The World of Science

THE WORLD OF SCIENCE

An Anthology for Writers

Gladys Garner Leithauser
Marilynn Powe Bell

The University of Michigan at Dearborn

Nadine Cowan Dyer

Consultant in Interdisciplinary Studies

Holt, Rinehart and Winston, Inc.
New York   Chicago   San Francisco   Philadelphia
Montreal   Toronto   London   Sydney   Tokyo

Senior Acquisitions Editor: Charlyce Jones Owen
Developmental Editor: Kate Morgan
Senior Project Editor: Lester A. Sheinis
Production Manager: Annette Mayeski/Stefania Taflinska
Design Supervisor: Robert Kopelman

Library of Congress Cataloging-in-Publication Data

The World of Science.

  Bibliography: p.
  1. Science. I. Leithauser, Gladys G. II. Bell,
Marilynn P.
Q171.W86 1987   500   86–12133
ISBN 0-03-006117-2

Holt, Rinehart and Winston, Inc.
The Dryden Press
Saunders College Publishing

Credits

1. "A Child's Garden of Science," reprinted from *The American Scholar*, Volume 52, Number 3, Summer, 1983. Copyright © 1983 by the United Chapters of Phi Beta Kappa. By permission of the publishers.
2. "A Scientific Apprenticeship," from Chapter 2 in *Disturbing the Universe* by Freeman Dyson. Copyright © 1979 by Freeman J. Dyson. By permission of Harper & Row, Publishers, Inc.
3. "The Portrait of a Scientist," from *Arrowsmith* by Sinclair Lewis, copyright 1925 by Harcourt Brace Jovanovich, Inc.; renewed 1953 by Michael Lewis. Reproduced by permission of the publisher.
4. "Flight into the Forest," reprinted by permission of the publishers from *In a Patch of Fireweed* by Bernd Heinrich, Cambridge, Mass.: Harvard University Press, Copyright © 1984 by the President and Fellows of Harvard College.
5. "Lost in the Twentieth Century," reproduced, with permission, from the Annual Review of Biochemistry, Volume 32, © 1963 by Annual Reviews, Inc.
6. "How Can I Tell If I Am Cut Out to Be a Research Worker?" from Chapter 2 in

(**Credits** continue on page 516.)

PREFACE

The World of Science: An Anthology for Writers focuses on the human side of scientific discovery. It presents the inner as well as the outer world of science, introducing students of all disciplines to science not as an abstract methodology, but as a creative activity. Offering insights and ideas that students may not realize they lack, these writings introduce readers to both tradition and controversy.

One organizing principle of this collection is that student writers often do better work when their readings reflect their special interests. Yet anthologies of such readings are rare for students in the natural sciences. Thomas Kuhn, science historian and author of *The Structure of Scientific Revolutions*, states that "the single most striking feature of education in the natural sciences is that, to an extent totally unknown in other creative fields, it is conducted entirely through textbooks. . . . There are no collections of 'readings' in the natural sciences." Many science instructors agree that the pressure of learning the masses of facts in these textbooks prevents students from first-hand encounters with the great thinkers and investigators of their own fields. *The World of Science* provides such encounters.

Importance, depth of insight, the ability to stimulate a sense of discovery—these criteria have determined our selections. We have chosen the familiar and the unfamiliar, the idealistic and the utilitarian, the rational and the romantic. They invite readers to explore the worlds of scientific thought and to experience what Martin Gardner calls the "Aha Moment." We include excerpts and complete selections from technical papers, scientific and philosophical books and essays, autobiography, fiction, and poetry. We place here Aristotle, Lucretius, Copernicus, and Darwin with Robert Frost, Albert Einstein, James Thurber, and Stephen Jay Gould.

We have grouped the selections into three major parts, each divided into units. Part One, "The Inquiring Attitude," presents the scientist's way of looking at the external world. The opening unit, "The Individual Perspective: Scientists in the Making," seeks to illuminate some of the ways in which individuals have discovered their aptitude for special fields. Unit Two, "From

Perspective to Method: Induction and Intuition," offers varying opinions on the nature of the scientific method. Unit Three, "The Broader Perspective: Science as an Element in Culture," analyzes the increasing role of science since its efflorescence in the sixteenth century. In this unit, the selections develop the historical perspective of the rise of science, the dialectical relation of science to society, and the influence of science on the humanities in the twentieth century.

Part Two, "The Disciplined Approach to Knowledge," offers works from several disciplines: mathematics, physics, chemistry, geology, the life sciences, and medicine. The section reflects the diverse development of the scientific disciplines.

Part Three, "The Intensifying Vision," opens with "On Science Applied," a unit that examines the "child" of science—technology. Unit Ten, "On the Nature of Things," offers observations, investigations, and conjectures on the external world. The final unit, "On the Issues of the World," presents writings in which scientists address the urgent issues of our time.

Our intent is that the organization of this anthology will make this collection also a writing guide. Biographical headnotes provide some of the "external factors" that set the pieces in context and provide information about the authors' original audiences and purposes. Questions and Topics for Discussion and Writing direct attention both to content and style, helping students to discover and develop their own topics for essays or research projects. The Suggestions for Further Reading at the end of each unit include a selected bibliography to guide further study and research. These lists of readings suggest pieces that would provide profitable comparison and contrast with the selections.

This pedagogical structure supports the overall purpose of the book: to communicate to students the experience and fascination of science. A Rhetorical Contents follows the regular Contents. For the instructor who desires additional material, an Instructor's Resource Book accompanies *The World of Science*. This Resource Book provides a number of supplemental topics and themes that give direction for discussion and writing, or for more extensive reading and research, and offers sample answers to selected questions.

We believe that our book will help students "identify" with scientific giants who might otherwise seem remote and to inquire profitably about the very nature and limitations of science itself.

Acknowledgments

This anthology grew from our experience as teachers of composition for students in the natural sciences. As we collected essays and discussed them in class, we developed the conviction that students need more in scientific reading than just principles and facts; they profit from direct experience of scientists' intellectual excitement and personal angles of vision.

A number of people supported us in this project. During the early stages, the following persons gave enthusiastic encouragement: Christopher Dahl, Eugene Grewe, Reginald Witherspoon, Dianne Martin, Robert Rohm, William Linn, David Emerson, Mary Powe, Louise and Richard Randall, and Dick Darios. Thoughtful suggestions for selections came from Victor Wong, Elizabeth Crowell, John Harrington, Bernd Heinrich, William Simpson, Kenneth Angyal, Brad Leithauser, Kenneth Blackwell, Harry Ruja, Lee Eisler, Margaret Moran, Edward Sayles, Paul Zitzewitz, Lucy Merritt, Gail Leithauser, and Sue Fromm Marshall. As the anthology developed, the following people gave practical assistance: Sue Falconer, with typing; Eugene Dyer and Mark Leithauser, with artwork; and John Devlin, with an annotated bibliography of physics.

We are particularly grateful to Shirley Smith, Director of The University of Michigan-Dearborn Library, for steady support and for the haven of a library study room. Special thanks for assistance go also to the library reference staff: Barbara Lukasiewicz, Judy Stuck, Caroline Smith, Robert Kelly, and Virginia Coots. For reference materials from other institutions, we thank Carol Schmidt and Earlene Pawley.

Barbara Loy provided painstaking record-keeping, particularly on the permissions log. David Sosnowski, Virginia Clark, Raymond Duda, and Jan McKae contributed careful research assistance and preparation of a number of biographical headnotes. Others who gave valued help with the biographical material include Richard Turner, Tija Spitsberg, Lois Breitmeyer, and Grace Stewart. A number of people contributed variously: Allie McCarter, Michael Reese, Ted-Larry Pebworth, Neil Leithauser, Hilde Junkermann, Lawrence Berkove, Patricia Hernlund. For assistance with Questions and Topics for Discussion and Writing, we thank Ayotunde Oyinsan, Gregory Loselle, and Barbara Hardt.

Thoughtful criticism came from Mary Jo Salter and Jean Owen. This anthology is better, too, for Norma Merry, who contributed ideas, materials, and equanimity. Others also smoothed the path: Cecilia Benner, Blinn Rush, Rosella Sanderson, and Alene Huguley.

Nadine Cowan Dyer has acted throughout the text as an interdisciplinary consultant. We are grateful, in particular, for her many vital study questions, some of the most illuminating in the text, and for her breadth of vision in making thematic connections.

For their expertise, we thank our editors at Holt, Rinehart and Winston: Lester A. Sheinis, Senior Project Editor; Kate Morgan, Developmental Editor; Kathleen Nevils, Copy Editor; and Charlyce Jones Owen, Senior Acquisitions Editor, English.

We wish to thank the following reviewers for their help: Victoria Aarons, Trinity University; Laureen Belmont, North Idaho College; Helen Dale, University of Wisconsin-Eau Claire; Robert Forman, St. John's University; Carol Freeman, University of California, Santa Cruz; William Ibbs, Loop College; and Christopher Thaiss, George Mason University.

Finally, we thank our understanding families: We credit the children, Carmen, Michael, and Markail Bell, who shared "Mom" with the computer, the typewriter, the file cabinet, and the telephone. Especially, we remember the good humor, encouragement, and support from our husbands, Clarence Bell and Harold Leithauser.

GGL
MPB

CONTENTS

PART TWO The Disciplined Approach to Knowledge 165

UNIT 4 Mathematics: Language in the Search for Certainty 167

UNIT 5 Physics: The Classical and the New 192

UNIT 6 Chemistry and Geology: Searching Out Matter 237

RHETORICAL CONTENTS

Scientific Method

Language and Diction

Analogy

Classification and Analysis

Cause and Effect

The World of Science

INTRODUCTION

Modern culture is so steeped in practical applications of scientific knowledge that many people confuse impressive technological achievements with the activity of science itself. They see science as a kind of cornucopia spilling out the fruits of research as ingenious inventions. Quick to claim a connection with this perceived beneficence, advertisers often make science the source, or at least the certifier, of their wares: "Scientific tests" prove the worth and effectiveness of their products. This book centers, in contrast, not on technological feats and marvels, but on the experience of the scientific quest. The experience is one that geologist John Harrington calls "the lure of the hunt," and much of the lure is strategy, an emphasis on ideas more than on facts.

Readers witness the scientific "hunters" pursuing their quest while rigorously and soberly controlling the search by systematic procedures. The reward for their effort is challenge, complexity, intellectual adventure—and sometimes exhilaration. *The World of Science* invites readers to share, through records and reports, the scientists' experiences and patterns of thought as they analyze problems, describe and investigate phenomena, and envision possibilities. The book offers its readers the opportunity to explore what Bertrand Russell has called the "expanding mental universe" as well as the physical one.

An organizing principle of this collection, then, is to demonstrate science as a human activity. "Science is a way of thinking," says Carl Sagan, "much more than it is a body of knowledge." This "way of thinking" is an increasingly systematized one that has led—especially in the last three centuries—to a vast in-

1

crease in generalized knowledge, a phenomenon often spoken of as the "knowl-edge explosion." Like Sagan, many scientists place the purpose of science less in the accumulation of facts than in the development of new ways of thinking about them. By its applications, this expanding human activity, of course, has made our world both a more fearsome and a more rewarding one.

To be considered science, the activity of investigating nature must meet at least two requirements: It must be faithful to evidence gathered by a strict methodology of observation or experiment, and it must make rational use of inference. Although science may involve serendipitous discoveries, it is tradi-tionally viewed as a disciplined inquiry into nature. Scientists may be viewed as persons who first note "That's odd," or who ask "Why?" (in Immanuel Kant's words, "putting Nature to the question") and then try to answer the question by systematic steps. In this traditional view, the modern scientist begins by observing many facts, goes on to construct theories, and then tests them by carefully devised and controlled experiments. It is this method that establishes results as generalizations that deserve—in contrast to the everyday generaliza-tions of common sense—to be called scientific "laws."

This systematic means for getting from observations to facts to laws is known as *the inductive method.* It is generally regarded as the hallmark that distin-guishes science from non-science or pseudo-science. Taking its name from the form of reasoning known as *induction*, the line of inference that proceeds from an instance to the universal (or "the particular to the general"), the inductive method stands in contrast to the deductive method. Highly regarded as a sys-tem of acquiring knowledge since the demonstrations of Aristotle, *deduction* is the process that discovers truth by logical reasoning from true premises ("the general to the particular").

Induction—although as old as human history—became the core of modern scientific method during the Renaissance. For his insistence on inductive logic, Sir Francis Bacon, the fifteenth-century English philosopher and statesman, is often called the "father of the scientific method." He was concerned with the tendency of scholars to emphasize deductive method over observation and with their failure to go to nature first-hand. He, therefore, argued against accepting premises on the basis of authority alone and for individual classification, experi-mentation, and inference. For over a century, traditional explanation has held that the vast increase in science since Bacon's day can be attributed to the inductive method.

Recently, however, some theorists have advanced the view that in addition to deduction and induction, a third approach—intuitive "guessing"—may come first in scientific thinking. Then the detailed and systematic procedure called for by the "rules" follows. Karl Popper, especially in *The Logic of Scien-tific Discovery,* is a leader against inductivism, the philosophical view that the scientific method is synonymous with induction. He and similar thinkers see induction as concerned with testing hypotheses, not with formulating them; in the same vein, they point out that deduction from premises (as in mathematics) does not help the reasoner obtain the premises in the first place.

Mario Bunge, philosopher and theoretical physicist, in *Intuition and Science,* calls the concept that these two separate, standardized methods are sufficient to explain the scientific method a "widespread fable" and a "cartoon of scientific work." He states: "Whoever has done something in science knows that the scientist, whether mathematician, naturalist, or sociologist, makes use of *all* psychical mechanisms, and that he cannot control them all and cannot always ascertain which has functioned in every case. . . . When we do not know exactly which of the . . . mechanisms has played a part . . . we tend to say that it has all been the work of *intuition*. Intuition is the collections of odds and ends where we place all the intellectual mechanisms which we do not know how to analyze or even name with precision." In assigning a role to intuition, such thinkers make science a more individualized human endeavor.

Nevertheless, we must be careful, warns Bunge, not to make intuition a philosophy, for it is "fertile to the extent that it is refined and waited on by reason." The scientific "vision" must be subjected to analysis and testing; scientific activity must construct theories and methods.

Developing similar ideas, Sir Peter Brian Medawar, in *The Art of the Soluble,* terms the scientific method a "hypothetico–deductive" conception. Through observations, the investigator creates a "working guess" or hypothesis, a procedure that is inductive; through the preparation of tests, the investigator checks his or her tentative answer, a procedure that is deductive. The conception reconciles the contradictory notions of the scientist as creator, for whom induction is preceded by the "imaginative leap"—a romantic view—and that of the scientist as critic—a classical view. The proper view, Medawar declares, is that the scientific process of thought employs both views as "successive and complementary episodes of thought." Discovery, a function of the imagination, co-exists with proof, a function of the reason; conjecture co-exists with refutation.

This kind of thinking about scientific method is part of the justification for grounding this book in the context of science as human activity. To those who may have judged science as a dehumanizing activity or an expression only of skepticism, the selections of this collection will offer surprise—even reassurance. These writings illustrate that, as a human activity, science is also a social one. What distinguishes science from individual and species systems of prediction, says Jacob Bronowski in *The Common Sense of Science,* is "at bottom this, that it is a method which is shared by the whole society consciously and at one time."

These writings, gleaned from a wide range of vantage points, help readers to understand what the scientific vision offers of special validity and vitality to the whole activity of human learning.

PART ONE

The Inquiring Attitude

Unit 1

THE INDIVIDUAL PERSPECTIVE: SCIENTISTS IN THE MAKING

The purest science is still biographical.

Henry David Thoreau

Jeremy Bernstein

Jeremy Bernstein (b. 1929) can claim the dual role of physicist and writer. His activities as a writer spring from his being what he calls a "science watcher," that is, one who strives to interpret "the scientific experience in all its manifestations."

Bernstein entered Harvard University at the age of 17 with no clear idea of what he wanted to do with his life: "Becoming a scientist was the last thing I had in mind. I was a near cipher in high school science and, although good at various kinds of mathematics, I always regarded such things as 'subjects' rather than disciplines that could arouse the passionate concerns of serious people. Nonetheless I had a random, and rather romantic, curiosity about Einstein and his theory of relativity. . . ."

After receiving his master's in Mathematics from Harvard, he became "hooked on modern physics," and went on to obtain the doctorate in Physics from the same institution. He has been a physicist at Los Alamos and a member of the Institute for Advanced Study in Princeton; he is currently a professor of physics at Stevens Institute of Technology.

Bernstein's distinguished second career as a writer has produced such books as *Einstein* and *Experiencing Science;* a profile of the Bell Laboratories; many articles for *The New Yorker* magazine; and an ongoing column, *Out of My Mind* (for *The American Scholar*), from which comes "A Child's Garden of Science."

A Child's Garden of Science

Jeremy Bernstein

Over a two-year period from 1973 to 1975, I conducted a series of interviews with the Nobel-Prize-winning physicist I. I. Rabi, which culminated in a *New Yorker* profile that was published in October 1975. Rabi, who was born on July 29, 1898, in Galicia, was then in his midseventies. He was, and is, one of the wisest people I have ever known. One of the things that we discussed was the aging process in scientists. It is a cliché that the great scientists appear to do their best work when they are extremely young and that many scientists seem to burn out by the time they are fifty or even earlier. I asked Rabi why he thought this was, and here is his answer:

> I think it must be basically neurological or physiological. The mind ceases to operate with the same richness and associations. The information-retrieval part sort of goes, along with the interconnections. I know that when I was in my late teens and early twenties the world was just a Roman candle—rockets all the time. The world was aglow. You lose that sort of thing as time goes on. It's the sort of thing that you want to hang on to if you can. And physics is such an out-of-the-world thing. It's not like history or poetry, or even painting. In them you never really lose contact with the world—it's right there before you. But physics is an otherworldly thing. It requires a taste for things unseen, even unheard of—a high degree of abstraction and a sort of profound innate philosophy. These faculties die off somehow when you grow up. You see them in children, who are fantastically interested in making things and in asking "Why? Why? Why?" Then, at a certain age, the children just become adults and are no longer very deeply interested in anything, except in the process of making a living and in sex and power. Money. Otherwise, they're not terribly interested. Profound curiosity happens when they are young. I think physicists are the Peter Pans of the human race. They never grow up, and they keep their curiosity.

The phrase "physicists [and for "physicists" one can substitute, I think, "scientists"] are the Peter Pans of the human race" has haunted me ever since I first heard it from Rabi. I believe that it is true, but it raises another question— the question that is the subject of this essay—namely, why is it that certain children become scientists? What is the triggering mechanism in childhood that starts some child on the road to where he, or she, becomes an Einstein, a Feynman, or a Rabi? In short, if one looks at the lives of scientists, what was the first childhood experience that they themselves can connect to what they later became? There is, very likely, an enormous literature on this subject, of which I am blissfully ignorant. But what I have been doing for the past twenty years is to ask many of the scientists I have interviewed just this question and, where, as in the case of Einstein, I had no opportunity to ask this question myself, I have tried to find out the answer by reading autobiographical statements. This has led me to read the first chapters of an awful lot of autobiographies.

One conclusion that I have come to is that there appears to be a clear distinction between mathematically oriented scientists and the rest. The triggering experiences of, say, experimental physicists are so diverse that they do not seem to fall into any obvious pattern. I will give a few examples later. But of mathematicians, there seems to be a universal law that I can summarize with a slight distortion of that celebrated bit of dialogue from *Die Fledermaus*. My version reads, "So young and already a mathematician." (The late Wolfgang Pauli—a Viennese noted for his acerbic wit, which was often applied to his fellow physicists—once remarked apropos of one of them, "So young and already so unknown." The original, in case one has not heard it recently, was "So jung und schon ein Prinz.") I really believe, having observed a good deal of it, that mathematical genius—even high talent—is "hard-wired." I do not have the foggiest idea where in the brain, or how, this hardware is hooked up—after all, brain scientists do not even know where, if anywhere, in the brain memory is located. But I am quite sure that it is there. We can all learn algebra, and many of us can learn calculus, but we cannot be taught to be Euler, Gauss, or Von Neumann. Even *they* cannot be taught to be Euler, Gauss, or Von Neumann.

Let me give a few examples of early mathematical experience that involve people I have spoken to myself. Hans Bethe, one of the premier contemporary mathematical physicists, when I asked him if he had any early mathematical memories, answered, "Oh yes—many. I was interested in numbers from a very early age. When I was five, I said to my mother on a walk one day, 'Isn't it strange that if a zero comes at the end of a number it means a lot but if it is at the beginning of a number it doesn't mean anything?' And one day when I was about four, Richard Ewald, a professor of physiology, who was my father's boss, asked me on the street, 'What is .5 divided by 2?' I answered, 'Dear Uncle Ewald, that I don't know,' but the next time I saw him I ran to him and said, 'Uncle Ewald, it's .25.' I knew about decimals then. When I was seven I learned about powers, and filled a whole book with the powers of two or three."

I not long ago asked the same question of Stanislaw Ulam. Ulam, who was born in Poland in 1909 and who has recently retired from Los Alamos, is the very distinguished mathematician who, among many other things, invented what is called the "Monte Carlo method" of doing approximate numerical calculations. (Ulam, along with Edward Teller, discovered what is known as the "secret of the hydrogen bomb.") He told me that when he was about ten he began attending lectures on the theory of relativity. In his delightful book *Adventures of a Mathematician,* he writes, "I did not really understand any of the details, but I had a good idea of the main thrust of the theory. Almost like learning a language in childhood, one develops the ability to speak it without knowing anything about grammar." (There is a school of thought, as is well known, that claims, and I think correctly, that grammar is also "hard-wired"; but that is a subject for another essay.) Ulam concluded: "I understood the schema of special relativity and even some of the consequences without being able to verify the details mathematically."

A little later in this chapter of his book Ulam writes: "I had mathematical curiosity very early. My father had in his library a wonderful series of German paperback books—*Reklam*, they were called. One was Euler's *Algebra*. I looked at it when I was perhaps ten or eleven, and it gave me a mysterious feeling. The symbols looked like magic signs; I wondered whether one day I could understand them. This probably contributed to the development of my mathematical curiosity. I discovered by myself how to solve quadratic equations. I remember that I did this by an incredible concentration and almost painful and not-quite conscious effort. What I did amounted to completing the square in my head without paper or pencil."

The computer scientist and artificial-intelligence expert Marvin Minsky told me that many of his earliest memories are mathematical. At some very young age he learned that there were both positive and negative numbers. "I thought to myself," he said, "that is very nice, but maybe there are three kinds of numbers—a, b and c. I tried for days to find a number system with three bases that looked like arithmetic. Many years later I realized that there isn't one. At the time I didn't have the courage to add a fourth base number that works and gives you the complex number system."

I once asked Freeman Dyson if he could recall *his* earliest mathematical memories. He told me that, among them, was a time when he was still being put down for naps in the afternoon—he was not exactly sure of the age, but less than ten—he began adding up numbers like $1 + \frac{1}{2} + \frac{1}{4} + \frac{1}{8} + \ . \ . \ .$ and realized that this series was adding up to 2. In other words, he had discovered for himself the notion of the convergent infinite series.

To these examples I would like to add a few more that I have culled from various biographies and autobiographies. Emilio Segrè wrote a lovely biography of his lifelong friend and colleague Fermi, called *Enrico Fermi, Physicist*. Early in the book, Segrè writes: "Fermi never told me how he first became acquainted with mathematics; it is possible that a friend of his father introduced him to the subject. . . . Fermi told me that one of his great intellectual efforts was his attempt to understand—at the age of ten—what was meant by the statement that the equation $x^2 + y^2 = r^2$ represents a circle. Someone must have stated the fact to him, but he had to rediscover its meaning by himself."

Recently I heard an interview that Richard Feynman, perhaps the most brilliant theoretical physicist of his generation, gave on "Nova." He was asked about his earliest memories. He replied, "When I was just a little kid, very small in a high chair, [my father] brought home a lot of tiles, little bathroom tiles—seconds of different colors. . . . We played with them, setting them out like dominoes, I mean vertically, on my high chair . . . so they tell me this anyway . . . and when we'd got them all set up I would push one end so that they would all go down. Then after a while I'd help to set them up in a more complicated way . . . two white tiles and a blue tile, two white tiles and a blue tile . . . and when my mother complained 'Leave the poor child alone, if he wants to put a blue tile, let him put a blue tile,' he said, 'No, I want to show him what patterns are like and how interesting they are as it's a kind of mathematics.'" And so it is.

Einstein never thought of himself as much of a mathematician, although, by almost any standards, he was as powerful a mathematician as he needed to be when it came time to invent the general theory of relativity. Einstein created, or re-created, the so-called "tensor calculus," a highly non-trivial affair. But the earliest scientific memory that he recalls, for the purpose of his wonderful "Autobiographical Notes," was not mathematical at all. When he was four or five years old he remembers being shown a compass. He was fascinated by the way the compass needle behaved, especially that it pointed north. He wrote, "That this needle behaved in such a determined way did not at all fit into the nature of events, which could find a place in the unconscious world of concepts (effect connected with direct 'touch'). I can still remember—or at least I believe I can remember—that this experience made a deep and lasting impression upon me. Something deeply hidden had to be behind things."

This sort of trigger experience—the experience of wanting to know how something mechanical or electrical actually works, often to the point of wanting to re-create it by building it oneself is, as far as I can make out, the characteristic of the less mathematically inclined scientists. Many scientists, of course, and Einstein is one, have both sorts of early memory. (Einstein invented a proof of the Pythagorean theorem by himself when he was twelve.) It is interesting to me that the non-mathematical trigger experience of Einstein involved magnetism, since electromagnetism was to play so important a part in his early creative life. I was also interested to learn that Ernst Mach, whose great book *The Science of Mechanics* (1893) had such an influence on the young Einstein, had a triggering experience that was mechanical. (By his own admission Mach was never much of a mathematician.) This is the story as recounted in John T. Blackmore's biography *Ernst Mach*.

There was a turning point in my fifth year. Up to that time I represented to myself everything I did not understand—a pianoforte, for instance—as simply a motley assemblage of the most wonderful things, to which I ascribed the sound of the notes. That the pressed key struck the chord with the hammer did not occur to me. Then one day I saw a windmill.

We [Ernst and his sister Octavia] had to bring a message to the miller. Upon our arrival the mill had just begun to work. The terrible noise frightened me, but did not hinder me from watching the teeth of the shaft which meshed with the gear of the grinding mechanism and moved on one tooth after another. This sight remained until I reached a more mature level, and in my opinion, raised my childlike thinking from the level of the wonder-believing savage to causal thinking; from now on, in order to understand the unintelligible, I no longer imagined magic things in the background but traced in a broken toy the cord or lever which had caused the effect.

Before I comment on this passage, I would like to give one more example, and this from Isaac Newton. As far as I know, it is not known what Newton's first scientific intimations were. I am not aware, for example, that he showed any special fascination for numbers when he was a child, although in adult life, as we all know, he became a prodigious mathematician. One of Newton's younger contemporaries, William Stukeley, collected information about Newton that he published in a memoir. He, like Newton, was from Lincolnshire, and in the preparation of his memoir, published in 1752, twenty-five years after Newton's death, he visited Newton's boyhood home, including the study where Newton did his great work when he came down from Cambridge in the plague year of 1665. Stukeley reports in his memoir:

> Every one that knew Sir Isaac, or have heard of him, recount the pregnancy of his parts when a boy, his strange inventions, and extraordinary inclination for mechanics. That instead of playing among the other boys, when from school, he always busied himself in making knick-knacks and models of wood in many kinds. For which purposes he had got little saws, hatchets, hammers, and a whole shop of tools, which he would use with great dexterity. In particular they speak of his making a wooden clock. About this time a new windmill [shades of Mach] was set up near Grantham, in the way to Gunnerby, which is now demolished, this country chiefly using water mills. Our lad's imitating spirit was soon excited, and by frequently prying into the fabric of it, as they were making it, he became master enough to make a very perfect model thereof, and it was said to be as clean and curious a piece of workmanship as the original. This sometimes he would set upon the house-top where he lodged, and clothing it with sail-cloth, the wind would really take it; but what was most extraordinary in its composition was, that he put a mouse into it, which he called the miller, and that mouse made the mill turn round when he pleased.

So young and already a mechanic.

What are we to make of all of this? It would appear that all children, to various degrees and with various aptitudes, go through this "scientific phase," but that very few of them become scientists. The reasons for the latter are clearly many, not the least of which have to do with environmental opportunities. An Isaac Newton born in a mountain village in Nepal might well have grown up to be the village curiosity, or even a great religious figure, but would, one can safely assume, be unlikely to have invented the law of universal gravitation. But many children are exposed to similar environments, and some of them grow up to rob banks and some to own them.

Perhaps some clue about what makes certain children scientists is to be found in the quotation from Mach. He had, he tells us, this sudden realization as a child that causal explanation could make the unintelligible understandable. In his "Autobiographical Notes" Einstein describes the tension created when some experience causes us, quite spontaneously, to "wonder" about its meaning and cause. He speaks of this as a "conflict" and he writes, "Whenever such a conflict is experienced hard and intensively it reacts back upon our thought world in a decisive way. The development of this thought world is in a certain sense a continuous flight from 'wonder.'"

"Flight" here is a very strong word. It suggests, at least to me, that these scientific children learn very early that causal explanation is a great aid and comfort in the constant flight we all make from the unintelligible. Causal explanation helps us to flee from the terrors of the unknown, and these terrors can loom very large in the mind of a child. In this respect, here is what Rabi told me when I asked him if some childhood experience turned him toward science: "Yes," he said, "a very profound one. One time, I was walking down the street—looked right down the street, which faced east. The moon was just rising. And it scared the hell out of me! Absolutely scared the hell out of me." Not long afterward Rabi began reading through—in alphabetical order—all of the books in the children's section of the Brooklyn Public Library branch that was near his home. First he began with fiction, reading from Alcott through Trowbridge. Then he came to the science shelf and began with astronomy—a little book on astronomy—which changed his life. It was his first encounter with scientific explanation and, as he told me, "When you have the astronomical explanation, the rising of the moon becomes a sort of non-event."

Questions and Topics for Discussion and Writing

1. Robert Louis Stevenson wrote a collection of poems called *A Child's Garden of Verses.* In the light of your reading of Bernstein's essay, what do you think is the significance of Bernstein's title, which alludes to Stevenson's book?
2. What do you understand Bernstein to mean by a "triggering mechanism"? To express this experience, suggest another image, one that has psychological validity for you.
3. What is being said of scientists when they are called the "Peter Pans of the human race"?
4. What is suggested about scientists when they are metaphorically described as "hard-wired"? How does Bernstein challenge the accuracy of this metaphor?
5. Bernstein says *parenthetically* that there exists a theory that "grammar is also 'hard-wired'" (a theory developed by the American linguist Noam Chomsky). Why do you think he adds this statement in the context of the discussion about scientists?
6. Summarize the conclusions about the "unknown" drawn by the different scientists listed and discussed.

 7. Interview two people you believe to be mathematically inclined. Write a short essay relating their early mathematical experiences.
 8. Analyzing a "triggering mechanism" in your own life, write a brief essay, emulating Bernstein's use of specific details of narrative.
 9. Determine as best you can why you decided to study science. Was there one "triggering mechanism," or a number of causes?
10. Discuss the implication of the concluding two sentences of the second paragraph as they relate to his later statement: "It would appear that all children . . . go through this 'scientific phase,' but that very few of them become scientists."
11. How does the word *flight*, together with the concluding quotation, serve to summarize the author's conclusion about what makes certain children linger in the "scientific phase" to become scientists? What do you think Rabi meant by the term *non-event*? How can a "non-event" become important?

Freeman Dyson

"It is one of the special beauties of science," writes Freeman Dyson (b. 1923), "that points of view which seem diametrically opposed turn out later, in a broader perspective, to be both right."

This vision that great ideas sometimes come in diametrically opposed pairs is at the basis of Dyson's belief that he can work in two "separate worlds." These worlds comprise his present roles, that of theoretical physicist at the Princeton Institute of Advanced Study, and that of military expert as a member of the Jason Project, the group of scientists who work on technical problems for the United States government. This belief in reconciling two worlds is expressed in *Weapons and Hope*, which won the Book Critics Circle Award for 1984.

In answer to why a scientist, "a peaceful theoretician," would write about weapons and war, Dyson writes, "I am possessed by an immodest hope that I may improve mankind's chances of escaping the horrors of nuclear holocaust if I can help these two worlds to understand and listen to each other."

Dyson is widely recognized for his work on the theory of quantum electrodynamics and for his speculations on the possibility that extraterrestrial civilizations exist. In *Disturbing the Universe*, his autobiography, Dyson describes himself as a mathematically inclined child who came to feel during his teen years that "every hour not spent doing mathematics was a tragic waste."

Born in Crowthorne, England, Dyson attended mathematics-oriented Cambridge University. During World War II, he interrupted his undergraduate studies to work as a civilian scientist and adviser at the British Bomber Command with the Royal Air Force, where his expertise in mathematics was used to study the causes of the heavy loss of bombers on night missions. In 1945, he graduated from Cambridge with a degree in Physics.

Financed by a Commonwealth Fund Fellowship, he came to Cornell University, where he studied with Hans Bethe and Richard P. Feynman to "learn physics from the Americans." He soon joined the faculty at the Institute for Advanced Study at Princeton and, in 1957, became an American citizen.

"A Scientific Apprenticeship," from *Disturbing the Universe*, recalls the period at Cornell University when he worked with Robert Oppenheimer and other scientists who had spent the war years on military projects. This dynamic group created what Dyson calls the "postwar flowering of physics."

A Scientific Apprenticeship

Freeman Dyson

I was lucky to arrive at Cornell at that particular moment. Nineteen forty-seven was the year of the postwar flowering of physics, when new ideas and new experiments were sprouting everywhere from seeds that had lain dormant through the war. The scientists who had spent the war years at places like Bomber Command headquarters and Los Alamos came back to the universities impatient to get started again in pure science. They were in a hurry to make up for the years they had lost, and they went to work with energy and enthusiasm. Pure science in 1947 was starting to hum. And right in the middle of the renascence of pure physics was Hans Bethe.

At that time there was a single central unsolved problem that absorbed the attention of a large fraction of physicists. We called it the quantum electrodynamics problem. The problem was simply that there existed no accurate theory to describe the everyday behavior of atoms and electrons emitting and absorbing light. Quantum electrodynamics was the name of the missing theory. It was called quantum because it had to take into account the quantum nature of light, electro because it had to deal with electrons, and dynamics because it had to describe forces and motions. We had inherited from the prewar generation of physicists, Einstein and Bohr and Heisenberg and Dirac, the basic ideas for such a theory. But the basic ideas were not enough. The basic ideas could tell you roughly how an atom would behave. But we wanted to be able to calculate the behavior exactly. Of course it often happens in science that things are too complicated to be calculated exactly, so that one has to be content with a rough qualitative understanding. The strange thing in 1947 was that even the simplest and most elementary objects, hydrogen atoms and light quanta, could not be accurately understood. Hans Bethe was convinced that a correct and exact theory would emerge if we could figure out how to calculate consistently using the old prewar ideas. He stood like Moses on the mountain showing us the promised land. It was for us students to move in and make ourselves at home there.

A few months before I arrived at Cornell, two important things had happened. First, there were some experiments at Columbia University in New York which measured the behavior of an electron a thousand times more accurately than it had been measured before. This made the problem of creating an accurate theory far more urgent and gave the theorists some accurate numbers which they had to try to explain. Second, Hans Bethe himself did the first theoretical calculation that went substantially beyond what had been done before the war. He calculated the energy of an electron in an atom of hydrogen and found an answer agreeing fairly well with the Columbia measurement. This showed that he was on the right track. But his calculation was still a pastiche of old ideas held together by physical intuition. It had no firm mathematical basis. And it was not even consistent with Einstein's principle of relativity. That was how things stood in September when I joined Hans's group of students.

The problem that Hans gave me was to repeat his calculation of the electron energy with the minimum changes that were needed to make it consistent with Einstein. It was an ideal problem for somebody like me, who had a good mathematical background and little knowledge of physics. I plunged in and filled hundreds of pages with calculations, learning the physics as I went along. After a few months I had an answer, again agreeing near enough with Columbia. My calculation was still a pastiche. I had not improved on Hans's calculation in any fundamental sense. I came no closer than Hans had come to a basic understanding of the electron. But those winter months of calculation had given me skill and confidence. I had mastered the tools of my trade. I was now ready to start thinking.

As a relaxation from quantum electrodynamics, I was encouraged to spend a few hours a week in the student laboratory doing experiments. These were not real research experiments. We were just going through the motions, repeating famous old experiments, knowing beforehand what the answers ought to be. The other students grumbled at having to waste their time doing Mickey Mouse experiments. But I found the experiments fascinating. In all my time in England I had never been let loose in a laboratory. All these strange objects that I had read about, crystals and magnets and prisms and spectroscopes, were actually there and could be touched and handled. It seemed like a miracle when I measured the electric voltage produced by light of various colors falling on a metal surface and found that Einstein's law of the photoelectric effect is really true. Unfortunately I came to grief on the Millikan oil drop experiment. Millikan was a great physicist at the University of Chicago who first measured the electric charge of individual electrons. He made a mist of tiny drops of oil and watched them float around under his microscope while he pulled and pushed them with strong electric fields. The drops were so small that some of them carried a net electric charge of only one or two electrons. I had my oil drops floating nicely, and then I grabbed hold of the wrong knob to adjust the electric field. They found me stretched out on the floor, and that finished my career as an experimenter.

I never regretted my brief and almost fatal exposure to experiments. This experience brought home to me as nothing else could the truth of Einstein's remark, "One may say the eternal mystery of the world is its comprehensibility." Here was I, sitting at my desk for weeks on end, doing the most elaborate and sophisticated calculations to figure out how an electron should behave. And here was the electron on my little oil drop, knowing quite well how to behave without waiting for the result of my calculation. How could one seriously believe that the electron really cared about my calculation, one way or the other? And yet the experiments at Columbia showed that it did care. Somehow or other, all this complicated mathematics that I was scribbling established rules that the electron on the oil drop was bound to follow. We know that this is so. Why it is so, why the electron pays attention to our mathematics, is a mystery that even Einstein could not fathom.

At our daily lunches with Hans we talked endlessly about physics, about the technical details and about the deep philosophical mysteries. On the whole, Hans was more interested in details than in philosophy. When I raised philosophical questions he would often say, "You ought to go and talk to Oppy about that." Oppy was Robert Oppenheimer, then newly appointed as director of the Institute for Advanced Study at Princeton. Sometime during the winter, Hans spoke with Oppy about me and they agreed that after my year at Cornell I should go for a year to Princeton. I looked forward to working with Oppy but I was also a bit scared. Oppy was already a legendary figure. He had been the originator and leader of the bomb project at Los Alamos. Hans had worked there under him as head of the Theoretical Division. Hans had enormous respect for Oppy. But he warned me not to expect an easy life at Princeton. He said Oppy did not suffer fools gladly and was sometimes hasty in deciding who was a fool.

One of our group of students at Cornell was Rossi Lomanitz, a rugged character from Oklahoma who lived in a dilapidated farmhouse outside Ithaca and was rumored to be a Communist. Lomanitz was never at Los Alamos, but he had worked with Oppy on the bomb project in California before Los Alamos was started. Being a Communist was not such a serious crime in 1947 as it became later. Seven years later, when Oppy was declared to be a Security Risk, one of the charges against him was that he had tried to stop the army from drafting Lomanitz. Mr. Robb, the prosecuting attorney at the trial, imputed sinister motives in Oppy's concern for Lomanitz. Oppy replied to Robb, "The relations between me and my students were not that I stood at the head of a class and lectured." That remark summed up exactly what made both Hans and Oppy great teachers. In 1947 security hearings and witch hunts were far from our thoughts. Rossi Lomanitz was a student just like the rest of us. And Oppy was the great national hero whose face could be seen ornamenting the covers of *Time* and *Life* magazines.

I knew before I came to Cornell that Hans had been at Los Alamos. I had not known beforehand that I would find a large fraction of the entire Los Alamos gang, with the exception of Oppy, reassembled at Cornell. Hans had been at Cornell before the war, and when he returned he found jobs for as many as possible of the bright young people he had worked with at Los Alamos. So we had at Cornell Robert Wilson, who had been head of experimental physics at Los Alamos, Philip Morrison, who had gone to the Mariana Islands to take care of the bombs that were used at Hiroshima and Nagasaki, Dick Feynman, who had been in charge of the computing center, and many others. I was amazed to see how quickly and easily I fitted in with this bunch of weaponeers whose experience of the war had been so utterly different from my own. There was endless talk about the Los Alamos days. Through all the talk shone a glow of pride and nostalgia. For every one of these people, the Los Alamos days had been a great experience, a time of hard work and comradeship and deep happiness. I had the impression that the main reason they were happy to be at Cornell was that the Cornell physics department still retained something of the Los Alamos atmosphere. I, too, could feel the vivid presence of this atmosphere. It was youth, it was exuberance, it was informality, it was a shared ambition to do great things together in science without any personal jealousies or squabbles over credit. Hans Bethe and Dick Feynman did, many years later, receive well-earned Nobel Prizes, but nobody at Cornell was grabbing for prizes or for personal glory.

The Los Alamos people did not speak in public about the technical details of bombs. It was surprisingly easy to talk around that subject without getting onto dangerous ground. Only once I embarrassed everybody at the lunch table by remarking in all innocence, "It's lucky that Eddington proved it's impossible to make a bomb out of hydrogen." There was an awkward silence and the subject of conversation was abruptly changed. In those days the existence of any thoughts about hydrogen bombs was a deadly secret. After lunch one of the students took me aside and told me in confidence that unfortunately Eddington was wrong, that a lot of work on hydrogen bombs had been done at Los Alamos, and would I please never refer to the subject again. I was pleased that they trusted me enough to let me in on the secret. After that I felt I was really one of the gang.

Many of the Los Alamos veterans were involved in political activities aimed at educating the public about the nuclear facts of life. The main thrust of their message was that the American monopoly of nuclear weapons could not last, and that in the long run the only hope of survival would lie in a complete surrender of all nuclear activities to a strong international authority. Philip Morrison was especially eloquent in spreading this message. Oppy had been saying the same thing more quietly to his friends inside the government. But by 1948 it was clear that the chance of establishing an effective international authority on the basis of the wartime Soviet-American alliance had been missed. The nuclear arms race had begun, and the idea of international control could at best be a long-range dream.

Our lunchtime conversations with Hans were often centered on Los Alamos and on the moral questions surrounding the development and use of the bomb. Hans was troubled by these questions. But few of the other Los Alamos people were troubled. It seemed that hardly anybody had been troubled until after Hiroshima. While the work was going on, they were absorbed in scientific details and totally dedicated to the technical success of the project. They were far too busy with their work to worry about the consequences. In June 1945 Oppy had been a member of the group appointed by Henry Stimson to advise him about the use of the bombs. Oppy had supported Stimson's decision to use them as they were used. But Oppy did not at that time discuss the matter with any of his colleagues at Los Alamos. Not even with Hans. That responsibility he bore alone.

In February 1948 *Time* magazine published an interview with Oppy in which appeared his famous confession, "In some sort of crude sense, which no vulgarity, no humor, no overstatement can quite extinguish, the physicists have known sin; and this is a knowledge which they cannot lose." Most of the Los Alamos people at Cornell repudiated Oppy's remark indignantly. They felt no sense of sin. They had done a difficult and necessary job to help win the war. They felt it was unfair of Oppy to weep in public over their guilt when anybody who built any kind of lethal weapons for use in war was equally guilty. I understood the anger of the Los Alamos people, but I agreed with Oppy. The sin of the physicists at Los Alamos did not lie in their having built a lethal weapon. To have built the bomb, when their country was engaged in a desperate war against Hitler's Germany, was morally justifiable. But they did not just build the bomb. They enjoyed building it. They had the best time of their lives while building it. That, I believe, is what Oppy had in mind when he said they had sinned. And he was right.

After a few months I was able to identify the quality that I found strange and attractive in the American students. They lacked the tragic sense of life which was deeply ingrained in every European of my generation. They had never lived with tragedy and had no feeling for it. Having no sense of tragedy, they also had no sense of guilt. They seemed very young and innocent although most of them were older than I was. They had come through the war without scars. Los Alamos had been for them a great lark. It left their innocence untouched. That was why they were unable to accept Oppy's statement as expressing a truth about themselves.

For Europeans the great turning point of history was the First World War, not the Second. The first war had created that tragic mood which was a part of the air we breathed long before the second war started. Oppy had grown up immersed in European culture and had acquired the tragic sense. Hans, being a European, had it too. The younger native-born Americans, with the exception of Dick Feynman, still lived in a world without shadows. Things are very different now, thirty years later. The Vietnam war produced in American life the same fundamental change of mood that the First World War produced in Europe. The young Americans of today are closer in spirit to the Europeans than to the Americans of thirty years ago. The age of innocence is now over for all of us.

Dick Feynman was in this respect, as in almost every other respect, an exception. He was a young native American who had lived with tragedy. He had loved and married a brilliant, artistic girl who was dying of TB. They knew she was dying when they married. When Dick went to work at Los Alamos, Oppy arranged for his wife to stay at a sanitarium in Albuquerque so that they could be together as much as possible. She died there, a few weeks before the war ended.

As soon as I arrived at Cornell, I became aware of Dick as the liveliest personality in our department. In many ways he reminded me of Frank Thompson. Dick was no poet and certainly no Communist. But he was like Frank in his loud voice, his quick mind, his intense interest in all kinds of things and people, his crazy jokes, and his disrespect for authority. I had a room in a student dormitory and sometimes around two o'clock in the morning I would wake up to the sound of a strange rhythm pulsating over the silent campus. That was Dick playing his bongo drums.

Dick was also a profoundly original scientist. He refused to take anybody's word for anything. This meant that he was forced to rediscover or reinvent for himself almost the whole of physics. It took him five years of concentrated work to reinvent quantum mechanics. He said that he couldn't understand the official version of quantum mechanics that was taught in textbooks, and so he had to begin afresh from the beginning. This was a heroic enterprise. He worked harder during those years than anybody else I ever knew. At the end he had a version of quantum mechanics that he could understand. He then went on to calculate with his version of quantum mechanics how an electron should behave. He was able to reproduce the result that Hans had calculated using orthodox theories a little earlier. But Dick could go much further. He calculated with his own theory fine details of the electron's behavior that Hans's method could not touch. Dick could calculate these things far more accurately, and far more easily, than anybody else could. The calculation that I did for Hans, using the orthodox theory, took me several months of work and several hundred sheets of paper. Dick could get the same answer, calculating on a blackboard, in half an hour.

So this was the situation which I found at Cornell. Hans was using the old cookbook quantum mechanics that Dick couldn't understand. Dick was using his own private quantum mechanics that nobody else could understand. They were getting the same answers whenever they calculated the same problems. And Dick could calculate a whole lot of things that Hans couldn't. It was obvious to me that Dick's theory must be fundamentally right. I decided that my main job, after I finished the calculation for Hans, must be to understand Dick and explain his ideas in a language that the rest of the world could understand.

In the spring of 1948, Hans and Dick went to a select meeting of experts arranged by Oppy at a lodge in the Pocono Mountains to discuss the quantum electrodynamics problem. I was not invited because I was not yet an expert. The Columbia experimenters were there, and Niels Bohr, and various other important physicists. The main event of the meeting was an eight-hour talk by Julian Schwinger, a young professor at Harvard who had been a student of Oppy's. Julian, it seemed, had solved the main problem. He had a new theory of quantum electrodynamics which explained all the Columbia experiments. His theory was built on orthodox principles and was a masterpiece of mathematical technique. His calculations were extremely complicated, and few in the audience stayed with him all the way through the eight-hour exposition. But Oppy understood and approved everything. After Julian had finished, it was Dick's turn. Dick tried to tell the exhausted listeners how he could explain the same experiments much more simply using his own unorthodox methods. Nobody understood a word that Dick said. At the end Oppy made some scathing comments and that was that. Dick came home from the meeting very depressed.

During the last months of my time at Cornell I made an effort to see as much of Dick as possible. The beautiful thing about Dick was that you did not have to be afraid you were wasting his time. Most scientists when you come to talk with them are very polite and let you sit down, and only after a while you notice from their bored expressions or their fidgety fingers that they are wishing you would go away. Dick was not like that. When I came to his room and he didn't want to talk he would just shout, "Go away, I'm busy," without even turning his head. So I would go away. And next time when I came and he let me sit down, I knew he was not just being polite. We talked for many hours about his private version of physics and I began finally to get the hang of it.

The reason Dick's physics was so hard for ordinary people to grasp was that he did not use equations. The usual way theoretical physics was done since the time of Newton was to begin by writing down some equations and then to work hard calculating solutions of the equations. This was the way Hans and Oppy and Julian Schwinger did physics. Dick just wrote down the solutions out of his head without ever writing down the equations. He had a physical picture of the way things happen, and the picture gave him the solutions directly with a minimum of calculation. It was no wonder that people who had spent their lives solving equations were baffled by him. Their minds were analytical; his was pictorial. My own training, since the far-off days when I struggled with Piaggio's differential equations, had been analytical. But as I listened to Dick and stared at the strange diagrams that he drew on the blackboard, I gradually absorbed some of his pictorial imagination and began to feel at home in his version of the universe.

The essence of Dick's vision was a loosening of all constraints. In orthodox physics you say, Suppose an electron is in this state at a certain time, then you calculate what it will do next by solving a certain differential equation, and from the solution of the equation you calculate what it will be doing at some later time. Instead of this, Dick said simply, the electron does whatever it likes. The electron goes all over space and time in all possible ways. It can even go backward in time whenever it chooses. If you start with an electron in this state at a certain time and you want to see whether it will be in some other state at another time, you just add together contributions from all the possible histories of the electron that take it from this state to the other. A history of the electron is any possible path in space and time, including paths zigzagging forward and back in time. The behavior of the electron is just the result of adding together all the histories according to some simple rules that Dick worked out. And the same trick works with minor changes not only for electrons but for everything else—atoms, baseballs, elephants and so on. Only for baseballs and elephants the rules are more complicated.

This sum-over-histories way of looking at things is not really so mysterious, once you get used to it. Like other profoundly original ideas, it has become slowly absorbed into the fabric of physics, so that now after thirty years it is difficult to remember why we found it at the beginning so hard to grasp. I had the enormous luck to be there at Cornell in 1948 when the idea was newborn, and to be for a short time Dick's sounding board. I witnessed the concluding stages of the five-year-long intellectual struggle by which Dick fought his way through to his unifying vision. What I saw of Dick reminded me of what I heard Keynes say of Newton six years earlier: "His peculiar gift was the power of holding continuously in his mind a purely mental problem until he had seen straight through it. I fancy his pre-eminence is due to his muscles of intuition being the strongest and most enduring with which a man has ever been gifted."

In that spring of 1948 there was another memorable event. Hans received a small package from Japan containing the first two issues of a new physics journal, *Progress of Theoretical Physics,* published in Kyoto. The two issues were printed in English on brownish paper of poor quality. They contained a total of six short articles. The first article in issue No. 2 was called "On a Relativistically Invariant Formulation of the Quantum Theory of Wave Fields," by S. Tomonaga of Tokyo University. Underneath it was a footnote saying, "Translated from the paper . . . (1943) appeared originally in Japanese." Hans gave me the article to read. It contained, set out simply and lucidly without any mathematical elaboration, the central idea of Julian Schwinger's theory. The implications of this were astonishing. Somehow or other, amid the ruin and turmoil of the war, totally isolated from the rest of the world, Tomonaga had maintained in Japan a school of research in theoretical physics that was in some respects ahead of anything existing anywhere else at that time. He had pushed on alone and laid the foundations of the new quantum electrodynamics, five years before Schwinger and without any help from the Columbia experiments. He had not, in 1943, completed the theory and developed it as a practical tool. To Schwinger rightly belongs the credit for making the theory into a coherent mathematical structure. But Tomonaga had taken the first essential step. There he was, in the spring of 1948, sitting amid the ashes and rubble of Tokyo and sending us that pathetic little package. It came to us as a voice out of the deep.

A few weeks later, Oppy received a personal letter from Tomonaga describing the more recent work of the Japanese physicists. They had been moving ahead fast in the same direction as Schwinger. Regular communications were soon established. Oppy invited Tomonaga to visit Princeton, and a succession of Tomonaga's students later came to work with us at Princeton and at Cornell. When I met Tomonaga for the first time, a letter to my parents recorded my immediate impression of him: "He is more able than either Schwinger or Feynman to talk about ideas other than his own. And he has enough of his own too. He is an exceptionally unselfish person." On his table among the physics journals was a copy of the New Testament.

Questions and Topics for Discussion and Writing

1. In recalling his experiments as a graduate student, Dyson concerns himself on one level of his essay with education. What distinction does he make between a scientist and a teacher? What does he imply about Bethe as a science teacher?
2. The third and fourth paragraphs, in particular, are clear, well organized, and readable. Analyze the elements of the paragraphs and determine what makes them successful.
3. In the sixth paragraph, Dyson personifies electrons and quotes Einstein's paradoxical statement "One may say the eternal mystery of the world is its comprehensibility." What is the effect of such techniques in a discussion of natural law?
4. What does Dyson say was the terrible effect, the "sin," of the scientists during war time? Is Oppenheimer justified in using such a strong term? Write an extended definition of "sin" or of "innocence."
5. Many psychologists have studied the effects of what they call "group think," the phenomenon that keeps members of a group from making critical objections to a course of action on which the group has embarked. Consult a library source on "group think" and write an essay applying it in this context, or in another that interests you (for example, the Vietnamese conflict, Watergate, mass suicide).
6. A major and continuing theme in American literature, especially in works by writers such as Nathaniel Hawthorne and Henry James, is one that contrasts the "innocence" of Americans with the greater "sophistication" of Europeans. How does Dyson treat this theme as a part of his distinction between European and American scientists during the two World Wars and the Vietnamese War? Do you agree that Americans exhibit a more "tragic view" of life since the Vietnamese War?
7. Dyson labels Dick Feynman a "profoundly original scientist." What qualities in Feynman has Dyson observed to justify this judgment?
8. According to Dyson, what is a "pictorial imagination" and a "muscle of intuition"? Why are such faculties said to be better than an "analytic mind" for a scientist?

Sinclair Lewis

Sinclair Lewis (1885–1951), the son, grandson, and nephew of physicians, was a keen and perceptive writer who gained international fame during the 1920s for his novels satirizing—often caricaturing—everyday American life. Born in Sauk Centre, Minnesota, and reared under the influence of a puritanical, stern father and a sometimes distant stepmother, Lewis was an awkward child, homely and solitary. At seventeen, he enrolled at Yale, where he edited the *Yale Literary Magazine.*

In 1925, he published *Arrowsmith,* a novel for which he was later offered the Pulitzer Prize. He declined the award, saying that such prizes tend to "legislate" taste. Some critics speculate, however, that he was trying to punish the Pulitzer committee for not giving him the prize for *Main Street* (1920) or *Babbitt* (1922). In 1930, he became the first American to win the Nobel Prize for Literature, which cited his "talent for painting life and creating character types." Lewis accepted the Nobel Prize because he felt it honored the writer's entire body of work.

The most compelling theme of *Arrowsmith* is the necessity for an obsessive struggle to maintain standards in medical science. The chief character, Martin Arrowsmith, a physician who becomes a research bacteriologist, pursues research to the exclusion of everything else. He cannot be satisfied with less than truth. Lewis, often concerned with the conflict between the Establishment and the individual, focuses in *Arrowsmith* on the role of the scientist as idealist. In his refusal to compromise the standards of science, Martin Arrowsmith remains victorious and thus reflects the high view of his creator, Lewis.

Portrait of a Scientist

Sinclair Lewis

I

Professor Max Gottlieb was about to assassinate a guinea pig with anthrax germs, and the bacteriology class were nervous.

They had studied the forms of bacteria, they had handled Petri dishes and platinum loops, they had proudly grown on potato slices the harmless red cultures of *Bacillus prodigiosus,* and they had come now to pathogenic germs and the inoculation of a living animal with swift disease. These two beady-eyed guinea pigs, chittering in a battery jar, would in two days be stiff and dead.

Martin had an excitement not free from anxiety. He laughed at it, he remembered with professional scorn how foolish were the lay visitors to the laboratory, who believed that sanguinary microbes would leap upon them from the mysterious centrifuge, from the benches, from the air itself. But he was conscious that in the cotton-plugged test-tube between the instrument-bath and the bichloride jar on the demonstrator's desk were millions of fatal anthrax germs.

The class looked respectful and did not stand too close. With the flair of technique, the sure rapidity which dignified the slightest movement of his hands, Dr. Gottlieb clipped the hair on the belly of a guinea pig held by the assistant. He soaped the belly with one flicker of a hand-brush, he shaved it and painted it with iodine.

(And all the while Max Gottlieb was recalling the eagerness of his first students, when he had just returned from working with Koch and Pasteur, when he was fresh from enormous beer seidels and Korpsbrüder and ferocious arguments. Passionate, beautiful days! *Die goldene Zeit!* His first classes in America, at Queen City College, had been awed by the sensational discoveries in bacteriology; they had crowded about him reverently; they had longed to know. Now the class was a mob. He looked at them—Fatty Pfaff in the front row, his face vacant as a doorknob; the co-eds emotional and frightened; only Martin Arrowsmith and Angus Duer visibly intelligent. His memory fumbled for a pale blue twilight in Munich, a bridge and a waiting girl, and the sound of music.)

He dipped his hands in the bichloride solution and shook them—a quick shake, fingers down, like the fingers of a pianist above the keys. He took a hypodermic needle from the instrument-bath and lifted the test-tube. His voice flowed indolently, with German vowels and blurred w's:

"This, gentlemen, iss a twenty-four-hour culture of *Bacillus anthracis.* You will note, I am sure you will have noted already, that in the bottom of the tumbler there was cotton to keep the tube from being broken. I cannot advise breaking tubes of anthrax germs and afterwards getting the hands into the culture. You *might* merely get anthrax boils—"

The class shuddered.

Gottlieb twitched out the cotton plug with his little finger, so neatly that the medical student who had complained, "Bacteriology is junk; urinalysis and blood tests are all the lab stuff we need to know," now gave him something of the respect they had for a man who could do card tricks or remove an appendix in seven minutes. He agitated the mouth of the tube in the Bunsen burner, droning, "Everytime you take the plug from a tube, flame the mouth of the tube. Make that a rule. It is a necessity of the technique, and technique, gentlemen, is the beginning of all science. It iss also the least-known thing in science."

The class was impatient. Why didn't he get on with it, on to the entertainingly dreadful moment of inoculating the pig?

(And Max Gottlieb, glancing at the other guinea pig in the prison of its battery jar, meditated, "Wretched innocent! Why should I murder him, to teach Dummköpfe? It would be better to experiment on that fat young man.")

He thrust the syringe into the tube, he withdrew the piston dextrously with his index finger, and lectured:

"Take one half c.c. of the culture. There are two kinds of M.D.'s—those to whom c.c. means cubic centimeter and those to whom it means compound cathartic. The second kind are more prosperous."

(But one cannot convey the quality of it: the thin drawl, the sardonic amiability, the hiss of the s's, the d's turned into blunt and challenging t's.)

The assistant held the guinea pig close; Gottlieb pinched up the skin of the belly and punctured it with a quick down thrust of the hypodermic needle. The pig gave a little jerk, a little squeak, and the co-eds shuddered. Gottlieb's wise fingers knew when the peritoneal wall was reached. He pushed home the plunger of the syringe. He said quietly, "This poor animal will now soon be dead as Moses." The class glanced at one another uneasily. "Some of you will think that it does not matter; some of you will think, like Bernard Shaw, that I am an executioner and the more monstrous because I am cool about it; and some of you will not think at all. This difference in philosophy iss what makes life interesting."

While the assistant tagged the pig with a tin disk in its ear and restored it to the battery jar, Gottlieb set down its weight in a notebook, with the time of inoculation and the age of the bacterial culture. These notes he reproduced on the blackboard, in his fastidious script, murmuring, "Gentlemen, the most important part of living is not the living but pondering upon it. And the most important part of experimentation is *quantitative* notes—in ink. I am told that a great many clever people feel they can keep notes in their heads. I have often observed with pleasure that such persons do not have heads in which to keep their notes. This iss very good, because thus the world never sees their results and science is not encumbered with them. I shall now inoculate the second guinea pig, and the class will be dismissed. Before the next lab hour I shall be glad if you will read Pater's 'Marius the Epicurean,' to derive from it the calmness which is the secret of laboratory skill."

II

As they bustled down the hall, Angus Duer observed to a brother, Digam, "Gottlieb is an old laboratory plug; he hasn't got any imagination; he sticks here instead of getting out into the world and enjoying the fight. But he certainly is handy. Awfully good technique. He might have been a first-rate surgeon, and made fifty thousand dollars a year. As it is, I don't suppose he gets a cent over four thousand!"

Ira Hinkley walked alone, worrying. He was an extraordinarily kindly man, this huge and bumbling parson. He reverently accepted everything, no matter how contradictory to everything else, that his medical instructors told him, but this killing of animals—he hated it. By a connection not evident to him he remembered that the Sunday before, in the slummy chapel where he preached during his medical course, he had exalted the sacrifice of the martyrs and they had sung of the blood of the lamb, the fountain filled with blood drawn from Emmanuel's veins, but this meditation he lost, and he lumbered toward Digamma Pi in a fog of pondering pity.

Clif Clawson, walking with Fatty Pfaff, shouted, "Gosh, ole pig certainly did jerk when Pa Gottlieb rammed that needle home!" and Fatty begged, "Don't! Please!"

But Martin Arrowsmith saw himself doing the same experiment and, as he remembered Gottlieb's unerring fingers, his hands curved in imitation.

III

The guinea pigs grew drowsier and drowsier. In two days they rolled over, kicked convulsively, and died. Full of dramatic expectation, the class reassembled for the necropsy. On the demonstrator's table was a wooden tray, scarred from the tacks which for years had pinned down the corpses. The guinea pigs were in a glass jar, rigid, their hair ruffled. The class tried to remember how nibbling and alive they had been. The assistant stretched out one of them with thumb-tacks. Gottlieb swabbed its belly with a cotton wad soaked in lysol, slit it from belly to neck, and cauterized the heart with a red-hot spatula—the class quivered as they heard the searing of the flesh. Like a priest of diabolic mysteries, he drew out the blackened blood with a pipette. With the distended lungs, the spleen and kidneys and liver, the assistant made wavy smears on glass slides which were stained and given to the class for examination. The students who had learned to look through the microscope without having to close one eye were proud and professional, and all of them talked of the beauty of identifying the bacillus, as they twiddled the brass thumb-screws to the right focus and the cells rose from cloudiness to sharp distinctness on the slides before them. But they were uneasy, for Gottlieb remained with them that day, stalking behind them, saying nothing, watching them always, watching the disposal of the remains of the guinea pigs, and along the benches ran nervous rumors about a bygone student who had died from anthrax infection in the laboratory.

IV

There was for Martin in these days a quality of satisfying delight; the zest of a fast hockey game, the serenity of the prairie, the bewilderment of great music, and a feeling of creation. He woke early and thought contentedly of the day; he hurried to his work, devout, unseeing.

The confusion of the bacteriological laboratory was ecstasy to him—the students in shirt-sleeves, filtering nutrient gelatine, their fingers gummed from the crinkly gelatine leaves; or heating media in an autoclave like a silver howitzer. The roaring Bunsen flames beneath the hot-air ovens, the steam from the Arnold sterilizers rolling to the rafters, clouding the windows, were to Martin lovely with activity, and to him the most radiant things in the world were rows of test-tubes filled with watery serum and plugged with cotton singed to a coffee brown, a fine platinum loop leaning in a shiny test-glass, a fantastic hedge of tall glass tubes mysteriously connecting jars, or a bottle rich with gentian violet stain.

He had begun, perhaps in youthful imitation of Gottlieb, to work by himself in the laboratory at night. . . . The long room was dark, thick dark, but for the gas-mantle behind his microscope. The cone of light cast a gloss on the bright brass tube, a sheen on his black hair, as he bent over the eyepiece. He was studying trypanosomes from a rat—an eight-branched rosette stained with polychrome methylene blue; a cluster of organisms delicate as a narcissus, with their purple nuclei, their light blue cells, and the thin lines of the flagella. He was excited and a little proud; he had stained the germs perfectly, and it is not easy to stain a rosette without breaking the petal shape. In the darkness, a step, the weary step of Max Gottlieb, and a hand on Martin's shoulder. Silently Martin raised his head, pushed the microscope toward him. Bending down, a cigarette stub in his mouth—the smoke would have stung the eyes of any human being—Gottlieb peered at the preparation.

He adjusted the gas light a quarter inch, and mused, "Splendid! You have craftsmanship. Oh, there is an art in science—for a few. You Americans, so many of you—all full with ideas, but you are impatient with the beautiful dullness of long labors. I see already—and I watch you in the lab before— perhaps you may try the trypanosomes of sleeping sickness. They are very, very interesting, and very, very ticklish to handle. It is quite a nice disease. In some villages in Africa, fifty per cent of the people have it, and it is invariably fatal. Yes, I think you might work on the bugs."

Which, to Martin, was getting his brigade in battle.

"I shall have," said Gottlieb, "a little sandwich in my room at midnight. If you should happen to work so late, I should be very pleast if you would come to have a bite."

Diffidently, Martin crossed the hall to Gottlieb's immaculate laboratory at midnight. On the bench were coffee and sandwiches, curiously small and excellent sandwiches, foreign to Martin's lunchroom taste.

Gottlieb talked till Clif had faded from existence and Angus Duer seemed but an absurd climber. He summoned forth London laboratories, dinners on frosty evenings in Stockholm, walks on the Pincio with sunset behind the dome of San Pietro, extreme danger and overpowering disgust from excreta-smeared garments in an epidemic at Marseilles. His reserve slipped from him and he talked of himself and of his family as though Martin were a contemporary.

The cousin who was a colonel in Uruguay and the cousin, a rabbi, who was tortured in a pogrom in Moscow. His sick wife—it might be cancer. The three children—the youngest girl, Miriam, she was a good musician, but the boy, the fourteen-year-old, he was a worry; he was a saucy, he would not study. Himself, he had worked for years on the synthesis of antibodies; he was at present in a blind alley, and at Mohalis there was no one who was interested, no one to stir him, but he was having an agreeable time massacring the opsonin theory, and that cheered him.

"No, I have done nothing except be unpleasant to people that claim too much, but I have dreams of real discoveries some day. And— No. Not five times in five years do I have students who understand craftsmanship and precision and maybe some big imagination in hypotheses. I t'ink perhaps you may have them. If I can help you—So!

"I do not t'ink you will be a good doctor. Good doctors are fine—often they are artists—but their trade, it is not for us lonely ones that work in labs. Once, I took an M.D. label. In Heidelberg that was—Herr Gott, back in 1875! I could not get much interested in bandaging legs and looking at tongues. I was a follower of Helmholtz—what a wild blithering young fellow! I tried to make researches into the physics of sound—I was bad, most unbelievable, but I learned that in this vale of tears there is nothing certain but the quantitative method. And I was a chemist—a fine stink-maker was I. And so into biology and much trouble. It has been good. I have found one or two things. And if sometimes I feel an exile, cold—I had to get out of Germany one time for refusing to sing *Die Wacht am Rhein* and trying to kill a cavalry captain—he was a stout fellow—I had to choke him—you see I am boasting, but I was a lifely *Kerl* thirty years ago! Ah! So!

"There is but one trouble of a philosophical bacteriologist. Why should we destroy these amiable pathogenic germs? Are we too sure, when we regard these oh, most unbeautiful young students attending Y.M.C.A.'s and singing dinkle-songs and wearing hats with initials burned into them—iss it worth while to protect them from the so elegantly functioning *Bacillus typhosus* with its lovely flagella? You know, once I asked Dean Silva would it not be better to let loose the pathogenic germs on the world, and so solve all economic questions. But he did not care for my met'od. Oh, well, he is older than I am; he also gives, I hear, some dinner parties with bishops and judges present, all in nice clothes. He would know more than a German Jew who loves Father Nietzsche and Father Schopenhauer (but damn him, he was teleological-minded!) and Father Koch and Father Pasteur and Brother Jacques Loeb and Brother Arrhenius. Ja! I talk foolishness. Let us go look at your slides and so good-night."

When he had left Gottlieb at his stupid brown little house, his face as reticent as though the midnight supper and all the rambling talk had never happened, Martin ran home, altogether drunk.

Questions and Topics for Discussion and Writing

1. Are you able to judge from this excerpt that the novel is, as some have termed it, a "warning to scientists"? Consider the impact of the opening sentence. What stylistic device does Lewis employ?
2. Consider Lewis's description of Gottlieb's actions while he clipped the hair on the

       belly of a guinea pig. What is the effect of Lewis's analogy of Gottlieb's actions to those of a pianist and a magician? What is the effect of Gottlieb's struggling to recall a love experience and the sound of music?

3.   How does Gottlieb's statement "This poor animal will now soon be dead as Moses" convey Lewis's belief about the conflict between science and religion?

4.   Contrast the different reactions of Angus Duer, Digam, Ira Hinkley, Clif Clawson, and Martin Arrowsmith to observations of Gottlieb's actions.

5.   How does Lewis's language in Section IV indicate that Martin has experienced an almost mystical impulse that will make him a scientist? Lewis uses *oxymorons*, figures of speech that combine two contradictory terms (for example, "beautiful dullness" and "nice disease"). How do these oxymorons help to externalize Gottlieb's attitude toward science and its relationship to humanity?

6.   According to Gottlieb, what are the traits desirable in a real scientist?

7.   What do you make of the term *philosophical bacteriologist*?

8.   After researching briefly the ideas and contributions offered by Nietzsche, Schopenhauer, Koch, Pasteur, Loeb, and Arrhenius, consider why Gottlieb calls the first four "Father" and the last two "Brother."

Bernd Heinrich

Bernd Heinrich (b. 1940) was shaped by unusual childhood experiences. With his parents and sisters, he spent his early years in deep forests, living on what could be foraged at a time after World War II when starvation threatened many. Born on a large farm in what is now Poland, Heinrich was barely aware of the serenity of his early life before he and his family "were caught between Communists, Nazis, and British and American bombers."

His father, an enterprising amateur biologist, supplemented the family's livelihood before and after the war years by selling to American museums rare birds, insects, and small mammals, which the three children helped to catch. Heinrich credits his "Huck Finn existence" with leading him toward his career of biologist.

A United States citizen since 1958, Heinrich received his B.A. and M.S. degrees from the University of Maine and his Ph.D. from the University of California in 1970, advancing to a professorship at the UC–Berkeley. From there, he joined the zoology faculty at the University of Vermont, where he currently teaches.

Heinrich's published research in zoology is marked by originality. His *Bumblebee Economics* (1978) contributes new understanding to insect physiology. It reveals how bumblebees and some other insects regulate an elevated body temperature regardless of the temperature of the air, an ability previously believed possible only in birds and mammals. The book uses the metaphor of economics to explain Heinrich's findings on the flexible foraging strategy of bumblebees. Looking at bees from an ecological perspective of plants, he sees them as collectors and distributors of wealth and considers their co-evolution with flowers. "Even though bees are free agents," Heinrich writes, "an order ensues out of their combined actions."

Heinrich's individuality also shows in his personal activities. At age 40, he wrote in a recent letter, "I hold the U.S. national record (for any age) for 100 km., 200 km., 100 miles, and 24-hour run." He also holds some world running records for his age.

Heinrich's vigorous personal viewpoint creates a message on ecology: "I feel strongly about *conservation*—in terms of *ecology*—and I am not too sympathetic with the misty-eyed who focus on *individual* furry, feathered creatures to the exclusion of the *environment*, where the real issues are." Heinrich sees overpopulation as "the root of most of our problems" and deplores the fact that the "rich boyhood" he recollects is out of reach for most modern American youngsters, largely because of the failure of lawmakers to develop an "ecological conscience." Heinrich adds, however, that this failure also comes through lack of understanding: Legislators prohibit, for example, the keeping of a pet crow (a "protected" bird), when a genuine "ecological conscience" would protect, not the crow, but "the fields and forests that breed crows."

"Flight into the Forest" is taken from Heinrich's autobiographical account, *In a Patch of Fireweed.* One reviewer, who sees the title as suggesting "randomness and unpredictability," describes Heinrich as having "double vision—the ecologist's perception of struggle and the physiologist's awareness of integration." "Flight into the Forest" includes Heinrich's original drawings, of which he writes, "Drawing is a way for me to get reacquainted with my feelings for the subject and focuses my attention on what I see as beautiful—in the same way that motivates my research and my writing."

Heinrich explains that this book "is not really for scholars. . . . My hope is to capture here some of the feelings of science that I have had to leave out of my [scientific] writings; the sounds and sights, the endless chores and happy accidents, the obsessions, the wonder of it all."

Flight into the Forest
Bernd Heinrich

We lived deep in the forest for five years. We had no work and hardly ever any money. The civilization that seals most of us off from the stark reality of existence had broken down. We were totally immersed in nature. Like most animals, our major concern was with finding food. Especially I remember foraging for berries.

It was late in the afternoon. I must have picked a million raspberries already. Ulla, Mamusha, and Marianne were picking too. We worked fast to clear out the key patch we had discovered before any of the other refugees from the nearby towns might find it. It was boring to pick raspberries (and later in the summer blueberries), but occasionally I would find a big fat green caterpillar with red spots and tufts of short bristles and would take it back to our hut under the tall spruce trees by the brook. I would feed it fresh raspberry leaves every day, watch it spin a cocoon out of brown silk, and at long last see a beautiful silkmoth emerge. That was exciting and compensated somewhat for the tedious chore of picking berries all day long.

I didn't like picking berries because I had to move so slowly, from bush to bush. I much preferred picking mushrooms, when I could run at will through the damp forest, feeling the soft green moss under my bare feet. Every place in the forest was different. The orange Rehfusschen (*Chanterelles*) grew under deep-shaded spruces on dark brown needles, and I might see kinglets hopping among the branches above. Chocolate brown Steinpilze (*Boletus*) grew under huge beech trees along slopes that led down to the heather bog. The Birkenpilze (*Leccium*) with their scabrous stalks grew near birches when the first leaves began to fall and the ferns turned yellow. There were never too many mushrooms in any one place, so I had a chance to move around and to be by myself for many hours. When there is a space, the mind fills it. I filled mine with the sights, sounds, and smells of the woods.

Fishing for brown trout was the most fun of all, even though I had to spend most of the time on my belly to do it. I would lie on the edge of the stream and reach over and underneath the undercut banks. When my fingertips touched the smooth skin of a fish, I reached forward bit by bit until I had felt enough and could visualize in my mind the shape and size of the trout. Then I'd grab quickly as hard as I could right behind its gills. I had to hide each fish so that neither the gamekeeper nor the British occupation soldiers who sometimes also fished there (with hook and line) would know I was poaching "their" fish. I liked the soldiers. They sometimes gave me chocolate bars when they were out hunting. Also I found many of their cigarette butts from which I carefully took the tobacco for the grown-ups; cigarettes were scarce and available only on the black market.

I had no playmates and never owned a toy. Yet I didn't feel deprived. Who needs toys, after having seen caterpillars from up close and knowing they can turn into moths? In the spring I watched the brilliant red peacock butterfly with its yellow and blue eyespots on the wings, as it sunned itself after coming out of hibernation. Its spiny black caterpillars would later feed on the nettles growing along the brook. We, too, fed on the nettles, which make a good "spinach" when boiled.

I watched the warblers singing and fluttering in the shady forest as the green, almost transparent beech leaves were opening in the spring after a warm rain. I found their domed nests on the ground among the dead leaves, and I saw the

tiny speckled eggs. In the gnarled willow trees by the brook I saw a tiny long-tailed titmouse disappear into a fork of a tree. The bird had entered a nest covered with lichens, and the nest looked exactly like the lichen-covered tree itself. I climbed up and examined the lichens held together by spiderwebs collected by the birds. Soft feathers protruded from the entrance hole, and with my fingers I felt tiny eggs within the nest. I especially liked the bright blue titmouse. It was a compact little bird that was very inquisitive and like an acrobat, as easily hanging from branches as perching on them. It raised its young in old woodpecker nest holes.

In the fall the beechnuts and acorns ripened, and we had another harvest. We sold the acorns to the farmers in the village, who fed them to their pigs, and we sold the beechnuts to be pressed for oil and made into margarine. Picking beechnuts was worse than picking berries because they were so small, and because Mamusha did not like it when I stopped to pop one into my mouth.

The British occupation soldiers who came to the forest to hunt were not all good stalkers and trackers. But they had a lot of ammunition. They sometimes wounded game that got away and died later. One morning, when there was frost on the leaves, I saw ravens circling near the bend in the sandy road where I walked to go to school in the village of Hamfelde. I moved into the dense spruce thicket where the ravens went down, and there I found a half-eaten boar with coarse black hair and long tusks. The guts had been pulled out and eaten by the birds, but it still had fat on its hide and a little muscle. I got help, and we

Blue titmouse

hauled the remains home to our cabin, where we cut the fat off and fried it. I'd never tasted anything so delicious.

Boar were not scarce in the Hahnheide, but Marianne and I were sometimes less afraid of them than of people. I particularly remember one incident as we were coming back from the village carrying a fresh loaf of bread with thick, flaky crust. The aroma of that bread was overwhelming, and we succumbed, nibbling off some of the flakes of crust. They were so good that we pried off a few more, then pieces of the crust itself. We felt guilty for our misdeed, and walked slower and slower. Eventually dusk arrived, and we were still only at the bend in the dirt road where there was a slight rise. And at that place, looking up, we saw a man standing, staring at us. We panicked and bolted into the underbrush. I then led the way through the woods back onto the road far ahead, almost up to the footpath through the spruces that led us to the safety of our hut.

Early one spring, shortly after the snow was gone, we found ourselves without food: we had eaten all of our dried berries and mushrooms, and there wouldn't be new ones for a long time. Papa and I walked into the forest. He didn't seem to know where he was going, and he didn't talk at all. We came to a stand of beech trees, but there were no nuts under the dead leaves on the ground. The new leaves weren't yet out, and the sun still shone through the branches. We walked around all morning and found nothing. When we got tired, we sat down in the sun at the base of a huge tree. It was quiet, with only a dog yapping in the distance. Suddenly Papa sprang up, as though stung by a bee, and ran off in the direction of the sound of the dog. When he came back he was carrying over his shoulders a deer that was bloody with a torn throat. He had heard and recognized the cry that a deer makes when it dies. We picked our way carefully through the thickets back to the cabin, away from the dirt roads where the soldiers might see us and take the deer away.

Later Marianne and I found the third, and best, large game animal we were to have in our five years in the Hahnheide. One of our daily chores was to collect branches for the fire. Occasionally we heard a boar crashing through the thickets, and in the fall we heard elk bulls bellowing at night. One day we saw what looked like a huge patch of brown fur in a thicket near the brook. We came closer, and it didn't move: a dead elk. We ran back to the cabin and told Mamusha, Papa, and Ulla, who did not believe us at first. But they did investigate and could hardly believe our good fortune. We hid the elk by covering it with branches and waited for nightfall.

As soon as it was dark we dragged the elk into a spruce thicket, where we could work by candlelight and not be seen from the nearby road. Through most of the night we were busy skinning it, cutting it up, and carrying the pieces to the cabin. Of course we had no refrigerator. But we could not afford to let such a treasure spoil, nor could we afford to be seen with it. Our solution was to hang some of it in the chimney, where it was preserved by the smoke and hidden as well.

We were not permitted to own firearms, but slingshots were allowed. I made one using a forked branch and rubber from a discarded inner tube. I recall the intense pleasure of stalking birds in the woods, and occasionally killing one. Having a bird in my hand gave me a chance to see details of eyes, feet, and colors and patterns of feathers that could never be imagined. We also trapped some birds using the hairs from our horses' tails. The long, strong hairs were made into stiff lasso-like loops that were pegged onto a board. Birds hopping on the board, trying to pick up grain, were entangled in the snares and held fast. We hunted mice and shrews with pit traps dug in the forest. We fried the bodies of the birds and rodents, bones and all, in the grease we got from the wild boar the ravens had shown us, and there is, in my recollection, no greater treat than mice fried to a brown crisp, except possibly pork chops (which at that time I did not remember having eaten).

We used not only the meat of the animals we caught. We removed fleas from the mammals and sold them to Dr. Rothschild, a flea specialist in London. Each kind of mammal hosted different species of parasites. Merchandising fleas didn't make us rich, but every pfennig helped. Also, Mamusha skinned the birds and mammals and made study skins that we sold to the New York and

Chicago museums of natural history. However, at this time there was no mail service from Germany to other countries. Again we had incredible luck. Papa had long known Piet Hart Nibbrig, the general manager of the Bols Liquor Company in the Netherlands. Piet and his wife Nela had been frequent guests at Borowke before the war. As Hollanders, they were allowed to drive across the German border. In order to help us now, Piet picked up our merchandise and brought it to Holland, whence it was mailed to New York.

I enjoyed the hunting and trapping. It was always interesting. Before digging a pit trap for a small mammal, we searched for tiny tunnels under leaves, matted grass, and moss. It intrigued me that, as if by magic, various kinds of small furry animals had built amazingly intricate mazes and were hidden from our view, probably aware of our presence. Whatever it was they were doing, it was aimed at survival.

To make a good pit trap for mice, one must make sure that the sides of the pit are smooth and that no piece of root hangs down to afford a foothold. Mice caught in pits we made in the Hahnheide often tried to burrow down when they could not escape upward. We put a large clump of moss in one corner in each pit to give them a false sense of safety, so that they would not try to escape. Animals would hide under the moss rather than try to jump, climb, or dig out. I always ran ahead of Papa, to be first to lift the moss to see what kind of field, wood, or deer mouse or shrew might be underneath.

While we were building and tending our pit traps to catch mice, Papa showed me some small conical depressions in loose sand. They were pit traps made by an insect larva, the ant lion. These traps are designed to catch ants, beetles, caterpillars, centipedes, spiders, and other prey. The ant lion itself lies hidden in the sand at the bottom of the pit; there it flings up sand to line the walls with loose, treacherous footing. If a victim blunders past the rim, it slips down to the bottom of the pit.

I liked to stretch out on the ground and drop red wood ants into the ant lions' pit traps, watching the ant-lion larvae hurl sand up the sides of the pit. The loose sand fell back down, perpetuating small sandslides that kept the ant slip-

Shrew, with remnants of its last meal

ping downward, until the ant lion could reach it and pull it under the sand at the bottom. Little did I know then that thirty years later I would again be feeding ant lions in order to learn the details of their prey-catching behavior.

Questions and Topics for Discussion and Writing

1. What comparison or contrast does Heinrich make between conflict in civilization (war) and the struggle in nature?
2. How did his return to nature and his struggle for survival provide the basis for his choice of a *particular scientific career*?
3. How does the epigrammatic sentence in the third paragraph, "Where there is a space, the mind fills it," serve both to introduce and to conclude the author's meaning throughout the essay?
4. How does the author's telling of beautiful sights in nature and his relating of what seem to the reader to be repulsive acts such as eating "mice fried to a brown crisp" and "boar remains" express his acceptance of the indifference of nature?
5. Find specific sentences that reveal the author's early attraction to the life and death process.
6. Find passages in the essay that you believe are particularly effective.
7. How does the statement "We saw a man standing, staring at us" convey the author's feeling that he had become a part of the animal life in the forest?
8. How does the characterization of bees as "independent entrepreneurs" and "free agents" (*Bumblebee Economics*) resemble or differ from other views of insect activities? For example, you might study Petrunkevitch's essay "The Spider and the Wasp" or Whittemore's poem "The Tarantula."
9. How would you characterize Heinrich's prose style? Objective, poetic, practical, persuasive—do any of these terms apply?

Albert Szent-Györgyi

Nobel-Laureate Albert Szent-Györgyi (b. 1893), a native of Budapest, Hungary, is another of the scientists who are hard to confine to one neat compartment of science. On his mother's side a fourth-generation scientist, he learned early that "only intellectual values were worth striving for, artistic or scientific creation being the highest aim."

In pursuit of this aim, Szent-Györgyi found his way through several different fields of science. He began with anatomy and histology, but felt the need to expand his inquiry to living materials and therefore turned to physiology. When the complexity of physiology proved "overwhelming," he shifted to pharmacology, then bacteriology, physicochemistry, and chemistry, always seeking "simpler" units. His frequent change of research subjects provided him with an overall view of the problems in the life sciences. Szent-Györgyi has had what justly may be considered many different careers.

In addition to his dedication to science and its standards, he has been a voice in the political arena of our time, speaking of freedom and the responsibilities of the scientist to his fellow human beings. Szent-Györgyi calls this political activism the "external course" of his life, science being the "internal."

Szent-Györgyi came to the United States in 1947 to establish the Marine Biological Laboratory at Woods Hole, Massachusetts, now a world-famed institution. Many biologists, oceanographers, and students have received training and contributed to the unique ambiance—at once scientific and humane—of that research center, which encourages both scientific innovation and respect for nature. Through research and teaching, Szent-Györgyi in his later years has continued to influence the direction of science. The following excerpt is from "Lost in the Twentieth Century," which details the extraordinary scientific and political adventures that have formed his life.

Lost in the Twentieth Century

Albert Szent-Györgyi

Overlooking my case history, I find a complete dichotomy. On the one hand, my inner story is exceedingly simple, if not indeed dull: my life has been devoted to science and my only real ambition has been to contribute to it and live up to its standards. In complete contradiction to this, the external course has been rather bumpy. I finished school in feudal Hungary as the son of a wealthy landowner and I had no worries about my future. A few years later I find myself working in Hamburg, Germany, with a slight hunger edema. In 1942 I find myself in Istanbul, involved in secret diplomatic activity with a setting fit for a cheap and exciting spy story. Shortly after, I get a warning that Hitler had ordered the Governor of Hungary to appear before him, screaming my name at the top of his voice and demanding my delivery. Arrest warrants were passed out even against members of my family. In my pocket I find a Swedish passport, having been made a full Swedish citizen on the order of the King of Sweden—I am "Mr. Swenson," my wife, "Mrs. Swenson." Sometime later I find myself in Moscow, treated in the most royal fashion by the Government (with caviar three times a day), but it does not take long before I am declared "a traitor of the people" and I play the role of the villain on the stages of Budapest. At the same time, I am refused entrance to the USA for my Soviet sympathies. Eventually, I find peace at Woods Hole, Massachusetts, working in a solitary corner of the Marine Biological Laboratory. After some nerve-racking complications, due to McCarthy, things straightened out, but the internal struggle is not completely over. I am troubled by grave doubts about the usefulness of scientific endeavor and have a whole drawer filled with treatises on politics and their relation to science, written for myself with the sole purpose of clarifying my mind, and finding an answer to the question: will science lead to the elevation or destruction of man, and has my scientific endeavor any sense?

All this, in itself, would have no interest. There are many who did more for science, were braver, suffered more agony and even paid the penalty of death. What may lend interest to my story is that it reflects the turbulence of our days. So to give sense to my story I will have to start by asking: why all this trouble and what is its relation to science? . . .

We are living in the middle of the transition from the prescientific to the scientific thinking, hence the "tumult." We still have God on our lips and our coins, but no more in our hearts. If we are taken ill we may still pray, but we take penicillin alongside. We pray for peace but heap up H-bombs for safety. We preach Christ and talk "overkill." This world is symbolized for me by the colossal statue of Christ, standing on a hill in Spain, stretching out His Arms to mankind, and wearing on His Head an enormous lightning conductor to protect Him, should the Almighty Father try to smite Him by lightning. We find the new expanding Universe a rather cold place and do not dare to abandon the old one. The trouble is that the two worlds cannot be mixed and the father inquisitor was right when he said to Galileo that "your teaching and the teaching of the church cannot exist side by side." We cannot build, unpunished, H-bombs by science either, and then run them with the XVIII Century egotistic, narrow, sentimental, and deceitful political thinking. It makes no sense to shoot astronauts out into space to reach other stars and erect ten-foot concrete walls to separate man from man. In its own time prescientific thinking did build a stable world, but science has irretrievably undermined the acquiescence in misery as the attribute of human existence, and has undermined the old hierarchies of gods, princes, barons, haves and have-nots, well-fed and hungry, developed and underdeveloped.

There is no way back, and we have to face squarely, the free choice between undreamed of wealth and dignity, and self-destruction which science has offered. My problem is: to what is science leading, and whether science can build a world in which man can feel, once more, at home? I will attempt to answer these questions at the end, after having given my case history. . . .

On my Mother's side, I am the fourth generation of scientists. My Father was interested only in farming and so my Mother's influence prevailed. Music filled the house and the conversation at the table roamed about the intellectual achievements of the entire world. Politics and finance had no place in our thoughts. I am a scientist, myself, because at an early age I learned that only intellectual values were worth striving for, artistic or scientific creation being the highest aim. I strongly believe that we establish the coordinates of our evaluation at a very early age. What we do later depends on this scale of values which mostly cannot be changed later. We are somewhat like Dr. Lorenz' goose which has hatched at the foot of a chair and recognized the chair as its mother all its later life. This is important for education, in case we are not intending to produce only "corporation men" with their intellectual crew cuts.

I must have been a very dull child. Nothing happened to me. I read no books and needed private tutoring to pass my exams. Around puberty, something changed and I became a voracious reader and decided to become a scientist. My uncle, a noted histologist (M. Lenhossek), who dominated our family and was a precocious child himself, violently protested, seeing no future for such a dull youngster in science. When his opinion gradually improved, he consented to my going into cosmetics. Later, he even considered my becoming a dentist. When I finished high school with top marks, he admitted the possibility of my becoming a proctologist (specialist of anus and rectum; he had haemorrhoids). So my first scientific paper, written in the first year of my medical studies, dealt with the epithelium of the anus. I started science on the wrong end, but soon I shifted to the vitreous body, the fibrillar fine structure I explored with new methods.

I have mentioned this early history of mine because it suggests that no final judgment should be made of children at too early an age.

I must have achieved some reputation as a histologist when, as a third-year medical student, I became increasingly discontent with morphology which told me little about life. So, I shifted to physiology but had to break my studies for compulsory military service. World War I found me in uniform.

Centuries-old tradition told us Hungarians to ask no question when we were called upon to fight. I did accordingly, but during the first three years of the war I was gradually overcome by a burning desire to return to science. At the same time I became increasingly disgusted with the moral turpitude of military service. I could see clearly that we had lost the war and that we were being sacrificed senselessly by a ruling clique; the best service I could do for my country was to stay alive. So, one day, when in the field, I took my gun and shot myself through the bone of my arm. With all the deeply ingrained tradition this was quite difficult to do and it was also the more dangerous road. Anyway, it took me back to the capital where I got my M.D., after which I continued my service in a bacteriological laboratory of the army. Here, I got into trouble but once, when I objected to experiments, dangerous to life, done on Italian prisoners of war. Since the man responsible for these experiments had two stars more than I had, I was punished, and sent to the North Italian swamps where tropical malaria made life expectancy very short. A few weeks later the war collapsed and so I pulled out alive and returned to the laboratory.

I wanted to understand life but found the complexity of physiology overwhelming. So I shifted to pharmacology where, at least, one of the partners, the drug, was simple. This, I found, did not relieve the difficulty. So, I went into bacteriology, but found bacteria too complex, too. I shifted on, to physicochemistry and then to chemistry, that is, to molecules, the smallest units in those days. Ten years ago I found molecules too complex and shifted to electrons, hoping to have reached bottom. But Nature has no bottom: its most basic principle is "organization." If Nature puts two things together she produces something new with new qualities, which cannot be expressed in terms of qualities of the components. When going from electrons and protons to atoms, from here to molecules, molecular aggregates, etc., up to the cell or the whole animal, at every level we find something new, a new breathtaking vista. Whenever we separate two things, we lose something, something which may have been the most essential feature. So now, at 68, I am to work my way up again following electrons in their motion through more extensive systems, hoping to arrive, someday, at an understanding of the cellular level of organization. So the internal course of my life made a smooth sinusoid curve; not so the external course.

After the War, I became assistant at the pharmacological laboratory of the newly founded University in Pozsony, an old Hungarian town. A few months later Pozsony was given, by the Versailles Treaty, to Czechoslovakia (it is now called Bratislava) and we had to clear out. We saved our scientific equipment not without danger, getting it one night, dressed as workmen, through the closely guarded gates of the campus. Meanwhile, in Hungary, the communists took over, which meant a complete loss of all my belongings. At the very last moment, I rescued one thousand English pounds. These I shared with my Mother, whom I visited at Budapest. For such a visit the wintry Danube had to be crossed in a small overcrowded boat at night, at a point where there were no Czech patrols, who shot at sight. In my company was a nun, Sister Angelica, who was deadly frightened and clung to me desperately. On my return I had to spend a night in the snow and arrived in Pozsony with a grave pneumonia. I probably owe my life to the devoted nursing of Sister Angelica. After this, I took my wife and child and steered west. The English pounds allowed me to live, very modestly, for a little while, during which time I wanted to gratify my desire to do research. . . .

Now, I thought myself capable of tackling a biochemical problem. I embarked on biological oxidations. At that time a violent controversy raged between O. Warburg and H. Wieland and their followers. The former thought that oxygen activation was the most essential feature of respiration, while Wieland put H-activation in the fore. I could show that both processes were involved. I simply knocked out O_2 activation (and with it, respiration) by cyanide and then added methylene blue to the minced tissue. The dye restored respiration, replacing O_2 activation. It was reduced by activated H and then reoxidized spontaneously. During these experiments I became fascinated by the succino and citrocodehydrogenase. These dehydrogenases differed from other dehydrogenases by being bound to structure, and "structure" had to mean something very important. They could not possibly be just ordinary metabolic enzymes, they had to have some general catalytic role. If this was so, then the whole of respiration had to be inhibited once the succino-dehydrogenase was inactivated, which could be done by malonic acid, as shown earlier by Quastel. So I added malonic acid to the minced tissue, and respiration stopped. This proved that succinic acid (and citric acid) had to have some general catalytic activity and could not be simply metabolites, as thought before. These ideas were later completed by Krebs and are the foundation of the so-called "Krebs cycle." It was partly this discovery of the C_4 dicarboxylic acid catalysis which was honored later by the Nobel prize. . . .

One day a nice young American-born Hungarian, J. Swirbely, came to Szeged to work with me. When I asked him what he knew he said he could find out whether a substance contained Vitamin C. I still had a gram or so of my hexuronic acid. I gave it to him to test for vitaminic activity. I told him that I expected he would find it identical with Vitamin C. I always had a strong hunch that this was so but never had tested it. I was not acquainted with animal tests in this field and the whole problem was, for me, too glamourous, and vitamins were, to my mind, theoretically uninteresting. "Vitamin" means that one has to eat it. What one has to eat is the first concern of the chef, not the scientist.

Anyway, Swirbely tested hexuronic acid. A full test took two months but after one month the result was evident: hexuronic acid was Vitamin C. We made no secret of this and finished the test which left no doubt about the identity. So, we (Haworth and I) rebaptized hexuronic acid to "ascorbic acid."

There we were. Ascorbic acid seemed medically most important but there was none of it, and none of the available vegetable sources allowed big-scale preparation. Adrenals were not available, in quantity, in Hungary. As it happened, Szeged is the center of the paprika (red pepper) industry. Paprika was not available at Cambridge. I once saw it on the market but the vendor cautioned me that it was poisonous. One night we had fresh red pepper for supper. I did not feel like eating it and thought of a way out. Suddenly it occurred to me that this was practically the only plant I had never tested. I took it to the laboratory and about midnight I knew that it was a treasure chest of vitamin C, containing 2 mg per gram. A few weeks later I had kilograms of crystalline Vitamin C which I distributed all over the world among researchers who wanted to work on it. This soon made complete analysis and synthesis possible. I received my Nobel prize partly for this work which also led to another unexpected discovery. When I still had only impure but highly concentrated solutions of ascorbic acid we tried my extracts in cases of Henochs' Purpura. In scurvy there is a great capillary fragility causing subcutaneous bleeding, so it seemed logical to try my extracts in purpura (subcutaneous bleeding). They worked. When I had crystalline ascorbic acid we tried it again, expecting a still stronger action. It did nothing. Evidently, my impure extract contained an additional substance responsible for the action. I guessed that it might be "flavones" which did the trick. My guess proved right. I isolated the flavones from "paprika" and they cured purpura. I called this group of substances Vitamin "P." I used the letter *P* because I was not quite sure that it was a vitamin. The alphabet was occupied only up to F so there was ample time to eliminate "P" without causing trouble if the vitamin nature became disproved. . . .

I felt I had now enough experience for attacking some more complex biological process, which could lead me closer to the understanding of life. I chose muscle contraction. With its violent physical, chemical, and dimensional changes, muscle is an ideal material to study. If one embarks on such a new field one usually does not know where to begin. There is one thing one can always do, and this I did: repeat the work of old masters. I repeated what W. Kühne did a hundred years earlier. I extracted myosin with strong potassium chloride (KCl) and kept my eyes open. With my associate, I. Banga, we observed that if the extraction was prolonged, a more sticky extract was obtained without extracting much more protein. We soon found that this change was due to the appearance of a new protein "actin," isolated in a very elegant piece of work by my pupil, F. Straub, while I "crystallized" myosin. Myosin, evidently, was a contractile protein, but the trouble was that *in vitro* it would do nothing. A contractile protein should contract wherever it is. So we made threads of the highly viscous new complex of actin and myosin, "actomyosin," and added boiled muscle juice. The threads contracted. To see them contract for the first time, and to have reproduced *in vitro* one of the oldest signs of life, motion, was perhaps the most thrilling moment of my life. A little cookery soon showed that what made it contract was ATP and ions. My conclusion, that muscle contraction was essentially an interaction of actomyosin and ATP, was soon strongly attacked, so I developed (later at Woods Hole) the method of glycerination, and glycerinated (extracted with diluted glycerol at low temperature) the psoas muscle of the rabbit. This method is now widely used for conservation of biological material such as sperm. On addition of ATP, my glycerinated muscle contracted, developing the same tension as it developed maximally *in vivo*. This satisfied me and I was sure that in a few weeks' time the whole problem of muscle contraction would be cleared up, but ten years later I still did not understand muscle, which made me conclude that something had to be missing from our basic ideas, something that was essential for the understanding of energy transformation. So I left muscle to find what this something is. This took me, gradually, into my present field, that of electronic dimensions and mobility.

As a temporary president of my university at Szeged, I tried to put into action the ideas picked up in the west. I created an intense cultural life among students which culminated in our producing Hamlet, and producing it well. But my democratic ideas brought me more and more into conflict with the rising tide of fascism. It was not I who went into politics. Politics came into our lives and when books were burned and my Jewish friends were prosecuted I had to say "yes" or "no." I said "no" and when later, during World War II a group of leading Hungarians came, secretly, to me and asked me to do something to save Hungary from Germany's grip, I went, under cover of an alleged lecture, to Istanbul to get in touch with the British and American diplomats to see what could be done. This was a risky undertaking, for German-occupied territory had to be crossed and Istanbul was the spying center, with highly developed techniques, and I was a newcomer in this business. I felt that I could be more useful if I did not go merely as a private individual to Istanbul and took a chance. I went to our Prime Minister, Mr. M. Kallay, and told him about my plans. Outwardly, Mr. Kallay was a Nazi, but I suspected that he was a good Hungarian, waiting for his chance to bring his country over to the other side. My guess was right. Instead of having me arrested he asked me to represent him and convey certain messages to the Allies. In Istanbul I succeeded in getting in touch with the head of the British Secret Service, making with him detailed plans which soon had the blessing of London. What made these dealings exciting was that, till the end, I could not know for certain whether I was dealing with the British, or the German Secret Service. This I could only find out later, when crossing German territory. Not being arrested on my return, I was finally sure that it was the British to whom I had talked.

Unfortunately, the secret of my mission leaked out, and I could not set up a secret wireless station which was essential for my plans. I was placed under house arrest. Hitler demanded my delivery. Later, when he occupied Hungary, I avoided final arrest by the Gestapo only by an inch, owing my escape more to good luck than ability. Arrest would have meant a very painful death. Even my daughter had to go into hiding, an arrest warrant having been issued also against her. Working against Hitler and living underground was full of colors which were not always pleasant. I expected to be killed so I wrote up my observations on muscle, which I did not want to be lost. I sent them for publication in the *Acta Scandinavica* to my friend Hugo Theorell. Not knowing where I was, he corroborated acceptance by wire, "care of Swedish Legation, Budapest." Fate would have it that at that time I was actually hiding at the Swedish Legation, and so Theorell's wire gave me away. The Gestapo immediately searched the surrounding houses for subterranean exits from the Swedish Legation, which served as a warning. Also, a hint from a friendly German diplomat made it evident that arrest was imminent. So Per Anger, the head of the Legation, smuggled me out in the back of his car the next night. Shortly after, the Nazis broke into the Swedish Legation, searched, robbed, and practically destroyed it. Then followed a series of exciting situations shared by my wife. At the end, we had to part, hiding together becoming too risky. Two of my hiding places were destroyed by bombs shortly after I left them, and, in the end I could avoid arrest only by hiding in the vicinity of the Soviet lines where the Gestapo did not dare to come.

The profound disgust we felt for Nazism made us guilty of a fatal sin in politics—wishful thinking. It made us believe that after Hitler was finished all we had to do to bring on the great golden age of peace was to show good will towards the Soviets. It is true that in the short communist period of Hungary, after World War I, the Communists behaved very badly, but that was long ago. A new world was to come. This was a most tragic error with fatal consequences. From my hiding places I contacted Governor Horthy, who was still the master of the situation. We met in secret, and I offered my services as an envoy to the Allies to prepare Hungary's joining them. He seemed to accept but when he noticed my friendly disposition towards the Soviets he edged out of the room and I never saw him again. I can reproach only myself for this failure. I should have taken Horthy's mentality into account. He hated Russia and feared it.

Personally, I did not expect a better treatment from the Soviets than I had expected from Hitler, having given my heavy golden Nobel Medal to Finland when the Soviets declared war on her, and this medal meant more than just gold. So I was not surprised when, after the "liberation" of Budapest, a Soviet patrol, with an English-speaking major at its head, came searching for me. I gave myself up. To my surprise the patrol did not come to arrest me but to bring me to safety on Molotov's personal order. I refused to go along, not wanting to leave my wife's big family in the very dangerous situation then prevailing in the Capital. So the whole family was taken to safety, while my wife and I were taken to Malinowski's headquarters where we were fed back to life with utmost care and consideration. Later, I was invited to Moscow where I spent two months and attended the Centennial Celebration of the Academy, finishing up with a trip to Armenia.

I went to Moscow with the hope of seeing Stalin. What made me want to meet him was the fact that the Soviet Army in Hungary behaved very badly. Near my home town a Hungarian regiment laid down its arms, not wanting to fight for Hitler. The whole regiment was crowded into a small prison where it was soon exterminated by typhus fever. In Budapest the ends of streets were suddenly closed by Soviet soldiers and all the younger men were herded together. Their documents were taken away, which wiped out their identity. About 30–40,000 men were arrested this way and then herded to Czegled, a nearby camp where there was no food and poor sanitation. Dysentery and typhoid began to decimate them. The screams could be heard from long distances. Those who were left were herded into trains, the doors of which were sealed; nobody knew where they went. We could not guess, at that time, that these people were simply taken to Russia as slaves, the whole transaction recalling the darkest days of African slave trade. With our wishful thinking we tried to find excuses for the Soviet atrocities. We even tried to find excuses for the individual misbehavior of Soviet soldiers; war is a beastly business, and makes beasts of men. So, I went to Moscow with the hope of being able to tell Stalin what was going on in Hungary, that we Hungarians wanted to be friends with the Soviet but couldn't be if he did not end this rule. I asked for an interview and was taken into the Foreign Office before Mr. Decanozov, who had to find out what I wanted from Stalin. Mr. Decanozov must have been a very high official because he was later executed together with Beria. He asked me what I wanted. I told him. His reaction was unexpected: he began to shout. At this moment I felt that what I thought to be the overzeal of local commanders was all planned in Moscow. Going home, I still continued working for an understanding with the Soviets. If we had to live together, we had better understand each other. The Russian people are a fine people whom one cannot help liking once one knows them. . . .

The sole general interest in this story is that it sheds a vivid light on the turbulence of our days, showing the conflict between my scientific world and prescientific surroundings which were immiscible. Looking back gives me the feeling of frustration. Resisting Hitler, building academies, research schools, living for years with a finger on the trigger instead of fingering test tubes—and all this to see the part of the world I worked for trodden down as a colony, and to see mankind on the brink of extinction. The idea of being killed for my ideas never frightened me. At one time it even seemed natural. But to have spent so much life and energy in vain is depressing, and I have to ask myself, as so many other scientists must do: has research any sense? Should science not be stopped till man reaches the maturity necessary to deal with the forces which science creates, without the danger of self-destruction?

In a way, the question has no sense, for scientific progress cannot be stopped. Human curiosity cannot be quenched. The question is, rather: does scientific progress offer a way out? To this question my answer is an emphatic "yes." . . .

Questions and Topics for Discussion and Writing

1. Teachers of composition often point out that titles provide the first opportunity for writers to influence their readers. How do you respond to Szent-Györgyi's title? Who is "lost"?

2. The noted writer Gertrude Stein, after World War I, declared to Ernest Hemingway, "You are all a lost generation." Does reading this declaration modify your response to the title? How does Szent-Györgyi's original purpose in writing make him a part of the "lost generation"?

3. Review the many historical events, countries, and people referred to by Szent-Györgyi. Since use of so many references is called *cataloguing*, a technique that gives epic stature to certain literary works, one could conclude that Szent-Györgyi gives epic stature to scientific endeavor. Write a paragraph in which you agree or disagree with such judgment.

4. According to Szent-Györgyi, why did he become a scientist? Szent-Györgyi's life was adventurous in both the scientific and nonscientific areas. Do you think this dual adventurousness is coincidental? Do you see a consistent pattern of behavior?

5. Scientists are sometimes characterized as being "optimists"—perhaps routinely and too easily so. How would you relate Szent-Györgyi to this label?
6. What is the importance to scientists of Szent-Györgyi's assertion that "Nature has no bottom"?
7. Literary existentialism offers a view of the human being in a fragmented world, one in which traditional values and religious beliefs have been destroyed. How does Szent-Györgyi relate science to this existential situation? Find words and phrases that illustrate his attitude.
8. In his conclusion, Szent-Györgyi uses the phrase "the turbulence of our days." Using that phrase as your title, write a well-developed essay: Provide a strong controlling idea, specific examples, and illustrations. Use concrete supporting details.
9. Considering that the human race is "lost in the twentieth century," Szent-Györgyi states that scientific progress offers hope for humanity. What is it?

Peter Brian Medawar

Sir Peter Medawar (b. 1915) is a British biologist, medical scientist, and writer. Together with F. M. Burnet, Medawar in 1960 received the Nobel Prize in Physiology or Medicine for his discovery of acquired immunological tolerance. His writings on both theoretical and practical aspects of science have won him an international audience of readers. Because biology, in particular, has expanded dramatically in the past three decades, Medawar's mastery in presenting the issues has made his audiences especially loyal.

Born in Rio de Janeiro of English parents, Medawar obtained degrees from Marlborough College and Oxford University, England, and became a professor of zoology and a researcher. The onset of World War II turned his biological research to a medical direction: investigating how the problem of severe burns might be alleviated by preventing the rejection of skin grafts. In collaborative work with Burnet and later with graduate students, Medawar showed that the immunological reaction acquired when homographs were used was a problem that could be solved in principle; the discovery, an impetus for clinical research in tissue transplantation, has led to contemporary triumphs in the field of organ transplantation.

In his writings, as in his research, Medawar invigorates thinking. In works such as *Induction and Intuition in Scientific Thought* (1969), a lucid, reflective, small book on the nature of the scientific method, he stimulates the reader to consider theoretical matters. He also challenges the reader by strong opinions: In *The Art of the Soluble*, for example, he asserts that one cause of the division between scientists and humanists—the "Two Cultures" phenomenon—and even between "pure" and applied scientists is the feeling that practical applications of science tend to put their practitioners "in trade" and are, therefore, not quite respectable. Because this feeling manifests what Medawar considers a "by-product" of "our Anglo-Saxon attitudes to research" (a "very English" kind of class consciousness that entails a reverence for purity and, therefore, for "Pure Science"), he addresses his English audience in particular. But the address suggests that an examination of attitudes may be worthwhile anywhere that an automatic distrust of scientists appears.

Medawar worked as a professor of zoology at several English universities until 1962, when he became director of the National Institute for Medical Research in London. Since 1984, he has been a Medical Research Grant Holder at the Clinical Research Centre of Middlesex, England.

Medawar's works, animated by wit and verve, concern the larger issues of science and tend to be philosophical. Other important books are *The Uniqueness of the Individual, The Hope of Progress, The Limits of Science,* and, with J. S. Medawar, *The Life Science* and *Aristotle to Zoos: A Philosophical Dictionary of Biology.*

The selection "How Can I Tell If I Am Cut Out to Be a Research Worker?" is from *Advice to a Young Scientist* and provides an informative preview for readers debating their suitability for a career in the sciences.

How Can I Tell If I Am Cut Out to Be a Research Worker?

P. B. Medawar

People who believe themselves cut out for a scientific life are sometimes dismayed and depressed by, in Sir Francis Bacon's words, "The subtilty of nature, the secret recesses of truth, the obscurity of things, the difficulty of experiment, the implication of causes and the infirmity of man's discerning power, being men no longer excited, either out of desire or hope, to penetrate farther."

There is no certain way of telling in advance if the daydreams of a life dedicated to the pursuit of truth will carry a novice through the frustration of seeing experiments fail and of making the dismaying discovery that some of one's favorite ideas are groundless.

Twice in my life I have spent two weary and scientifically profitless years seeking evidence to corroborate dearly loved hypotheses that later proved to be groundless; times such as these are hard for scientists—days of leaden gray skies bringing with them a miserable sense of oppression and inadequacy. It is my recollection of these bad times that accounts for the earnestness of my advice to young scientists that they should have more than one string to their bow and should be willing to take no for an answer if the evidence points that way.

It is especially important that no novice should be fooled by old-fashioned misrepresentations about what a scientific life is like. Whatever it may have been alleged to be, it is in reality exciting, rather passionate and—in terms of hours of work—a very demanding and sometimes exhausting occupation. It is also likely to be tough on a wife or husband and children who have to live with an obsession without the compensation of being possessed by it themselves.

A novice must stick it out until he discovers whether the rewards and compensations of a scientific life are for him commensurate with the disappointments and the toil; but if once a scientist experiences the exhilaration of discovery and the satisfaction of carrying through a really tricky experiment—once he has felt that deeper and more expansive feeling Freud has called the "oceanic feeling" that is the reward for any real advancement of the understanding—then he is hooked and no other kind of life will do.

MOTIVES

What about the motives for becoming a scientist in the first place? This is the kind of subject upon which psychologists might be expected to make some pronouncement. Love of finicky detail was said by Lou Andreas Salomé to be one of the outward manifestations of—uh—"anal erotism," but scientists in general are not finicking, nor, luckily, do they often have to be. Conventional wisdom has always had it that curiosity is the mainspring of a scientist's work. This has always seemed an inadequate motive to me; *curiosity* is a nursery word. "Curiosity killed the cat" is an old nanny's saying, though it may have been that same curiosity which found a remedy for the cat on what might otherwise have been its deathbed.

Most able scientists I know have something for which "exploratory impulsion" is not too grand a description. Immanuel Kant spoke of a "restless endeavor" to get at the truth of things, though in the context of the not wholly convincing argument that nature would hardly have implanted such an ambition in our breasts if it had not been possible to gratify it. A strong sense of unease and dissatisfaction always goes with lack of comprehension. Laymen feel it, too; how otherwise can we account for the relief they feel when they learn that some odd and disturbing phenomenon can be explained? It cannot be the explanation itself that brings relief, for it may easily be too technical to be widely understood. It is not the knowledge itself, but the satisfaction of knowing that something is known. The writings of Francis Bacon and of Jan Amos Comenius—two of the philosophic founders of modern science whose writings I shall often refer to—are suffused by the imagery of light. Perhaps the restless unease I am writing of is an adult equivalent of that childish fear of the dark that can be dispelled, Bacon said, only by kindling a light in nature.

I am often asked, "What made *you* become a scientist?" But I can't stand far enough away from myself to give a really satisfactory answer, for I cannot distinctly remember a time when I did not think that a scientist was the most exciting possible thing to be. Certainly I had been stirred and persuaded by the writings of Jules Verne and H. G. Wells and also by the not necessarily posh encyclopedias that can come the way of lucky children who read incessantly and who are forever poring over books. Works of popular science helped, too: sixpenny—in effect, dime—books on stars, atoms, the earth, the oceans, and suchlike. I was literally afraid of the dark, too—and if my conjecture in the paragraph above is right, that may also have helped.

AM I BRAINY ENOUGH TO BE A SCIENTIST?

An anxiety that may trouble some novices, and perhaps particularly some women because of the socially engendered habit—not often enough corrected—of self-depreciation, is whether they have brains enough to do well in science. It is an anxiety they could well spare themselves, for one does not need to be terrifically brainy to be a good scientist. An antipathy or a total indifference to the life of the mind and an impatience of abstract ideas can be taken as contraindications, to be sure, but there is nothing in experimental science that calls for great feats of ratiocination or a preternatural gift for deductive reasoning. Common sense one cannot do without, and one would be the better for owning some of those old-fashioned virtues that seem unaccountably to have fallen into disrepute. I mean application, diligence, a sense of purpose, the power to concentrate, to persevere and not be cast down by adversity—by finding out after long and weary inquiry, for example, that a dearly loved hypothesis is in large measure mistaken.

An Intelligence Test

For full measure I interpolate an intelligence test, the performance of which will differentiate between common sense and the dizzily higher intellections that scientists are sometimes thought to be capable of or to need. To many eyes, some of the figures (particularly the holy ones) of El Greco's paintings seem unnaturally tall and thin. An ophthalmologist who shall be nameless surmised that they were drawn so because El Greco suffered a defect of vision that made him *see* people that way, and as he saw them, so he would necessarily draw them.

Can such an interpretation be valid? When putting this question, sometimes to quite large academic audiences, I have added, "Anyone who can see *instantly* that this explanation is nonsense and is nonsense for philosophic rather than aesthetic reasons is undoubtedly bright. On the other hand, anyone who still can't see it is nonsense even when its nonsensicality is explained must be rather dull." The explanation is epistemological—that is, it has to do with the theory of knowledge.

Suppose a painter's defect of vision was, as it might easily have been, diplopia—in effect, seeing everything double. If the ophthalmologist's explanation were right, then such a painter would paint his figures double; but if he did so, then when he came to inspect his handiwork, would he not see all the figures fourfold and maybe suspect that something was amiss? If a defect of vision is in question, the only figures that could seem natural (that is, representational) to the painter must seem natural to us also, even if we ourselves suffer defects of vision; if some of El Greco's figures seem unnaturally tall and thin, they appear so because this was El Greco's intention.

I do not wish to undervalue the importance of intellectual skills in science, but I would rather undervalue them than overrate them to a degree that might frighten recruits away. Different branches of science call for rather different abilities, anyway, but after deriding the idea that there is any such thing as *the* scientist, I must not speak of "science" as if it were a single species of activity. To collect and classify beetles requires abilities, talents and incentives quite different from, I do not say inferior to, those that enter into theoretical physics or statistical epidemiology. The pecking order within science—a most complicated *snobismus*—certainly rates theoretical physics above the taxonomy of beetles, perhaps because in the collection and classification of beetles the order of nature is thought to spare us any great feat of judgment or intellection: is not there a slot waiting for each beetle to fit into?

Any such supposition is merely inductive mythology, however, and an experienced taxonomist or paleobiologist will assure a beginner that taxonomy well done requires great deliberation, considerable powers of judgment and a flair for the discernment of affinities that can come only with experience and the will to acquire it.

At all events scientists do not often think of themselves as brilliantly brainy people—and some, at least, like to avow themselves rather stupid. This is a transparent affectation, though—unless some uneasy recognition of the truth tempts them to fish for reassurance. Certainly very many scientists are not intellectuals. I myself do not happen to know any who are Philistines unless—in a very special sense—it is being a Philistine to be so overawed by the judgments of literary and aesthetic critics as to take them far more seriously than they deserve.

Because so many experimental sciences call for the use of manipulative skills, it is part of conventional wisdom to declare that a predilection for or proficiency at mechanical or constructive play portends a special aptitude for experimental science. A taste for Baconian experimentation is often thought significant, too— for example, an insistent inner impulsion to find out what happens when several ounces of a mixture of sulfur, saltpeter and finely powdered charcoal is ignited. We cannot tell if the successful prosecution of such an experiment genuinely portends a successful research career because only they become scientists who don't find out. To devise some means of ascertaining whether or not these conventional beliefs hold water is work for sociologists of science. I do not feel, though, that a novice need be turned away from science by clumsiness or an inability to mend radio sets or bikes. These skills are not instinctual; they can be learned, as dexterity can be. A trait surely incompatible with a scientific career is to regard manual work as undignified or inferior, or to believe that a scientist has achieved success only when he packs away test tubes and culture dishes, turns off the Bunsen burner, and sits at a desk dressed in collar and tie. Another scientifically disabling belief is to expect to be able to carry out experimental research by issuing instructions to lesser mortals who scurry hither and thither to do one's bidding. What is disabling about this belief is the failure to realize that experimentation is a form of thinking as well as a practical expression of thought.

Opting Out

The novice who tries his hand at research and finds himself indifferent to or bored by it should leave science without any sense of self-reproach or misdirection.

This is easy enough to say, but in practice the qualifications required of scientists are so specialized and time-consuming that they do not qualify him to take up any other occupation; this is especially a fault of the current English scheme of education and does not apply with the same force in America, whose experience of *general* university education is so much greater than our own.[1]

A scientist who pulls out may regret it all his life or he may feel liberated; if the latter, he probably did well to quit, but any regret he felt would be well-founded, for several scientists have told me with an air of delighted wonderment how very satisfactory it is that they should be paid—perhaps even adequately paid—for work that is so absorbing and deeply pleasurable as scientific research.

Questions and Topics for Discussion and Writing

1.   Sir Francis Bacon, often called "the father of the scientific method," insisted that we examine nature first-hand, then put inductive logic to work on the facts so gained. Despite his famous dictum "Knowledge is power," Bacon also believed

---

[1] The great wave of university building that in England transformed city colleges into the civic universities happened about 1890–1910, but in America the mountain-building epoch of university evolution happened about a hundred years ago.

that the experiments of fruit were less important than the experiments of light. Comment on Medawar's quotation from Bacon at the beginning of the essay in terms of Medawar's own use of the imagery of light.

2. How do you define *curiosity*? How does Medawar believe curiosity relates to the scientist's calling? Why does Medawar find the term *curiosity* inadequate?

3. Why does Medawar include an "intelligence test" and what does it contribute to his essay?

4. What does Medawar mean by "Philistines" and why does he use the term?

5. Why does Medawar deride the idea of "*the* scientist" or "science" as a single species of activity?

6. How do you react to Medawar's declaration that "experimentation is a form of thinking as well as a practical expression of thought"?

7. As a science student, you have certain expectations about science. How, if in any way, has this essay changed your expectations?

SUGGESTIONS FOR FURTHER READING

Bernstein, Jeremy. *Experiencing Science: Profiles in Discovery.* New York: Basic Books, 1978.

Blackwell, Elizabeth. *Pioneer Work in Opening the Medical Profession to Women: Autobiography.* New York: Schocken Books, 1977.

Bohr, Niels. *The Man, His Science, and the World They Changed.* New York: Alfred A. Knopf, 1966.

Curie, Eve. *Madame Curie: A Biography.* Trans. Vincent Sheean. New York: Doubleday, 1937.

Elsasser, Walter M. *Memoirs of a Physicist in the Atomic Age.* New York: Science History Publications, 1978.

Feynman, Richard P. *Surely You're Joking, Mr. Feynman! Adventures of a Curious Character.* As told to Ralph Leighton. Ed. Edward Hutchings. New York: W. W. Norton & Company, 1985.

Frisch, Otto R. *What Little I Remember.* Cambridge: Cambridge University Press, 1979.

Gamow, George. *My World Line: An Informal Autobiography.* New York: Viking Press, 1970.

Infeld, Leopold. *Quest: An Autobiography.* New York: Chelsea Publishing Company, 1980.

Medawar, Peter. *Memoir of a Thinking Radish.* New York: Oxford University Press, 1986.

Russell, Bertrand. *The Autobiography of Bertrand Russell. 3 Vols. Vol I: 1872–1914.* Boston: Little, Brown, 1967.

Ulam, S. M. *Adventures of a Mathematician.* New York: Scribner, 1976.

Wiener, Norbert. *Ex-Prodigy: My Childhood and Youth.* Cambridge: MIT Press, 1964.

Unit 2

FROM PERSPECTIVE TO METHOD: INDUCTION AND INTUITION

Thomas Henry Huxley

Known to his Victorian contemporaries as "Darwin's Bulldog" for his tenacious championing of the controversial theory of evolution, Thomas Henry Huxley (1825–1895) was a naturalist, paleontologist, noted teacher–lecturer, and eminent humanist. A man of action, Huxley set forth tirelessly from his busy laboratory in downtown London as a defender of science in general and the work of Darwin in particular.

Huxley developed his career as scientist, writer, and orator despite early obstacles. The seventh child in a family with scant financial resources, Huxley had only two years of regular school training, from the ages of eight to ten. These years, to judge from his brief remarks in his *Autobiography*, were evidently times of more hardship than help to him. During adolescence, Huxley taught himself science, logic, and German. His true desire—to become a mechanical engineer—was thwarted, and he was apprenticed at fifteen to medical practitioners. His experiences in this field occurred in poverty-ridden areas, including the dock region of London.

This training led Huxley to a scholarship to Charing Cross School of Medicine, then to a position in the Royal Navy as assistant surgeon on the frigate *Rattlesnake*. Four years' exploration in the South Seas followed, permitting Huxley to collect marine specimens. His scientific research gained him admission to the Royal Society on his return and to a post as a professor of paleontology.

Huxley's lectures popularizing science, especially those addressed to the average intelligent person, attracted wide audiences. He campaigned for a place for science in the liberal arts curricula of the day. In dramatic encounters, Huxley debated evolution with clerics, in particular the noted Bishop Wilberforce. In these activities, Huxley showed himself exceptional, yet a man of his times. Although an agnostic, he kept the earnest Victorian moral sense and strenuously worked to advance the scientific method because he believed it to be concerned with truth finding.

This effort to spread rational thinking and scientific techniques marks T. H. Huxley as part of a tradition that runs from the Renaissance through the Enlightenment. His link in scientific tradition is personal as well: He was the grandfather of Julian Huxley, twentieth-century biologist and writer, and Aldous Huxley, author of the science-fiction novel *Brave New World* (1932).

In the selection below, Huxley's chief purpose is to ground his listeners in the scientific method. The occasion was one of the "lectures for workingmen," which Huxley used to advance the cause of technical education and preparation for "practical life."

We All Use the Scientific Method

Thomas H. Huxley

The method of scientific investigation is nothing but the expression of the necessary mode of working of the human mind. It is simply the mode at which all phenomena are reasoned about, rendered precise and exact. There is no more difference, but there is just the same kind of difference, between the mental operations of a man of science and those of an ordinary person, as there is between the operations and methods of a baker or of a butcher weighing out his goods in common scales, and the operations of a chemist in performing a difficult and complex analysis by means of his balance and finely-graduated weights. It is not that the action of the scales in the one case, and the balance in the other, differ in the principles of their construction or manner of working; but the beam of one is set on an infinitely finer axis than the other, and of course turns by the addition of a much smaller weight.

You will understand this better, perhaps, if I give you some familiar example. You have all heard it repeated, I dare say, that men of science work by means of induction and deduction, and that by the help of these operations, they, in a sort of sense, wring from Nature certain other things, which are called natural laws, and causes, and that out of these, by some cunning skill of their own, they build up hypotheses and theories. And it is imagined by many, that the operations of the common mind can be by no means compared with these processes, and that they have to be acquired by a sort of special apprenticeship to the craft. To hear all these large words, you would think that the mind of a man of science must be constituted differently from that of his fellow men; but if you will not be frightened by terms, you will discover that you are quite wrong, and that all these terrible apparatus are being used by yourselves every day and every hour of your lives.

There is a well-known incident in one of Molière's plays, where the author makes the hero express unbounded delight on being told that he had been talking prose during the whole of his life. In the same way, I trust that you will take comfort, and be delighted with yourselves, on the discovery that you have been acting on the principles of inductive and deductive philosophy during the same period. Probably there is not one here who has not in the course of the day had occasion to set in motion a complex train of reasoning, of the very same kind, though differing of course in degree, as that which a scientific man goes through in tracing the causes of natural phenomena.

A very trivial circumstance will serve to exemplify this. Suppose you go into a fruiterer's shop, wanting an apple,—you take up one, and, on biting it, you find it is sour; you look at it, and see that it is hard and green. You take up another one, and that too is hard, green, and sour. The shopman offers you a third; but, before biting it, you examine it, and find that it is hard and green, and you immediately say that you will not have it, as it must be sour, like those that you have already tried.

Nothing can be more simple than that, you think; but if you will take the trouble to analyse and trace out into its logical elements what has been done by the mind, you will be greatly surprised. In the first place, you have performed the operation of induction. You found that, in two experiences, hardness and greenness in apples went together with sourness. It was so in the first case, and it was confirmed by the second. True, it is a very small basis, but still it is enough to make an induction from; you generalise the facts, and you expect to find sourness in apples where you get hardness and greenness. You found upon that a general law, that all hard and green apples are sour; and that, so far as it goes, is a perfect induction. Well, having got your natural law in this way, when you are offered another apple which you find is hard and green, you say, "All hard and green apples are sour; this apple is hard and green, therefore this apple is sour." That train of reasoning is what logicians call a syllogism, and has all its various parts and terms—its major premise, its minor premise, and its conclusion. And, by the help of further reasoning, which, if drawn out, would have to be exhibited in two or three other syllogisms, you arrive at your final determination, "I will not have that apple." So that, you see, you have, in the first place, established a law by induction, and upon that you have founded a deduction, and reasoned out the special conclusion of the particular case. Well now, suppose, having got your law, that at some time afterwards, you are discussing the qualities of apples with a friend: you will say to him, "It is a very curious thing—but I find that all hard and green apples are sour!" Your friend says to you, "But how do you know that?" You at once reply, "Oh, because I have tried them over and over again, and have always found them to be so." Well, if we were talking science instead of common sense, we should call that an experimental verification. And, if still opposed, you go further, and say, "I have heard from the people in Somersetshire and Devonshire, where a large number of apples are grown, that they have observed the same thing. It is also found to be the case in Normandy, and in North America. In short, I find it to be the universal experience of mankind wherever attention has been directed to the subject." Whereupon, your friend, unless he is a very unreasonable man, agrees with you, and is convinced that you are quite right in the conclusion you have drawn. He believes, although perhaps he does not know he believes it, that the more extensive verifications are,—that the more frequently experiments have been made, and results of the same kind arrived at,—that the more varied the conditions under which the same results are attained, the more certain is the ultimate conclusion, and he disputes the question no further. He sees that the experiment has been tried under all sorts of conditions, as to time, place, and people, with the same result; and he says with you, therefore, that the law you have laid down must be a good one, and he must believe it.

In science we do the same thing;—the philosopher exercises precisely the same faculties, though in a much more delicate manner. In scientific inquiry it becomes a matter of duty to expose a supposed law to every possible kind of verification, and to take care, moreover, that this is done intentionally, and not left to a mere accident, as in the case of the apples. And in science, as in common life, our confidence in a law is in exact proportion to the absence of variation in the result of our experimental verifications. For instance, if you let go your grasp of an article you may have in your hand, it will immediately fall to the ground. That is a very common verification of one of the best established laws of nature—that of gravitation. The method by which men of science establish the existence of that law is exactly the same as that by which we have established the trivial proposition about the sourness of hard and green apples. But we believe it in such an extensive, thorough, and unhesitating manner because the universal experience of mankind verifies it, and we can verify it ourselves at any time; and that is the strongest possible foundation on which any natural law can rest.

So much, then, by way of proof that the method of establishing laws in science is exactly the same as that pursued in common life. Let us now turn to another matter (though really it is but another phase of the same question), and that is, the method by which, from the relations of certain phenomena, we prove that some stand in the position of causes towards the others.

I want to put the case clearly before you, and I will therefore show you what I mean by another familiar example. I will suppose that one of you, on coming down in the morning to the parlour of your house, finds that a tea-pot and some spoons which had been left in the room on the previous evening are gone,—the window is open, and you observe the mark of a dirty hand on the window-frame, and perhaps, in addition to that, you notice the impress of a hob-nailed shoe on the gravel outside. All these phenomena have struck your attention instantly, and before two seconds have passed you say, "Oh, somebody has broken open the window, entered the room, and run off with the spoons and the tea-pot!" That speech is out of your mouth in a moment. And you will probably add, "I know there has; I am quite sure of it!" You mean to say exactly what you know; but in reality you are giving expression to what is, in all essential particulars, an hypothesis. You do not *know* it at all; it is nothing but an hypothesis rapidly framed in your own mind. And it is an hypothesis founded on a long train of inductions and deductions.

What are those inductions and deductions, and how have you got at this hypothesis? You have observed, in the first place, that the window is open; but by a train of reasoning involving many inductions and deductions, you have probably arrived long before at the general law—and a very good one it is—that windows do not open of themselves; and you therefore conclude that something has opened the window. A second general law that you have arrived at in the same way is, that tea-pots and spoons do not go out of a window spontaneously, and you are satisfied that, as they are not now where you left them, they have been removed. In the third place, you look at the marks on the window-sill, and the shoe-marks outside, and you say that in all previous experience the former kind of mark has never been produced by anything else but the hand of a human being; and the same experience shows that no other animal but man at present wears shoes with hob-nails in them such as would produce the marks in the gravel. I do not know, even if we could discover any of those "missing links" that are talked about, that they would help us to any other conclusion! At any rate the law which states our present experience is strong enough for my present purpose. You next reach the conclusion, that as these kinds of marks have not been left by any other animals than men, or are liable to be formed in any other way than by a man's hand and shoe, the marks in question have been formed by a man in that way. You have, further, a general law, founded on observation and experience, and that, too, is, I am sorry to say, a very universal and unimpeachable one,—that some men are thieves; and you assume at once from all these premises—and that is what constitutes your hypothesis—that the man who made the marks outside and on the window-sill, opened the window, got into the room, and stole your tea-pot and spoons. You have now arrived at a *vera causa;*—you have assumed a cause which, it is plain, is competent to produce all the phenomena you have observed. You can explain all these phenomena only by the hypothesis of a thief. But that is a hypothetical conclusion, of the justice of which you have no absolute proof at all; it is only rendered highly probable by a series of inductive and deductive reasonings.

I suppose your first action, assuming that you are a man of ordinary common sense, and that you have established this hypothesis to your own satisfaction, will very likely be to go off for the police, and set them on the track of the burglar, with the view to the recovery of your property. But just as you are starting with this object, some person comes in, and on learning what you are about, says, "My good friend, you are going on a great deal too fast. How do you know that the man who really made the marks took the spoons? It might have been a monkey that took them, and the man may have merely looked in afterwards." You would probably reply, "Well, that is all very well, but you see it is contrary to all experience of the way tea-pots and spoons are abstracted; so that, at any rate, your hypothesis is less probable than mine." While you are talking the thing over in this way, another friend arrives, one of that good kind of people that I was talking of a little while ago. And he might say, "Oh, my dear sir, you are certainly going on a great deal too fast. You are most presumptuous. You admit that all these occurrences took place when you were fast asleep, at a time when you could not possibly have known anything about what was taking place. How do you know that the laws of Nature are not suspended during the night? It may be that there has been some kind of supernatural interference in this case." In point of fact, he declares that your hypothesis is one of which you cannot at all demonstrate the truth, and that you are by no means sure that the laws of Nature are the same when you are asleep as when you are awake.

Well, now, you cannot at the moment answer that kind of reasoning. You feel that your worthy friend has you somewhat at a disadvantage. You will feel perfectly convinced in your own mind, however, that you are quite right, and you say to him, "My good friend, I can only be guided by the natural probabilities of the case, and if you will be kind enough to stand aside and permit me to pass, I will go and fetch the police." Well, we will suppose that your journey is successful, and that by good luck you meet with a policeman; that eventually the burglar is found with your property on his person, and the marks correspond to his hand and to his boots. Probably any jury would consider those facts a very good experimental verification of your hypothesis, touching the cause of the abnormal phenomena observed in your parlour, and would act accordingly.

Now, in this suppositious case, I have taken phenomena of a very common kind, in order that you might see what are the different steps in an ordinary process of reasoning, if you will only take the trouble to analyse it carefully. All the operations I have described, you will see, are involved in the mind of any man of sense in leading him to a conclusion as to the course he should take in order to make good a robbery and punish the offender. I say that you are led, in that case, to your conclusion by exactly the same train of reasoning as that which a man of science pursues when he is endeavouring to discover the origin and laws of the most occult phenomena. The process is, and always must be, the same; and precisely the same mode of reasoning was employed by Newton and Laplace in their endeavours to discover and define the causes of the movements of the heavenly bodies, as you, with your own common sense, would employ to detect a burglar. The only difference is, that the nature of the inquiry being more abstruse, every step has to be most carefully watched, so that there may not be a single crack or flaw in your hypothesis. A flaw or crack in many of the hypotheses of daily life may be of little or no moment as affecting the general correctness of the conclusions at which we may arrive; but, in a scientific inquiry, a fallacy, great or small, is always of importance, and is sure to be in the long run constantly productive of mischievous, if not fatal results.

Do not allow yourselves to be misled by the common notion that an hypothesis is untrustworthy simply because it is an hypothesis. It is often urged, in respect to some scientific conclusion, that, after all, it is only an hypothesis. But what more have we to guide us in nine-tenths of the most important affairs of daily life than hypotheses, and often very ill-based ones? So that in science, where the evidence of an hypothesis is subjected to the most rigid examination, we may rightly pursue the same course. You may have hypotheses and hypotheses. A man may say, if he likes, that the moon is made of green cheese: that is an hypothesis. But another man, who has devoted a great deal of time and attention to the subject, and availed himself of the most powerful telescopes and the results of the observations of others, declares that in his opinion it is probably composed of materials very similar to those of which our own earth is made up: and that is also only an hypothesis. But I need not tell you that there is an enormous difference in the value of the two hypotheses. That one which is based on sound scientific knowledge is sure to have a corresponding value; and that which is a mere hasty random guess is likely to have but little value. Every great step in our progress in discovering causes has been made in exactly the same way as that which I have detailed to you. A person observing the occurrence of certain facts and phenomena asks, naturally enough, what process, what kind of operation known to occur in Nature applied to the particular case, will unravel and explain the mystery? Hence you have the scientific hypothesis; and its value will be proportionate to the care and completeness with which its basis had been tested and verified. It is in these matters as in the commonest affairs of practical life: the guess of the fool will be folly, while the guess of the wise man will contain wisdom. In all cases, you see that the value of the result depends on the patience and faithfulness with which the investigator applies to his hypothesis every possible kind of verification.

Questions and Topics for Discussion and Writing

1. This selection is taken from one of Huxley's "lectures for workingmen." What evidence do you find in the piece that indicates that it was first presented as an oral communication? What methods of development do you think were chosen for that audience in particular?
2. Huxley obviously believes that a thorough education should ground students in the inductive method. He seems to stress that this method gives a practical prepara-

tion for life. Do you agree that the study of a rational subject matter and method can generate a mind that deals well with rational materials and modes? Do you think that working people need science to make their work more effective? In short, does Huxley imply that the scientific method is the key to nature?

3.   Compare Huxley's attempt to mediate between the demands of the new, technical world and the virtues of the old education with C. P. Snow's efforts to reconcile the "Two Cultures" in his essay (Unit 3).

4.   This piece presents an extended definition of the inductive method. Construct your own extended definition.

5.   In the fifth paragraph, Huxley uses a "train of reasoning" that is a *syllogism*. Look up the word in an unabridged dictionary. What is the major premise, the minor premise, and the conclusion of Huxley's syllogism?

6.   Again comparing Huxley's essay with Snow's, would Huxley seem to be one of the "Philistines," or does he qualify and soften his views with liberalism? Some writers have called Huxley a "statesman of culture." Comment on your own reactions to Huxley's attitude and purpose.

7.   Is Huxley eloquent? What do his metaphors emphasize? Is he persuasive? If so, how?

8.   To explain the inductive method, Huxley uses a homely illustration of the taste testing of green apples. Do you see a connection between the development of his subject here and his vigorous defense of the evolutionary hypothesis?

Arthur Conan Doyle

The creator of Sherlock Holmes, Arthur Conan Doyle (1859–1930), was born in Scotland to Anglo-Norman parents and was educated at Edinburgh University and at the Royal Infirmary in Edinburgh. During his years as a medical student, he met Joseph Bell, an Infirmary surgeon whose deductions and diagnostic intuitions impressed him greatly. In his autobiography, Doyle says that it was Bell's method that suggested those "dazzling deductive displays of Holmes that amazed and, when they became too personal, deeply offended the good Dr. Watson."

After his graduation in 1881, Doyle opened a medical practice but with little success. While waiting for the patients to arrive, he tried his hand at writing. His early short stories earned him varying amounts of money—most of them small. In 1887 his fortunes changed. According to a biographer, he had been thinking of writing a story about a "detective whose work would embody the most modern scientific methods, who would not be as 'showy and superficial' as Edgar Allan Poe's Dupin, or 'a miserable bungler' like Gaboriau's Lecoq."

The first name that he gave his detective was "Sherringford Holmes," but the name was changed to Sherlock Holmes before the publication of *A Study in Scarlet*, which was an immediate success. This success opened his eyes to the possibilities of his detective fiction. Sherlock Holmes appears in 56 short stories and 3 other novels. In 1893, Doyle killed off Holmes. Sherlock Holmes's huge and enthusiastic magazine following protested: Young men wore mourning bands to work, and elderly women wrote to tell Doyle that he was a "brute." In 1901, Doyle brought Holmes back in *The Hound of the Baskervilles*, a posthumous tale. A living, breathing Holmes finally appeared again in 1903 in *The Adventures of the Empty House*.

Among Doyle's numerous works are *The Sign of the Four, Adventures of Sherlock Holmes, The Memoirs of Sherlock Holmes,* and *The Valley of Fear.* The following section from the detective novel *A Study in Scarlet* establishes the "Holmesian-and-Watsonian-pattern" and is a basis for reading the stories of Sherlock Holmes. Like Thomas Huxley, Doyle may be seen as a "popularizer" of logic, for here he puts into fiction a demonstration of reasoning by inference in order to entertain as he instructs.

The Science of Deduction

Arthur Conan Doyle

We met next day as he had arranged, and inspected the rooms at No. 221B, Baker Street, of which he had spoken at our meeting. They consisted of a couple of comfortable bedrooms and a single large airy sitting-room, cheerfully furnished, and illuminated by two broad windows. So desirable in every way were the apartments, and so moderate did the terms seem when divided between us, that the bargain was concluded upon the spot, and we at once entered into possession. That very evening I moved my things round from the hotel, and on the following morning Sherlock Holmes followed me with several boxes and portmanteaux. For a day or two we were busily employed in unpacking and laying out our property to the best advantage. That done, we gradually began to settle down and to accommodate ourselves to our new surroundings.

Holmes was certainly not a difficult man to live with. He was quiet in his ways, and his habits were regular. It was rare for him to be up after ten at night, and he had invariably breakfasted and gone out before I rose in the morning. Sometimes he spent his day at the chemical laboratory, sometimes in the dissecting-rooms, and occasionally in long walks, which appeared to take him into the lowest portions of the city. Nothing could exceed his energy when the working fit was upon him; but now and again a reaction would seize him, and for days on end he would lie upon the sofa in the sitting-room, hardly uttering a word or moving a muscle from morning to night. On these occasions I have noticed such a dreamy, vacant expression in his eyes, that I might have suspected him of being addicted to the use of some narcotic, had not the temperance and cleanliness of his whole life forbidden such a notion.

As the weeks went by, my interest in him and my curiosity as to his aims in life gradually deepened and increased. His very person and appearance were such as to strike the attention of the most casual observer. In height he was rather over six feet, and so excessively lean that he seemed to be considerably taller. His eyes were sharp and piercing, save during those intervals of torpor to which I have alluded; and his thin, hawk-like nose gave his whole expression an air of alertness and decision. His chin, too, had the prominence and squareness which mark the man of determination. His hands were invariably blotted with ink and stained with chemicals, yet he was possessed of extraordinary delicacy of touch, as I frequently had occasion to observe when I watched him manipulating his fragile philosophical instruments.

The reader may set me down as a hopeless busybody, when I confess how much this man stimulated my curiosity, and how often I endeavoured to break through the reticence which he showed on all that concerned himself. Before pronouncing judgment, however, be it remembered how objectless was my life, and how little there was to engage my attention. My health forbade me from venturing out unless the weather was exceptionally genial, and I had no friends who would call upon me and break the monotony of my daily existence. Under these circumstances I eagerly hailed the little mystery which hung around my companion, and spent much of my time in endeavouring to unravel it.

He was not studying medicine. He had himself, in reply to a question, confirmed Stamford's opinion upon that point. Neither did he appear to have pursued any course of reading which might fit him for a degree in science or any other recognized portal which would give him an entrance into the learned world. Yet his zeal for certain studies was remarkable, and within eccentric limits his knowledge was so extraordinarily ample and minute that his observations have fairly astounded me. Surely no man would work so hard or attain such precise information unless he had some definite end in view. Desultory readers are seldom remarkable for the exactness of their learning. No man burdens his mind with small matters unless he has some very good reason for doing so.

His ignorance was as remarkable as his knowledge. Of contemporary literature, philosophy and politics he appeared to know next to nothing. Upon my quoting Thomas Carlyle, he inquired in the naïvest way who he might be and what he had done. My surprise reached a climax, however, when I found incidentally that he was ignorant of the Copernican Theory and of the composition of the Solar System. That any civilized human being in this nineteenth-century should not be aware that the earth travelled round the sun appeared to be to me such an extraordinary fact that I could hardly realize it.

"You appear to be astonished," he said, smiling at my expression of surprise. "Now that I do know it I shall do my best to forget it."

"To forget it!"

"You see," he explained, "I consider that a man's brain originally is like a little empty attic, and you have to stock it with such furniture as you choose. A fool takes in all the lumber of every sort that he comes across, so that the knowledge which might be useful to him gets crowded out, or at best is jumbled up with a lot of other things, so that he has a difficulty in laying his hands upon it. Now the skilful workman is very careful indeed as to what he takes into his brain-attic. He will have nothing but the tools which may help him in doing his work, but of these he has a large assortment, and all in the most perfect order. It is a mistake to think that that little room has elastic walls and can distend to any extent. Depend upon it there comes a time when for every addition of knowledge you forget something that you knew before. It is of the highest importance, therefore, not to have useless facts elbowing out the useful ones."

"But the Solar System!" I protested.

"What the deuce is it to me?" he interrupted impatiently: "you say that we go round the sun. If we went round the moon it would not make a penny-worth of difference to me or to my work."

I was on the point of asking him what that work might be, but something in his manner showed me that the question would be an unwelcome one. I pondered over our short conversation, however, and endeavoured to draw my deductions from it. He said that he would acquire no knowledge which did not bear upon his object. Therefore all the knowledge which he possessed was such as would be useful to him. I enumerated in my own mind all the various points upon which he had shown me that he was exceptionally well-informed. I even took a pencil and jotted them down. I could not help smiling at the document when I had completed it. It ran in this way:

SHERLOCK HOLMES—his limits

1. Knowledge of Literature.—Nil.
2.   "   " Philosophy.—Nil.
3.   "   " Astronomy.—Nil.
4.   "   " Politics.—Feeble.
5.   "   " Botany.—Variable. Well up in belladonna, opium, and poisons generally. Knows nothing of practical gardening.
6.   "   " Geology.—Practical, but limited. Tells at a glance different soils from each other. After walks has shown me splashes upon his trousers, and told me by their colour and consistence in what part of London he had received them.
7. Knowledge of Chemistry.—Profound.
8.   "   " Anatomy.—Accurate, but unsystematic.
9.   "   " Sensational Literature.—Immense. He appears to know every detail of every horror perpetrated in the century.

10.   Plays the violin well.
11.   Is an expert singlestick player, boxer, and swordsman.
12.   Has a good practical knowledge of British law.

When I had got so far in my list I threw it into the fire in despair. "If I can only find what the fellow is driving at by reconciling all these accomplishments, and discovering a calling which needs them all," I said to myself, "I may as well give up the attempt at once."

I see that I have alluded above to his powers upon the violin. These were very remarkable, but as eccentric as all his other accomplishments. That he could play pieces, and difficult pieces, I knew well, because at my request he has played me some of Mendelssohn's Lieder, and other favourites. When left to himself, however, he would seldom produce any music or attempt any recognized air. Leaning back in his arm-chair of an evening, he would close his eyes and scrape carelessly at the fiddle which was thrown across his knee. Sometimes the chords were sonorous and melancholy. Occasionally they were fantastic and cheerful. Clearly they reflected the thoughts which possessed him, but whether the music aided those thoughts, or whether the playing was simply the result of a whim or fancy, was more than I could determine. I might have rebelled against those exasperating solos had it not been that he usually terminated them by playing in quick succession a whole series of my favourite airs as a slight compensation for the trial upon my patience.

During the first week or so we had no callers, and I had begun to think that my companion was as friendless a man as I was myself. Presently, however, I found that he had many acquaintances, and those in the most different classes of society. There was one little sallow, rat-faced, dark-eyed fellow, who was introduced to me as Mr. Lestrade, and who came three or four times in a single week. One morning a young girl called, fashionably dressed, and stayed for half an hour or more. The same afternoon brought a grey-headed, seedy visitor, looking like a Jew pedlar, who appeared to me to be much excited, and who was closely followed by a slip-shod elderly woman. On another occasion an old white-haired gentleman had an interview with my companion; and on another, a railway porter in his velveteen uniform. When any of these nondescript individuals put in an appearance, Sherlock Holmes used to beg for the use of the sitting-room, and I would retire to my bedroom. He always apologized to me for putting me to this inconvenience. "I have to use this room as a place of business," he said, "and these people are my clients." Again I had an opportunity of asking him a point-blank question, and again my delicacy prevented me from forcing another man to confide in me. I imagined at the time that he had some strong reason for not alluding to it, but he soon dispelled the idea by coming round to the subject of his own accord.

It was upon the 4th of March, as I have good reason to remember, that I rose somewhat earlier than usual, and found that Sherlock Holmes had not yet finished his breakfast. The landlady had become so accustomed to my late habits that my place had not been laid nor my coffee prepared. With the unreasonable petulance of mankind I rang the bell and give a curt intimation that I was ready. Then I picked up a magazine from the table and attempted to while away the time with it, while my companion munched silently at his toast. One of the articles had a pencil mark at the heading, and I naturally began to run my eye through it.

Its somewhat ambitious title was "The Book of Life," and it attempted to show how much an observant man might learn by an accurate and systematic examination of all that came in his way. It struck me as being a remarkable mixture of shrewdness and of absurdity. The reasoning was close and intense, but the deductions appeared to me to be far-fetched and exaggerated. The writer claimed by a momentary expression, a twitch of a muscle or a glance of an eye, to fathom a man's inmost thoughts. Deceit, according to him, was an impossibility in the case of one trained to observation and analysis. His conclusions were as infallible as so many propositions of Euclid. So startling would his results appear to the uninitiated that until they learned the processes by which he had arrived at them they might well consider him as a necromancer.

"From a drop of water," said the writer, "a logician could infer the possibility of an Atlantic or a Niagara without having seen or heard of one or the other. So all life is a great chain, the nature of which is known whenever we are shown a single link of it. Like all other arts, the Science of Deduction and Analysis is one which can only be acquired by long and patient study, nor is life long enough to allow any mortal to attain the highest possible perfection in it. Before turning to those moral and mental aspects of the matter which present the greatest difficulties, let the inquirer begin by mastering more elementary problems. Let him on meeting a fellow-mortal, learn at a glance to distinguish the history of the man, and the trade or profession to which he belongs. Puerile as such an exercise may seem, it sharpens the faculties of observation, and teaches one where to look and what to look for. By a man's finger-nails, by his coat-sleeve, by his boot, by his trouser-knees, by the callosities of his forefinger and thumb, by his expression, by his shirtcuffs—by each of these things a man's calling is plainly revealed. That all united should fail to enlighten the competent inquirer in any case is almost inconceivable."

"What ineffable twaddle!" I cried, slapping the magazine down on the table; "I never read such rubbish in my life."

"What is it?" asked Sherlock Holmes.

"Why, this article," I said, pointing at it with my egg-spoon as I sat down to my breakfast. "I see that you have read it since you have marked it. I don't deny that it is smartly written. It irritates me though. It is evidently the theory of some arm-chair lounger who evolves all these neat little paradoxes in the seclusion of his own study. It is not practical. I should like to see him clapped down in a third-class carriage on the Underground, and asked to give the trades of all his fellow-travellers. I would lay a thousand to one against him."

"You would lose your money," Holmes remarked calmly. "As for the article, I wrote it myself."

"You!"

"Yes; I have a turn both for observation and for deduction. The theories which I have expressed there, and which appear to you to be so chimerical, are really extremely practical—so practical that I depend upon them for my bread and cheese."

"And how?" I asked involuntarily.

"Well, I have a trade of my own. I suppose I am the only one in the world. I'm a consulting detective, if you can understand what that is. Here in London we have lots of Government detectives and lots of private ones. When these fellows are at fault, they come to me, and I manage to put them on the right scent. They lay all the evidence before me, and I am generally able, by the help of my knowledge of the history of crime, to set them straight. There is a strong family resemblance about misdeeds, and if you have all the details of a thousand at your finger ends, it is odd if you can't unravel the thousand and first. Lestrade is a well-known detective. He got himself into a fog recently over a forgery case, and that was what brought him here."

"And these other people?"

"They are mostly sent on by private inquiry agencies. They are all people who are in trouble about something, and want a little enlightening. I listen to their story, they listen to my comments, and then I pocket my fee."

"But do you mean to say," I said, "that without leaving your room you can unravel some knot which other men can make nothing of, although they have seen every detail for themselves?"

"Quite so. I have a kind of intuition that way. Now and again a case turns up which is a little more complex. Then I have to bustle about and see things with my own eyes. You see I have a lot of special knowledge which I apply to the problem, and which facilitates matters wonderfully. Those rules of deduction laid down in that article which aroused your scorn are invaluable to me in practical work. Observation with me is second nature. You appeared to be surprised when I told you, on our first meeting, that you had come from Afghanistan."

"You were told, no doubt."

"Nothing of the sort. I *knew* you came from Afghanistan. From long habit the train of thoughts ran so swiftly through my mind that I arrived at the conclusion without being conscious of intermediate steps. There were such steps, however. The train of reasoning ran: "Here is a gentleman of a medical type, but with the air of a military man. Clearly an army doctor then. He has just come from the tropics, for his face is dark, and that is not the natural tint of his skin, for his wrists are fair. He has undergone hardship and sickness, as his haggard face says clearly. His left arm has been injured. He holds it in a stiff and unnatural manner. Where in the tropics could an English army doctor have seen much hardship and got his arm wounded? Clearly in Afghanistan." The whole train of thought did not occupy a second. I then remarked that you came from Afghanistan, and you were astonished."

"It is simple enough as you explain it," I said, smiling. "You remind me of Edgar Allan Poe's Dupin. I had no idea that such individuals did exist outside of stories."

Questions and Topics for Discussion and Writing

1. Although Watson states about Holmes, "Neither did he appear to have pursued any course of reading which might fit him for a degree in science . . . ," the descriptions of the activity of his mind make him out to be a scientist. Find evidence of this characterization.
2. Would scientists today "consider [with Holmes] that a man's brain originally is like a little empty attic, and you have to stock it with such furniture as you choose"? Would you agree?
3. How would certain scientists and teachers today challenge Holmes's statement (paraphrased by Watson) that "he would acquire no knowledge which did not bear upon his object. Therefore all the knowledge which he possessed was such as would be useful to him"? Do you agree with Holmes?
4. How does the adage "Practice makes perfect" describe Holmes's method of "observation and deduction," his "train of thought"?
5. Contrast Watson's observation of the message bearer with that of Holmes. How does Holmes's description of his thought process reflect what he says about his "train of thought"?
6. Is Holmes's description a demonstration of *deductive* or *inductive* reasoning—or both?
7. Some critics attribute Holmes's popularity to the "fun" of watching reason at work. He is a "role model" for us to see our mental powers in operation. Taking as your theme the power of play as a demonstration of reasoning, write a short story to illustrate it or an essay to discuss it. Provide details, incidents, and characters.

Robert M. Pirsig

In his autobiographical *Zen and the Art of Motorcycle Maintenance: An Inquiry into Values*, Robert Pirsig (b. 1928) presents modern readers with what seems an incongruous pair of subjects. Such seeming incompatibility suggests what Samuel Johnson called "*discordia concors . . .* a combination of dissimilar images" having a negative effect on art. When Pirsig's book appeared in 1974, however, many of his readers became almost cultish in their enthusiastic response to it.

Zen, a Japanese development of Chinese Buddhism, is a religion to be experienced, not explained. It advocates meditation on nature, combined with certain exercises, as a way to enlightenment—a blessed release from reflective thought and social conformism. The motorcycle, on the other hand, is a contemporary device, a two-wheeled device for one or two riders, which provides more motorized travel power and speed than does the conventional automobile. One segment of the American population, in the 60s and 70s, would conceivably have given credence to Pirsig's linking of two unlike concepts. Many in this group had given up the auto, a symbol of corporate authority, for the cycle, at the same time reinvestigating the contemplative life, as suggested in Zen and other antirationalist philosophies. Still other readers are attracted to the book as significant to the spirit of our times.

Pirsig elaborates on the connections between such disparate subjects as Nature, capitalized as the state of the human condition as well as the environment, and motorcycle maintenance because to know about each of them as much as we can will help us solve some of life's problems. The book tells the tale of two journeys—Pirsig's and his young son's motorcycle trip from the Midwest to California and Pirsig's mental journey as he re-examines his life and his relationships. he manages to combine humor with some serious and illuminating discussion about being as meticulous about our relationships, the caring about them, as we are about our consumer products. Thus both journeys become ways to explore the concept of authentic individuality.

Pirsig, a native of Minnesota, received his bachelor's and master's degrees from the University of Minnesota. He has been a Guggenheim Fellow.

In the passage that follows, Pirsig discusses the application of scientific thinking to techniques of motorcycle maintenance. Stating that a mechanic must "design an experiment properly," he emphasizes the mental aspects of physical problem solving.

Reflections on the Scientific Method

Robert M. Pirsig

Now we follow the Yellowstone Valley right across Montana. It changes from Western sagebrush to Midwestern cornfields and back again, depending on whether it's under irrigation from the river. Sometimes we cross over bluffs that take us out of the irrigated area, but usually we stay close to the river. We pass by a marker saying something about Lewis and Clark. One of them came up this way on a side excursion from the Northwest Passage.

Nice sound. Fits the Chautauqua. We're really on a kind of Northwest Passage too. We pass through more fields and desert and the day wears on.

I want to pursue further now that same ghost that Phaedrus pursued—rationality itself, that dull, complex, classical ghost of underlying form.

This morning I talked about hierarchies of thought—the system. Now I want to talk about methods of finding one's way through these hierarchies—logic.

Two kinds of logic are used, inductive and deductive. Inductive inferences start with observations of the machine and arrive at general conclusions. For example, if the cycle goes over a bump and the engine misfires, and then goes over another bump and the engine misfires, and then goes over another bump and the engine misfires, and then goes over a long smooth stretch of road and there is no misfiring, and then goes over a fourth bump and the engine misfires again, one can logically conclude that the misfiring is caused by the bumps. That is induction: reasoning from particular experiences to general truths.

Deductive inferences do the reverse. They start with general knowledge and predict a specific observation. For example, if, from reading the hierarchy of facts about the machine, the mechanic knows the horn of the cycle is powered exclusively by electricity from the battery, then he can logically infer that if the battery is dead the horn will not work. That is deduction.

Solution of problems too complicated for common sense to solve is achieved by long strings of mixed inductive and deductive inferences that weave back and forth between the observed machine and the mental hierarchy of the machine found in the manuals. The correct program for this interweaving is formalized as scientific method.

Actually I've never seen a cycle-maintenance problem complex enough really to require full-scale formal scientific method. Repair problems are not that hard. When I think of formal scientific method an image sometimes comes to mind of an enormous juggernaut, a huge bulldozer—slow, tedious, lumbering, laborious, but invincible. It takes twice as long, five times as long, maybe a dozen times as long as informal mechanics' techniques, but you know in the end you're going to *get* it. There's no fault isolation problem in motorcycle maintenance that can stand up to it. When you've hit a really tough one, tried everything, racked your brain and nothing works, and you know that this time Nature has really decided to be difficult, you say, "Okay, Nature, that's the end of the *nice* guy," and you crank up the formal scientific method.

For this you keep a lab notebook. Everything gets written down, formally, so that you know at all times where you are, where you've been, where you're going and where you want to get. In scientific work and electronics technology this is necessary because otherwise the problems get so complex you get lost in them and confused and forget what you know and what you don't know and have to give up. In cycle maintenance things are not that involved, but when confusion starts it's a good idea to hold it down by making everything formal and exact. Sometimes just the act of writing down the problems straightens out your head as to what they really are.

The logical statements entered into the notebook are broken down into six categories: (1) statement of the problem, (2) hypotheses as to the cause of the problem, (3) experiments designed to test each hypothesis, (4) predicted results of the experiments, (5) observed results of the experiments and (6) conclusions from the results of the experiments. This is not different from the formal arrangement of many college and high-school lab notebooks but the purpose here is no longer just busywork. The purpose now is precise guidance of thoughts that will fail if they are not accurate.

The real purpose of scientific method is to make sure Nature hasn't misled you into thinking you know something you don't actually know. There's not a mechanic or scientist or technician alive who hasn't suffered from that one so much that he's not instinctively on guard. That's the main reason why so much scientific and mechanical information sounds so dull and so cautious. If you get careless or go romanticizing scientific information, giving it a flourish here and there, Nature will soon make a complete fool out of you. It does it often enough anyway even when you don't give it opportunities. One must be extremely careful and rigidly logical when dealing with Nature: one logical slip and an entire scientific edifice comes tumbling down. One false deduction about the machine and you can get hung up indefinitely.

In Part One of formal scientific method, which is the statement of the problem, the main skill is in stating absolutely no more than you are positive you know. It is much better to enter a statement "Solve Problem: Why doesn't cycle work?" which sounds dumb but is correct, than it is to enter a statement "Solve Problem: What is wrong with the electrical system?" when you don't absolutely *know* the trouble is *in* the electrical system. What you should state is "Solve Problem: What is wrong with cycle?" and *then* state as the first entry of Party Two: "Hypothesis Number One: The trouble is in the electrical system." You think of as many hypotheses as you can, then you design experiments to test them to see which are true and which are false.

This careful approach to the beginning questions keeps you from taking a major wrong turn which might cause you weeks of extra work or can even hang you up completely. Scientific questions often have a surface appearance of dumbness for this reason. They are asked in order to prevent dumb mistakes later on.

Part Three, that part of formal scientific method called experimentation, is sometimes thought of by romantics as all of science itself because that's the only part with much visual surface. They see lots of test tubes and bizarre equipment and people running around making discoveries. They do not see the experiment as part of a larger intellectual process and so they often confuse experiments with demonstrations, which look the same. A man conducting a gee-whiz science show with fifty thousand dollars' worth of Frankenstein equipment is not doing anything scientific if he knows beforehand what the results of his efforts are going to be. A motorcycle mechanic, on the other hand, who honks the horn to see if the battery works is informally conducting a true scientific experiment. He is testing a hypothesis by putting the question to nature. The TV scientist who mutters sadly, "The experiment is a failure; we have failed to achieve what we had hoped for," is suffering mainly from a bad scriptwriter. An experiment is never a failure solely because it fails to achieve predicted results. An experiment is a failure only when it also fails adequately to test the hypothesis in question, when the data it produces don't prove anything one way or another.

Skill at this point consists of using experiments that test only the hypothesis in question, nothing less, nothing more. If the horn honks, and the mechanic concludes that the whole electrical system is working, he is in deep trouble. He has reached an illogical conclusion. The honking horn only tells him that the battery and horn are working. To design an experiment properly he has to think very rigidly in terms of what directly causes what. This you know from the hierarchy. The horn doesn't make the cycle go. Neither does the battery, except in a very indirect way. The point at which the electrical system *directly* causes the engine to fire is at the spark plugs, and if you don't test here, at the output of the electrical system, you will never really know whether the failure is electrical or not.

To test properly the mechanic removes the plug and lays it against the engine so that the base around the plug is electrically grounded, kicks the starter lever and watches the spark-plug gap for a blue spark. If there isn't any he can conclude one of two things: (a) there is an electrical failure or (b) his experiment is sloppy. If he is experienced he will try it a few more times, checking connections, trying every way he can think of to get that plug to fire. Then, if he can't get it to fire, he finally concludes that *a* is correct, there's an electrical failure, and the experiment is over. He has proved that his hypothesis is correct.

In the final category, conclusions, skill comes in stating no more than the experiment has proved. It hasn't proved that when he fixes the electrical system the motorcycle will start. There may be other things wrong. But he does know that the motorcycle isn't going to run until the electrical system is working and he sets up the next formal question: "Solve Problem: What is wrong with the electrical system?"

He then sets up hypotheses for these and tests them. By asking the right questions and choosing the right tests and drawing the right conclusions the mechanic works his way down the echelons of the motorcycle hierarchy until he has found the exact specific cause or causes of the engine failure, and then he changes them so that they no longer cause the failure.

An untrained observer will see only physical labor and often get the idea that physical labor is mainly what the mechanic does. Actually the physical labor is the smallest and easiest part of what the mechanic does. By far the greatest part of his work is careful observation and precise thinking. That is why mechanics sometimes seem so taciturn and withdrawn when performing tests. They don't like it when you talk to them because they are concentrating on mental images, hierarchies, and not really looking at you or the physical motorcycle at all. They are using the experiment as part of a program to expand their hierarchy of knowledge of the faulty motorcycle and compare it to the correct hierarchy in their mind. They are looking at underlying form.

Questions and Topics for Discussion and Writing

1. As the section begins, Pirsig speaks of the "Chautauqua," which he describes elsewhere as "an old-time series of popular talks intended to edify and entertain." Look up "Chautauqua" and explain why the method it embodies is a suitable one for Pirsig's purpose in this narrative.
2. Why do you think Pirsig speaks of his "Northwest Passage"? Is he speaking metaphorically? Explain.
3. Pirsig speaks of the character Phaedrus. The name is that of one of the men who conducted conversations with Socrates in Plato's *Dialogues*. It is also the name Pirsig gives to his youthful self, a brilliant but troubled student of both science and Oriental thought. In this context, what do you take Pirsig to mean when he writes that Phaedrus pursued the "ghost" of rationality itself, "that dull, complex, classical ghost of underlying form"?
4. Pirsig discusses the theory of logic for the benefit of mechanics. In what way does he make his discussion a worthwhile and valid one for this audience? Explain.
5. Compare Pirsig's purpose and method with those of Thomas Huxley in "We All Use the Scientific Method" (Unit 2).
6. How does Pirsig distinguish experiments from demonstrations in science?
7. What does Pirsig see as the value of writing as part of scientific procedure?
8. When, according to Pirsig, is an experiment a failure?
9. Pirsig offers an image as an analogy for the formal scientific method. What is that image? Does it increase understanding for you? Why, or why not?

John Updike

John Updike (b. 1932) divides his creative time among the demands of being a novelist, poet, and critic. Pennsylvania-born, he was educated at Harvard and the Ruskin School of Drawing and Fine Art in Oxford, England. His name is often associated with *The New Yorker* magazine, where he was a staff member from 1955 to 1957 and for which he has written stories, essays, reviews, and poems. To see him primarily in this role, however, is to limit his works, in a sense, to their obvious cleverness and vividness, when the body of his work demands a larger evaluation.

Prolific in his writing, Updike has created the well-known trilogy *Rabbit Run* 1960), *Rabbit Redux* (1971), and *Rabbit Is Rich* (1981), in which the rather ordinary "Rabbit" Angstrom portrays the drives, hopes, and problems of a typical contemporary American. In addition, together with other novels, volumes of poetry, and juvenile books, Updike has added to his reputation with several books of critical commentary, including the best-selling *Hugging the Shore* (1983).

A number of themes vie for attention in his work; religion, although subtly presented as a sense of mystery and value in our lives, is probably the most significant. Science also plays a role, usually serious or philosophical. In "V. B. Nimble, V. B. Quick," Updike creates comedy from his reading of a newspaper item and his own sense that a serious subject can be viewed in a lighthearted mood.

V. B. Nimble, V. B. Quick

John Updike

SCIENCE, PURE AND APPLIED, by V. B. Wigglesworth, F.R.S., Quick Professor of Biology in the University of Cambridge.
A talk listed in the B.B.C. Radio Times of February 2, 1955.

V. B. Wigglesworth wakes at noon,
Washes, shaves, and very soon
Is at the lab; he reads his mail,
Swings a tadpole by the tail,
Undoes his coat, removes his hat,

Dips a spider in a vat
Of alkaline, phones the press,
Tells them he is F.R.S.,
Subdivides six protocells,
Kills a rat by ringing bells,

Writes a treatise, edits two
Symposia on "Will Man Do?,"
Gives a lecture, audits three,
Has the Sperm Club in for tea,
Pensions off an aging spore,

Cracks a test tube, takes some pure
Science and applies it, finds
His hat, adjusts it, pulls the blinds,
Instructs the jellyfish to spawn,
And, by one o'clock, is gone.

Questions and Topics for Discussion and Writing

1. Updike's poem carries the echo of the nursery rhyme "Jack Be Nimble, Jack Be Quick." Why do you think Updike employs such an allusion in treating this subject?
2. The abbreviation "F.R.S." stands for "Fellow of the Royal Society." What dimension does this abbreviation add to the portrait of V. B. Wigglesworth and to the poem?
3. Consider the item in the series "takes some pure/ Science and applies it . . . ." Characterize Updike's attitude toward those who profess to practice "pure Science." What is your understanding of the term? Do you think it is possible for someone to be a "pure scientist"?
4. Write a paragraph discussing Updike's portrayal of an "image" of the scientist. (You may wish to refer to Holton's essay "Modern Science and the Intellectual Tradition: The Seven Images of Science" in Unit 3.)

Carl Sagan

"By profession, I'm a planetary astronomer," declares Carl Sagan. "My job is to examine other worlds. It's invigorating, exciting, even magical work for me. And in the last 20 years, the United States and the Soviet Union have accomplished something stunning and historic—the close-up examination of all those points of light, from Mercury to Saturn, that moved our ancestors to wonder and to science."

As an astronomer, researcher, lecturer, Professor of Astronomy and Space Sciences at Cornell University, and advisor and consultant to NASA and the National Academy of Science, Carl Sagan (b. 1934) is known for his multiple, often controversial, perspectives on life in the universe. One insistent theme of Sagan's is that humans are not alone in the cosmos. For his work, he has received many honors: Among them are the NASA medals for Exceptional Scientific Achievement (1972) and Distinguished Public Service (1977) and the Joseph Priestley Award (1975) for "distinguished contributions to the welfare of mankind."

Sagan is perhaps best known as the host and writer of *Cosmos*, a television series and book, and has written popular pieces on planetary physics, biology, extraterrestrial life, and the romance of science. He has served as editor in chief of *Icarus*, the leading professional astronomical journal. He was one of the driving forces behind the *Mariner, Viking*, and *Voyager* expeditions. Winner of the Pulitzer Prize (for *Dragons of Eden*), Sagan is author of *The Cosmic Connection, Mars and the Mind of Man, Intelligent Life in the Universe*, and, in 1985, *Contact: A Novel. Broca's Brain*, from which the following essay is taken, deals with a number of themes: "the joys and social consequences of the scientific endeavor; borderline or pop science; the not entirely different subject of religious doctrine; the exploration of the planets; and the search for extraterrestrial life."

In "Can We Know the Universe?" Sagan shares his enthusiasm and excitement over what he calls "that vast universe in which we are embedded like a grain of sand in a cosmic ocean."

Can We Know the Universe?
Reflections on a Grain of Salt
Carl Sagan

Nothing is rich but the inexhaustible wealth of nature. She shows us only surfaces, but she is a million fathoms deep.

Ralph Waldo Emerson

Science is a way of thinking much more than it is a body of knowledge. Its goal is to find out how the world works, to seek what regularities there may be, to penetrate to the connections of things—from subnuclear particles, which may be the constituents of all matter, to living organisms, the human social community, and thence to the cosmos as a whole. Our intuition is by no means an infallible guide. Our perceptions may be distorted by training and prejudice or merely because of the limitations of our sense organs, which, of course, perceive directly but a small fraction of the phenomena of the world. Even so straightforward a question as whether in the absence of friction a pound of lead falls faster than a gram of fluff was answered incorrectly by Aristotle and almost everyone else before the time of Galileo. Science is based on experiment, on a willingness to challenge old dogma, on an openness to see the universe as it really is. Accordingly, science sometimes requires courage—at the very least the courage to question the conventional wisdom.

Beyond this the main trick of science is to *really* think of something: the shape of clouds and their occasional sharp bottom edges at the same altitude everywhere in the sky; the formation of a dewdrop on a leaf; the origin of a name or a word—Shakespeare, say, or "philanthropic"; the reason for human social customs—the incest taboo, for example; how it is that a lens in sunlight can make paper burn; how a "walking stick" got to look so much like a twig; why the Moon seems to follow us as we walk; what prevents us from digging a hole down to the center of the Earth; what the definition is of "down" on a spherical Earth; how it is possible for the body to convert yesterday's lunch into today's muscle and sinew; or how far is up—does the universe go on forever, or if it does not, is there any meaning to the question of what lies on the other side?

Some of these questions are pretty easy. Others, especially the last, are mysteries to which no one even today knows the answer. They are natural questions to ask. Every culture has posed such questions in one way or another. Almost always the proposed answers are in the nature of "Just So Stories," attempted explanations divorced from experiment, or even from careful comparative observations.

But the scientific cast of mind examines the world critically as if many alternative worlds might exist, as if other things might be here which are not. Then we are forced to ask why what we see is present and not something else. Why are the Sun and the Moon and the planets spheres? Why not pyramids, or cubes, or dodecahedra? Why not irregular, jumbly shapes? Why so symmetrical, worlds? If you spend any time spinning hypotheses, checking to see whether they make sense, whether they conform to what else we know, thinking of tests you can pose to substantiate or deflate your hypotheses, you will find yourself doing science. And as you come to practice this habit of thought more and more you will get better and better at it. To penetrate into the heart of the thing—even a little thing, a blade of grass, as Walt Whitman said—is to experience a kind of exhilaration that, it may be, only human beings of all the beings on this planet can feel. We are an intelligent species and the use of our intelligence quite properly gives us pleasure. In this respect the brain is like a muscle. When we think well, we feel good. Understanding is a kind of ecstasy.

But to what extent can we *really* know the universe around us? Sometimes this question is posed by people who hope the answer will be in the negative, who are fearful of a universe in which everything might one day be known. And sometimes we hear pronouncements from scientists who confidently state that everything worth knowing will soon be known—or even is already known— and who paint pictures of a Dionysian or Polynesian age in which the zest for intellectual discovery has withered, to be replaced by a kind of subdued languor, the lotus eaters drinking fermented coconut milk or some other mild hallucinogen. In addition to maligning both the Polynesians, who were intrepid explorers (and whose brief respite in paradise is now sadly ending), as well as the inducements to intellectual discovery provided by some hallucinogens, this contention turns out to be trivially mistaken.

Let us approach a much more modest question: not whether we can know the universe or the Milky Way Galaxy or a star or a world. Can we know, ultimately and in detail, a grain of salt? Consider one microgram of table salt, a speck just barely large enough for someone with keen eyesight to make out without a microscope. In that grain of salt there are about 10^{16} sodium and chlorine atoms. This is a 1 followed by 16 zeros, 10 million billion atoms. If we wish to know a grain of salt, we must know at least the three-dimensional positions of each of these atoms. (In fact, there is much more to be known—for example, the nature of the forces between the atoms—but we are making only a modest calculation.) Now, is this number more or less than the number of things which the brain can know?

How much *can* the brain know? There are perhaps 10^{11} neurons in the brain, the circuit elements and switches that are responsible in their electrical and chemical activity for the functioning of our minds. A typical brain neuron has perhaps a thousand little wires, called dendrites, which connect it with its fellows. If, as seems likely, every bit of information in the brain corresponds to one of these connections, the total number of things knowable by the brain is no more than 10^{14}, one hundred trillion. But this number is only one percent of the number of atoms in our speck of salt.

So in this sense the universe is intractable, astonishingly immune to any human attempt at full knowledge. We cannot on this level understand a grain of salt, much less the universe.

But let us look a little more deeply at our microgram of salt. Salt happens to be a crystal in which, except for defects in the structure of the crystal lattice, the position of every sodium and chlorine atom is predetermined. If we could shrink ourselves into this crystalline world, we would see rank upon rank of atoms in an ordered array, a regularly alternating structure—sodium, chlorine, sodium, chlorine, specifying the sheet of atoms we are standing on and all the sheets above us and below us. An absolutely pure crystal of salt could have the position of every atom specified by something like 10 bits of information.[1] This would not strain the information-carrying capacity of the brain.

If the universe had natural laws that governed its behavior to the same degree of regularity that determines a crystal of salt, then, of course, the universe would be knowable. Even if there were many such laws, each of considerable complexity, human beings might have the capability to understand them all. Even if such knowledge exceeded the information-carrying capacity of the brain, we might store the additional information outside our bodies—in books, for example, or in computer memories—and still, in some sense, know the universe.

Human beings are, understandably, highly motivated to find regularities, natural laws. The search for rules, the only possible way to understand such a vast and complex universe, is called science. The universe forces those who live in it to understand it. Those creatures who find everyday experience a muddled jumble of events with no predictability, no regularity, are in grave peril. The universe belongs to those who, at least to some degree, have figured it out.

It is an astonishing fact that there *are* laws of nature, rules that summarize conveniently—not just qualitatively but quantitatively—how the world works. We might imagine a universe in which there are no such laws, in which the 10^{80} elementary particles that make up a universe like our own behave with utter and uncompromising abandon. To understand such a universe we would need a brain at least as massive as the universe. It seems unlikely that such a universe

---

[1] Chlorine is a deadly poison gas employed on European battlefields in World War I. Sodium is a corrosive metal which burns upon contact with water. Together they make a placid and un-poisonous material, table salt. Why each of these substances has the properties it does is a subject called chemistry, which requires more than 10 bits of information to understand.

could have life and intelligence, because beings and brains require some degree of internal stability and order. But even if in a much more random universe there were such beings with an intelligence much greater than our own, there could not be much knowledge, passion or joy.

Fortunately for us, we live in a universe that has at least important parts that are knowable. Our common-sense experience and our evolutionary history have prepared us to understand something of the workaday world. When we go into other realms, however, common sense and ordinary intuition turn out to be highly unreliable guides. It is stunning that as we go close to the speed of light our mass increases indefinitely, we shrink toward zero thickness in the direction of motion, and time for us comes as near to stopping as we would like. Many people think that this is silly, and every week or two I get a letter from someone who complains to me about it. But it is a virtually certain consequence not just of experiment but also of Albert Einstein's brilliant analysis of space and time called the Special Theory of Relativity. It does not matter that these effects seem unreasonable to us. We are not in the habit of traveling close to the speed of light. The testimony of our common sense is suspect at high velocities.

Or consider an isolated molecule composed of two atoms shaped something like a dumbbell—a molecule of salt, it might be. Such a molecule rotates about an axis through the line connecting the two atoms. But in the world of quantum mechanics, the realm of the very small, not all orientations of our dumbbell molecule are possible. It might be that the molecule could be oriented in a horizontal position, say, or in a vertical position, but not at many angles in between. Some rotational positions are forbidden. Forbidden by what? By the laws of nature. The universe is built in such a way as to limit, or quantize, rotation. We do not experience this directly in everyday life; we would find it startling as well as awkward in sitting-up exercises, to find arms outstretched from the sides or pointed up to the skies permitted but many intermediate positions forbidden. We do not live in the world of the small, on the scale of 10^{-13} centimeters, in the realm where there are twelve zeros between the decimal place and the one. Our common-sense intuitions do not count. What does count is experiment—in this case observations from the far infrared spectra of molecules. They show molecular rotation to be quantized.

The idea that the world places restrictions on what humans might do is frustrating. Why *shouldn't* we be able to have intermediate rotational positions? Why *can't* we travel faster than the speed of light? But so far as we can tell, this is the way the universe is constructed. Such prohibitions not only press us toward a little humility; they also make the world more knowable. Every restriction corresponds to a law of nature, a regularization of the universe. The more restrictions there are on what matter and energy can do, the more knowledge human beings can attain. Whether in some sense the universe is ultimately knowable depends not only on how many natural laws there are that encompass widely divergent phenomena, but also on whether we have the openness and the intellectual capacity to understand such laws. Our formulations of the regularities of nature are surely dependent on how the brain is built, but also, and to a significant degree, on how the universe is built.

For myself, I like a universe that includes much that is unknown and, at the same time, much that is knowable. A universe in which everything is known would be static and dull, as boring as the heaven of some weak-minded theologians. A universe that is unknowable is no fit place for a thinking being. The ideal universe for us is one very much like the universe we inhabit. And I would guess that this is not really much of a coincidence.

Questions and Topics for Discussion and Writing

1. How does the scientific "cast of mind" differ from that of people in other fields in its attempt to answer the questions about nature and the universe?
2. Interpret what Sagan means by saying, "Understanding is a kind of ecstasy."
3. How does Sagan define science? In a paragraph, write your definition.
4. Discuss the significance and warning to humanity of Sagan's saying, "The universe belongs to those who, at least to some degree, have figured it out."
5. Although Sagan quotes Emerson as a lead to his essay, his title alludes to a poem by the mystical poet William Blake, "Auguries of Innocence." One line of the poem expresses the poet's belief that it is possible to know the whole by observing a part: "To see eternity in a grain of sand." Does Sagan conclude that it is possible to know the universe by "reflecting on a grain of salt"?
6. Some philosophers have used the metaphor of the "grain of sand" to express the size of the earth in the universe. Critics of these philosophers hold that the reader is "humbled," "humiliated," even "terrified" by such a picture, which presents a "perverted sense of values." When Sagan reflects on the grain of salt, do *you* feel insignificant or diminished before the vastness of the universe? Explain.
7. What is implied by Sagan in his concluding paragraph when he challenges "some weak-minded theologians," yet adds: "The ideal universe for us is one very much like the universe we inhabit. And I would guess that this is not really much of a coincidence"? Does his statement assert a religious belief?

Loren Eiseley

Loren Eiseley (1907–1977), American anthropologist and human paleontologist, won recognition as both a scientist and a literary artist. Born in eastern Nebraska, Eiseley seems always to have kept his sense of the American prairie, with its wide horizons and its loneliness, as part of his individualistic heritage. His mother's deafness and paranoia deepened the isolation of their household: Always suspicious and jealous of any social bonds he might form, she drove his young friends from the Eiseley home. Without friends, Eiseley grew up in what one biographer called "unnatural silence," taking refuge in books, solitary fields, and caves. Recounting his personal experiences, Eiseley remembers her one redeeming feature, that she used an artist's eye to observe nature. A second compensation is that his father was a loving parent. The once-itinerant actor and hardware salesman with a well-trained voice demonstrated for his son the beauty of the spoken word. As one family friend recalls, "He grew up on his father's reading from the Bible and Shakespeare and never heard 'fussing' or angry talk. That, and his profound contemplation of nature, influenced his remarkable style."

One of Eiseley's earliest, most significant memories is being held in his father's arms to observe the passage of Halley's comet in 1910. Eiseley writes: "'If you live to be an old man,' he said carefully, fixing my eyes on the midnight spectacle, 'you will see it again. It will come back in seventy-five years. . . .' I was destined to recall the incident all my life."

Eiseley sensed very early his affinity for physical anthropology. As a young boy in Lincoln, Nebraska, he frequently visited the university museum and there became fascinated by the collection of skulls. In 1925 he entered the University of Nebraska, but did not graduate until 1933. According to his biographers, this period encompasses "the most important and difficult events of his life—the Depression and Dust Bowl, his father's death, his growing bitterness toward his mother, his bout with tuberculosis, his drifting and dangerous days riding the rails, the first real recognition of his writing ability, and the first archeological field work." Rejected, although he wanted desperately to be a part of the war effort, Eiseley taught anatomy to premedical reservists in Kansas.

For many years the Benjamin Franklin Professor of Anthropology at the University of Pennsylvania in Philadelphia, Eiseley was also Curator of Early Man in the University Museum. Always a fossil hunter (his term was "bone hunter"), he remained preoccupied with the themes of time and change. His works include the dichotomies of "time and eternity, permanence and change, memory and oblivion, mythical and real time, relative and absolute time."

In addition to the many honors he has received as a scientist, Eiseley has also been much anthologized for his complex, meditative, interpretive essays. He was elected in 1971 to the National Institute of Arts and Letters, an event attesting to his remarkable literary gifts and his stylistic accomplishment. He has written articles popularizing science, essays and books on the history of science, technical papers in anthropology, and imagistic poetry. Of his poetry, one critic observes, "Eiseley is a poet, of course; but most of his poetry, like Thoreau's, is found in his prose."

Eiseley's impulse to explore uncharted territory has led him to write many books, among them *The Immense Journey, Darwin's Century, All the Strange Hours* (his autobiography), *The Firmament of Time, Francis Bacon and the Modern Dilemma, The Unexpected Universe, Notes of an Alchemist* (a book of poems), *The Innocent Assassins* (a book of poems), *The Invisible Pyramid, Darwin and the Mysterious Mr. X,* and *The Star Thrower,* from which the following selection is taken. "Science and the Sense of the Holy" presents Eiseley's insights into what lies "behind the visible."

Science and the Sense of the Holy

Loren Eiseley

I

When I was a young man engaged in fossil hunting in the Nebraska badlands I was frequently reminded that the ravines, washes, and gullies over which we wandered resembled the fissures in a giant exposed brain. The human brain contains the fossil memories of its past—buried but not extinguished moments—just as this more formidable replica contained deep in its inner stratigraphic convolutions earth's past in the shape of horned titanotheres and stalking, dirk-toothed cats. Man's memory erodes away in the short space of a lifetime. Jutting from the coils of the earth brain over which I clambered were the buried remnants, the changing history, of the entire age of mammals—millions of years of vanished daylight with their accompanying traces of volcanic outbursts and upheavals. It may well be asked why this analogy of earth's memory should so preoccupy the mind of a scientist as to have affected his entire outlook upon nature and upon his kinship with—even his concern for—the plant and animal world about him.

Perhaps the problem can best be formulated by pointing out that there are two extreme approaches to the interpretation of the living world. One was expressed by Charles Darwin at the age of twenty-eight; one by Sigmund Freud in his mature years. Other men of science have been arrayed on opposite sides of the question, but the eminence of these two scholars will serve to point up a controversy that has been going on since science arose, sometimes quietly, sometimes marked by vitriolic behavior, as when a certain specialist wedded to his own view of the universe hurled his opponent's book into his wastebasket only to have it retrieved and cherished by a graduate student who became a lifelong advocate of the opinions reviled by his mentor. Thus it is evident that, in the supposed objective world of science, emotion and temperament may play a role in our selection of the mental tools with which we choose to investigate nature.

Charles Darwin, at a time when the majority of learned men looked upon animals as either automatons or creatures created merely for human exploitation, jotted thoughtfully into one of his early journals upon evolution the following observation:

"If we choose to let conjecture run wild, then animals, our fellow brethren in pain, disease, suffering and famine—our slaves in the most laborious works, our companions in our amusements—they may partake of our origin in one common ancestor—we may be all netted together."

What, we may now inquire, is the world view here implied, one way in which a great scientist looked upon the subject matter that was to preoccupy his entire working life? In spite of the fact that Darwin was, in his later years, an agnostic, in spite of confessing he was "in thick mud" so far as metaphysics was concerned, the remark I have quoted gives every sign of that feeling of awe, of dread of the holy playing upon nature, which characterizes the work of a number of naturalists and physicists down even to the present day. Darwin's remark reveals an intuitive sensitivity to the life of other creatures about him, an attitude quite distinct from that of the laboratory experimentalist who is hardened to the infliction of pain. In addition, Darwin's final comment that we may be all netted together in one gigantic mode of experience, that we are in a mystic sense one single diffuse animal, subject to joy and suffering beyond what we endure as individuals, reveals a youth drawn to the world of nature by far more than just the curiosity to be readily satisfied by the knife or the scalpel.

If we turn to Sigmund Freud by way of contrast we find an oddly inhibited reaction. Freud, though obviously influenced by the elegant medical experimenters of his college days, groped his way alone, and by methods not subject to quantification or absolute verification, into the dark realms of the subconscious. His reaction to the natural world, or at least his feelings and intuitions about it, are basically cold, clinical, and reserved. He of all men recognized what one poet has termed "the terrible archaeology of the brain." Freud states that "nothing once constructed has perished, and all the earlier stages of development have survived alongside the latest." But for Freud, convinced that childhood made the man, adult reactions were apt to fall under the suspicion of being childhood ghosts raised up in a disguised fashion. Thus, insightful though he could be, the very nature of his study of man tended to generate distrust of that outgoing empathy we observed in the young Darwin. "I find it very difficult to work with these intangible qualities," confessed Freud. He was suspicious of their representing some lingering monster of childhood, even if reduced in size. Since Freud regarded any type of religious feeling—even the illuminative quality of the universe—as an illusion, feelings of awe before natural phenomena such as that manifested by Darwin were to him basically remnants of childhood and to be dismissed accordingly.

In *Civilization and Its Discontents* Freud speaks with slight condescension of a friend who claimed a sensation of eternity, something limitless, unbounded—"oceanic," to use the friend's expression. The feeling had no sectarian origin, no assurance of immortality, but implied just such a sense of awe as might lie at the root of the religious impulse. "I cannot," maintained Freud, "discover this 'oceanic' impulse in myself." Instead he promptly psychoanalyzes the feeling of oneness with the universe into the child's pleasure ego which holds to itself all that is comforting; in short, the original ego, the infant's ego, included everything. Later, by experience, contended Freud, our adult ego becomes only a shrunken vestige of that far more extensive feeling which "expressed an inseparable connection . . . with the external world."

In essence, then, Freud is explaining away one of the great feelings characteristic of the best in man by relegating it to a childhood atavistic survival in
adult life. The most highly developed animals, he observes, have arisen from
the lowest. Although the great saurians are gone, the dwarfed crocodile remains. Presumably if Freud had completed the analogy he would have been
forced to say that crocodilian adults without awe and with egos shrunken safely
into their petty concerns represented a higher, more practical evolutionary
level than the aberrant adult who persists in feelings of wonder before which
Freud recoiled with a nineteenth-century mechanist's distaste, although not
without acknowledging that this lurking childlike corruption might be widespread. He chose to regard it, however, as just another manifestation of the
irrational aspect of man's divided psyche.

Over six decades before the present, a German theologian, Rudolf Otto,
had chosen for his examination what he termed *The Idea of the Holy (Das
Heilige)*. Appearing in 1917 in a time of bitterness and disillusionment, his book
was and is still widely read. It cut across denominational divisions and spoke to
all those concerned with that *mysterium tremendum,* that very awe before the
universe which Freud had sighed over and dismissed as irrational. I think it
safe to affirm that Freud left adult man somewhat shrunken and misjudged—
misjudged because some of the world's scientists and artists have been deeply
affected by the great mystery, less so the child at one's knee who frequently has
to be disciplined to what in India has been called the "opening of the heavenly
eye."

Ever since man first painted animals in the dark of caves he has been re-
sponding to the holy, to the numinous, to the mystery of being and becoming,
to what Goethe very aptly called "the weird portentous." Something inexpress-
ible was felt to lie behind nature. The bear cult, circumpolar in distribution and
known archaeologically to extend into Neanderthal times, is a further and most
ancient example. The widespread beliefs in descent from a totemic animal,
guardian helpers in the shapes of animals, the concept of the game lords who
released or held back game to man are all part of a variety of a sanctified,
reverent experience that extends from the beautiful rock paintings of South
Africa to the men of the Labradorean forests or the Plains Indian seeking by
starvation and isolation to bring the sacred spirits to his assistance. All this is
part of the human inheritance, the wonder of the world, and nowhere does that
wonder press closer to us than in the guise of animals which, whether supernat-
urally as in the caves of our origins or, as in Darwin's sudden illumination,
perceived to be, at heart, one form, one awe-inspiring mystery, seemingly
diverse and apart but derived from the same genetic source. Thus the *mysterium* arose not by primitive campfires alone. Skins may still prickle in a modern
classroom.

In the end, science as we know it has two basic types of practitioners. One is
the educated man who still has a controlled sense of wonder before the univer-

sal mystery, whether it hides in a snail's eye or within the light that impinges on that delicate organ. The second kind of observer is the extreme reductionist who is so busy stripping things apart that the tremendous mystery has been reduced to a trifle, to intangibles not worth troubling one's head about. The world of the secondary qualities—color, sound, thought—is reduced to illusion. The *only* true reality becomes the chill void of ever-streaming particles.

If one is a biologist this approach can result in behavior so remarkably cruel that it ceases to be objective but rather suggests a deep grain of sadism that is not science. To list but one example, a recent newspaper article reported that a great urban museum of national reputation had spent over a half-million dollars on mutilating experiments with cats. The experiments are too revolting to chronicle here and the museum has not seen fit to enlighten the public on the knowledge gained at so frightful a cost in pain. The cost, it would appear, lies not alone in animal suffering but in the dehumanization of those willing to engage in such blind, random cruelty. The practice was defended by museum officials, who in a muted show of scientific defense maintained the right to study what they chose "without regard to its demonstrable practical value."

This is a scientific precept hard to override since the days of Galileo, as the official well knew. Nevertheless, behind its seamless façade of probity many terrible things are and will be done. Blaise Pascal, as far back as the seventeenth century, foresaw our two opposed methods. Of them he said: "There are two equally dangerous extremes, to shut reason out, and to let nothing else in." It is the reductionist who, too frequently, would claim that the end justifies the means, who would assert reason as his defense and let that *mysterium* which guards man's moral nature fall away in indifference, a phantom without reality.

"The whole of existence frightens me," protested the philosoper Søren Kierkegaard; "from the smallest fly to the mystery of the Incarnation, everything is unintelligible to me, most of all myself." By contrast, the evolutionary reductionist Ernst Haeckel, writing in 1877, commented that "the cell consists of matter . . . composed chiefly of carbon with an admixture of hydrogen, nitrogen and sulphur. These component parts, properly united, produce the soul and body of the animated world, and suitably nourished become man. With this single argument the mystery of the universe is explained, the Deity annulled and a new era of infinite knowledge ushered in." Since these remarks of Haeckel's, uttered a hundred years ago, the genetic alphabet has scarcely substantiated in its essential intricacy Haeckel's carefree dismissal of the complexity of life. If anything, it has given weight to Kierkegaard's wary statement or at least heightened the compassionate wonder with which we are led to look upon our kind.

"A conviction akin to religious feeling of the rationality or intelligibility of the world lies behind all scientific work of a high order," says Albert Einstein. Here once more the eternal dichotomy manifests itself. Thoreau, the man of literature, writes compassionately, "Shall I not have intelligence with the earth? Am

I not partly leaves and vegetable mould myself?" Or Walt Whitman, the poet, protests in his *Song of Myself:* "whoever walks a furlong without sympathy walks to his own funeral drest in a shroud."

> "Magnifying and applying come I"—he thunders—
> "Outbidding at the start the old cautious hucksters . . .
> Not objecting to special revelations, considering a curl of smoke or a hair
> on the back of my hand just as curious as any revelation."

Strange, is it not, that so many of these voices are not those of children, but those of great men—Newton playing on the vast shores of the universe, or Whitman touched with pity or Darwin infused with wonder over the clambering tree of life. Strange, that all these many voices should be dismissed as the atavistic yearnings of an unreduced childlike ego seeking in "oceanic" fashion to absorb its entire surroundings, as though in revolt against the counting house, the laboratory, and the computer.

II

Not long ago in a Manhattan art gallery there were exhibited paintings by Irwin Fleminger, a modernist whose vast lawless Martianlike landscapes contain cryptic human artifacts. One of these paintings attracted my eye by its title: "Laws of Nature." Here in a jumbled desert waste without visible life two thin laths had been erected a little distance apart. Strung across the top of the laths was an insubstantial string with even more insubstantial filaments depending from the connecting cord. The effect was terrifying. In the huge inhuman universe that constituted the background, man, who was even more diminished by his absence, had attempted to delineate and bring under natural law an area too big for his comprehension. His effort, his "law," whatever it was, denoted a tiny measure in the midst of an ominous landscape looming away to the horizon. The frail slats and dangling string would not have sufficed to fence a chicken run.

The message grew as one looked. With all the great powers of the human intellect we were safe, we understood, in degree, a space between some slats and string, a little gate into the world of infinitude. The effect was crushing and it brought before one that sense of the "other" of which Rudolf Otto spoke, the sense beyond our senses, unspoken awe, or, as the reductionist would have it, nothing but waste. There the slats stood and the string drooped hopelessly. It was the natural law imposed by man, but outside its compass, again to use the words of Thoreau, was something terrific, not bound to be kind to man. Not man's at all really—a star's substance totally indifferent to life or what laws life might concoct. No man would greatly extend that trifling toy. The line sagged hopelessly. Man's attempt had failed, leaving but an artifact in the wilderness.

Perhaps, I thought, this is man's own measure. Perhaps he has already gone. The crepitation at my spine increased. I felt the mood of the paleolithic artists, lost in the mysteries of birth and coming, as they carved pregnant beasts in the dark of caves and tried by crayons to secure the food necessarily wrung from similar vast landscapes. Their art had the same holy quality that shows in the ivory figurines, the worship before the sacred mother who brought man mysteriously into the limited world of the cave mouth.

The numinous then is touched with superstition, the reductionist would say, but all the rituals suggest even toward hunted animals a respect and sympathy leading to ceremonial treatment of hunted souls; whereas by contrast in the modern world the degradation of animals in experiments of little, or vile, meaning, were easily turned to the experimental human torture practiced at Dachau and Buchenwald by men dignified with medical degrees. So the extremes of temperament stand today: the man with reverence and compassion in his heart whose eye ranges farther than the two slats in the wilderness, and the modern vandal totally lacking in empathy for life beyond his own, his sense of wonder reduced to a crushing series of gears and quantitative formula, the educated vandal without mercy or tolerance, the collecting man that I once tried to prevent from killing an endangered falcon, who raised his rifle, fired, and laughed as the bird tumbled at my feet. I suppose Freud might have argued that this was a man of normal ego, but I, extending my childlike mind into the composite life of the world, bled accordingly.

Perhaps Freud was right, but let us look once more at this brain that in many distinguished minds has agonized over life and the mysterious road by which it has come. Certainly, as Darwin recognized, it was not the tough-minded, logical inductionists of the early nineteenth century who in a deliberate distortion of Baconian philosophy solved the problem of evolution. Rather, it was what Darwin chose to call "speculative" men, men, in other words, with just a touch of the numinous in their eye, a sense of marvel, a glimpse of what was happening behind the visible, who saw the whole of the living world as though turning in a child's kaleidoscope.

Among the purely human marvels of the world is the way the human brain after birth, when its cranial capacity is scarcely larger than that of a gorilla or other big anthropoid, spurts onward to treble its size during the first year of life. The human infant's skull will soar to a cranial capacity of 950 cubic centimeters while the gorilla has reached only 380 cubic centimeters. In other words, the human brain grows at an exponential rate, a spurt which carries it almost to adult capacity at the age of fourteen.

This clever and specifically human adaptation enables the human offspring successfully to pass the birth canal like a reasonably small-headed animal, but in a more larval and helpless condition than its giant relatives. The brain burgeons after birth, not before, and it is this fact which enables the child, with proper care, to assimilate all that larger world which will be forever denied to its relative the gorilla. The big anthropoids enjoy no such expansion. Their brains grow without exponential quickening into maturity. Somewhere in the far past of man something strange happened in his evolutionary development. His skull has enhanced its youthful globularity; he has lost most of his body hair and what remains grows strangely. He demands, because of his immature emergence into the world, a lengthened and protected childhood. Without prolonged familial attendance he would not survive, yet in him reposes the capacity for great art, inventiveness, and his first mental tool, speech, which creates his humanity. He is without doubt the oddest and most unusual evolutionary product that this planet has yet seen.

The term applied to this condition is neoteny, or pedomorphism. Basically the evolutionary forces, and here "forces" stands for complete ignorance, seem to have taken a roughhewn ordinary primate and softened and eliminated the adult state in order to allow for a fantastic leap in brain growth. In fact, there is a growing suspicion that some, at least, of the African fossils found and ascribed to the direct line of human ascent in eastern Africa may never, except for bipedalism and some incipient tool-using capacities, have taken the human road at all.

Some with brains that seem to have remained at the same level through long ages have what amounts quantitatively to no more than an anthropoid brain. Allowing for upright posture and free use of the hand, there is no assurance that they spoke or in any effective way were more than well-adapted bipedal apes. Collateral relatives, in short, but scarcely to be termed men. All this is the more remarkable because their history can now be traced over roughly five if not six million years—a singularly unprogressive period for a creature destined later to break upon the world with such remarkable results after so long a period of gestation.

Has something about our calculations gone wrong? Are we studying, however necessarily, some bipedal cousins but not ancestral men? The human phylogeny which we seemed well on the way to arranging satisfactorily is now confused by a multiplicity of material contended over by an almost equal number of scholars. Just as a superfluity of flying particles is embarrassing the physicist, so man's evolution, once thought to be so clearly delineated, is showing signs of similar strain. A skull from Lake Rudolf with an estimated capacity of 775 cubic centimeters or even 800 and an antiquity ranging into the three-million-year range is at the human Rubicon, yet much younger fossils are nowhere out of the anthropoid range.

Are these all parts of a single variable subhumanity from which we arose, or are some parts of this assemblage neotenous of brain and others not? The scientific exchanges are as stiff with politeness as exchanges between enemies on the floor of the Senate. "Professor so-and-so forgets the difficult task of restoring to its proper position a frontal bone trampled by cattle." A million years may be covertly jumped because there is nothing to be found in it. We must never lose sight of one fact, however: it is by neotenous brain growth that we came to be men, and certain of the South African hominids to which we have given such careful attention seem to have been remarkably slow in revealing such development. Some of them, in fact, during more years than present mankind has been alive seem to have flourished quite well as simple grassland apes.

Why indeed should they all have become men? Because they occupied the same ecological niche, contend those who would lump this variable assemblage. But surely paleontology does not always so bind its deliberations. We are here dealing with a gleam, a whisper, a thing of awe in the mind itself, that oceanic feeling which even the hardheaded Freud did not deny existed though he tried to assign it to childhood.

With animals whose precise environment through time may overlap, extinction may result among contending forms; it can and did happen among men. But with the first stirrings of the neotenous brain and its superinduced transformation of the family system a new type of ecological niche had incipiently appeared—a speaking niche, a wondering niche which need not have been first manifested in tools but in family organization, in wonder over what lay over the next hill or what became of the dead. Whether man preferred seeds or flesh, how he regarded his silent collateral relatives, may not at first have induced great competition. Only those gifted with the pedomorphic brain would in some degree have fallen out of competition with the real. It would have been their danger and at the same time their beginning triumph. They were starting to occupy, not a niche in nature, but an invisible niche carved into thought which in time would bring them suffering, superstition, and great power.

It cannot, in the beginning, be recognized clearly because it is not a matter of molar teeth and seeds, or killer instincts and ill-interpreted pebbles. Rather it was something happening in the brain, some blinding, irradiating thing. Until the quantity of that gray matter reached the threshold of human proportions no one could be sure whether the creature saw with a human eye or looked upon life with even the faint stirrings of some kind of religious compassion.

The new niche in its beginnings is invisible; it has to be inferred. It does not lie waiting to be discovered in a pebble or a massive molar. These things are important in the human dawn but so is the mystery that ordained that mind should pass the channel of birth and then grow like a fungus in the night—grow and convolute and overlap its older buried strata, while a 600-pound gorilla retains by contrast the cranial content of a very small child. When man cast off his fur and placed his trust in that remarkable brain linked by neural pathways to his tongue he had potentially abandoned niches for dreams. Henceforth the world was man's niche. All else would live by his toleration—even the earth from which he sprang. Perhaps this is the hardest, most expensive lesson the layers of the fungus brain have yet to learn: that man is not as other creatures and that without the sense of the holy, without compassion, his brain can become a gray stalking horror—the deviser of Belsen.

Its beginning is not the only curious thing about that brain. There are some finds in South Africa dating into immediately post-glacial times that reveal a face and calvaria more "modern" in appearance, more pedomorphic, than that of the average European. The skull is marked by cranial capacities in excess of

1700 cubic centimeters—big brained by any standards. The mastoids are child-like, the teeth reduced, the cranial base foreshortened. These people, variously termed Boskopoid or Strandlooper, have, in the words of one anthropologist, "the amazing cranium to face ratio of almost five to one. In Europeans it is about three to one. Face size has been modernized and subordinated to brain growth." In a culture still relying on coarse fare and primitive implements, the face and brain had been subtly altered in the direction science fiction writers like to imagine as the direction in which mankind is progressing. Yet here the curious foetalization of the human body seems to have outrun man's cultural status, though in the process giving warning that man's brain could still pass the straitened threshold of its birth.

How did these people look upon the primitive world into which they found themselves precipitated? History gives back no answer save that here there flourished striking three-dimensional art—the art of the brother animal seen in beauty. Childlike, Freud might have muttered, with childlike dreams, rushed into conflict with the strong, the adult and shrunken ego, the ego that gets what it wants. Yet how strangely we linger over this lost episode of the human story, its pathos, its possible meaning. From whence did these people come? We are not sure. We are not even sure that they derive from one of the groups among the ruck of bipedal wandering apes long ago in Kenya that reveal some relation-ship to ourselves. Their development was slow, if indeed some of them took that road, the strange road to the foetalized brain that was to carry man outside of the little niche that fed him his tuberous, sandy diet.

We thought we were on the verge of solving the human story, but now we hold in our hands gross jaws and delicate, and are unsure of our direction save that the trail is longer than we had imagined twenty years ago. Yet still the question haunts us, the numinous, the holy in man's mind. Early man laid gifts beside the dead, but then in the modern unbelieving world, Ernst Haeckel's world, a renowned philosopher says, "The whole of existence frightens me," or another humbler thinker says, "In the world there is nothing to explain the world" but remembers the gold eyes of the falcon thrown brutally at his feet. He shivers while Freud says, "As for me I have never had such feelings." They are a part of childhood, Freud argues, though there are some who counter that in childhood—yes, even Freud might grant it—the man is made, the awe persists or is turned off by blows or the dullness of unthinking parents. One can only assert that in science, as in religion, when one has destroyed human won-der and compassion, one has killed man, even if the man in question continues to go about his laboratory tasks.

III

Perhaps there is one great book out of all American literature which best ex-presses the clash between the man who has genuine perception and the one who pursues nature as ruthlessly as a hunted animal. I refer to *Moby Dick*, whose narrator, Ishmael, is the namesake of a Biblical wanderer. Every literate person knows the story of Moby Dick and his pursuit by the crazed Captain Ahab who had yielded a leg to the great albino whale. It is the whale and Ahab who dominate the story. What does the whale represent? A symbol of evil, as some critics have contended? Fate, destiny, the universe personified, as other scholars have protested?

Moby Dick is "all a magnet," remarks Ahab cryptically at one moment. "And be he agent or be he principal I will wreak my hate upon him." Here, reduced to the deck of a whaler out of Nantucket, the old immortal questions resound, the questions labeled science in our era. Nothing is to go unchallenged. Thrice, by different vessels, Ahab is warned away from his contemplated conquest. The whale does not pursue Ahab, Ahab pursues the whale. If there is evil represented in the white whale it cannot be personalized. The evils of self-murder, of megalomania, are at work in a single soul calling up its foreordained destruction. Ahab heartlessly brushes aside the supplications of a brother captain to aid in the search for his son, lost somewhere in a boat in the trail of the white whale's passing. Such a search would only impede the headlong fury of the pursuit.

In Ahab's anxiety to "strike through the mask," to confront "the principal," whether god or destiny, he is denuding himself of all humanity. He has forgotten his owners, his responsibility to his crew. His single obsession, the hidden obsession that lies at the root of much Faustian overdrive in science, totally possesses him. Like Faust he must know, if the knowing kills him, if naught lies beyond. "All my means are sane," he writes, like Haeckel and many another since. "My motive and my object mad."

So it must have been in the laboratories of the atom breakers in their first heady days of success. Yet again on the third day Starbuck, the doomed mate, cries out to Ahab, "Desist. See. Moby Dick seeks thee not. It is thou, thou, that madly seekest him." This then is not the pursuit of evil. It is man in his pride that the almighty gods will challenge soon enough. Not for nothing is Moby Dick a white snow hill rushing through Pacific nights. He carries upon his brow the inscrutability of fate. Agent or principal, Moby Dick presents to Ahab the mystery he may confront but never conquer. There is no harpoon tempered that will strike successfully the heart of the great enigma.

So much for the seeking peg-legged man without heart. We know he launched his boats and struck his blows and in the fury of returning vengeance lost his ship, his comrades, and his own life. If, indeed, he pierced momentarily the mask of the "agent," it was not long enough to tell the tale. But what of the sometimes silent narrator, the man who begins the book with the nonchalant announcement, "Call me Ishmael," the man whose Biblical namesake had every man's hand lifted against him? What did he tell? After all, Moby Dick is his book.

Ishmael, in contrast to Ahab, is the wondering man, the acceptor of all races and their gods. In contrast to the obsessed Ahab he paints a magnificent picture of the peace that reigned in the giant whale schools of the 1840s, the snuffling and grunting about the boats like dogs, loving and being loved, huge mothers gazing in bliss upon their offspring. After hours of staring in those peaceful depths, "Deep down," says Ishmael, "there I still bathe in eternal mildness of joy." The weird, the holy, hangs undisturbed over the whales' huge cradle. Ishmael knows it, others do not.

At the end, when Ahab has done his worst and the *Pequod* with the wounded whale is dragged into the depths amidst shrieking seafowl, it is Ishmael, buoyed up on the calked coffin of his cannibal friend Queequeg, who survives to tell the tale. Like Whitman, like W. H. Hudson, like Thoreau, Ishmael, the wanderer, has noted more of nature and his fellow men than has the headstrong pursuer of the white whale, whether "agent" or "principal," within the universe. The tale is not of science, but it symbolizes on a gigantic canvas the struggle between two ways of looking at the universe: the magnification by the poet's mind attempting to see all, while disturbing as little as possible, as opposed to the plunging fury of Ahab with his cry, "Strike, strike through the mask, whatever it may cost in lives and suffering." Within our generation we have seen the one view plead for endangered species and reject the despoliation of the earth; the other has left us lingering in the shadow of atomic disaster. Actually, the division is not so abrupt as this would imply, but we are conscious as never before in history that there is an invisible line of demarcation, an ethic that science must sooner or later devise for itself if mankind is to survive. Herman Melville glimpsed in his huge mythology of the white beast that was nature's agent something that only the twentieth century can fully grasp.

It may be that those childlike big-brained skulls from Africa are not of the past but of the future, man, not, in Freud's words, retaining an atavistic child's ego, but pushing onward in an evolutionary attempt to become truly at peace with the universe, to know and enjoy the sperm-whale nursery as did Ishmael, to paint in three dimensions the beauty of the world while not to harm it.

Yesterday, wandering along a railroad spur line, I glimpsed a surprising sight. All summer long, nourished by a few clods of earth on a boxcar roof, a sunflower had been growing. At last, the car had been remembered. A train was being made up. The box car with its swaying rooftop inhabitant was coupled in. The engine tooted and slowly, with nodding dignity, my plant began to travel.

Throughout the summer I had watched it grow but never troubled it. Now it lingered and bowed a trifle toward me as the winds began to touch it. A light not quite the sunlight of this earth was touching the flower, or perhaps it was the watering of my aging eye—who knows? The plant would not long survive its journey but the flower seeds were autumn-brown. At every jolt for miles they would drop along the embankment. They were travelers—travelers like Ishmael and myself, outlasting all fierce pursuits and destined to re-emerge into future autumns. Like Ishmael, I thought, they will speak with the voice of the one true agent. "I only am escaped to tell thee."

Questions and Topics for Discussion and Writing

1. Analyze the first paragraph of this essay and determine how it is related to the rest of the work, especially to the sixth paragraph.
2. Although many famous people are referred to in the essay, because of the allusion in the title to Rudolf Otto's work *The Idea of the Holy*, a classic work on mysticism, Otto looms as an influence of supreme importance. Why do you think Eiseley changed the word "Idea" to "Sense"?
3. Summarize in your own words Eiseley's conclusions about Charles Darwin and Sigmund Freud.
4. What docs Eiseley mean by "world view"?
5. How does Eiseley define mysticism?
6. Eiseley distinguishes between two basic types of practitioners of science: the one who experiences a "terrible" sense of wonder about the universe and the "extreme reductionist." What does Eiseley mean by "terrible"? What is his judgment of the reductionist?
7. How does Eiseley use art—painting and literature—as a focus of discussion? Considering his concluding identification of himself with Ishmael, the narrator of *Moby Dick*, explain Eiseley's idea of the value of literature to scientists.
8. Can you find specific examples in the news today of what Eiseley terms "educated vandals"?
9. What are the importance and implications of Eiseley's essay to teachers of biological science?
10. Since scientists are involved in the survival of the human race, what is Eiseley's warning to them?

Isaac Asimov

The influential science-fiction writer and science popularizer Isaac Asimov (b. 1920) is the son of a Jewish candy-store keeper. The family came to the United States from Russia in 1923, and Asimov became a naturalized citizen in 1928. He was educated at Columbia University, graduating at the age of nineteen and remaining for further work in chemistry. World War II interrupted his studies, and he began work as a chemist at the United States Navy Yard in Philadelphia, later serving as a member of the armed forces. He returned to Columbia in 1946, earning his Ph.D. in 1948. A year later he accepted a faculty position at Boston University School of Medicine, where he has remained.

Asimov is best known for his science fiction. His first acquaintance with the genre came through *Amazing Stories,* a magazine he began reading in his father's store when he was nine. According to Asimov, the encounter with the magazine was fateful: It interested him permanently not only in science fiction but also in writing and in science itself. His first sale was the story "Marooned off Vesta," which appeared in the 1939 issue of *Amazing Stories*. Since then, while remaining a practicing biochemist, he has written hundreds of stories, which have tended to popularize and humanize robots.

Asimov's books on biochemistry, mathematics, physics, astronomy, the Bible, Shakespeare, and a range of nonscientific subjects number over 200, allowing him the material for *Opus 200* (1979), an anthology of works from his second hundred books. His works include *The Stars, Of Time and Space and Other Things, ABC's of the Earth, I, Robot*, the *Foundation* trilogy, *The Gods Themselves*, and *The Autobiography of Isaac Asimov, 1920–1954*.

"The Eureka Phenomenon" is Asimov's contribution to the dialogue on what constitutes scientific method. The essay presents vivid examples of "the flash of deep insight"—the "workings of involuntary reasoning" often called scientific intuition. Asimov demonstrates his ability to present complex information in a style that engages the general reader.

The Eureka Phenomenon

Isaac Asimov

In the old days, when I was writing a great deal of fiction, there would come, once in a while, moments when I was stymied. Suddenly, I would find I had written myself into a hole and could see no way out. To take care of that, I developed a technique which invariably worked.

It was simply this—I went to the movies. Not just any movie. I had to pick a movie which was loaded with action but which made no demands on the intellect. As I watched, I did my best to avoid any conscious thinking concerning my problem, and when I came out of the movie I knew exactly what I would have to do to put the story back on the track.

It never failed.

In fact, when I was working on my doctoral dissertation, too many years ago, I suddenly came across a flaw in my logic that I had not noticed before and that knocked out everything I had done. In utter panic, I made my way to a Bob Hope movie—and came out with the necessary changes in point of view.

It is my belief, you see, that thinking is a double phenomenon like breathing.

You can control breathing by deliberate voluntary action: you can breathe deeply and quickly, or you can hold your breath altogether, regardless of the body's needs at the time. This, however, doesn't work well for very long. Your chest muscles grow tired, your body clamors for more oxygen, or less, and you relax. The automatic involuntary control of breathing takes over, adjusts it to the body's needs and unless you have some respiratory disorder, you can forget about the whole thing.

Well, you can think by deliberate voluntary action, too, and I don't think it is much more efficient on the whole than voluntary breath control is. You can deliberately force your mind through channels of deductions and associations in search of a solution to some problem and before long you have dug mental furrows for yourself and find yourself circling round and round the same limited pathways. If those pathways yield no solution, no amount of further conscious thought will help.

On the other hand, if you let go, then the thinking process comes under automatic involuntary control and is more apt to take new pathways and make erratic associations you would not think of consciously. The solution will then come while you *think* you are *not* thinking.

The trouble is, though, that conscious thought involves no muscular action and so there is no sensation of physical weariness that would force you to quit. What's more, the panic of necessity tends to force you to go on uselessly, with each added bit of useless effort adding to the panic in a vicious cycle.

It is my feeling that it helps to relax, deliberately, by subjecting your mind to material complicated enough to occupy the voluntary faculty of thought, but superficial enough not to engage the deeper involuntary one. In my case, it is an action movie; in your case, it might be something else.

I suspect it is the involuntary faculty of thought that gives rise to what we call "a flash of intuition," something that I imagine must be merely the result of unnoticed thinking.

Perhaps the most famous flash of intuition in the history of science took place in the city of Syracuse in third-century B.C. Sicily. Bear with me and I will tell you the story—

About 250 B.C., the city of Syracuse was experiencing a kind of Golden Age. It was under the protection of the rising power of Rome, but it retained a king of its own and considerable self-government; it was prosperous; and it had a flourishing intellectual life.

The king was Hieron II, and he had commissioned a new golden crown from a goldsmith, to whom he had given an ingot of gold as raw material. Hieron, being a practical man, had carefully weighed the ingot and then weighed the crown he received back. The two weights were precisely equal. Good deal!

But then he sat and thought for a while. Suppose the goldsmith had sub- tracted a little bit of the gold, not too much, and had substituted an equal weight of the considerably less valuable copper. The resulting alloy would still have the appearance of pure gold, but the goldsmith would be plus a quantity of gold over and above his fee. He would be buying gold with copper, so to speak, and Hieron would be neatly cheated.

Hieron didn't like the thought of being cheated any more than you or I would, but he didn't know how to find out for sure if he had been. He could scarcely punish the goldsmith on mere suspicion. What to do?

Fortunately, Hieron had an advantage few rulers in the history of the world could boast. He had a relative of considerable talent. The relative was named

Archimedes and he probably had the greatest intellect the world was to see prior to the birth of Newton.

Archimedes was called in and was posed the problem. He had to determine whether the crown Hieron showed him was pure gold, or was gold to which a small but significant quantity of copper had been added.

If we were to reconstruct Archimedes' reasoning, it might go as follows. Gold was the densest known substance (at that time). Its density in modern terms is 19.3 grams per cubic centimeter. This means that a given weight of gold takes up less volume than the same weight of anything else! In fact, a given weight of pure gold takes up less volume than the same weight of *any* kind of impure gold.

The density of copper is 8.92 grams per cubic centimeter, just about half that of gold. If we consider 100 grams of pure gold, for instance, it is easy to calculate it to have a volume of 5.18 cubic centimeters. But suppose that 100 grams of what looked like pure gold was really only 90 grams of gold and 10 grams of copper. The 90 grams of gold would have a volume of 4.66 cubic centimeters, while the 10 grams of copper would have a volume of 1.12 cubic centimeters; for a total value of 5.78 cubic centimeters.

The difference between 5.18 cubic centimeters and 5.78 cubic centimeters is quite a noticeable one, and would instantly tell if the crown were of pure gold, or if it contained 10 per cent copper (with the missing 10 per cent of gold tucked neatly in the goldsmith's strongbox).

All one had to do, then, was measure the volume of the crown and compare it with the volume of the same weight of pure gold.

The mathematics of the time made it easy to measure the volume of many simple shapes: a cube, a sphere, a cone, a cylinder, any flattened object of simple regular shape and known thickness, and so on.

We can imagine Archimedes saying, "All that is necessary, sire, is to pound that crown flat, shape it into a square of uniform thickness, and then I can have the answer for you in a moment."

Whereupon Hieron must certainly have snatched the crown away and said, "No such thing. I can do that much without you; I've studied the principles of mathematics, too. This crown is a highly satisfactory work of art and I won't have it damaged. Just calculate its volume without in any way altering it."

But Greek mathematics had no way of determining the volume of anything with a shape as irregular as the crown, since integral calculus had not yet been invented (and wouldn't be for two thousand years, almost). Archimedes would have had to say, "There is no known way, sire, to carry through a non-destructive determination of volume."

"Then think of one," said Hieron testily.

And Archimedes must have set about thinking of one, and gotten nowhere. Nobody knows how long he thought, or how hard, or what hypotheses he considered and discarded, or any of the details.

What we do know is that, worn out with thinking, Archimedes decided to visit the public baths and relax. I think we are quite safe in saying that Archimedes had no intention of taking his problem to the baths with him. It would be ridiculous to imagine he would, for the public baths of a Greek metropolis weren't intended for that sort of thing.

The Greek baths were a place for relaxation. Half the social aristocracy of the town would be there and there was a great deal more to do than wash. One steamed one's self, got a massage, exercised, and engaged in general socializing. We can be sure that Archimedes intended to forget the stupid crown for a while.

One can envisage him engaging in light talk, discussing the latest news from Alexandria and Carthage, the latest scandals in town, the latest funny jokes at the expense of the country-squire Romans—and then he lowered himself into a nice hot bath which some bumbling attendant had filled too full.

The water in the bath slopped over as Archimedes got in. Did Archimedes notice that at once, or did he sigh, sink back, and paddle his feet awhile before noting the water-slop? I guess the latter. But, whether soon or late, he noticed, and that one fact, added to all the chains of reasoning his brain had been working on during the period of relaxation when it was unhampered by the comparative stupidities (even in Archimedes) of voluntary thought, gave Archimedes his answer in one blinding flash of insight.

Jumping out of the bath, he proceeded to run home at top speed through the streets of Syracuse. He did *not* bother to put on his clothes. The thought of Archimedes running naked through Syracuse has titillated dozens of generations of youngsters who have heard this story, but I must explain that the ancient Greeks were quite lighthearted in their attitude toward nudity. They thought no more of seeing a naked man on the streets of Syracuse, than we would on the Broadway stage.

And as he ran, Archimedes shouted over and over, "I've got it! I've got it!" Of course, knowing no English, he was compelled to shout it in Greek, so it came out, *"Eureka! Eureka!"*

Archimedes' solution was so simple that anyone could understand it—once Archimedes explained it.

If an object that is not affected by water in any way, is immersed in water, it is bound to displace an amount of water equal to its own volume, since two objects cannot occupy the same space at the same time.

Suppose, then, you had a vessel large enough to hold the crown and suppose it had a small overflow spout set into the middle of its side. And suppose further that the vessel was filled with water exactly to the spout, so that if the water level were raised a bit higher, however slightly, some would overflow.

Next, suppose that you carefully lower the crown into the water. The water level would rise by an amount equal to the volume of the crown, and that volume of water would pour out the overflow and be caught in a small vessel. Next, a lump of gold, known to be pure and exactly equal in weight to the crown, is also immersed in the water and again the level rises and the overflow is caught in a second vessel.

If the crown were pure gold, the overflow would be exactly the same in each case, and the volume of water caught in the two small vessels would be equal. If, however, the crown were of alloy, it would produce a larger overflow than the pure gold would and this would be easily noticeable.

What's more, the crown would in no way be harmed, defaced, or even as much as scratched. More important, Archimedes had discovered the "principle of buoyancy."

And was the crown pure gold? I've heard that it turned out to be alloy and that the goldsmith was executed, but I wouldn't swear to it.

How often does this "Eureka phenomenon" happen? How often is there this flash of deep insight during a moment of relaxation, this triumphant cry of "I've got it! I've got it!" which must surely be a moment of the purest ecstasy this sorry world can afford?

I wish there were some way we could tell. I suspect that in the history of science it happens *often;* I suspect that very few significant discoveries are made by the pure technique of voluntary thought; I suspect that voluntary thought may possibly prepare the ground (if even that), but that the final touch, the real inspiration, comes when thinking is under involuntary control.

But the world is in a conspiracy to hide the fact. Scientists are wedded to reason, to the meticulous working out of consequences from assumptions to the careful organization of experiments designed to check those consequences. If a certain line of experiments ends nowhere, it is omitted from the final report. If an inspired guess turns out to be correct, it is *not* reported as an inspired guess. Instead, a solid line of voluntary thought is invented after the fact to lead up to the thought, and that is what is inserted in the final report.

The result is that anyone reading scientific papers would swear that *nothing* took place but voluntary thought maintaining a steady clumping stride from origin to destination, and that just can't be true.

It's such a shame. Not only does it deprive science of much of its glamour (how much of the dramatic story in Watson's *Double Helix* do you suppose got into the final reports announcing the great discovery of the structure of DNA?[1]), but it hands over the important process of "insight," "inspiration," "revelation" to the mystic.

The scientist actually becomes ashamed of having what we might call a revelation, as though to have one is to betray reason—when actually what we call revelation in a man who has devoted his life to reasoned thought, is after all merely reasoned thought that is not under voluntary control.

Only once in a while in modern times do we ever get a glimpse into the workings of involuntary reasoning, and when we do, it is always fascinating. Consider, for instance, the case of Friedrich August Kekule von Stradonitz.

In Kekule's time, a century and a quarter ago, a subject of great interest to

---

[1] I'll tell you, in case you're curious. None! [Asimov's note]. How Francis Crick and James Watson discovered the molecular structure of this vital substance is told in Watson's autobiographical book, *The Double Helix.*

chemists was the structure of organic molecules (those associated with living tissue). Inorganic molecules were generally simple in the sense that they were made up of few atoms. Water molecules, for instance, are made up of two atoms of hydrogen and one of oxygen (H_2O). Molecules of ordinary salt are made up of one atom of sodium and one of chlorine (NaCl), and so on.

Organic molecules, on the other hand, often contained a large number of atoms. Ethyl alcohol molecules have two carbon atoms, six hydrogen atoms, and an oxygen atom (C_2H_6O); the molecule of ordinary cane sugar is $C_{12}H_{22}O_{11}$, and other molecules are even more complex.

Then, too, it is sufficient, in the case of inorganic molecules generally, merely to know the kinds and numbers of atoms in the molecule; in organic molecules, more is necessary. Thus, dimethyl ether has the formula C_2H_6O, just as ethyl alcohol does, and yet the two are quite different in properties. Apparently, the atoms are arranged differently within the molecules—but how to determine the arrangements?

In 1852, an English chemist, Edward Frankland, had noticed that the atoms of a particular element tended to combine with a fixed number of other atoms. This combining number was called "valence." Kekule in 1858 reduced this notion to a system. The carbon atom, he decided (on the basis of plenty of chemical evidence) had a valence of four; the hydrogen atom, a valence of one; and the oxygen atom, a valence of two (and so on).

Why not represent the atoms as their symbols plus a number of attached dashes, that number being equal to the valence? Such atoms could then be put together as though they were so many Tinker Toy units and "structural formulas" could be built up.

It was possible to reason out that the structural formula of ethyl alcohol was

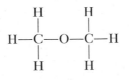

while that of dimethyl ether was

$$
\begin{array}{ccc}
H & & H \\
| & & | \\
H-C-O-C-H \\
| & & | \\
H & & H \\
\end{array}
$$

In each case, there were two carbon atoms, each with four dashes attached; six hydrogen atoms, each with one dash attached; and an oxygen atom with two dashes attached. The molecules were built up of the same components, but in different arrangements.

Kekule's theory worked beautifully. It has been immensely deepened and elaborated since his day, but you can still find structures very much like Kekule's Tinker Toy formulas in any modern chemical textbook. They represent oversimplifications of the true situation, but they remain extremely useful in practice even so.

The Kekule structures were applied to many organic molecules in the years after 1858 and the similarities and contrasts in the structures neatly matched similarities and contrasts in properties. The key to the rationalization of organic chemistry had, it seemed, been found.

Yet there was one disturbing fact. The well-known chemical benzene wouldn't fit. It was known to have a molecule made up of equal numbers of carbon and hydrogen atoms. Its molecular weight was known to be 78 and a single carbon-hydrogen combination had a weight of 13. Therefore, the benzene molecule had to contain six carbon-hydrogen combinations and its formula had to be C_6H_6.

But that meant trouble. By the Kekule formulas, the hydrocarbons (molecules made up of carbon and hydrogen atoms only) could easily be envisioned as chains of carbon atoms with hydrogen atoms attached. If all the valences of the carbon atoms were filled with hydrogen atoms, as in "hexane," whose molecule looks like this—

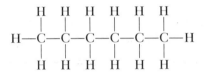

the compound is said to be saturated. Such saturated hydrocarbons were found to have very little tendency to react with other substances.

If some of the valences were not filled, unused bonds were added to those connecting the carbon atoms. Double bonds were formed as in "hexene"—

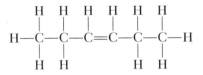

Hexene is unsaturated, for that double bond has a tendency to open up and add other atoms. Hexene is chemically active.

When six carbons are present in a molecule, it takes fourteen hydrogen atoms to occupy all the valence bonds and make it inert—as in hexane. In hexene, on the other hand, there are only twelve hydrogens. If there were still fewer hydrogen atoms, there would be more than one double bond; there might even be triple bonds, and the compound would be still more active than hexene.

Yet benzene, which is C_6H_6 and has eight fewer hydrogen atoms than hexane, is *less* active than hexene, which has only two fewer hydrogen atoms than hexane. In fact, benzene is even less active than hexane itself. The six hydrogen atoms in the benzene molecule seem to satisfy the six carbon atoms to a greater extent than do the fourteen hydrogen atoms in hexane.

For heaven's sake, why?

This might seem unimportant. The Kekule formulas were so beautifully suitable in the case of so many compounds that one might simply dismiss benzene as an exception to the general rule.

Science, however, is not English grammar. You can't just categorize something as an exception. If the exception doesn't fit into the general system, then the general system must be wrong.

Or, take the more positive approach. An exception can often be made to fit into a general system, provided the general system is broadened. Such broadening generally represents a great advance and for this reason, exceptions ought to be paid great attention.

For some seven years, Kekule faced the problem of benzene and tried to puzzle out how a chain of six carbon atoms could be completely satisfied with as few as six hydrogen atoms in benzene and yet be left unsatisfied with twelve hydrogen atoms in hexene.

Nothing came to him!

And then one day in 1865 (he tells the story himself) he was in Ghent, Belgium, and in order to get to some destination, he boarded a public bus. He was tired and, undoubtedly, the droning beat of the horses' hooves on the cobblestones, lulled him. He fell into a comatose half-sleep.

In that sleep, he seemed to see a vision of atoms attaching themselves to each other in chains that moved about. (Why not? It was the sort of thing that constantly occupied his waking thoughts.) But then one chain twisted in such a way that head and tail joined, forming a ring—and Kekule woke with a start.

To himself, he must surely have shouted "Eureka," for indeed he had it. The six carbon atoms of benzene formed a ring and not a chain, so that the structural formula looked like this:

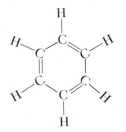

To be sure, there were still three double bonds, so you might think the molecule had to be very active—but now there was a difference. Atoms in a ring might be expected to have different properties from those in a chain and double bonds in one case might not have the properties of those in the other. At least, chemists could work on that assumption and see if it involved them in contradictions.

It didn't. The assumption worked excellently well. It turned out that organic molecules could be divided into two groups: aromatic and aliphatic. The former had the benzene ring (or certain other similar rings) as part of the structure and the latter did not. Allowing for different properties within each group, the Kekule structures worked very well.

For nearly seventy years, Kekule's vision held good in the hard field of actual chemical techniques, guiding the chemist through the jungle of reactions that led to the synthesis of more and more molecules. Then, in 1932, Linus Pauling applied quantum mechanics to chemical structure with sufficient subtlety to explain just why the benzene ring was so special and what had proven correct in practice proved correct in theory as well.

Other cases? Certainly.

In 1764, the Scottish engineer James Watt was working as an instrument maker for the University of Glasgow. The university gave him a model of a Newcomen steam engine, which didn't work well, and asked him to fix it. Watt fixed it without trouble, but even when it worked perfectly, it didn't work well. It was far too inefficient and consumed incredible quantities of fuel. Was there a way to improve that?

Thought didn't help, but a peaceful, relaxed walk on a Sunday afternoon did. Watt returned with the key notion in mind of using two separate chambers, one for steam only and one for cold water only, so that the same chamber did not have to be constantly cooled and reheated to the infinite waste of fuel.

The Irish mathematician William Rowan Hamilton worked up a theory of "quaternions" in 1843 but couldn't complete that theory until he grasped the fact that there were conditions under which $p \times q$ was *not* equal to $q \times p$. The necessary thought came to him in a flash one time when he was walking to town with his wife.

The German physiologist Otto Loewi was working on the mechanism of nerve action, in particular, on the chemicals produced by nerve endings. He awoke at 3 A.M. one night in 1921 with a perfectly clear notion of the type of experiment he would have to run to settle a key point that was puzzling him. He wrote it down and went back to sleep. When he woke in the morning, he found he couldn't remember what his inspiration had been. He remembered he had written it down, but he couldn't read his writing.

The next night, he woke again at 3 A.M. with the clear thought once more in mind. This time, he didn't fool around. He got up, dressed himself, went straight to the laboratory and began work. By 5 A.M. he had proved his point and the consequences of his findings became important enough in later years so that in 1936 he received a share in the Nobel prize in medicine and physiology.

How very often this sort of thing must happen, and what a shame that scientists are so devoted to their belief in conscious thought that they so consistently obscure the actual methods by which they obtain their results.

Questions and Topics for Discussion and Writing

1. Asimov begins his essay with an account of personal experience that he calls the "Eureka phenomenon." He then develops the essay by recounting experiences of several renowned scientists. What is the value of his beginning his work in this way?
2. What is the "Eureka phenomenon"?
3. Asimov writes a single-sentence paragraph that reads, "It is my belief, you see, that thinking is a double phenomenon like breathing." Why do you think he sets off this statement as a separate paragraph?
4. A basis for Zen Buddhism is that a release from voluntary and rational thought allows room for spontaneous, intuitive thought. Does such experience parallel what Asimov says when he writes of using rational thought, "If those pathways yield no solution, no amount of further conscious thought will help. On the other hand, if you let go, then the thinking process comes under automatic involuntary control and is more apt to take new pathways and make erratic associations you would not think of consciously"?
5. What fault does Asimov find with scientific papers?
6. Asimov regrets that science "hands over the important process of 'insight,' 'inspiration,' 'revelation' to the mystic." He then defines "revelation." What is his definition?
7. What does Asimov mean when he says that "Science, however, is not English grammar"?
8. Why did Asimov choose to relate experiences of scientists in so many fields?

SUGGESTIONS FOR FURTHER READING

Ackermann, Robert. *Philosophy of Science*. New York: Western Publishing Company, 1970.

Beveridge, W. I. B. *The Art of Scientific Investigation*. New York: W. W. Norton Company, 1957.

Blake, Ralph M., Curt J. Ducasse, and Edward H. Madden. *Theories of Scientific Method: The Renaissance Through the Nineteenth Century*. Ed. Edward H. Madden. Seattle: University of Washington Press, 1960.

Bunge, Mario. *Causality: The Place of the Causal Principle in Modern Science*. Cambridge: Harvard University Press, 1959.

————. *Intuition and Science*. Englewood Cliffs: Prentice-Hall, 1962.

Dewey, John. *How We Think*. Boston: D. C. Heath, 1933.

Medawar, P. B. *Induction and Intuition in Scientific Thought*. Philadelphia: American Philosophical Society, 1969.

Poincaré, Jules Henri. *Science and Hypothesis*. New York: Dover Publications, 1952.

Popper, K. *The Logic of Scientific Discovery*. New York: Scientific Editions, 1961.

Rosenthal-Schneider, Ilse. *Reality and Scientific Truth: Discussions with Einstein, von Laue, Planck*. Detroit: Wayne State University Press, 1980.

Russell, Bertrand. *Mysticism and Logic and Other Essays*. London: Allen and Unwin, 1917.

———. *The Scientific Outlook*. New York: W. W. Norton, 1931.

Salk, Jonas. *Anatomy of Reality: Merging of Intuition and Reason*. New York: Columbia University Press, 1983.

Scientific American. September 1958 (entire issue devoted to creative aspects of the scientific method).

Unit 3

THE BROADER PERSPECTIVE: SCIENCE AS AN ELEMENT IN CULTURE

Bertrand Russell

Bertrand Russell (the third Earl Russell), British philosopher, mathematician, logician, and social reformer, is one of the architects of our modern world. His life was extraordinarily long (1872–1970), and he frequently reappraised and adjusted his ideas. The grandson of Lord John Russell, twice Queen Victoria's Prime Minister, Russell spent his childhood and adolescence during a period of great stability in world affairs. At the start of the twentieth century, however, Russell began to grapple with difficult new ideas in mathematics and philosophy. In maturity, he faced the shocks and destruction of the two World Wars and reacted as a philosopher and a pacifist. Finally, in old age, he recognized the peril of nuclear armaments for the world and engaged himself until his death in the issues of the peace movement. With these changing interests, Russell became an international lecturer as well as a writer.

Many commentators see Russell as a titan, controversial and complex. Such complexity has inspired contradictory descriptions: He is sometimes called an "aristocratic rebel" for his liberal politics, his three efforts as a political candidate for Parliament, and his activity as a pacifist, for which he was twice imprisoned. He is sometimes described as "a romantic rationalist." One biographer labeled him a "passionate sceptic."

Russell's achievements, together with his diversity, are attested to by the Order of Merit and the Nobel Prize for Literature in 1950. The citation reads:

"In recognition for his varied and significant writing in which he champions humanitarian ideals and freedom of thought."

The search for freedom and the pursuit of truth are continuing threads through Russell's full life, as is his intention to "become a scientific philosopher." These themes—with other perennial human themes such as love, marriage, brotherhood, parenthood, and education—are to be seen in his hundreds of newspaper essays (most of them for American audiences), as well as in his major philosophical works. Among almost a hundred major works are *Principia Mathematica* (with Alfred North Whitehead); *The Problems of Philosophy; Mysticism and Logic; The Scientific Outlook; The Analysis of Mind; An Analysis of Matter;* several contributions to the popular literature on science, including *The ABC of Relativity* and *The ABC of Atoms;* and, in his late years, the three-volume *Autobiography.*

"The Rise of Science," which follows, is excerpted from his widely read *A History of Western Philosophy.* It reveals Russell's fascination with the role of ideas in human history, especially the emergence of the scientific attitude.

The Rise of Science

Bertrand Russell

Almost everything that distinguishes the modern world from earlier centuries is attributable to science, which achieved its most spectacular triumphs in the seventeenth century. The Italian Renaissance, though not medieval, is not modern; it is more akin to the best age of Greece. The sixteenth century, with its absorption in theology, is more medieval than the world of Machiavelli. The modern world, so far as mental outlook is concerned, begins in the seventeenth century. No Italian of the Renaissance would have been unintelligible to Plato or Aristotle; Luther would have horrified Thomas Aquinas, but would not have been difficult for him to understand. With the seventeenth century it is different: Plato and Aristotle, Aquinas and Occam, could not have made head or tail of Newton.

The new conceptions that science introduced profoundly influenced modern philosophy. Descartes, who was in a sense the founder of modern philosophy, was himself one of the creators of seventeenth-century science. Something must be said about the methods and results of astronomy and physics before the mental atmosphere of the time in which modern philosophy began can be understood.

Four great men—Copernicus, Kepler, Galileo, and Newton—are preeminent in the creation of science. Of these, Copernicus belongs to the sixteenth century, but in his own time he had little influence.

Copernicus (1473–1543) was a Polish ecclesiastic, of unimpeachable orthodoxy. In his youth he travelled in Italy, and absorbed something of the atmosphere of the Renaissance. In 1500 he had a lectureship or professorship of mathematics in Rome, but in 1503 he returned to his native land, where he was a canon of Frauenburg. Much of his time seems to have been spent in combating the Germans and reforming the currency, but his leisure was devoted to astronomy. He came early to believe that the sun is at the centre of the universe, and that the earth has a twofold motion: a diurnal rotation, and an annual revolution about the sun. Fear of ecclesiastical censure led him to delay publication of his views, though he allowed them to become known. . . .

The atmosphere of Copernicus's work is not modern; it might rather be described as Pythagorean. He takes it as axiomatic that all celestial motions must be circular and uniform, and like the Greeks he allows himself to be influenced by aesthetic motives. There are still epicycles in his system, though their centres are at the sun, or, rather, near the sun. The fact that the sun is not exactly in the centre marred the simplicity of his theory. He does not seem to have known of Aristarchus's heliocentric theory, but there is nothing in his speculations that could not have occurred to a Greek astronomer. What was important in his work was the dethronement of the earth from its geometrical pre-eminence. In the long run, this made it difficult to give to man the cosmic importance assigned to him in the Christian theology, but such consequences of his theory would not have been accepted by Copernicus, whose orthodoxy was sincere, and who protested against the view that his theory contradicted the Bible.

There were genuine difficulties in the Copernican theory. The greatest of these was the absence of stellar parallax. If the earth at any one point of its orbit is 186,000,000 miles from the point at which it will be in six months, this ought to cause a shift in the apparent positions of the stars, just as a ship at sea which is due north from one point of the coast will not be due north from another. No parallax was observed, and Copernicus rightly inferred that the fixed stars must be very much more remote than the sun. It was not till the nineteenth century that the technique of measurement became sufficiently precise for stellar parallax to be observed, and then only in the case of a few of the nearest stars.

Another difficulty arose as regards falling bodies. If the earth is continually rotating from west to east, a body dropped from a height ought not to fall to a point vertically below its starting-point, but to a point somewhat further west, since the earth will have slipped away a certain distance during the time of the fall. To this difficulty the answer was found by Galileo's law of inertia, but in the time of Copernicus no answer was forthcoming.

There is an interesting book by E. A. Burtt, called *The Metaphisical Foundations of Modern Physical Science* (1925), which sets forth with much force the many unwarrantable assumptions made by the men who founded modern science. He points out quite truly that there were in the time of Copernicus no known facts which compelled the adoption of his system, and several which militated against it. "Contemporary empiricists, had they lived in the sixteenth century, would have been the first to scoff out of court the new philosophy of the universe." The general purpose of the book is to discredit modern science by suggesting that its discoveries were lucky accidents springing by chance from superstitions as gross as those of the Middle Ages. I think this shows a misconception of the scientific attitude: it is not *what* the man of science believes that distinguishes him but *how* and *why* he believes it. His beliefs are tentative, not dogmatic; they are based on evidence, not on authority or intuition. Copernicus was right to call his theory a hypothesis; his opponents were wrong in thinking new hypotheses undesirable.

The men who founded modern science had two merits which are not necessarily found together: immense patience in observation, and great boldness in framing hypotheses. The second of these merits had belonged to the earliest Greek philosophers; the first existed, to a considerable degree, in the later astronomers of antiquity. But no one among the ancients, except perhaps Aristarchus, possessed both merits, and no one in the Middle Ages possessed either. Copernicus, like his great successors, possessed both. He knew all that could be known, with the instruments existing in his day, about the apparent motions of the heavenly bodies on the celestial sphere, and he perceived that the diurnal rotation of the earth was a more economical hypothesis than the revolution of all the celestial spheres. According to modern views, which regard all motion as relative, simplicity is the only gain resulting from his hypothesis, but this was not his view or that of his contemporaries. As regards the earth's annual revolution, there was again a simplification, but not so notable a one as in the case of the diurnal rotation. Copernicus still needed epicycles, though fewer than were needed in the Ptolemaic system. It was not until Kepler discovered his laws that the new theory acquired its full simplicity.

Apart from the revolutionary effect on cosmic imagination, the great merits of the new astronomy were two: first, the recognition that what had been believed since ancient times might be false; second, that the test of scientific truth is patient collection of facts, combined with bold guessing as to laws binding the facts together. Neither merit is so fully developed in Copernicus as in his successors, but both are already present in a high degree in his work.

Some of the men to whom Copernicus communicated his theory were German Lutherans, but when Luther came to know of it, he was profoundly shocked. "People give ear," he said, "to an upstart astrologer who strove to show that the earth revolves, not the heavens or the firmament, the sun and the moon. Whoever wishes to appear clever must devise some new system, which of all systems is of course the very best. This fool wishes to reverse the entire science of astronomy; but sacred Scripture tells us that Joshua commanded the sun to stand still, and not the earth." Calvin, similarly, demolished Copernicus with the text: "The world also is stablished, that it cannot be moved" (Ps. XCIII, 1), and exclaimed: "Who will venture to place the authority of Copernicus above that of the Holy Spirit?" Protestant clergy were at least as bigoted as Catholic ecclesiastics; nevertheless there soon came to be much more liberty of speculation in Protestant than in Catholic countries, because in Protestant countries the clergy had less power. . . .

Copernicus was not in a position to give any conclusive evidence in favour of his hypothesis, and for a long time astronomers rejected it. The next astronomer of importance was Tycho Brahe (1546–1601), who adopted an intermediate position: he held that the sun and moon go round the earth, but the planets go round the sun. As regards theory he was not very original. He gave, however, two good reasons against Aristotle's view that everything above the moon is unchanging. One of these was the appearance of a new star in 1572, which was found to have no daily parallax, and must therefore be more distant than the moon. The other reason was derived from observation of comets, which were also found to be distant. The reader will remember Aristotle's doctrine that change and decay are confined to the sublunary sphere; this, like everything else that Aristotle said on scientific subjects, proved an obstacle to progress.

The importance of Tycho Brahe was not as a theorist, but as an observer, first under the patronage of the king of Denmark, then under the Emperor Rudolf II. He made a star catalogue, and noted the positions of the planets throughout many years. Towards the end of his life Kepler, then a young man, became his assistant. To Kepler his observations were invaluable.

Kepler (1571–1630) is one of the most notable examples of what can be achieved by patience without much in the way of genius. He was the first important astronomer after Copernicus to adopt the heliocentric theory, but Tycho Brahe's data showed that it could not be quite right in the form given to it by Copernicus. He was influenced by Pythagoreanism, and more or less fancifully inclined to sun-worship, though a good Protestant. These motives no doubt gave him a bias in favour of the heliocentric hypothesis. His Pythagoreanism also inclined him to follow Plato's *Timaeus* in supposing that cosmic significance must attach to the five regular solids. He used them to suggest hypotheses to his mind; at last, by good luck, one of these worked.

Kepler's great achievement was the discovery of his three laws of planetary motion. Two of these he published in 1609, and the third in 1619. His first law states: The planets describe elliptic orbits, of which the sun occupies one focus. His second law states: The line joining a planet to the sun sweeps out equal areas in equal times. His third law states: The square of the period of revolution of a planet is proportioned to the cube of its average distance from the sun. Something must be said in explanation of the importance of these laws. . . .

[Here Russell explains the importance of Kepler's three laws, pointing out in what ways they were an "effort of emancipation from tradition." He also points out that, although the laws could not be completely proved in Kepler's time, they soon found "decisive confirmation."]

Galileo (1564–1642) is the greatest of the founders of modern science, with the possible exception of Newton. He was born on about the day on which

Michelangelo died, and he died in the year in which Newton was born. I commend these facts to those (if any) who still believe in metempsychosis. He is important as an astronomer, but perhaps even more as the founder of dynamics.

Galileo first discovered the importance of *acceleration* in dynamics. "Acceleration" means change of velocity, whether in magnitude or direction; thus a body moving uniformly in a circle has at all times an acceleration towards the centre of the circle. In the language that had been customary before his time, we might say that he treated uniform motion in a straight line as alone "natural," whether on earth or in the heavens. It had been thought "natural" for heavenly bodies to move in circles, and for terrestrial bodies to move in straight lines; but moving terrestrial bodies, it was thought, would gradually cease to move if they were let alone. Galileo held, as against this view, that every body, if let alone, will continue to move in a straight line with uniform velocity; any change, either in the rapidity or the direction of motion, requires to be explained as due to the action of some "force." This principle was enunciated by Newton as the "first law of motion." It is also called the law of inertia. I shall return to its purport later, but first something must be said as to the detail of Galileo's discoveries.

Galileo was the first to establish the law of falling bodies. This law, given the concept of "acceleration," is of the utmost simplicity. It says that, when a body is falling freely, its acceleration is constant, except in so far as the resistance of the air may interfere; further, the acceleration is the same for all bodies, heavy or light, great or small. The complete proof of this law was not possible until the air pump had been invented, which was about 1654. After this, it was possible to observe bodies falling in what was practically a vacuum, and it was found that feathers fell as fast as lead. What Galileo proved was that there is no measurable difference between large and small lumps of the same substance. Until his time it had been supposed that a large lump of lead would fall much quicker than a small one, but Galileo proved by experiment that this is not the case. Measurement, in his day, was not such an accurate business as it has since become; nevertheless he arrived at the true law of falling bodies. If a body is falling freely in a vacuum, its velocity increases at a constant rate. At the end of the first second, its velocity will be 32 feet per second; at the end of another second, 64 feet per second; at the end of the third, 96 feet per second; and so on. The acceleration, i.e., the rate at which the velocity increases, is always the same; in each second, the increase of velocity is (approximately) 32 feet per second.

Galileo also studied projectiles, a subject of importance to his employer, the duke of Tuscany. It had been thought that a projectile fired horizontally will move horizontally for a while, and then suddenly begin to fall vertically. Galileo showed that, apart from the resistance of the air, the horizontal velocity would remain constant, in accordance with the law of inertia, but a vertical velocity would be added, which would grow according to the law of falling bodies. To find out how the projectile will move during some short time, say a second, after it has been in flight for some time, we proceed as follows: First, if it were not falling, it would cover a certain horizontal distance, equal to that which it covered in the first second of its flight. Second, if it were not moving horizontally, but merely falling, it would fall vertically with a velocity proportional to the time since the flight began. In fact, its change of place is what it would be if it first moved horizontally for a second with the initial velocity, and then fell vertically for a second with a velocity proportional to the time during which it has been in flight. A simple calculation shows that its consequent course is a parabola, and this is confirmed by observation except in so far as the resistance of the air interferes.

The above gives a simple instance of a principle which proved immensely fruitful in dynamics, the principle that, when several forces act simultaneously, the effect is as if each acted in turn. This is part of a more general principle called the parallelogram law. Suppose, for example, that you are on the deck of a moving ship, and you walk across the deck. While you are walking the ship has moved on, so that, in relation to the water, you have moved both forward and across the direction of the ship's motion. If you want to know where you will have got to in relation to the water, you may suppose that first you stood still while the ship moved, and then, for an equal time, the ship stood still while you walked across it. The same principle applies to forces. This makes it possible to work out the total effect of a number of forces, and makes it feasible to analyse physical phenomena, discovering the separate laws of the several forces to which moving bodies are subject. It was Galileo who introduced this immensely fruitful method.

In what I have been saying, I have tried to speak, as nearly as possible, in the language of the seventeenth century. Modern language is different in important respects, but to explain what the seventeenth century achieved it is desirable to adopt its modes of expression for the time being. . . .

[Here Russell explains that the law of inertia clarifies puzzles that the Copernican system had left unanswered. He describes Galileo's correspondence with Kepler and Galileo's making of a telescope, with his consequent discoveries.]

Newton (1642–1727) achieved the final and complete triumph for which Copernicus, Kepler, and Galileo had prepared the way. Starting from his three laws of motion—of which the first two are due to Galileo—he proved that Kepler's three laws are equivalent to the proposition that every planet, at every moment, has an acceleration towards the sun which varies inversely as the square of the distance from the sun. He showed that accelerations towards the earth and the sun, following the same formula, explain the moon's motion, and that the acceleration of falling bodies on the earth's surface is again related to that of the moon according to the inverse square law. He defined "force" as the cause of change of motion, i.e., of acceleration. He was thus able to enunciate his law of universal gravitation: "Every body attracts every other with a force directly proportional to the product of their masses and inversely proportional to the square of the distance between them." From this formula he was able to deduce everything in planetary theory: the motions of the planets and their satellites, the orbits of comets, the tides. It appeared later that even the minute departures from elliptical orbits on the part of the planets were deducible from Newton's law. The triumph was so complete that Newton was in danger of becoming another Aristotle, and imposing an insuperable barrier to progress. In England, it was not till a century after his death that men freed themselves from his authority sufficiently to do important original work in the subjects of which he had treated.

The consequence of the scientific work we have been considering was that the outlook of educated men was completely transformed. At the beginning of the century, Sir Thomas Browne took part in trials for witchcraft; at the end, such a thing would have been impossible. In Shakespeare's time, comets were still portents; after the publication of Newton's *Principia* in 1687, it was known that he and Halley had calculated the orbits of certain comets, and that they were as obedient as the planets to the law of gravitation. The reign of law had established its hold on men's imaginations, making such things as magic and sorcery incredible. In 1700 the mental outlook of educated men was completely modern; in 1600, except among a very few, it was still largely medieval.

. . . I shall try to state briefly the philosophical beliefs which appeared to follow from seventeenth-century science, and some of the respects in which modern science differs from that of Newton.

The first thing to note is the removal of almost all traces of animism from the laws of physics. The Greeks, though they did not say so explicitly, evidently considered the power of movement a sign of life. To common-sense observation it seems that animals move themselves, while dead matter only moves when impelled by an external force. The soul of an animal, in Aristotle, has various functions, and one of them is to move the animal's body. The sun and planets, in Greek thinking, are apt to be gods, or at least regulated and moved by gods. Anaxagoras thought otherwise, but was impious. Democritus thought otherwise, but was neglected, except by the Epicureans, in favour of Plato and Aristotle. Aristotle's forty-seven or fifty-five unmoved movers are divine spirits, and are the ultimate source of all the motion in the universe. Left to itself, any inanimate body would soon become motionless; thus the operation of soul on matter has to be continuous if motion is not to cease.

All this was changed by the first law of motion. Lifeless matter, once set moving, will continue to move for ever unless stopped by some external cause. Moreover the external causes of change of motion turned out to be themselves material, whenever they could be definitely ascertained. The solar system, at any rate, was kept going by its own momentum and its own laws; no outside interference was needed. There might still seem to be need of God to set the mechanism working; the planets, according to Newton, were originally hurled by the hand of God. But when He had done this, and decreed the law of gravitation, everything went on by itself without further need of divine intervention. When Laplace suggested that the same forces which are now operative might have caused the planets to grow out of the sun, God's share in the course of nature was pushed still further back. He might remain as Creator, but even that was doubtful, since it was not clear that the world had a beginning in time. Although most of the men of science were models of piety, the outlook suggested by their work was disturbing to orthodoxy, and the theologians were quite justified in feeling uneasy.

Another thing that resulted from science was a profound change in the conception of man's place in the universe. In the medieval world, the earth was the centre of the heavens, and everything had a purpose concerned with man. In the Newtonian world, the earth was a minor planet of a not specially distinguished star; astronomical distances were so vast that the earth, in comparison, was a mere pin-point. It seemed unlikely that this immense apparatus was all designed for the good of certain small creatures on this pin-point. Moreover purpose, which had since Aristotle formed an intimate part of the conception of science, was now thrust out of scientific procedure. Any one might still believe that the heavens exist to declare the glory of God, but no one could let this belief intervene in an astronomical calculation. The world might have a purpose, but purposes could no longer enter into scientific explanations.

The Copernican theory should have been humbling to human pride, but in fact the contrary effect was produced, for the triumphs of science revived human pride. The dying ancient world had been obsessed with a sense of sin, and had bequeathed this as an oppression to the Middle Ages. To be humble before God was both right and prudent, for God would punish pride. Pestilences, floods, earthquakes, Turks, Tartars, and comets perplexed the gloomy centuries, and it was felt that only greater and greater humility would avert these real or threatened calamities. But it became impossible to remain humble when men were achieving such triumphs:

Nature and Nature's laws lay hid in night.
God said "Let Newton be," and all was light.

And as for damnation, surely the Creator of so vast a universe had something better to think about than sending men to hell for minute theological errors. Judas Iscariot might be damned, but not Newton, though he were an Arian.

There were of course many other reasons for self-satisfaction. The Tartars had been confined to Asia, and the Turks were ceasing to be a menace. Comets had been humbled by Halley, and as for earthquakes, though they were still formidable, they were so interesting that men of science could hardly regret them. Western Europeans were growing rapidly richer, and were becoming lords of all the world: they had conquered North and South America, they were powerful in Africa and India, respected in China and feared in Japan. When to all this were added the triumphs of science, it is no wonder that the men of the seventeenth century felt themselves to be fine fellows, not the miserable sinners that they still proclaimed themselves on Sundays.

There are some respects in which the concepts of modern theoretical physics differ from those of the Newtonian system. To begin with, the conception of "force," which is prominent in the seventeenth century, has been found to be superfluous. "Force," in Newton, is the cause of change of motion, whether in magnitude or direction. The notion of cause is regarded as important, and force is conceived imaginatively as the sort of thing that we experience when we push or pull. For this reason it was considered an objection to gravitation that it acted at a distance, and Newton himself conceded that there must be some medium by which it was transmitted. Gradually it was found that all the equations could be written down without bringing in forces. What was observable was a certain relation between acceleration and configuration; to say that this relation was brought about by the intermediacy of "force" was to add nothing to our knowledge. Observation shows that planets have at all times an acceleration towards the sun, which varies inversely as the square of their distance from it. To say that this is due to the "force" of gravitation is merely verbal, like saying that opium makes people sleep because it has a dormitive virtue. The modern physicist, therefore, merely states formulae which determine accelerations, and avoids the word "force" altogether. "Force" was the faint ghost of the vitalist view as to the causes of motions, and gradually the ghost has been exorcized.

Until the coming of quantum mechanics, nothing happened to modify in any degree what is the essential purport of the first two laws of motion, namely this: that the laws of dynamics are to be stated in terms of accelerations. In this respect, Copernicus and Kepler are still to be classed with the ancients; they sought laws stating the shapes of the orbits of the heavenly bodies. Newton made it clear that laws stated in this form could never be more than approximate. The planets do not move in *exact* ellipses, because of the perturbations caused by the attractions of other planets. Nor is the orbit of a planet ever exactly repeated, for the same reason. But the law of gravitation, which deals with accelerations, was very simple, and was thought to be quite exact until two hundred years after Newton's time. When it was emended by Einstein, it still remained a law dealing with accelerations.

It is true that the conservation of energy is a law dealing with velocities, not accelerations. But in calculations which use this law it is still accelerations that have to be employed.

As for the changes introduced by quantum mechanics, they are very profound, but still, to some degree, a matter of controversy and uncertainty.

There is one change from the Newtonian philosophy which must be mentioned now, and that is the abandonment of absolute space and time. . . . Newton believed in a space composed of points, and a time composed of instants, which had an existence independent of the bodies and events that occupied them. As regards space, he had an empirical argument to support his view, namely that physical phenomena enable us to distinguish absolute rotation. If the water in a bucket is rotated, it climbs up the sides and is depressed in the centre; but if the bucket is rotated while the water is not, there is no such effect. Since his day, the experiment of Foucault's pendulum has been devised, giving what has been considered a demonstration of the earth's rotation. Even on the most modern views, the question of absolute rotation presents difficulties. If all motion is relative, the difference between the hypothesis that the earth rotates and the hypothesis that the heavens revolve is purely verbal; it is no more than the difference between "John is the father of James" and "James is the son of John." But if the heavens revolve, the stars move faster than light, which is considered impossible. It cannot be said that the modern answers to this difficulty are completely satisfying, but they are sufficiently satisfying to cause almost all physicists to accept the view that motion and space are purely relative. This, combined with the amalgamation of space and time into space-time, has considerably altered our view of the universe from that which resulted from the work of Galileo and Newton. But of this, as of quantum theory, I will say no more at this time.

Questions and Topics for Discussion and Writing

1. Why is Russell's chapter important to students of science, mathematics, philosophy, and religion? Cite several specific examples to support your reply.
2. Why, according to Russell, is it important for the student of modern science to study the scientific achievements of the seventeenth century? Outline.
3. Draw a diagram showing the names of contributors to the evolution of modern science stemming from Copernicus. Draw another showing the evolution of various sciences and technological inventions evolving from the astronomy of the seventeenth century. For additional information, consult some of the suggested further readings at the end of this unit.
4. What, according to Russell, is the true scientific attitude?
5. Considering the qualities that make Copernicus a true scientist, what is the great tribute Russell pays to him?
6. What are the two great merits of the new astronomy begun by Copernicus?
7. What is Russell's judgment of the general and prolonged acceptance of Aristotle as a scientific authority?
8. How did Copernicus's, Brahe's, and Kepler's scientific contributions illustrate the "simplification" of which Russell speaks?
9. What is the implication of Russell's stating that Galileo was born on approximately the day Michelangelo died, and that he died in the year Newton was born, adding "I commend these facts to those (if any) who still believe in metempsychosis [the transformation of souls]"? Is he simply being humorous or does he, perhaps, note reinforcement for his general thesis?

10. What is the great tribute that Russell pays to Galileo?
11. List Galileo's discoveries. Why is his "law of falling bodies" of "utmost simplicity"?
12. Russell declares in the essay: "Modern language is different in important respects, but to explain what the seventeenth century achieved it is desirable to adopt its modes of expression for the time being." How does the statement prepare readers for problems they might have in comprehending this discussion of seventeenth-century science?
13. Awarded the Nobel Prize for literature in 1950, Russell has also been acclaimed by such outstanding critics as T. S. Eliot for his clear, logical, "easy-to-read" prose. Find examples of his writing in this chapter that support such recognition.

Francis Bacon

The environment of Francis Bacon (1561–1626) introduced him early to government service. His father was lord keeper of the Great Seal of England, and Bacon in turn advanced to that office, then in 1618 to Lord Chancellor, being created Baron Verulam. But this high office was his for only a few years. His enemies accused him of receiving bribes in chancery suits—a charge Bacon did not contest, pleading, however, that gifts had never influenced his judgment. Nevertheless, he was thereafter excluded from political power.

The close of his political career allowed Bacon to expand on his deeper interests: He began to write on history, philosophy, and science. Earlier (1599), with a popular book of essays, he had established himself as a literary figure. The essay, a form of inquiry recently invented by the French writer Montaigne, served Bacon's purposes well. He developed the form to explore many subjects in true Renaissance manner.

The proper pursuit of science was Bacon's subject in *Novum Organon*, published in 1620. Bacon's thesis in this work was that he had developed a new instrument of thought, or organon. His new method, he believed, would replace Aristotle's logic of the syllogism, still everywhere in use, but bankrupt as an approach to the science of the day.

Bacon proposed a new version of induction (the form of logic that proceeds from the particular to the general). While Aristotle himself had used induction, the method until Bacon's time had been practiced by simple enumeration. Bacon offered a set of prescriptions for collecting and tabulating exhaustive lists of qualities of things being investigated; the lists categorized items with a quality, items without it, and items possessing it to varying degrees. Bacon was confident that his version of induction would increase the power of the investigator: It would become a method of discovery.

For statistical problems, Bacon's method is adequate. But, in actuality, scientists do not obtain hypotheses in this fashion, although Bacon believed that they would. His method of investigation works well after the hypotheses have already been formulated.

Bacon's real contribution was to insist, for the list making he prescribed, on observations drawn from nature. This empirical approach generated the power he had surmised, despite his inadequate, mechanical explanation. The impulse to base investigation on observations, not traditional rationalism, did indeed lead to discoveries Bacon glimpsed.

For his contribution, Bacon is sometimes called "the father of the scientific method"— too sweeping a label. On the other hand, his commitment to experimental evidence is beyond question. Ironically, it cost him his life. Investigating the preservation of food by freezing, he caught a fatal chill while stuffing chickens with snow. Among Bacon's admirers is anthropologist Loren Eiseley, who wrote about Bacon as "the man who saw through time."

The selection below displays a small example of Bacon's rhetorical skill. He defines Science through a fable.

Sphinx or Science

Francis Bacon

Sphinx, says the story, was a monster combining many shapes in one. She had the face and voice of a virgin, the wings of a bird, the claws of a griffin. She dwelt on the ridge of a mountain near Thebes and infested the roads, lying in ambush for travellers, whom she would suddenly attack and lay hold of; and when she had mastered them, she propounded to them certain dark and perplexed riddles, which she was thought to have obtained from the Muses. And if the wretched captives could not at once solve and interpret the same, as they stood hesitating and confused she cruelly tore them to pieces. Time bringing no abatement of the calamity, the Thebans offered to any man who should expound the Sphinx's riddles (for this was the only way to subdue her) the sovereignty of Thebes as his reward. The greatness of the prize induced Oedipus, a man of wisdom and penetration, but lame from wounds in his feet, to accept the condition and make the trial: who presenting himself full of confidence and alacrity before the Sphinx, and being asked what kind of animal it was which was born four-footed, afterwards became two-footed, then three-footed, and at last four-footed again, answered readily that it was man; who at his birth and during his infancy sprawls on all four, hardly attempting to creep; in a little while walks upright on two feet; in later years leans on a walking-stick and so goes as it were on three; and at last in extreme age and decrepitude, his sinews all failing, sinks into a quadruped again, and keeps his bed. This was the right answer and gave him the victory; whereupon he slew the Sphinx; whose body was put on the back of an ass and carried about in triumph; while himself was made according to compact King of Thebes.

The fable is an elegant and a wise one, invented apparently in allusion to

Science; especially in its application to practical life. Science, being the wonder of the ignorant and unskilful, may be not absurdly called a monster. In figure and aspect it is represented as many-shaped, in allusion to the immense variety of matter with which it deals. It is said to have the face and voice of a woman, in respect of its beauty and facility of utterance. Wings are added because the sciences and the discoveries of science spread and fly abroad in an instant; the communication of knowledge being like that of one candle with another, which lights up at once. Claws, sharp and hooked, are ascribed to it with great elegance, because the axioms and arguments of science penetrate and hold fast the mind, so that it has no means of evasion or escape; a point which the sacred philosopher also noted: *The words of the wise are as goads, and as nails driven deep in.* Again, all knowledge may be regarded as having its station on the heights of mountains; for it is deservedly esteemed a thing sublime and lofty, which looks down upon ignorance as from an eminence, and has moreover a spacious prospect on every side, such as we find on hill-tops. It is described as infesting the roads, because at every turn in the journey or pilgrimage of human life, matter and occasion for study assails and encounters us. Again Sphinx proposes to men a variety of hard questions and riddles which she received from the Muses. In these, while they remain with the Muses, there is probably no cruelty; for so long as the object of meditation and inquiry is merely to know, the understanding is not oppressed or straitened by it, but is free to wander and expatiate, and finds in the very uncertainty of conclusion and variety of choice a certain pleasure and delight; but when they pass from the Muses to Sphinx, that is from contemplation to practice, whereby there is necessity for present action, choice, and decision, then they begin to be painful and cruel; and unless they be solved and disposed of, they strangely torment and worry the mind, pulling it first this way and then that, and fairly tearing it to pieces. Moreover the riddles of the Sphinx have always a twofold condition attached to them; distraction and laceration of mind, if you fail to solve them; if you succeed, a kingdom. For he who understands his subject is master of his end; and every workman is king over his work.

Now of the Sphinx's riddles there are in all two kinds; one concerning the nature of things, another concerning the nature of man; and in like manner there are two kinds of kingdom offered as the reward of solving them; one over nature, and the other over man. For the command over things natural—over bodies, medicines, mechanical powers, and infinite other of the kind—is the one proper and ultimate end of true natural philosophy; however the philosophy of the School,[1] content with what it finds, and swelling with talk, may neglect or spurn the search after realities and works. But the riddle proposed to Oedipus, by the solution of which he became King of Thebes, related to the nature of man; for whoever has a thorough insight into the nature of man may shape his fortune almost as he will, and is born for empire; as was well declared concerning the arts of the Romans,—

[1] "of the School." Of the speculative medieval Scholastics.

Be thine the art,
O Rome, with government to rule the nations,
And to know whom to spare and whom to abate,
And settle the condition of the world.

And therefore it fell out happily that Augustus Caesar, whether on purpose or by chance, used a Sphinx for his seal. For he certainly excelled in the art of politics if ever man did; and succeeded in the course of his life in solving most happily a great many new riddles concerning the nature of man, which if he had not dexterously and readily answered he would many times have been in imminent danger of destruction. The fable adds very prettily that when the Sphinx was subdued, her body was laid on the back of an ass: for there is nothing so subtle and abstruse, but when it is once thoroughly understood and published to the world, even a dull wit can carry it. Nor is that other point to be passed over, that the Sphinx was subdued by a lame man with club feet; for men generally proceed too fast and in too great a hurry to the solution of the Sphinx's riddles; whence it follows that the Sphinx has the better of them, and instead of obtaining the sovereignty by works and effects, they only distract and worry their minds with disputations.

Questions and Topics for Discussion and Writing

1. Bacon says that the fable of the Sphinx "is an elegant and a wise one, invented apparently in allusion to Science." In your view, is the fable suitably instructive and illuminating for the subject of science?
2. How, in Bacon's view, can science be called a "monster"?
3. In Paragraph 2, Bacon quotes "the sacred philosopher." This quotation is taken from the Old Testament, Ecclesiastes 12:11. Does this passage seem to you applicable to science? Why or why not?
4. Consult your dictionary—or, if possible the *Oxford English Dictionary (OED)*—for a definition of the word "Sphinx." After reading the essay, what characteristics of the Sphinx have you learned that would help you write an extended definition?
5. In one paragraph, summarize the substance of the story of Oedipus and the Sphinx.
6. What is Bacon's theme? What assumptions underlie his theme?
7. The fable has a lengthy history as a form of literature. You have probably read Aesop's *Fables*, written in antiquity, and perhaps George Orwell's *Animal Farm*, written in 1945. What advantages do you see in Bacon's employment of this form?
8. What is the function of the concept of the *riddle*, both in the original fable and in the context of allusion to science? Why, in your opinion, does Bacon point out the "two-fold" condition that the Sphinx attaches to the riddle?
9. What number and kinds of riddles does the Sphinx have?
10. Write a brief summary of Bacon's ideas about science. Use only the general and abstract ideas in the piece. What does Bacon gain by adding the notation of the fable to the original?

Jacob Bronowski

Jacob Bronowski (1908–1974), the Polish-born American statistician, teacher, inventor, administrator, philosopher, poet, historian, and statesman, is almost as well known for his writing as for his scientific accomplishments. A mathematician by training, Bronowski pioneered the development of the operations research method and served as Deputy Director and Fellow of the Salk Institute for Biological Studies in La Jolla, California. As Scientific Deputy to the British Chiefs of Staff Mission to Japan in 1945, he wrote the classic British study, "The Effects of the Atomic Bombs at Hiroshima and Nagasaki."

Bronowski describes his lifelong intention in these words: "My ambition has been to create a philosophy for the twentieth century which shall be all of one piece. There cannot be a decent philosophy, there cannot even be a decent science, without humanity. For me, the understanding of nature has as its goal the understanding of human nature, and of the human condition within nature." A reviewer describes the writer Bronowski as "an ideal of the Renaissance, at home equally in all the pursuits of knowledge and art, a good man and wise . . . seeking most deeply, in the new rationalism, the fulfillment of what is human. . . ." Bronowski's ideas provide young scientists who "yearn for a philosophy" with subject matter ranging from science, philosophy, literature, and the arts to the human social condition and its questions and problems.

As a Carnegie Visiting Professor at Massachusetts Institute of Technology in 1953, Bronowski delivered a series of lectures that addressed the concept of the "two cultures." Among his many published volumes are *Science and Human Values, The Poet's Defense, The Common Sense of Science, The Ascent of Man,* and *The Identity of Man, Nature, and Knowledge.*

In "The Creative Process," Bronowski makes a strong case for the "essential oneness of artistic and scientific creativity." In this definition of creativity, Bronowski studies the "nature of the scientific activity and the imaginative acts of understanding which exercise the creative mind."

The Creative Process

Jacob Bronowski

The most remarkable discovery made by scientists is science itself. The discovery must be compared in importance with the invention of cave-painting and of writing. Like these earlier human creations, science is an attempt to control our surroundings by entering into them and understanding them from inside. And like them, science has surely made a critical step in human development which cannot be reversed. We cannot conceive a future society without science.

I have used three words to describe these far-reaching changes: discovery, invention and creation. There are contexts in which one of these words is more appropriate than the others. Christopher Columbus discovered the West Indies, and Alexander Graham Bell invented the telephone. We do not call their achievements creations because they are not personal enough. The West Indies were there all the time; as for the telephone, we feel that Bell's ingenious thought was somehow not fundamental. The groundwork was there, and if not Bell then someone else would have stumbled on the telephone as casually as on the West Indies.

By contrast, we feel that *Othello* is genuinely a creation. This is not because *Othello* came out of a clear sky; it did not. There were Elizabethan dramatists before Shakespeare, and without them he could not have written as he did. Yet within their tradition *Othello* remains profoundly personal; and though every element in the play has been a theme of other poets, we know that the amalgam of these elements is Shakespeare's; we feel the presence of his single mind. The Elizabethan drama would have gone on without Shakespeare, but no one else would have written *Othello*.

There are discoveries in science like Columbus's, of something which was always there: the discovery of sex in plants, for example. There are tidy inventions like Bell's, which combine a set of known principles: the use of a beam of electrons as a microscope, for example. In this article I ask the question: Is there anything more? Does a scientific theory, however deep, ever reach the roundness, the expression of a whole personality that we get from *Othello*?

A fact is discovered, a theory is invented; is any theory ever deep enough for it to be truly called a creation? Most nonscientists would answer: No! Science, they would say, engages only part of the mind—the rational intellect—but creation must engage the whole mind. Science demands none of that groundswell of emotion, none of that rich bottom of personality, which fills out the work of art.

This picture by the nonscientist of how a scientist works is of course mistaken. A gifted man cannot handle bacteria or equations without taking fire from what he does and having his emotions engaged. It may happen that his emotions are immature, but then so are the intellects of many poets. When Ella Wheeler Wilcox died, having published poems from the age of seven, *The Times* of London wrote that she was "the most popular poet of either sex and of any age, read by thousands who never open Shakespeare." A scientist who is emotionally immature is like a poet who is intellectually backward: both produce work which appeals to others like them, but which is second-rate.

I am not discussing the second-rate, and neither am I discussing all that useful but commonplace work which fills most of our lives, whether we are chemists or architects. There are in my laboratory of the British National Coal Board about 200 industrial scientists—pleasant, intelligent, sprightly people who thoroughly earn their pay. It is ridiculous to ask whether they are creators who produce works that could be compared with *Othello*. They are men with the same ambitions as other university graduates, and their work is most like the work of a college department of Greek or of English. When the Greek departments produce a Sophocles, or the English departments produce a Shakespeare, then I shall begin to look in my laboratory for a Newton.

Literature ranges from Shakespeare to Ella Wheeler Wilcox, and science ranges from relativity to market research. A comparison must be of the best with the best. We must look for what is created in the deep scientific theories: in Copernicus and Darwin, in Thomas Young's theory of light and in William Rowan Hamilton's equations, in the pioneering concepts of Freud, of Bohr and of Pavlov.

The most remarkable discovery made by scientists, I have said, is science itself. It is therefore worth considering the history of this discovery, which was not made all at once but in two periods. The first period falls in the great age of Greece, between 600 B.C. and 300 B.C. The second period begins roughly with the Renaissance, and is given impetus at several points by the rediscovery of Greek mathematics and philosophy.

When one looks at these two periods of history, it leaps to the eye that they were not specifically scientific. On the contrary: Greece between Pythagoras and Aristotle is still, in the minds of most scholars, a shining sequence of classical texts. The Renaissance is still thought of as a rebirth of art, and only specialists are uncouth enough to link it also with what is at last being called, reluctantly, the Scientific Revolution. The accepted view of Greece and of the Renaissance is that they were the great creative periods of literature and art. Now that we recognize in them also the two periods in which science was born, we must surely ask whether this conjunction is accidental. Is it a coincidence that Phidias and the Greek dramatists lived in the time of Socrates? Is it a coincidence that Galileo shared the patronage of the Venetian republic with sculptors and painters? Is it a coincidence that, when Galileo was at the height of his intellectual power, there were published in England in the span of 12 years the following three works: the Authorized Version of the Bible, the First Folio of Shakespeare and the first table of logarithms?

The sciences and the arts have flourished together. And they have been fixed together as sharply in place as in time. In some way both spring from one civilization: the civilization of the Mediterranean, which expresses itself in action. There are civilizations which have a different outlook; they express themselves in contemplation, and in them neither science nor the arts are practiced as such. For a civilization which expresses itself in contemplation values no creative activity. What it values is a mystic immersion in nature, the union with what already exists.

The contemplative civilization we know best is that of the Middle Ages. It has left its own monuments, from the Bayeux Tapestry to the cathedrals; and characteristically they are anonymous. The Middle Ages did not value the cathedrals, but only the act of worship which they served. It seems to me that the works of Asia Minor and of India (if I understand them) have the same anonymous quality of contemplation, and like the cathedrals were made by craftsmen rather than by artists. For the artist as a creator is personal; he cannot drop his work and have it taken up by another without doing it violence. It may be odd to claim the same personal engagement for the scientist; yet in this the scientist stands to the technician much as the artist stands to the craftsman. It is at least remarkable that science has not flourished either in an anonymous age, such as the age of medieval crafts, or in an anonymous place, such as the craftsmanlike countries of the East.

The change from an outlook of contemplation to one of action is striking in the long transition of the Renaissance and the Scientific Revolution. The new men, even when they are churchmen, have ideals which are flatly opposed to the monastic and withdrawn ideals of the Middle Ages. Their outlook is active, whether they are artists, humanist scholars or scientists.

The new man is represented by Leonardo da Vinci, whose achievement has never, I think, been rightly understood. There is an obvious difference between Leonardo's painting and that of his elders—between, for example, an angel painted by him and one by Verrocchio. It is usual to say that Leonardo's angel is more human and more tender; and this is true, but it misses the point. Leonardo's pictures of children and of women are human and tender; yet the evidence is powerful that Leonardo liked neither children nor women. Why then did he paint them as if he were entering their lives? Not because he saw them as people, but because he saw them as expressive parts of nature. We do not understand the luminous and transparent affection with which Leonardo lingers on a head or a hand until we look at the equal affection with which he paints the grass and the flowers in the same picture.

To call Leonardo either a human or a naturalist painter does not go to the root of his mind. He is a painter to whom the detail of nature speaks aloud; for him, nature expresses herself in the detail. This is a view which other Renaissance artists had; they lavished care on perspective and on flesh tones because these seemed to them (as they had not seemed in the Bayeux Tapestry) to carry the message of nature. But Leonardo went further; he took this artist's vision into science. He understood that science as much as painting has to find the design of nature in her detail.

When Leonardo was born in 1452, science was still Aristotle's structure of cosmic theories, and the criticism of Aristotle in Paris and Padua was equally grandiose. Leonardo distrusted all large theories, and this is one reason why his experiments and machines have been forgotten. Yet he gave science what it most needed, the artist's sense that the detail of nature is significant. Until science had this sense, no one could care—or could think that it mattered—

how fast two unequal masses fall and whether the orbits of the planets are accurately circles or ellipses.

The power which the scientific method has developed has grown from a procedure which the Greeks did not discover: the procedure of induction. This procedure is useless unless it is followed into the detail of nature; its discovery therefore flows from Leonardo's vision.

Francis Bacon in 1620 and Christian Huygens in 1690 set down the intellectual bases of induction. They saw that it is not possible to reach an explanation of what happens in nature by deductive steps. Every explanation goes beyond our experience and thereby becomes a speculation. Huygens says, and philosophers have sheepishly followed him in this, that an explanation should therefore be called probable. He means that no induction is unique; there is always a set—an infinite set—of alternatives between which we must choose.

The man who proposes a theory makes a choice—an imaginative choice which outstrips the facts. The creative activity of science lies here, in the process of induction. For induction imagines more than there is ground for and creates relations which at bottom can never be verified. Every induction is a speculation and it guesses at a unity which the facts present but do not strictly imply.

To put the matter more formally: A scientific theory cannot be constructed from the facts by any procedure which can be laid down in advance, as if for a machine. To the man who makes the theory, it may seem as inevitable as the ending of *Othello* must have seemed to Shakespeare. But the theory is inevitable only to him; it is his choice, as a mind and as a person, among the alternatives which are open to everyone.

There are scientists who deny what I have said—that we are free to choose between alternative theories. They grant that there are alternative theories, but they hold that the choice between them is made mechanically. The principle of choice, in their view, is Occam's Razor: we choose, among the theories which fit the facts we know now, that one which is simplest. On this view, Newton's laws were the simplest theory which covered the facts of gravitation as they were then known; and general relativity is not a new conception but is the simplest theory which fits the additional facts.

This would be a plausible view if it had a meaning. Alas, it turns out to be a verbal deception, for we cannot define simplicity; we cannot even say what we mean by the simpler of two inductions. The tests which have been proposed are hopelessly artificial and, for example, can compare theories only if they can be expressed in differential equations of the same kind. Simplicity itself turns out to be a principle of choice which cannot be mechanized.

Of course every innovator has thought that his way of arranging the facts is particularly simple, but this is a delusion. Copernicus's theory in his day was not simple to others, because it demanded two rotations of the earth—a daily one and a yearly one—in place of one rotation of the sun. What made his theory seem simple to Copernicus was something else: an esthetic sense of unity. The motion of all the planets around the sun was both simple and beautiful to him, because it expressed the unity of God's design. The same thought has moved scientists ever since: that nature has a unity, and that this unity makes her laws seem beautiful in simplicity.

The scientist's demand that nature shall be lawful is a demand for unity. When he frames a new law, he links and organizes phenomena which were thought different in kind; for example, general relativity links light with gravitation. In such a law we feel that the disorder of nature has been made to reveal a pattern, and that under the colored chaos there rules a more profound unity.

A man becomes creative, whether he is an artist or a scientist, when he finds a new unity in the variety of nature. He does so by finding a likeness between things which were not thought alike before, and this gives him a sense both of richness and of understanding. The creative mind is a mind that looks for unexpected likenesses. This is not a mechanical procedure, and I believe that it engages the whole personality in science as in the arts. Certainly I cannot separate the abounding mind of Thomas Young (which all but read the Rosetta Stone) from his recovery of the wave theory of light, or the awkwardness of J. J. Thomson in experiment from his discovery of the electron. To me, William Rowan Hamilton drinking himself to death is as much part of his prodigal work as is any drunken young poet; and the childlike vision of Einstein has a poet's innocence.

When Max Planck proposed that the radiation of heat is discontinuous, he seems to us now to have been driven by nothing but the facts of experiment. But we are deceived; the facts did not go so far as this. The facts showed that the radiation is not continuous; they did not show that the only alternative is Planck's hail of quanta. This is an analogy which imagination and history brought into Planck's mind. So the later conflict in quantum physics between the behavior of matter as a wave and as a particle is a conflict between analogies, between poetic metaphors; and each metaphor enriches our understanding of the world without completing it.

In *Auguries of Innocence* William Blake wrote:

A dog starv'd at his Master's gate
Predicts the ruin of the State.

This seems to me to have the same imaginative incisiveness, the same understanding crowded into metaphor, that Planck had. And the imagery is as factual, as exact in observation, as that on which Planck built; the poetry would be meaningless if Blake used the words "dog," "master" and "state" less robustly than he does. Why does Blake say dog and not cat? Why does he say master and not mistress? Because the picture he is creating depends on our factual grasp of the relation between dog and master. Blake is saying that when the master's conscience no longer urges him to respect his dog, the whole society is in decay. This profound thought came to Blake again and again: that a morality expresses itself in what he called its Minute Particulars—that the moral detail is significant of a society. As for the emotional power of the couplet, it comes, I think, from the change of scale between the metaphor and its application: between the dog at the gate and the ruined state. This is why Blake, in writing it, seems to me to transmit the same excitement that Planck felt when he discovered, no, when he created, the quantum.

One of the values which science has made natural to us is originality; as I said earlier, in spite of appearances science is not anonymous. The growing tradition of science has now influenced the appreciation of works of art, so that we expect both to be original in the same way. We expect artists as well as scientists to be forward-looking, to fly in the face of what is established, and to create not what is acceptable but what will become accepted. One result of this prizing of originality is that the artist now shares the unpopularity of the scientist: the large public dislikes and fears the way that both of them look at the world.

As a more important result, the way in which the artist looks at the world has come close to the scientist's. For example, in what I have written science is pictured as preoccupied less with facts than with relations, less with numbers than with arrangement. This new vision, the search for structure, is marked throughout the other articles in this issue of *Scientific American;* and it is also marked in modern art. Abstract sculpture often looks like an exercise in topology, exactly because the sculptor shares the vision of the topologist.

In each of the articles which follow I find again my view, that a theory is the creation of unity in what is diverse by the discovery of unexpected likenesses. In all of them innovation is pictured as an act of imagination, a seeing of what others do not see; indeed. Dr. Pierce uses the phrase "creative observation," which would outrage many theoretical scientists, but which exactly describes the pioneer vision of Leonardo. And Dr. Eccles gives me almost a physical feeling of creation, as if the structure of a theory reproduces the pattern of interlacing paths engaged in the brain.

There is, however, one striking division in these articles, between those which treat the physical and those which treat the biological sciences. The physical scientists have more fun. Their theories are more eccentric; they live in a world in which the unexpected is everyday. This is a strange inversion of the way that we usually picture the dead and the living, and it reflects the age of these sciences. The physical sciences are old, and in that time the distance between fact and explanation has lengthened; their very concepts are unrealistic. The biological sciences are young, so that fact and theory look alike; the new entities which have been created to underlie the facts are still representational rather than abstract. One of the pleasant thoughts that these articles prompt is: How much more extravagant the biological sciences will become when they are as old as the physical sciences.

Questions and Topics for Discussion and Writing

1. What is Bronowski's definition of *science*?
2. Bronowski uses three words to characterize scientific activity. What are these words?
3. What does Bronowski consider a "genuine creation"?
4. What does Bronowski consider a general misapprehension about the faculty of mind used by scientists?
5. Bronowski says, "The most remarkable discovery made by scientists is science itself." What are the two periods distinguished by Bronowski as those in which scientists discovered science?
6. To what does Bronowski attribute the lack of scientific advancement during the Middle Ages?
7. What does Bronowski mean by saying that Leonardo da Vinci "took an artist's vision into science"?
8. What have scientists to gain from learning an artist's approach to nature?
9. What does Bronowski see as the role of the imagination in scientific progress?
10. Bronowski equates the inductive method in science with the creative process in artistic creation. He adds, "For induction imagines more than there is ground for, and creates relations which at bottom can never be verified." Why does Bronowski make induction the subject of the sentence and personify the method?
11. What is Bronowski's challenge to the deductive method as a means for constructing scientific theory?
12. Why, according to Bronowski, was Copernicus's theory "simple" to him but not to others of his day?
13. What is Bronowski's view of the devastating result of the contemporary stress on originality?
14. What is the "new vision" shared by modern science and modern art that Bronowski thinks history will deem characteristic of our age?
15. Many writers include a scholarly subtitle that states the thesis of their work. One writer's title reads, for example, "Can We Know the Universe? Reflections on a Grain of Salt." Write what you would consider a good subtitle for Bronowski's title "The Creative Process."

Gerald Holton

A professor of physics and of the history of science at Harvard University, Gerald Holton (b. 1922) is a naturalized citizen of the United States. Born in Germany, he was educated in Vienna and at Oxford, Wesleyan University, and Harvard University. Holton was the founding editor of *Daedalus* and has produced landmark works in science education by adding the history and philosophy of science to factual accounts of physics.

Holton's premise—that the philosophy and history of science have great inherent interest and add dimension to factual material—has been amply validated by the wide influence his writings have achieved. Holton urges writers of science essays to "celebrate" in science that which is worth celebrating; scientists must convey their sense of science as a "developing, sometimes inchoate private achievement that may have deep private consequences rather than to stress only the more usual sense of science as a finished institutional product." His pioneering ideas of curricula design are embodied in the texts *Concepts and Theories* and *Foundations of Modern Physical Science*. In 1962, he brought together a group of scientists and scholars who designed the nationally adopted course, Harvard Project Physics.

In *Thematic Origins of Scientific Thought*, Holton applies his breadth of knowledge by examining a number of specific themes in the sciences. He prepares for the "recognition that themata not merely belong to a pool of specifically scientific ideas but spring from the more general ground of the imagination."

In the middle of the twentieth century occurred what Freeman Dyson calls the "postwar flowering of physics." In "Modern Science and the Intellectual Tradition," Holton addresses this appreciation of science and warns that "underneath the euphoria there was an old, underlying disease that spelled trouble ahead. . . . A clearer understanding of what science is and is not will help make more explicit the bonds that could keep science in reciprocal contact with the rest of our culture." The following selection is excerpted from this essay and analyzes common views that the public holds of science and scientists. Here Holton reveals fallacies in the views and illuminates the role of science as a cultural element.

Modern Science and the Intellectual Tradition: The Seven Images of Science

Gerald Holton

PURE THOUGHT AND PRACTICAL POWER

Each person's image of the role of science may differ in detail from that of the next, but all public images are in the main based on one or more of seven positions. The first of these goes back to Plato and portrays science as an activity with double benefits: Science as pure thought helps the mind find truth, and science as power provides tools for effective action. In book 7 of the *Republic*,

Socrates tells Glaucon why the young rulers in the Ideal State should study mathematics: "This, then, is knowledge of the kind we are seeking, having a double use, military and philosophical; for the man of war must learn the art of number, or he will not know how to array his troops; and the philosopher also, because he has to rise out of the sea of change and lay hold of true being. . . . This will be the easiest way for the soul to pass from becoming to truth and being."

The main flaw in this image is that it omits a third vital aspect. Science has always had also a mythopoeic function—that is, it generates an important part of our symbolic vocabulary and provides some of the metaphysical bases and philosophical orientations of our ideology. As a consequence the methods of argument of science, its conceptions and its models, have permeated first the intellectual life of the time, then the tenets and usages of everyday life. All philosophies share with science the need to work with concepts such as space, time, quantity, matter, order, law, causality, verification, reality. Our language of ideas, for example, owes a great debt to statics, hydraulics, and the model of the solar system. These have furnished powerful analogies in many fields of study. Guiding ideas—such as conditions of equilibrium, centrifugal and centripetal forces, conservation laws, feedback, invariance, complementarity— enrich the general arsenal of imaginative tools of thought.

A sound image of science must embrace each of the three functions. However, usually only one of the three is recognized. For example, folklore often depicts the life of the scientist either as isolated from life and from beneficent action[1] or, at the other extreme, as dedicated to technological improvements.

ICONOCLASM

A second image of long standing is that of the scientist as iconoclast. Indeed, almost every major scientific advance has been interpreted—either triumphantly or with apprehension—as a blow against religion. To some extent science was pushed into this position by the ancient tendency to prove the existence of God by pointing to problems which science could not solve at the time. Newton thought that the regularities and stability of the solar system proved it "could only proceed from the counsel and dominion of an intelligent and powerful Being," and the same attitude governed thought concerning the earth's formation before the theory of geological evolution, concerning the descent of man before the theory of biological evolution, and concerning the origin of our galaxy before modern cosmology. The advance of knowledge therefore made inevitable an apparent conflict between science and religion. It is now clear how large a price had to be paid for a misunderstanding of both science and religion; to base religious beliefs on an estimate of what science cannot do is as foolhardy as it is blasphemous.

The iconoclastic image of science has, however, other components not ascribable to a misconception of its functions. For example, Arnold Toynbee charges science and technology with usurping the place of Christianity as the main source of our new symbols. Neo-orthodox theologians call science the "self-estrangement" of man because it carries him with idolatrous zeal along a dimension where no ultimate—that is, religious—concerns prevail. It is evident that these views fail to recognize the multitude of divergent influences that shape a culture, or a person. And on the other hand there is, of course, a group of scientists, though not a large one, which really does regard science as largely an iconoclastic activity. Ideologically they are, of course, descendants of Lucretius, who wrote on the first pages of *De rerum natura*, "The terror and darkness of mind must be dispelled not by the rays of the sun and glittering shafts of day, but by the aspect and the law of nature; whose first principle we shall begin by thus stating, nothing is ever gotten out of nothing by divine power." In our day this ancient trend has assumed political significance owing to the fact that in Soviet literature scientific teaching and atheistic propaganda are sometimes equated.

ETHICAL PERVERSION

The third image of science is that of a force which can invade, possess, pervert, and destroy man. The current stereotype of the soulless, evil scientist is the psychopathic investigator of science fiction or the nuclear destroyer—immoral if he develops the weapons he is asked to produce, traitorous if he refuses. According to this view, scientific morality is inherently negative. It causes the arts to languish, it blights culture, and when applied to human affairs, it leads to regimentation and to the impoverishment of life. Science is the serpent seducing us into eating the fruits of the tree of knowledge—thereby dooming us.

The fear behind this attitude is genuine but not confined to science; it is directed against all thinkers and innovators. Society has always found it hard to deal with creativity, innovation, and new knowledge. And since science assures a particularly rapid, and therefore particularly disturbing, turnover of ideas, it remains a prime target of suspicion.

Factors peculiar to our time intensify this suspicion. The discoveries of "pure" science often lend themselves readily to widespread exploitation through technology. The products of technology—whether they are better vaccines or better weapons—have the characteristics of frequently being very effective, easily made in large quantities, easily distributed, and very appealing. Thus we are in an inescapable dilemma—irresistibly tempted to reach for the fruits of science, yet, deep inside, aware that our metabolism may not be able to cope with this ever-increasing appetite.

Probably the dilemma can no longer be resolved, and this increases the anxiety and confusion concerning science. A current symptom is the popular identification of science with the technology of superweapons. The bomb is taking the place of the microscope, Wernher von Braun, the place of Einstein, as symbols for modern science and scientists. The efforts to convince people that science itself can give man only knowledge about himself and his environment, and occasionally a choice of action, have been largely unavailing. The scientist *as scientist* can take little credit or responsibility either for facts he discovers—for he did not create them—or for the uses others make of his discoveries, for he generally is neither permitted nor specially fitted to make these decisions. They are controlled by considerations of ethics, economics, or politics and therefore are shaped by the values and historical circumstances of the whole society.[2]

There are other evidences of the widespread notion that science itself cannot contribute positively to culture. Toynbee, for example, gives a list of "creative individuals," from Xenophon to Hindenburg and from Dante to Lenin, but does not include a single scientist. I cannot forego the remark that there is a significant equivalent on the level of casual conversation. For when the man in the street—or many an intellectual—hears that you are a physicist or mathematician, he will usually remark with a frank smile, "Oh, I never could understand that subject"; while intending this as a curious compliment, he betrays his intellectual dissociation from scientific fields. It is not fashionable to confess to a lack of acquaintance with the latest ephemera in literature or the arts, but one may even exhibit a touch of pride in professing ignorance of the structure of the universe or one's own body, of the behavior of matter or one's own mind.

THE SORCERER'S APPRENTICE

The last two views held that man is inherently good and science evil. The next image is based on the opposite assumption—that man cannot be trusted with scientific and technical knowledge. He has survived only because he lacked sufficiently destructive weapons; now he can immolate his world. Science, indirectly responsible for this new power, is here considered ethically neutral. But man, like the sorcerer's apprentice, can neither understand this tool nor control it. Unavoidably he will bring upon himself catastrophe, partly through his natural sinfulness, and partly through his lust for power, of which the pursuit of knowledge is a manifestation. It was in this mood that Pliny deplored the development of projectiles of iron for purposes of war: "This last I regard as the most criminal artifice that has been devised by the human mind; for, as if to bring death upon man with still greater rapidity, we have given wings to iron and taught it to fly. Let us, therefore, acquit Nature of a charge that belongs to man himself."

When science is viewed in this plane—as a temptation for the mischievous savage—it becomes easy to suggest a moratorium on science, a period of abstinence during which humanity somehow will develop adequate spiritual or social resources for coping with the possibilities of inhuman uses of modern technical results. Here I need point out only the two main misunderstandings implied in this recurrent call for a moratorium.

First, science of course is not an occupation, such as working in a store or on an assembly line, that one may pursue or abandon at will. For a creative scientist, it is not a matter of free choice what he shall do. Indeed it is erroneous to think of him as advancing toward knowledge; it is, rather, knowledge which advances towards him, grasps him, and overwhelms him. Even the most superficial glance at the life and work of a Kepler, a Dalton, or a Pasteur would clarify this point. It would be well if in his education each person were shown by example that the driving power of creativity is as strong and as sacred for the scientist as for the artist.

The second point can be put equally briefly. In order to survive and to progress, mankind surely cannot ever know too much. Salvation can hardly be thought of as the reward for ignorance. Man has been given his mind in order that he may find out where he is, what he is, who he is, and how he may assume the responsibility for himself which is the only obligation incurred in gaining knowledge.

Indeed, it may well turn out that the technological advances in warfare have brought us to the point where society is at last compelled to curb the aggressions that in the past were condoned and even glorified. Organized warfare and genocide have been practiced throughout recorded history, but never until now have even the war lords openly expressed fear of war. In the search for the causes and prevention of aggression among nations, we shall, I am convinced, find scientific investigations to be a main source of understanding.

ECOLOGICAL DISASTER

A change in the average temperature of a pond or in the salinity of an ocean may shift the ecological balance and cause the death of a large number of plants and animals. The fifth prevalent image of science similarly holds that while neither science nor man may be inherently evil, the rise of science happened, as if by accident, to initiate an ecological change that now corrodes the only conceivable basis for a stable society. In the words of Jacques Maritain, the "deadly disease" science set off in society is "the denial of eternal truth and absolute values."

The main events leading to this state are usually presented as follows. The abandonment of geocentric astronomy implied the abandonment of the conception of the earth as the center of creation and of man as its ultimate purpose. Then purposive creation gave way to blind evolution. Space, time, and certainty were shown to have no absolute meaning. All a priori axioms were discovered to be merely arbitrary conveniences. Modern psychology and anthropology led to cultural relativism. Truth itself has been dissolved into probabilistic and indeterministic statements. Drawing upon analogy with the sciences, liberal philosophers have become increasingly relativistic, denying either the necessity or the possibility of postulating immutable verities, and so have undermined the old foundations of moral and social authority on which a stable society must be built.

It should be noted in passing that many applications of recent scientific concepts outside science merely reveal ignorance about science. For example, relativism in nonscientific fields is generally based on farfetched analogies. Relativity theory, of course, does not find that truth depends on the point of view of the observer but, on the contrary, reformulates the laws of physics so that they hold good for every observer, no matter how he moves or where he stands. Its central meaning is that the most valued truths in science are wholly independent of the point of view. Ignorance of science is also the only excuse for adopting rapid changes within science as models for antitraditional attitudes outside science. In reality, no field of thought is more conservative than science. Each change necessarily encompasses previous knowledge. Science grows like a tree, ring by ring. Einstein did not prove the work of Newton wrong; he provided a larger setting within which some contradictions and asymmetries in the earlier physics disappeared.

But the image of science as an ecological disaster can be subjected to a more severe critique.[3] Regardless of science's part in the corrosion of absolute values, have those values really given us always a safe anchor? A priori absolutes abound all over the globe in completely contradictory varieties. Most of the horrors of history have been carried out under the banner of some absolutistic philosophy, from the Aztec mass sacrifices to the auto-da-fé of the Spanish Inquisition, from the massacre of the Huguenots to the Nazi gas chambers. It is far from clear that any society of the past did provide a meaningful and dignified life for more than a small fraction of its members. If, therefore, some of the new philosophies, inspired rightly or wrongly by science, point out that absolutes have a habit of changing in time and of contradicting one another, if they invite a re-examination of the bases of social authority and reject them when those bases prove false (as did the Colonists in this country), then one must not blame a relativistic philosophy for bringing out these faults. They were there all the time.

In the search for a new and sounder basis on which to build a stable world, science will be indispensable. We can hope to match the resources and structure of society to the needs and potentialities of people only if we know more about man. Already science has much to say that is valuable and important about human relationships and problems. From psychiatry to dietetics, from immunology to meteorology, from city planning to agricultural research, by far the largest part of our total scientific and technical effort today is concerned, indirectly or directly, with man—his needs, relationships, health, and comforts. Insofar as absolutes are to help guide mankind safely on the long and dangerous journey ahead, they surely should be at least strong enough to stand scrutiny against the background of developing factual knowledge.

SCIENTISM

While the last four images implied revulsion from science, scientism may be described as an addiction to science. Among the signs of scientism are the habit of dividing all thought into two categories, up-to-date scientific knowledge and nonsense; the view that the mathematical sciences and the large nuclear laboratory offer the only permissible models for successfully employing the mind or organizing effort; and the identification of science with technology, to which reference was made above.

One main source for this attitude is evidently the persuasive success of recent technical work. Another resides in the fact that we are passing through a period of revolutionary change in the nature of scientific activity—a change triggered by the perfecting and disseminating of the methods of basic research by teams of specialists with widely different training and interests. Twenty years ago the typical scientist worked alone or with a few students and colleagues. Today he usually belongs to a sizable group working under a contract with a substantial annual budget. In the research institute of one university more than 1500 scientists and technicians are grouped around a set of multimillion-dollar machines; the funds come from government agencies whose ultimate aim is national defense.

Everywhere the overlapping interests of basic research, industry, and the military establishment have been merged in a way that satisfies all three. Science has thereby become a large-scale operation with a potential for immediate and world-wide effects. The results are a splendid increase in knowledge, and also side effects that are analogous to those of sudden and rapid urbanization—a strain on communication facilities, the rise of an administrative bureaucracy, the depersonalization of some human relationships.

To a large degree, all this is unavoidable. The new scientific revolution will justify itself by the flow of new knowledge and of material benefits that will no doubt follow. The danger—and this is the point where scientism enters—is that the fascination with the *mechanism* of this successful enterprise may change the scientist himself and society around him. For example, the unorthodox, often withdrawn individual, on whom most great scientific advances have depended in the past, does not fit well into the new system. And society will be increasingly faced with the seductive urging of scientism to adopt generally what is regarded—often erroneously—as the pattern of organization of the new science. The crash program, the breakthrough pursuit, the megaton effect are becoming ruling ideas in complex fields such as education, where they may not be applicable.

MAGIC

Few nonscientists would suspect a hoax if it were suddenly announced that a stable chemical element lighter than hydrogen had been synthesized, or that a manned observation platform had been established at the surface of the sun. To most people it appears that science knows no inherent limitations. Thus, the seventh image depicts science as magic, and the scientist as wizard, *deus ex machina,* or oracle. The attitude toward the scientist on this plane ranges from terror to sentimental subservience, depending on what motives one ascribes to him.

IMPOTENCE OF THE MODERN INTELLECTUAL

The prevalence of these false images is a main source of the alienation between the scientific and nonscientific elements in our culture, and therefore the failure of image is important business for all of us. Now to pin much of the blame on the insufficient instruction in science which the general student receives at all levels is quite justifiable. I have implied the need, and most people nowadays seem to come to this conclusion anyway. But this is not enough. We must consider the full implications of the discovery that not only the man in the street but almost all of our intellectual leaders today know at most very little about science. And here we come to the central point underlying the analysis made above: the chilling realization that our intellectuals, for the first time in history, are losing their hold of understanding upon the world.

The wrong images would be impossible were they not anchored in two kinds of ignorance. One kind is ignorance on the basis level, that of *facts*—what biology says about life, what chemistry and physics say about matter, what astronomy says about the development and structure of our galaxy, and so forth. The nonscientist realizes that the old common-sense foundations of thought about the world of nature have become obsolete during the last two generations. The ground is trembling under his feet; the simple interpretations of solidity, permanence, and reality have been washed away, and he is plunged into the nightmarish ocean of four-dimensional continua, probability amplitudes, indeterminacies, and so forth. He knows only two things about the basic conceptions of modern science: that he does not understand them, and that he is now so far separated from them that he will never find out what they mean.

On the second level of ignorance, the contemporary intellectual knows just as little of the way in which the main facts from the different sciences fit together in a picture of the world taken as a whole. He has had to leave behind him, one by one, those great syntheses which used to represent our intellectual and moral home—the world view of the book of Genesis, of Homer, of Dante, of Milton, of Goethe. In the mid-20th century he finds himself abandoned in a universe which is to him an unsolvable puzzle on either the factual or the philosophical level. Of all the bad effects of the separation of culture and scientific knowledge, this feeling of bewilderment and basic homelessness is the most terrifying. Here is the reason, it seems to me, for the ineffectiveness and self-denigration of our contemporary intellectuals. Nor are the scientists themselves protected from this fate, for it has always been, and must always be, the job of the humanist to construct and disseminate the meaningful total picture of the world.

. . . Every great age has been shaped by intellectuals of the stamp of Hobbes, Locke, Berkeley, Leibnitz, Voltaire, Montesquieu, Rousseau, Kant, Jefferson, and Franklin—all of whom would have been horrified by the proposition that cultivated men and women could dispense with a good grasp of the scientific aspect of the contemporary world picture. This tradition is broken; very few intellectuals are now able to act as informed mediators. Meanwhile, as science moves every day faster and further from the bases of ordinary understanding, the gulf grows, and any remedial action becomes more difficult and more unlikely.

To restore science to reciprocal contact with the concerns of most men—to bring science into an orbit about us instead of letting it escape from our intellectual tradition—that is the great challenge that intellectuals face today.

REFERENCES AND NOTES

1. See, for example, the disturbing findings of M. Mead and R. Metraux, "Image of the scientist among high-school students," *Science* 126, 384 (1957). I have presented the approach in this middle section in the "Adventures of the Mind" series, *Saturday Evening Post*, 9 January 1960.
2. It is, however, also appropriate to say here that there has been only a moderate success in persuading the average scientist of the proposition that the privilege of freely pursuing a field of knowledge having large-scale secondary effects imposes on him, in his capacity as citizen, a proportionately larger burden of civic responsibility.
3. See, for example, C. Frankel, *The Case for Modern Man* (Beacon, Boston, 1959).

Questions and Topics for Discussion and Writing

1. Holton offers Plato's portrayal of the role of science as his first image of science. What is his challenge to Plato's definition of the role of science? In the title essay of his book *Mysticism and Logic*, Bertrand Russell blames Plato for saying that the ultimate end of science is to promote mind expansion to a realm outside nature. Does Holton offer the same challenge that Russell does?
2. What is the "mythopoeic function" of science to which Holton refers?
3. Holton says that a "sound image of science must embrace three functions." List these functions.
4. What is an iconoclast? Name the three sciences that brought the image of the scientist as an iconoclast.

5. How, according to Holton, did science come to be considered an enemy of religion?

6. Holton categorizes the third image of science as "ethical perversion." Summarize what Holton means by this term and restate his defense of science.

7. What, according to Holton, is the "inescapable dilemma" created by technology, the "child of science"?

8. According to Holton, how are the contributions of science to humanity limited?

9. Holton's sentence "Salvation can hardly be thought of as the reward for ignorance" is epigrammatic. How does this sentence present Holton's answer to those who pose the view that science is directly responsible for creating destructive weapons? What is the important contribution scientists can make toward a cessation of warfare?

10. How would Holton challenge the student who says that he or she intends to become a scientist?

11. In an essay, explain how education can prepare students for accepting the idea that "science is not an occupation"?

12. Many literary works characterize contemporary individuals in a fragmented world, one in which they find no absolute meaning and, as a consequence, must impose order on their own lives. How would Holton's statement in the section "Ecological Disaster" provide comfort and assurance to such individuals?

13. Define *scientism*. What does Holton see as the "danger" of scientism?

14. Holton says that we are passing through a revolutionary change in the nature of scientific activity. What is the nature of this change? How was it triggered? How will the new scientific revolution justify itself?

Margaret Mead

"Respected" and "controversial" are two words that best describe Margaret Mead (1901–1979), the American psychological anthropologist. The science of human beings and the relationship of the races in terms of physical character, origin, environment, and culture perhaps initially attracted her because the perimeters of anthropology are so flexible. Elastic as they are, however, they did not contain her. In her lifetime, neither oceans, continents, time-honored philosophies, prejudices, other scientists, nor institutions presented any barriers to her. She developed her own rules and ventured into both geographical and theoretical spheres where no other woman up to that time had dared to tread.

Mead's sense of family dominated her life. The oldest of five children, she grew up in an intellectually stimulating environment, her mother a sociologist, her father an economist. She spent her undergraduate years at Barnard College, where, as a senior, she changed her major from English to anthropology, and she went on to earn her master's and doctoral degrees from Columbia University.

She was to use her studies to promote social changes, particularly equality between the sexes. Choosing the island of Samoa, she made her first research trip in 1925. In 1926, she began a lifelong association with the American Museum of Natural History. She was later to make repeated trips to Samoa to document the changes in the society, and her *Coming of Age in Samoa* (1928) made Mead world famous before she was thirty.

By forty, Mead had field work behind her in Samoa, New Guinea, and Bali, six books, 95 other publications, and 25,000 annotated photographs to her credit. Among her best-known books are *Growing Up in New Guinea* (1930), *Male and Female* (1949), and the autobiographical *Blackberry Winter* (1972). During World War II, the American Museum granted her leave to work on various government projects in Washington.

An attack on her methods of work was made in 1983 by Australian anthropologist Derek Freeman. It led to worldwide forums, with some adverse criticism, but also to many defenses of her career.

The selection that follows, "The Role of the Scientist in Society," from *Anthropology: A Human Science*, reveals her preoccupation with human interaction and her ability to integrate knowledge from natural and social sciences with the humanities.

The Role of the Scientist in Society*

Margaret Mead

I want to discuss the problem of how we Americans today feel about the scientist and the scientist's role, and what significance that feeling has for the contribution the scientist can make to the contemporary world crisis. This seems relevant because I am addressing an audience of practitioners. However much you may be devoting some part of your lives to research, most of you are giving most of your time to applying scientific insights to the problems of individuals, or occasionally, of groups. As practitioners, your every word and tone of voice become significant in conveying to those with whom you work, as patients, as collaborators, as members of the general public, the meaning and the promise, or the threat and the limitations of science. Perhaps even more potently than the stylizations of the scientist in the press and radio, the stage and film, the way in which the practitioner stylizes his own role, and sees his own role, tends to build up in the mind of the layman either a faith or a distrust in science. And

*Reference notes have been omitted from this essay.—Eds.

to the degree that the practitioner sees the implications, the possible interpretations which may be placed upon his every act, he or she becomes the more aware and therefore the more effective as a communicator.

When science enters the realm of human relations, what will be the result? What, in fact, will happen to human relations and will they be seen as human at all? The problem can be approached from the standpoint of the scientist's picture of himself, as he sees himself mirrored in the conceptions of those around him. I would rather reverse this picture and explore some of the reasons why the layman entertains the various attitudes of fear, faith, hope, and distrust toward the scientist.

Central to this problem is the question of power. Science, as it has developed historically, has come to be associated with the idea of power, unlimited power over the forces of nature. Atomic discoveries have so enhanced this picture that it is safe to say that *power* is one of the first associations which the layman makes with the word *scientist*. A second association is the word *impersonal*. The scientist has been celebrated for his objectivity, his freedom from bias, his cool, aloof, impersonal—and almost by definition—inhuman behavior. This stereotype, frightening even when applied to someone who was experimenting with unrealizable entities, is extremely repellant when applied to human affairs. To treat another impersonally, coolly, aloofly, is to be lacking in warmth, in concern, in contact. The minute we are asked to think of the scientist in relation to human behavior, then this carefully built up picture of objectivity intervenes. The desire to make a split between this coolness and the human practitioner is seen in the contrast between the picture of the doctor—warm, and a little shabby, who sits by his patient's bedside—and the white-coated "scientist" who is pictured all alone with some shining piece of laboratory apparatus. One of the tasks of interpreting the meaning of science for human welfare becomes then to heal this split, to reunite the tired, friendly, country doctor, who knew and loved each patient as a person, and the cool impersonal man in a laboratory coat.

A second conflict centers around the fear that power over persons, even more than power over things, is blasphemous, is arrogating to man something which should be left to God. While this feeling is less strong in the United States, where man has come as an adult to deal with an unpatterned landscape, than it is in European countries where the works of man and the natural landscape both blend together in a past to which man adjusts, still the feeling is here. The phrase "playing God" comes readily to the lips when any specialist in human behavior seems too sure.

A third difficulty centers around the way in which the sciences of human behavior appear to restrict rather than expand the layman's sense of understanding and control of the world. The layman has no expectation of understanding, without expert help, the details of geological stratification, the movement of the stars, the wonders of embryology, or the operation of hormones and enzymes. Whatever the scientist discovers in these fields is felt to be added on to the layman's existing stock of knowledge, and the acquisition of wonderful new words like homeostasis, entropy, proton, adds to his sense of human dignity. But in the field of human relations each generation has characteristically thought of itself as well informed and well oriented. Parenthood and marriage, discipline and indulgence, love and hate, are matters which people think of themselves as "naturally" understanding. Every time a technical term, *affect* instead of love or hate, *ambivalence* for some simple phrase like mixed feelings, is developed in the sciences of human relations, the layman feels that part of his rightful inheritance as a social being has been snatched away from him, that what was simple and plain has been made mysterious and esoteric, that he is robbed of his dignity as a well-oriented human being.

This is perhaps especially likely to occur in a culture with a primary Protestant orientation, in which the insistence that each human being could read his Bible and deal directly with his God, without intervention of priest or sacrament, forms a natural background for a jealous guarding of individual choice and judgment in personal relations. The young Italian American graduate student who says "psychology is just the things my mother used to know, put in a way that no one can understand"; the jeering reaction of the press to the attempt of an educator to subsume all sorts of beatings, spankings, cuffings, ear boxings, and hand smackings under the heading of "manual discipline"—these are symptoms of this deep sense of loss and affront which the layman feels as the area of human relations is invaded, studied, classified, and labeled with new words which he must learn as he would have to learn the vocabulary of physics or chemistry.

It will probably be necessary to devise new educational methods which will set the student to wondering about human behavior first, before he is given any of this unwelcome knowledge, just as in the training of a natural scientist, the wondering curiosity of the student of natural history, of the child who holds a "cat's eye" in his hand and realizes that "this must have been alive," is the precursor of creative scientific curiosity. But too often expert knowledge in the field of human relations is offered to the layman and to the patient or client or student, not as an enlargement of a horizon which has first been opened up, but in response to what is technically called an expressed "need"—that is, a sense of individual inadequacy in solving one's psychological or social problems. If "need," a crying active awareness of trouble and inadequacy, is regarded as the appropriate setting for imparting expert knowledge in the field of personal relations, this practice is likely to reinforce the already existing sense of human outrage, that affairs which people should be able to think about and feel about as part of their adultness, must be handed over to specialists.

The position of the human relations scientist is further complicated by the current tendency to blame science and so scientists, for the plight to which our world has come. There is precedent in the history of human beliefs for expecting a cure from the one who causes the ailment, but when this occurs we classify it as "black magic." There are many primitive societies in which all disease, misfortune, and death are produced by men who are in special rapport with supernatural powers, and who may be persuaded by bribes, cajolery, threats or reversals of the behavior which induced them to start the train of evil, to undo what they have done. In such societies, power is to a degree undifferentiated, and reacted to with great ambivalence by those against whom it is exercised. (It is notable that witchcraft and black magic seem to increase in primitive societies in which the culture is disintegrating under contact with our civilization, and also where a village culture is giving place to an urban culture with the resulting atomization of the individual and increase in *anomie.*) A world in which the disintegration of all reliable values is attributed to the natural scientist, and the resulting disintegration of personality is then referred for treatment to the human relations scientist, is of course more complicated than a witch-ridden primitive society. Nevertheless useful parallels can be drawn. If the word scientist is used both for the men who discover the laws of thermodynamics and atomic fission and for the physician who must work day by day with individual breakdown, the possibility that the scientist will be seen as both the cause and the possible magical cure is very great. In this case, the success of the psychiatrist, especially with measures such as electric shock, drug or hypnotherapy, will also tend to be read backward and amalgamated with the attitudes toward the scientists who have produced the atomic bomb. The belief in the power of science will be increased, with a corresponding emphasis upon the malign nature of that power and an enhancement of the sense of individual helplessness of the layman.

The final and perhaps the central problem of the position of the human relations scientist in society concerns manipulation. Once the possibility of discovering and applying principles of human behavior is granted, what possible safeguard can society develop against the misuse of this power? The examples of recent decades in which a very little knowledge of human behavior has been used in commerce, in government, and in war to bemuse, befuddle, subjugate, corrupt, and disintegrate the minds of men, breed a very justified fear that a society with a real scientific grasp of human behavior would be a monstrous society in which no one would willingly live. It breeds the belief that it may be better to accept every human ill to which flesh is heir—disease, famine, war, insanity—than to risk the inevitable destruction of human dignity in a controlled world, in which those in absolute power have been absolutely corrupted by that power.

But while this fear is both justified and cogent, it is important to realize that the acceptance and incorporation of the science of human behavior is no longer a matter of choice. Atomic discoveries have introduced an order of urgency which the world has not hitherto faced, an urgency such that to neglect a single possible solution, no matter how difficult, becomes a treachery to the human race. In addition, the developments in domestic controls by totalitarian governments, in methods of opinion research and attitude testing by democratic governments, commercial undertakings, and in psychological warfare, especially of the "black" variety which introduces a final corrupting note in its denial of the source from which it emanates, have presented us with a degree of potentially destructive uses of the science of human behavior which we cannot eliminate by prayer or legislation or a refusal to face them. It is impossible to go back to an age of innocence; attempted returns to such earlier states invariably assume the unlovely aspect of political reaction and the oversimplifications of the near psychotic.

Our only course is to go forward and integrate the human sciences into the very fabric of our society. We must invent and introduce ethics and controls which will tie the hands of those with power, so that either by a self-denying ordinance or by carefully devised pressures, analogous to but much more complex than the "medical ethics" which have served mankind so well, or by controls which are in some way actually built into practice so that any manipulative behavior becomes self-defeating, manipulative behavior is impossible and human beings remain free in spite of having again eaten of the tree of knowledge. To do this, we need, most of all, a climate of opinion, a sense of the role of the scientist as the responsible expression of a new kind of civilization, a civilization to which disciplined self-awareness is the very breath of life. In developing such a climate of opinion, every practitioner, in every professional word and act, can contribute.

Questions and Topics for Discussion and Writing

1. After reading this essay, what statements can you make about the role of the scientist in society? Do your statements differ from Mead's? How?
2. In 1961, Mead began writing for *Redbook* on a number of topics. Read several back issues to get an idea of her subjects and themes. Who was her audience? Does she have a sense of the complexity of social issues? Establish your opinion with specific examples. How was Mead a pioneer of new ideas and social themes?
3. How do you think Mead would justify her "popularizing" her ideas in nonacademic publications?
4. Inventory this essay for words and phrases that support the view that Mead is an optimist.
5. Explain this statement: "But in the field of human relations each generation has characteristically thought of itself as well informed and well oriented." Use your own ideas and opinions.
6. Mead opens her autobiography, *Blackberry Winter*, with the following declaration: "I have spent most of my life studying the lives of other peoples, far-away peoples, so that Americans might better understand themselves." In what way could the study of very different cultures help Americans to "better understand themselves"?
7. This paper was presented in a symposium at the 1947 Annual Meeting of the American Orthopsychiatric Association to an audience concerned about problems relating to children. Why would such a seemingly unrelated topic be of interest to this audience? How does Mead justify her choice of subject?

Werner Heisenberg

The German nuclear physicist Werner Heisenberg (1907–1976) belongs in the special group of scientists whose work has contributed one of the major ideas of the century, thereby altering our entire world picture. Heisenberg is one of the co-founders of quantum mechanics, the system used to describe exceedingly small masses, an area where the system of Newtonian mechanics breaks down.

Heisenberg is best known as the author of the *uncertainty principle* (sometimes called the *principle of indeterminacy*). It states that the position and velocity of an object cannot with exactness be measured simultaneously, a limitation that becomes significant in the world of atoms and subatomic particles. The basis of the principle is still in dispute. Heisenberg's emphasis is on the fact that observation of the microphysical world entails interference with it.

The notion of uncertainty—which has been described as "this supremely negative assertion"—has modified not only the field of physics but also modern thought, with its implications for causality, determinism, and even the concept of miracles. Among Heisenberg's writings are *The Physical Principles of the Quantum Theory* (1930), *Physics and Philosophy* (1959), and the semi-autobiographical *Physics and Beyond* (1971).

Heisenberg received the Nobel Prize for physics in 1932. Although unsympathetic to the Nazis, he remained in Germany as a professor of physics during World War II, afterward becoming director of the Max Planck Institute for Physics and Astrophysics and promoting scientific research in Germany.

The following essay, "Tradition in Science," is an excerpt from a lecture on the role of tradition in influencing the choice of problems, the selection of method, and the formulation of concepts in science.

Tradition in Science

Werner Heisenberg

When we celebrate the 500th birthday of Copernicus, we do it because we believe that our present science is connected with his work; the direction that he had chosen for his research in astronomy still determines to some extent the scientific work of our time. We are convinced that our present problems, our methods, our scientific concepts are, at least partly, the result of a scientific tradition that accompanies or leads the way of science through the centuries. It is therefore natural to ask to what extent our present work is determined or influenced by tradition. Are the *problems* in which we are engaged freely cho-

sen according to our interest or inclination, or are they given to us by a historical process? To what extent can we select our scientific *methods* according to
the purpose? To what extent do we again follow a given tradition? Finally, how
free are we in choosing the *concepts* for formulating our questions?

Any scientific work can only be defined by formulating the questions that we
want to answer. But in order to formulate the questions we need concepts by
which we hope to get hold of the phenomena. These concepts usually are taken
from the past history of science; they suggest a possible picture of the phenomena. But if we are going to enter into a new realm of phenomena, these concepts may act as a collection of prejudices, which hamper progress rather than
foster it. Even then we have to use concepts, and we can't help falling back on
those given to us by tradition. Therefore, I will try to discuss the influence of
tradition in the selection of problems. . . .

To what extent are we bound by tradition in the selection of our problems?
When we look back into the history of science, we see that periods of intense
activity alternate with long periods of inactivity. In ancient Greece the philosophers started asking questions of principle with respect to the phenomena in
nature. There had been a considerable practical knowledge long before; great
skill had been developed in building houses, cutting and moving big stones,
constructing ships and so on; but it was first in the period after Pythagoras, that
this skill was supplemented by scientific inquiry. The relevance of mathematical relations in natural phenomena was discovered by Pythagoras and his pupils, and a great development both in mathematics, in astronomy and in natural
philosophy followed. The decline of Greek science after the Hellenistic period,
after Ptolemy, the last great astronomer, marked the beginning of a long period
of inactivity which lasted until the Renaissance in Italy.

During this period of stagnation again an admirable development of practical
knowledge led to a high civilization in the Arab countries; but it was not accompanied by a corresponding development in science, by a deeper understanding
of nature. More than a thousand years later, when humanism and renaissance
had shown the way to a more liberal trend of thought, when the explorers had
demonstrated the possibility of expansion on our earth, then a new activity in
science was inaugurated by the discoveries of Copernicus, Galileo and Kepler.
This activity has lasted until our present time, and we do not know whether it
will still continue for long, or will give way to a new period in which the interest
goes off into very different directions.

Looking back upon history in this way we see that we apparently have only
little freedom in the selection of our problems. We are bound up with historical
process. Our lives are part of this process and our choices seem to be restricted
to the decision whether or not we want to participate in a development that
takes place in our time, with or without our contribution. Without such a
favorable development, our activity would probably be lost.

If Einstein had lived in the 12th century, he would have had very little
chance to become a good scientist. Even within a fruitful period a scientist does

not have much choice in selecting his problems. On the contrary, one may say that a fruitful period is characterized by the fact that the problems are given, that we need not invent them.

This seems to be true in science as well as in art. In the 15th century when painters in the Netherlands discovered the possibility of portraying men as active members of society, many gifted people were attracted by this possibility and competed in solving the problem. In the 18th century, Haydn tried in his string quartets to express emotions that had appeared in the literature of his time, in the work of Rousseau and in Goethe's *Werther;* and, then, the musicians of the younger generation—Mozart, Beethoven, Schubert—gathered in Vienna to compete in the solution of this problem.

In our century the development of physics led Niels Bohr to the idea that Lord Rutherford's experiments on alpha rays, Max Planck's theory of radiation, and the facts of chemistry could be combined in a theory of the atom. And in the following years many young physicists went to Copenhagen in order to participate in the solution of this given problem. One cannot doubt that in the selection of problems the tradition, the historical development, plays an essential role.

This may also sometimes be true in a negative sense. It can happen that traditional themes have been exhausted and that the gifted people turn away from a field in which they see no more objects for their activity. After Thomas Aquinas, the philosophers got tired of the theological and philosophical problems of scholastics and turned to humanism. In our time the traditional themes of art seem to be exhausted. In 1972, one of the most popular yearly exhibitions of modern art in Germany, which is held in Kassel and called "Documenta," was a center of political propaganda rather than of art. On the outside of the building of the exhibition, the young artists had fixed a huge poster with the text: "Art is superfluous."

In a similar way we cannot exclude the possibility that after some time the themes of science and technology will be exhausted, that a younger generation will be tired of our rationalistic and pragmatic attitudes and will turn their interest to an entirely different activity. In the present situation, however, many problems still exist in pure and in applied science. No effort is needed to invent them, and they will be passed on from the teachers to their pupils.

In this connection it is important to emphasize the very great role of personal relations in the development of science or art. It need not only be the relationship between teacher and pupil, it may simply be personal friendship or respect between people working for the same goal. This is probably the most efficient instrument of tradition. Among the many examples which could be mentioned for this kind of tradition I will only recall some of the personal relationships which have shaped the history of physics in the first half of our present century.

Einstein was well acquainted with Planck; he corresponded with A. J. Sommerfeld about the theory of relativity and about quantum theory; he was a dear friend of Max Born, although he could never agree with him on the statistical interpretation of quantum theory; and he discussed with Niels Bohr the philosophical implications of the relations of uncertainty. A large part of the scientific analysis of those extremely difficult problems, arising out of relativity and quantum theory, was actually carried out in conversations between those who took an active part in the research.

Sommerfeld's school in Munich was a center of research in the early 1920s. Wolfgang Pauli, Gregor Wentzel, Otto Laporte, W. Lenz and many others belonged to this group, and we discussed almost daily the difficulties and paradoxes in the interpretation of recent experiments. When Sommerfeld received a letter from Einstein or Bohr, he read the important parts in a seminar and started at once a discussion on the critical problems. Niels Bohr held a close association with Lord Rutherford, Otto Hahn, and Lise Meitner, and he considered the continuous exchange of information between experiment and theory as a central task in the progress of physics. The enormous influence of Niels Bohr on the development of physics in his time was not primarily due to his papers, but to his way of discussing again and again with his partners the fundamental difficulties of quantum theory, which, as he knew, did not allow for any cheap solution.

When wave mechanics was introduced by Erwin Schrödinger, Bohr saw at once that this was a very important new aspect of quantum theory; but that a simple replacement of the electronic orbits in the atom by three-dimensional matter waves could not solve the real difficulties. Again, the only way of analyzing the problem seemed to be personal discussion with the author. Schrödinger was invited to Copenhagen. In two weeks of most intense discussions, the way was prepared for the later development in the interpretation of quantum theory, for Bohr's concept of complementarity, and for the relations of uncertainty. I need not enlarge upon these examples. It is obvious that personal relations play a decisive role in the progress of science and in the selection of problems.

There are, of course, other motives for the selection of problems; motives that have played their role in the history of science. The best known of these is the practical applicability of science. In ancient times the interest in astronomy and mathematics was stimulated by the fact that knowledge in these fields was helpful for navigation and for the surveying of land. Navigation played a very important role in the 15th century, when the explorers left Europe and the Mediterranean and sailed westward.

When Galileo defended the ideas of Copernicus he made use of a newly invented instrument, the telescope, thereby demonstrating that a practical tool may be helpful in the progress of science, and science may be helpful in leading to the invention of practical tools. Galileo and his followers were strongly interested in the practical side of science. They studied mechanical devices, for example, the mechanical clock; they invented optical instruments, and so on.

It has always been a tradition in science, guiding the activity of many generations, that science should be applied to practical purposes and that the practical application should be a check on the validity of the results and a justification for the efforts. The atomic physicists of the first half of our present century followed this old tradition of science when they looked for practical applications of atomic physics. It was, of course, extremely disappointing for them that the first practical application was for warfare. Still the fact that one now could transmute chemical elements into others in large quantities was justly considered as a real triumph of science.

Interest in the practical application of science is frequently misunderstood as the trivial attempt of the scientist to acquire economic wealth. It is true that this trivial motive does play a role, depending of course on the individuals. But this motive should not be overestimated. There is another much stronger motive that fascinates the good scientist in connection with the practical application, namely, to see that one has correctly understood nature.

I remember a conversation with Enrico Fermi after the war, a short time before the first hydrogen bomb was to be tested in the Pacific. We discussed this plan, and I suggested that one should perhaps abstain from such a test considering the biological and political consequences. Fermi replied, "But it is such a beautiful experiment." This is probably the strongest motive behind the applications of science; the scientist needs the confirmation from an impartial judge, from nature herself, that he has understood her structure. And he wants to see the effect of his effort.

From this attitude one can also easily understand the motives that determine the line of research for the individual scientist. Such a line of research is usually based on some theoretical ideas, on conjectures concerning the interpretation of the known phenomena, or on hopes for finding new ones. But which ideas are accepted? Experience teaches that it is usually not the consistency, the clarity of ideas, which makes them acceptable, but the hope that one can participate in their elaboration and verification. It is the wish for our own activity, the hope for results from our own efforts, that leads us on our way through science. This wish is stronger than our rational judgment about the merits of various theoretical ideas. In the early 1920s we knew that Bohr's theory of the atom could not be quite correct; but we guessed that it pointed in the right direction, and we hoped that we would be able some day to avoid the inconsistencies and to replace Bohr's theory by a more satisfactory picture. . . .

[At this point, Heisenberg discusses the effect of tradition on the scientific method, concluding that, despite our no longer founding science on a theological justification as seventeenth-century physicists did, we still "follow strictly the tradition inaugurated in the time of Galileo" because it has been highly successful in many fields. This tradition designs experiments "which idealize and isolate experience," thereby creating new phenomena and comparing these phenomena "with mathematical constructs, called natural laws."

Heisenberg then moves to the third part of his talk. To the effect of tradition in the selection of problems and in the scientific method, he adds the effect of tradition in the history of concepts. He begins with concepts from seventeenth-century astronomy, where the "new science" began, then moves on to concepts added by Newton, Faraday, and others, pointing out that, as in the case of the concept of the "ether," tradition can sometimes hinder understanding, as well as sometimes be a condition for progress. Heisenberg continues to our century and his own work in relativity and quantum mechanics, where language itself has acted as a part of tradition that limits. We cannot seem, for example, to get away from the concept of the "elementary particle," a notion that goes back 2500 years to the time of Democritus, even though the questions we raise according to that concept do not give us a "sensible answer." The expectations from that concept—the "old tradition"—still hold us.]

What is really needed is a change in fundamental concepts. We will have to abandon the philosophy of Democritos and the concept of fundamental elementary particles. We should accept instead the concept of fundamental symmetries, which is a concept out of the philosophy of Plato. Just as Copernicus and Galileo in their method abandoned the descriptive science of Aristotle and turned to the structural science of Plato, so we are probably forced in our concepts to abandon the atomic materialism of Democritos and to turn to the ideas of symmetry in the philosophy of Plato. Again we would return to a very old tradition. As I said before, such changes are extremely difficult. Even with the change many complicated details will have to be worked out, both experimentally and theoretically, in elementary particle physics; but I do not believe that there will be any spectacular breakthrough, except for this change in concepts.

After going through the three most important influences of tradition in science—those in the selection of problems, in the method, and in the concepts—I should perhaps, in conclusion, say a few words about the future development of science. Of course, I am not interested in futurology; but since we can scarcely work on other problems than those that are given to us by the historical process, we may ask where this process has led to new and interesting questions.

In physics I would like to mention astrophysics. In this field, the strange properties of the pulsars and the quasars, and perhaps also the gravitational waves, can be considered as a challenge. Then there is the new and wide field of molecular biology, where concepts of very different origin, namely, physical, chemical and biological concepts meet and produce a great wealth of interesting new problems. Finally, on the practical side, we have to solve the very urgent problems put by the deterioration of our environment. I have mentioned these points not in order to make predictions about the future, but in order to emphasize that we need not invent our problems. The scientific tradition, that is, the historical process, gives us many problems and encourages our efforts. And that is a sign for a very healthy state of affairs in science.

Questions and Topics for Discussion and Writing

1. Heisenberg's work was originally an address. Recorded here, it is an essay. An address, by its very nature, demands a more conversational tone than does an essay. Can you find evidence of conversational tone in this essay that suggests it was originally an address?

2. Because of Heisenberg's discussion of the importance of the role of tradition to scientists, he might have titled his work "Tradition and Science" instead of "Tradition in Science." What change in meaning would have been made by a change of the preposition "in" to the conjunction "and"?

3. What does Heisenberg see as the "movements" that provided a basis for astronomy?

4. A much discussed question involves "whether the man makes the time or the time makes the man." What does Heisenberg conclude about this question?

5. Heisenberg points to historical development to show that periods of intense activity in science alternate with long periods of inactivity. To what does he attribute the advance of science after such periods of inactivity?

6. If Heisenberg is considering such periods, what is his question about science in our time?

7. What is the role of education in the development of science?

8. List the motives that promote the selection of scientific problems.

9. What is the role of dialogue in scientific advance?

10. How does practical application of science serve as a check on theoretical science?

11. What are the major motives of scientists in the practical application of science?

12. Explain how Heisenberg differed from Fermi in considering the practical application of the hydrogen bomb.

13. What does Heisenberg offer as motives that determine the lines of research for the *individual* scientist?

14. Write a paragraph in which you relate the importance of the concluding paragraph to students who hope to enter scientific fields today.

Max Planck

Max Planck (1858–1947) looms as a central figure in the momentous re-evaluation of physics in the early twentieth century that paved the way for the modern development of the sciences. Nineteenth-century physics proceeded at a quiet, steady pace, confident that eventually humanity would discover and understand all the "secrets" of matter and energy.

At the turn of the century, new experiments, measuring techniques, and refined observations led scientists more and more to question hitherto accepted notions of the Newtonian universe and Euclidean geometry. The conventional scientific approach assumed that any "contradiction" to established laws could eventually be explained by harmonious applications of those laws, or it treated the contradictions as exceptions. The "new" would ultimately fit the "old." However, some scientists dared a bolder solution to these "contradictions": The old should be made to fit the new. So pervasive did this questioning and altering of the conclusions of classical physics regarding the universe become that the human race entered a period that George Gamow has called the "declassicalization" of physics. Planck, for example, applied his work on heat and physical chemistry in new ways to the problem of black-body radiation. He worked on the novel assumption that energy, rather than being given off continuously as physicists then believed, is given off in discrete bundles. In 1900, he developed one of the most fundamental of all constants, known as the Planck constant—which became a first step leading to quantum mechanics.

The "father of the quantum theory" in the minds of many, Planck first proposed the quantized character of energy as a temporary, makeshift solution to the problem of the ultraviolet catastrophe of black-body radiation, believing that eventually any contradictions would be reconciled by principles of classical physics. His quantum theory initiated revolutionary changes in the sciences, however, and brought him the Nobel Prize in 1919. Despite this recognition, he remained a conservative scientist and did not engage in the further, rapid development of quantum theory. It was Einstein, a personal friend, who extended quantum ideas to the radiation itself; others (Schrödinger and Heisenberg) developed quantum mechanics. Planck even spoke out against some of the new ideas he had helped to make possible with his theory.

Planck was born in Kiel, a part of Denmark passed to Germany in 1866 that later developed strategic significance in World Wars I and II. He received his degree in physics at age twenty-one and lectured at Munich, Kiel, and Berlin. Cultured and musically inclined, he almost followed a career in music. As a result of his firm belief in causality, his "world picture" of physics was rigorously deterministic, yet he did not extend this determinism to human action. He wrote on philosophy and religious problems as well as on scientific concerns and felt that no real opposition existed between science and religion. Highly respected in pre-Nazi Germany, he became President of the Kaiser Wilhelm Institute, which after World War II was renamed the Max Planck Institute. During Hitler's sway, Planck officially expressed resentment of Nazi treatment of the Jews, thereby enraging Hitler. Furthermore, his last son, the only surviving child, was executed for taking part in an anti-Hitler plot.

Despite Planck's Nobel Prize and his strong regard for the scientific enterprise, he retained some cynical distrust of organized science, owing, some suggest, to the little attention other scientists accorded his early work. (Even his quantum theory was initially rejected; it was Bohr's research in 1913 that helped to establish its validity.) This distrust is reflected in his *Scientific Autobiography* (1949), where he states that new ideas are accepted not because of their compelling truth but because "the opponents eventually die."

Phantom Problems in Science

Max Planck

The world is teeming with problems. Wherever man looks, some new problem crops up to meet his eye—in his home life as well as in his business or professional activity, in the realm of economics as well as in the field of technology, in the arts as well as in science. And some problems are very stubborn; they just refuse to let us in peace. Our agonized thinking of them may sometimes reach such a pitch that our thoughts haunt us throughout the day, and even rob us of sleep at night. And if by lucky chance we succeed in solving a problem, we experience a sense of deliverance, and rejoice over the enrichment of our knowledge. But it is an entirely different story, and an experience annoying as can be, to find after a long time spent in toil and effort, that the problem which has been preying on one's mind is totally incapable of any solution at all— either because there exists no indisputable method to unravel it, or because considered in the cold light of reason, it turns out to be absolutely void of all meaning—in other words, it is a *phantom problem,* and all that mental work and effort was expended on a mere nothing. There are many such phantom problems—in my opinion, far more than one would ordinarily suspect—even in the realm of science.

There is no better safeguard against such unpleasant experiences than to ascertain in each instance, and at the very outset, whether the problem under consideration is a genuine or meaningful one, and whether a solution for it is to be expected. In view of this situation I will cite and examine a number of problems, in order to see whether they happen to be mere phantom problems. By doing so, I may be able to render a genuinely useful service to some of you. My selection of these problems to be exhibited as specimens is not based on any systematic viewpoint, and even less can it lay a claim to completeness in any respect. Most of them are taken from the realm of science, because this is the field in which the relevant factors are the most clearly discernible. However, this consideration will not deter me from touching upon other fields, too, whenever I can reasonably surmise that the subject holds an interest for you.

I

In order to decide whether or not a given problem is truly meaningful, we must first of all examine closely the assumptions contained in its wording. In many instances, these alone will immediately reveal the problem under consideration to be a phantom problem. The matter is simplest when an error is lurking in the assumptions. In this case, of course, it is immaterial whether the erroneous assumption was introduced deliberately or has just escaped detection. A lucid example is the famous problem of perpetual motion, i.e., the problem of devising a periodically functioning apparatus which will perform mechanical work perpetually without any other change in nature. Since the existence of such an apparatus would contradict the principle of the conservation of energy, such an apparatus cannot possibly occur in nature, so that this problem is a phantom problem. Of course, one may raise the following argument: "The principle of the conservation of energy, after all, is an experimental law. Accordingly, although today it is considered to be universal and all-embracing, its validity may one day have to be restricted—and in fact, such a curtailment of its universal applicability has been sometimes suspected in nuclear physics—and the problem of perpetual motion would then suddenly become genuine. Its meaninglessness is, therefore, by no means absolute."

This counter-argument may actually acquire practical significance, as is demonstrated especially clearly by the example of a no less well known problem in chemistry: The ancient problem of changing base metal, for instance, mercury, into gold. Originally, prior to the birth of a scientific chemistry, this problem was considered to be pregnant with portentous meaning, and many a learned— and unlearned—mind was zealously occupied with it. But later, as the theory of chemical elements was developed and became universally accepted, the transmutation of metals turned into a phantom problem. In recent times, since the discovery of artificial radioactivity, the situation has again been reversed. The fact is that today it no longer seems to be fundamentally impossible to discover a process for removing a proton from the nucleus of the mercury atom and an electron from its shell. This operation would change the mercury atom into a gold atom. Therefore, at the present stage of science, the ancient quest of the alchemists no longer belongs to the class of phantom problems.

However, these examples must by no means be construed as indicating that the meaninglessness of a phantom problem is never absolute, but simply dependent on whether or not a certain theory is accepted as valid. There are also many phantom problems which are indubitably doomed to remain such forever. One of these, for instance, is the problem which used to keep many a great physicist busy for many years: the study of the mechanical properties of the luminiferous ether. The meaninglessness of this problem follows from its basic premise, which postulates that light vibrations are of a mechanical nature. This premise is erroneous, and must so remain forever.

Here is another example, taken from the field of physiology: It is a well-known fact that the convex lens of the human eye projects an inverted image on the retina. When we see a tower, its image appears on the retina with the top of the tower pointing downward. When this phenomenon was established, a number of scientists tried to detect in the human organ of sight that particular mechanism which supposedly re-inverts the image on the retina. This is a phantom problem, and never can be anything else, for it is based on an erroneous premise, for which there can be no possible proof—namely, that in the organ of vision the image of an object must be upright rather than inverted.

Far more difficult than those cases in which, as in the examples just cited, the assumptions are mistaken, are problems whose presuppositions contain no error, but are so vaguely worded that they must remain phantom ones because they are inadequately formulated. Yet, it so happens that it is just such cases with which we shall be chiefly preoccupied.

My first example is a phantom problem, for the triviality of which I beg your forgiveness. The room in which we now sit, has two side walls, a right-hand one and a left-hand one. To you, *this* is the right side, to me, sitting facing you, *that* is the right side. The problem is: Which side is in reality the right-hand one? I know that this question sounds ridiculous, yet I dare call it typical of an entire host of problems which have been, and in part still are, the subject of earnest and learned debates, except that the situation is not always quite so clear. It demonstrates, right at the very outset, what great caution must be exercised in using the word, *real*. In many instances, the word has any sense at all only when the speaker first defines clearly the point of view on which his considerations are based. Otherwise, the words, *real* or *reality*, are often empty and misleading.

Another example: I see a star shining in the sky. What is *real* in it? Is it the glowing substance, of which it is composed, or is it the sensation of light in my eyes? The question is meaningless so long as I do not state whether I am assuming a realistic or a positivistic point of view.

Still another example, this one from the realm of modern physics: When the behavior of a moving electron is studied through an electron microscope, the electron appears as a particle following a definite course. But when the electron is made to pass through a crystal, the image projected on the screen shows every characteristic of a refracted light wave. The question, whether the electron is in reality a particle, occupying a certain position in space at a certain time, or a wave, filling all of infinite space, will therefore constitute a phantom problem so long as we fail to stipulate which of the two viewpoints is applied in the study of the behavior of the electron.

The famous controversy between Newton's emission theory and Huygens' wave theory of light is also a phantom problem of science. For every decision for or against either of these two opposing theories will be a completely arbitrary one, depending on whether one accepts the point of view of the quantum theory or that of the classical theory.

II

In every one of the cases cited till now, we encountered a rather simple, easily appreciable situation. Now let us proceed to the consideration of a problem which has always been regarded as one of central importance because of its meaning to human life—the famous body-mind problem. In this case, first of all we must try to ascertain the meaning of our problem. For there are philosophers who claim that mental processes need not be accompanied by physical processes at all, but can take place totally independently from the latter. If this view is right, mental processes are subject to entirely different laws than those applying to physical processes. If so, then, the body-mind problems splits into two separate problems—the body problem and the mind problem—thus losing its meaning, and degenerating into a phantom problem. With this finding, the case may be considered as good as closed, and we need only concern ourselves with the reciprocal interaction of mental and physical processes. Experience shows that they are very closely influenced by each other. For instance, somebody asks me a question. His question is introduced by a physical process, the propagation of the sound waves of the spoken words which, emitted by him, hit my ears and are transmitted to my brain through the sensory nerve paths. They then cause mental processes to take place in my brain, namely, a reflection on the meaning of the words perceived, followed by a decision as to the content of the answer to be given. Then another physical process operates my motor nerves and my larynx, to transmit the answer to the questioner by means of the physical process of propagating sound waves through the air.

Now then, what is the nature of the interrelation of physical and mental processes? Are mental processes caused by physical ones? And if so, according to what laws? How can something material act on something immaterial, and *vice versa?* All these questions are difficult to answer. If we assume the existence of a causal interaction, a cause-and-effect relationship, between physical and mental processes, a continued, unrestricted applicability of the principle of the conservation of energy appears to be an indispensable premise. For one will not be disposed to sacrifice this universal foundation of exact science. But in that case, there would have to exist a numerically definite mechanical equivalent of psychic processes, as there is a definite equivalent of heat in thermodynamics, and there would be absolutely no method for measuring such a constant. For this reason, a solution has been attempted on the basis of the hypothesis that the mental forces contribute no perceptible energy to the physical processes, but act merely to liberate the latter, as a gentle breeze will start something that will grow into a mighty avalanche, or a tiny spark will blow up a huge powder magazine. However, this hypothesis does not solve the difficulty completely. Because in every case known to us, while the amount of energy expanded in liberating a process is very small in comparison with the energy released, yet it does exist, even though it may have just a microscopic magnitude. Even the very gentlest breeze and very tiniest spark possess an energy above zero—and this is what matters here.

However, it is well known that there are some forces which produce a perceptible effect without any expenditure whatever of energy. These are what we may call "guiding forces," such as, for instance, the resistance due to the rigidity of railroad rails which forces the wheels of a train to follow a pre-determined curve, without any expenditure of energy. An attempt might be made to ascribe a similar role to the mental forces in the guiding of physical processes along pre-determined paths in the human brain. But this, too, involves grave and insurmountable difficulties. For the modern science of brain physiology is based on the very premise that it is possible to achieve a satisfactory understanding of the laws of biological processes without postulating the intervention of any particular mental force. Such a hypothesis avoids also the theory of parallelism which, in contrast to the theory of interaction, assumes that mental and physical processes must, necessarily, run side by side, each according to its own laws, without interfering with each other. Of course, it still remains incomprehensible just how this reciprocal interdependence of two such fundamentally different occurrences is to be conceived, and whether it perhaps requires the assumption of some form of pre-established harmony. In this respect, the theory of parallelism, too, is hardly satisfactory.

And now, in order to get to the bottom of the matter, let us ask ourselves this question of basic significance: Just what do we know about mental processes? In what circumstances and in what sense may we speak of mental processes? Let us consider first where we come across mental processes in this world. We must take it for granted that members of the higher animal kingdom as well as human beings have emotions and sensations. But as we descend to the lower animals— where is the borderline where sensation ceases to exist? Has a worm any sensation of pain as it is crushed under our feet? And may plants be considered capable of some kind of sensation? There are botanists who are disposed to answer this question affirmatively. But such a theory can never be put to the test, let alone proved, and the wisest course seems to be not to venture any opinion in this regard. Along the entire ladder of evolution, from the lowest order of life up to Man, there is no point at which one can establish a discontinuity in the nature of mental processes.

It is nevertheless possible to specify a quite definite borderline, of decisive importance for all that follows. This is the borderline between the mental processes within other individuals and the mental processes within one's own Ego. For everybody experiences his own emotions and sensations directly. They just simply exist for him. But we do not experience directly the sensations of any other individuals, however certain their existence may be, and we can only infer them in analogy to our own sensations. To be sure, there are physicians who solemnly claim to be able to perceive the emotions and moods of their patients no less clearly than the latter themselves. But such a claim can never be proved indisputably. Its questionability becomes most striking if we think of certain specific instances. Even the most sensitive dentist cannot feel the piercing pains which his patient at times has to suffer under his treatment. He can ascertain them only indirectly, on the basis of the moans or squirming of the patient. Or, to speak of a more pleasant situation, such as for instance a banquet, however clearly one may sense the pleasure of one's neighbor over the taste of his favorite wine, it is something quite different from tasting it on one's own tongue. What *you* feel, think, want, only *you* can know as first-hand information. Other people can conclude it only indirectly, from your words, conduct, actions and mannerisms. When such physical manifestations are entirely absent, they have no basis whatever to enable them to know your momentary mental state.

This contrast between first-hand, or direct, and second-hand, or indirect, experience is a fundamental one. Since our primary aim is to gain direct, first-hand experience, we shall now discuss the interrelation of our mental and physical states.

First of all, we find that we may speak of conscious states only. To be sure, many processes, perhaps even the most decisive ones, must be taking place in the subconscious mind. But these are beyond the reach of scientific analysis. For there exists no science of the unconscious, or subconscious, mind. It would be a contradiction in terms, a self-contradiction. One does not know that which is subconscious. Therefore, all problems concerning the subconscious are phantom problems.

Let us therefore take a simple conscious process involving body and mind. I prick my hand with a needle, and feel a sensation of pain. The wound made by the pin is the physical element, the sensation of pain is the mental element of the process. The wound is seen, the pain is felt. Is there, then, an indisputable method of throwing light upon the interrelation of the two elements of this process? It is easy to realize that this is absolutely impossible. For there is nothing here upon which light is to be thrown. The visual perception of the wound and the feeling of the pain are elementary facts of experience, but they are as different in nature as knowledge and feeling. Therefore, the question as to their essential interrelation represents no meaningful problem—it is just a phantom problem.

It is obvious that the two occurrences, the pin-prick and the sensation of pain, can be examined and analyzed most thoroughly, in every detail. But such an analysis calls for two different methods, which mutually preclude each other. Each of the two corresponds to one of two different viewpoints.

Therefore, it will do no harm to say that the physical and the mental are in no way different from each other. They are the selfsame processes, only viewed from two diametrically opposite directions. This statement is the answer to the riddle, which has been inseparable from the theory of parallelism, namely, how one is to conceive the fact that two types of processes so different from each other as the physical and the mental, are so closely interlinked. The link has now been disclosed. At the same time, the body-mind problem has been recognized as another phantom problem.

Questions and Topics for Discussion and Writing

1. What, according to Planck, is a "phantom problem"? Why must scientists—and especially students considering careers as scientists—recognize the existence of "phantom problems"?

2. Study Planck's opening sentence. What rhetorical devices does he use to develop his controlling idea?
3. Consider Planck's statement that "the world is teeming with problems." Focusing on one of those problems, write an expository essay. Based on your readings, develop it with examples and illustrations (for example, nuclear waste, foreign policy, terrorism).
4. Why does Planck point particularly to the field of science for examples of phantom problems?
5. Why is a study of logic especially helpful to scientists?
6. What is the problem of language that faces scientists after assumptions are correctly made?
7. How does the first sentence of Paragraph 7 serve to illustrate and reinforce Planck's insistence on the need for clear and precise language?
8. List the examples of phantom problems used by Planck.
9. Evaluate Planck's argument that the unconscious is "beyond the reach of scientific analysis." Would you consider this statement a renunciation of psychology as a science?

C. P. Snow

Sir Charles Percy Snow was an English physicist and novelist. Born into a poor family, Snow (1905–1980) showed a bent for science and went on to a doctorate in physics from Cambridge University. His work for the British government during World War II occasioned his recognition and secured the knighthood he received in 1957.

While still pursuing his scientific career, Snow turned his talents toward fiction, using the setting of the scientific world and his experiences as an influential civil servant for many of his works. These special kinds of knowledge underlie the recurrent themes in Snow's works, which often deal with the problems and temptations of personal and professional power. *The Search* is an outstanding example of such a novel, portraying the excitement of a young scientist working in crystallography. Snow's powers of observation serve him in the craft of fiction as well as in the work of science.

In 1959, while delivering the Rede Lecture at Cambridge, Snow introduced the term *Two Cultures*, which has gained wide acceptance as representing the lack of communication between specialized groups in our modern society, particularly the scientists and the literary intellectuals. Actually, Snow was reviving an old controversy. The Victorians Thomas Henry Huxley and Matthew Arnold had engaged in an intellectual dispute over the roles of science versus literature and versus religion. Furthermore, Alfred North Whitehead, in *Science and the Modern World* (1927), had recognized this division and attempted a reconciliation. Somehow Snow's term and ideas caught the popular imagination, however, especially after F. R. Leavis, a professor and editor, challenged his views soon after Snow's lecture was published as *The Two Cultures and the Scientific Revolution.*

Snow has written perceptively on the esthetic joys of science, its moral responsibilities, and its possibilities. The excerpt that follows presents his thesis that society's intellectuals are divided, prevented from speaking to each other by the lack both of a common language and the motivation to understand the orientation of the other "culture."

The Two Cultures

C. P. Snow

I. THE TWO CULTURES

It is about three years since I made a sketch in print of a problem which had been on my mind for some time. It was a problem I could not avoid just because of the circumstances of my life. The only credentials I had to ruminate on the subject at all came through those circumstances, through nothing more than a set of chances. Anyone with similar experience would have seen much the same things and I think made very much the same comments about them. It just happened to be an unusual experience. By training I was a scientist: by vocation I was a writer. That was all. It was a piece of luck, if you like, that arose through coming from a poor home.

But my personal history isn't the point now. All that I need say is that I came to Cambridge and did a bit of research here at a time of major scientific activity. I was privileged to have a ringside view of one of the most wonderful creative periods in all physics. And it happened through the flukes of war—including meeting W. L. Bragg in the buffet on Kettering station on a very cold morning in 1939, which had a determining influence on my practical life—that I was able, and indeed morally forced, to keep that ringside view ever since. So for thirty years I have had to be in touch with scientists not only out of curiosity, but as part of a working existence. During the same thirty years I was trying to shape the books I wanted to write, which in due course took me among writers.

There have been plenty of days when I have spent the working hours with scientists and then gone off at night with some literary colleagues. I mean that literally. I have had, of course, intimate friends among both scientists and writers. It was through living among these groups and much more, I think, through moving regularly from one to the other and back again that I got occupied with the problem of what, long before I put it on paper, I christened to myself as the "two cultures." For constantly I felt I was moving among two groups—comparable in intelligence, identical in race, not grossly different in social origin, earning about the same incomes, who had almost ceased to communicate at all, who in intellectual, moral and psychological climate had little in common. . . .

I believe the intellectual life of the whole of Western society is increasingly being split into two polar groups. When I say the intellectual life, I mean to include also a large part of our practical life, because I should be the last person to suggest the two can at the deepest level be distinguished. I shall come back to the practical life a little later. Two polar groups: at one pole we have the literary intellectuals, who incidentally while no one was looking took to referring to themselves as "intellectuals" as though there were no others. I remember G. H. Hardy once remarking to me in mild puzzlement, some time in the 1930s: "Have you noticed how the word 'intellectual' is used nowadays? There seems to be a new definition which certainly doesn't include Rutherford or Eddington or Dirac or Adrian or me. It does seem rather odd, don't y' know."

Literary intellectuals at one pole—at the other scientists, and as the most representative, the physical scientists. Between the two a gulf of mutual incomprehension—sometimes (particularly among the young) hostility and dislike, but most of all lack of understanding. They have a curious distorted image of each other. Their attitudes are so different that, even on the level of emotion, they can't find much common ground. Non-scientists tend to think of scientists as brash and boastful. They hear Mr T. S. Eliot, who just for these illustrations we can take as an archetypal figure, saying about his attempts to revive verse-drama that we can hope for very little, but that he would feel content if he and his co-workers could prepare the ground for a new Kyd or a new Greene. That is the tone, restricted and constrained, with which literary intellectuals are at home: it is the subdued voice of their culture. Then they hear a much louder voice, that of another archetypal figure, Rutherford, trumpeting: "This is the heroic age of science! This is the Elizabethan age!" Many of us heard that, and a good many other statements beside which that was mild; and we weren't left in any doubt whom Rutherford was casting for the role of Shakespeare. What is hard for the literary intellectuals to understand, imaginatively or intellectually, is that he was absolutely right.

And compare "this is the way the world ends, not with a bang but a whimper"—incidentally, one of the least likely scientific prophecies ever made—compare that with Rutherford's famous repartee, "Lucky fellow, Rutherford, always on the crest of the wave." "Well, I made the wave, didn't I?"

The non-scientists have a rooted impression that the scientists are shallowly optimistic, unaware of man's condition. On the other hand, the scientists believe that the literary intellectuals are totally lacking in foresight, peculiarly unconcerned with their brother men, in a deep sense anti-intellectual, anxious to restrict both art and thought to the existential moment. And so on. Anyone with a mild talent for invective could produce plenty of this kind of subterranean back-chat. On each side there is some of it which is not entirely baseless. It is all destructive. Much of it rests on misinterpretations which are dangerous. . . .

The number 2 is a very dangerous number: that is why the dialectic is a dangerous process. Attempts to divide anything into two ought to be regarded with much suspicion. I have thought a long time about going in for further refinements: but in the end I have decided against. I was searching for something a little more than a dashing metaphor, a good deal less than a cultural map: and for those purposes the two cultures is about right, and subtilising any more would bring more disadvantages than it's worth.

At one pole, the scientific culture really is a culture, not only in an intellectual but also in an anthropological sense. That is, its members need not, and of course often do not, always completely understand each other; biologists more often than not will have a pretty hazy idea of contemporary physics; but there are common attitudes, common standards and patterns of behaviour, common approaches and assumptions. This goes surprisingly wide and deep. It cuts across other mental patterns, such as those of religion or politics or class.

Statistically, I suppose slightly more scientists are in religious terms unbelievers, compared with the rest of the intellectual world—though there are plenty who are religious, and that seems to be increasingly so among the young. Statistically also, slightly more scientists are on the Left in open politics—though again, plenty always have called themselves conservatives, and that also seems to be more common among the young. Compared with the rest of the intellectual world, considerably more scientists in this country and probably in the U.S. come from poor families. . . . Yet over a whole range of thought and behaviour, none of that matters very much. In their working, and in much of their emotional life, their attitudes are closer to other scientists than to non-scientists who in religion or politics or class have the same labels as themselves. If I were to risk a piece of shorthand, I should say that naturally they had the future in their bones. . . .

At the other pole, the spread of attitudes is wider. It is obvious that between the two, as one moves through intellectual society from the physicists to the literary intellectuals, there are all kinds of tones of feeling on the way. But I believe the pole of total incomprehension of science radiates its influence on all the rest. That total incomprehension gives, much more pervasively than we realise, living in it, an unscientific flavour to the whole "traditional" culture, and that unscientific flavour is often, much more than we admit, on the point of turning anti-scientific. The feelings of one pole become the anti-feelings of the other. If the scientists have the future in their bones, then the traditional culture responds by wishing the future did not exist. . . . It is the traditional culture, to an extent remarkably little diminished by the emergence of the scientific one, which manages the Western world.

This polarisation is sheer loss to us all. To us as people, and to our society. It is at the same time practical and intellectual and creative loss, and I repeat that it is false to imagine that those three considerations are clearly separable. . . .

As one would expect, some of the very best scientists had and have plenty of energy and interest to spare, and we came across several who had read everything that literary people talk about. But that's very rare. Most of the rest, when one tried to probe for what books they had read, would modestly confess, "Well, I've *tried* a bit of Dickens," rather as though Dickens were an extraordinarily esoteric, tangled and dubiously rewarding writer, something like Rainer Maria Rilke. In fact that is exactly how they do regard him: we thought that discovery, that Dickens had been transformed into the type-specimen of literary incomprehensibility, was one of the oddest results of the whole exercise.

But of course, in reading him, in reading almost any writer whom we should value, they are just touching their caps to the traditional culture. They have their own culture, intensive, rigorous, and constantly in action. This culture contains a great deal of argument, usually much more rigorous, and almost always at a higher conceptual level, than literary persons' arguments; even though the scientists do cheerfully use words in senses which literary persons don't recognise, the senses are exact ones, and when they talk about "subjective," "objective," "philosophy" or "progressive" . . . they know what they mean, even though it isn't what one is accustomed to expect.

Remember, these are very intelligent men. Their culture is in many ways an exacting and admirable one. It doesn't contain much art, with the exception, an important exception, of music. Verbal exchange, insistent argument. Long-playing records. Colour-photography. The ear, to some extent the eye. Books, very little, though perhaps not many would go so far as one hero, who perhaps I should admit was further down the scientific ladder than the people I've been talking about—who, when asked what books he read, replied firmly and confidently: "Books? I prefer to use my books as tools." It was very hard not to let the mind wander—what sort of tool would a book make? Perhaps a hammer? A primitive digging instrument?

Of books, though, very little. And of the books which to most literary persons are bread and butter, novels, history, poetry, plays, almost nothing at all. It isn't that they're not interested in the psychological or moral or social life. In the social life, they certainly are, more than most of us. In the moral, they are by and large the soundest group of intellectuals we have; there is a moral component right in the grain of science itself, and almost all scientists form their own judgments of the moral life. In the psychological they have as much interest as most of us, though occasionally I fancy they come to it rather late. It isn't that they lack the interests. It is much more that the whole literature of the traditional culture doesn't seem to them relevant to those interests. They are, of course, dead wrong. As a result, their imaginative understanding is less than it could be. They are self-impoverished.

But what about the other side? They are impoverished too—perhaps more seriously, because they are vainer about it. They still like to pretend that the traditional culture is the whole of "culture," as though the natural order didn't exist. As though the exploration of the natural order was of no interest in either its own value or its consequences. As though the scientific edifice of the physical world was not, in its intellectual depth, complexity and articulation, the most beautiful and wonderful collective work of the mind of man. Yet most non-scientists have no conception of that edifice at all. Even if they want to have it, they can't. It is rather as though, over an immense range of intellectual experience, a whole group was tone-deaf. Except that this tone-deafness doesn't come by nature, but by training, or rather the absence of training.

As with the tone-deaf, they don't know what they miss. They give a pitying chuckle at the news of scientists who have never read a major work of English literature. They dismiss them as ignorant specialists. Yet their own ignorance and their own specialisation is just as startling. A good many times I have been present at gatherings of people who, by the standards of the traditional culture, are thought highly educated and who have with considerable gusto been expressing their incredulity at the illiteracy of scientists. Once or twice I have been provoked and have asked the company how many of them could describe the Second Law of Thermodynamics. The response was cold: it was also negative. Yet I was asking something which is about the scientific equivalent of: *Have you read a work of Shakespeare's?*

I now believe that if I had asked an even simpler question—such as, What do you mean by mass, or acceleration, which is the scientific equivalent of saying, *Can you read?*—not more than one in ten of the highly educated would have felt that I was speaking the same language. So the great edifice of modern physics goes up, and the majority of the cleverest people in the Western world have about as much insight into it as their neolithic ancestors would have had. . . .

There seems then to be no place where the cultures meet. I am not going to waste time saying that this is a pity. It is much worse than that. Soon I shall come to some practical consequences. But at the heart of thought and creation we are letting some of our best chances go by default. The clashing point of two subjects, two disciplines, two cultures—of two galaxies, so far as that goes—ought to produce creative chances. In the history of mental activity that has been where some of the break-throughs came. The chances are there now. But they are there, as it were, in a vacuum, because those in the two cultures can't talk to each other. It is bizarre how very little of twentieth-century science has been assimilated into twentieth-century art. Now and then one used to find poets conscientiously using scientific expressions, and getting them wrong—there was a time when "refraction" kept cropping up in verse in a mystifying fashion, and when "polarised light" was used as though writers were under the illusion that it was a specially admirable kind of light.

Of course, that isn't the way that science could be any good to art. It has got to be assimilated along with, and as part and parcel of, the whole of our mental experience, and used as naturally as the rest.

This cultural divide is not just an English phenomenon: it exists all over the Western world.

Questions and Topics for Discussion and Writing

1. How does Snow define *culture*? What are the two cultures of which Snow speaks?
2. Why is Snow particularly qualified to discuss the polarity between the two cultures?
3. Snow says that he "was privileged to have a ringside view of one of the most wonderful creative periods in all physics." He adds that he was "morally forced" to keep his ringside seat. How do you interpret the last statement?
4. How does the humorous account of the lack of conversation when an Oxford don came to Cambridge help to call attention to the problem Snow sees? How does this account prepare readers for the gravity of his thesis: "I believe the intellectual life of the whole of Western society is increasingly being split into two polar groups"?
5. Write a short paragraph restating what Snow says are errors in judgment made by nonscientists and scientists about each other.
6. Snow says that both groups hold a "tragic" view of life. Explain what such a view entails. What is the "moral trap" that accompanies the insight into such a tragic view of life?
7. What, according to Snow, is the judgment of most of the famous twentieth-century writers about scientists?
8. What does Snow diagnose as a fault of scientists in various fields?
9. What are the kinds of "tones of feeling" that are on the point of turning the Western world "antiscientific"? Write an essay presenting your views on this subject.
10. Although the scientists' culture doesn't include much art, it does include one form. What is that form?
11. Snow says that "it is bizarre how very little of twentieth-century science has been assimilated into twentieth-century art." Yet, he asserts, science can aid art in two ways. What are these two ways? Write an essay explaining one way science can aid art—for example, in authenticating works and detecting forgeries. Consult your library.
12. In what way are scientists "self-impoverished"?
13. Snow creates a visual metaphor of facial expressions to show the condition of separation between scientists and nonscientists at the present time. How effective is this metaphor in summarizing his thesis?
14. What does Snow see to be a way out of the present dilemma?

SUGGESTIONS FOR FURTHER READING

Barzun, Jacques. *Science: The Glorious Entertainment.* New York: Harper & Row, 1964.

Bernal, John Desmond. *The Social Function of Science.* Cambridge: MIT Press, 1967.

Boorstin, Daniel J. *The Discoverers.* New York: Random House, 1983.

Bronowski, Jacob. *Science and Human Values.* New York: Harper, 1959.

Conant, James B. *Science and Common Sense.* New Haven: Yale University Press, 1951.

Finocchiaro, Maurice A. *History of Science as Explanation.* Detroit: Wayne State University Press, 1973.

Kuhn, Thomas S. *The Essential Tension*. Chicago: University of Chicago Press, 1977.
————. *The Structure of Scientific Revolutions*. Chicago: University of Chicago Press, 1970.
Oppenheimer, Robert. *Science and the Common Understanding*. New York: Simon & Schuster, 1954.
Ortega y Gasset, José. *The Modern Theme*. New York: Harper, 1961.
Ravetz, Jerome R. *Scientific Knowledge and Its Social Problems*. Oxford: Clarendon Press, 1971.
Reichenbach, Hans. *The Rise of Scientific Philosophy*. Berkeley: University of California Press, 1951.
Smithsonian Treasury of Twentieth Century Science. Ed. Webster Prentiss True. New York: Simon & Schuster, 1966.

PART TWO

The Disciplined Approach to Knowledge

Unit 4

MATHEMATICS: LANGUAGE IN THE SEARCH FOR CERTAINTY

G. H. Hardy

"A mathematician, like a painter or a poet, is a maker of patterns. If his patterns are more permanent than theirs, it is because they are made with *ideas*," asserts the British mathematician Godfrey Harold Hardy. The most notable pure mathematician of his time, Hardy (1877–1947) was born into a poor but enlightened family, both of his parents being mathematically minded. After showing himself precociously skillful with numbers, he progressed to Cambridge University, where he distinguished himself as a student and lecturer. Hardy was prodigious in publication, making important discoveries in every branch of mathematical analysis and eventually producing, alone or in collaboration, over 350 papers.

Many mathematicians attest to the positive influence that Hardy, as part of the "galaxy of talent" at Cambridge, exerted in attracting and helping to develop undergraduates. Norbert Wiener, for one, was Hardy's student. Hardy lists as personally significant events his collaboration with J. E. Littlewood, a colleague without parallel for fruitfulness in mathematics, and his discovery of the self-taught genius Srinivasa Ramanujan. Hardy removed him from a clerk's job in Madras, India, to Cambridge, where, after three years of spectacular accomplishment, the young protégé met an untimely death.

As Hardy's autobiography reveals, and as biographers agree, Hardy was dominated by his passion for pure mathematics. In a letter describing the logician Ludwig Wittgenstein, Hardy's fellow lecturer at Cambridge, Bertrand Russell wrote to a friend: "[Ludwig Wittgenstein] is the only man I have ever met with a real taste for philosophical scepticism; he is glad when it is *proved* that something can't be known. I told Hardy this, and Hardy said he himself would be glad to *prove* anything: 'If I could prove by logic that you would die in 5 minutes, I should be sorry you were going to die, but my sorrow would be very much mitigated by pleasure in the proof.'"

Despite the tenor of this amusing incident, Hardy proved himself a loyal friend to Russell. In 1916 Russell had been dismissed from his Cambridge lectureship for antiwar activities that had brought him into conflict with the British government. In 1919 a group of his fellows at Cambridge who had served in the armed forces during the war petitioned to have Russell's lectureship restored. Among them was Hardy; for various reasons, they were unsuccessful. Then, in the 1940s, Hardy wrote *Bertrand Russell at Trinity*, giving a full account of Russell's dismissal and concluding that a wrong remained to be righted; the book helped clear the way for Russell's return.

Hardy received a number of awards, including his election as a Fellow of the Royal Society in 1910. Of his more than dozen technical books, one is a popular account, *A Mathematician's Apology*, in which he presents his own eloquent defense of the art of mathematics.

A Mathematician's Apology

G. H. Hardy

There are many highly respectable motives which may lead men to prosecute research, but three which are much more important than the rest. The first (without which the rest must come to nothing) is intellectual curiosity, desire to know the truth. Then, professional pride, anxiety to be satisfied with one's performance, the shame that overcomes any self-respecting craftsman when his work is unworthy of his talent. Finally, ambition, desire for reputation, and the position, even the power or the money, which it brings. It may be fine to feel, when you have done your work, that you have added to the happiness or alleviated the sufferings of others, but that will not be why you did it. So if a mathematician, or a chemist, or even a physiologist, were to tell me that the driving force in his work had been the desire to benefit humanity, then I should not believe him (nor should I think the better of him if I did). His dominant motives have been those which I have stated, and in which, surely, there is nothing of which any decent man need be ashamed.

If intellectual curiosity, professional pride, and ambition are the dominant incentives to research, then assuredly no one has a fairer chance of gratifying them than a mathematician. His subject is the most curious of all—there is none in which truth plays such odd pranks. It has the most elaborate and the most fascinating technique, and gives unrivalled openings for the display of sheer professional skill. Finally, as history proves abundantly, mathematical achievement, whatever its intrinsic worth, is the most enduring of all.

We can see this even in semi-historic civilizations. The Babylonian and Assyrian civilizations have perished; Hammurabi, Sargon, and Nebuchadnezzar are empty names; yet Babylonian mathematics is still interesting, and the Babylonian scale of 60 is still used in astronomy. But of course the crucial case is that of the Greeks.

The Greeks were the first mathematicians who are still "real" to us today. Oriental mathematics may be an interesting curiosity, but Greek mathematics is the real thing. The Greeks first spoke a language which modern mathematicians can understand; as Littlewood said to me once, they are not clever schoolboys or "scholarship candidates," but "Fellows of another college." So Greek mathematics is "permanent," more permanent even than Greek literature. Archimedes will be remembered when Aeschylus is forgotten, because languages die and mathematical ideas do not. "Immortality" may be a silly word, but probably a mathematician has the best chance of whatever it may mean.

A mathematician, like a painter or a poet, is a maker of patterns. If his patterns are more permanent than theirs, it is because they are made with *ideas*. A painter makes patterns with shapes and colours, a poet with words. A painting may embody an "idea," but the idea is usually commonplace and unimportant. In poetry, ideas count for a good deal more; but, as Housman insisted, the importance of ideas in poetry is habitually exaggerated: "I cannot satisfy myself that there are any such things as poetical ideas. . . . Poetry is not the thing said but a way of saying it."

Not all the water in the rough rude sea
Can wash the balm from an anointed King.

Could lines be better, and could ideas be at once more trite and more false? The poverty of the ideas seems hardly to affect the beauty of the verbal pattern. A mathematician, on the other hand, has no material to work with but ideas, and so his patterns are likely to last longer, since ideas wear less with time than words.

The mathematician's patterns, like the painter's or the poet's, must be *beautiful*; the ideas, like the colours or the words, must fit together in a harmonious way. Beauty is the first test: there is no permanent place in the world for ugly mathematics. And here I must deal with a misconception which is still widespread (though probably much less so now than it was twenty years ago), what Whitehead has called the "literary superstition" that love of and aesthetic appreciation of mathematics is "a monomania confined to a few eccentrics in each generation."

It would be difficult now to find an educated man quite insensitive to the aesthetic appeal of mathematics. It may be very hard to *define* mathematical beauty, but that is just as true of beauty of any kind—we may not know quite what we mean by a beautiful poem, but that does not prevent us from recognizing one when we read it. Even Professor Hogben, who is out to minimize at all costs the importance of the aesthetic element in mathematics, does not venture to deny its reality. "There are, to be sure, individuals for whom mathematics exercises a coldly impersonal attraction. . . . The aesthetic appeal of mathematics may be very real for a chosen few." But they are "few," he suggests, and they feel "coldly" (and are really rather ridiculous people, who live in silly little university towns sheltered from the fresh breezes of the wide open spaces). In this he is merely echoing Whitehead's "literary superstition."

The fact is that there are few more "popular" subjects than mathematics. Most people have some appreciation of mathematics, just as most people can enjoy a pleasant tune; and there are probably more people really interested in mathematics than in music. Appearances may suggest the contrary, but there are easy explanations. Music can be used to stimulate mass emotion, while mathematics cannot; and musical incapacity is recognized (no doubt rightly) as mildly discreditable, whereas most people are so frightened of the name of mathematics that they are ready, quite unaffectedly, to exaggerate their own mathematical stupidity.

A very little reflection is enough to expose the absurdity of the "literary superstition." There are masses of chess-players in every civilized country—in Russia, almost the whole educated population; and every chess-player can recognize and appreciate a "beautiful" game or problem. Yet a chess problem is *simply* an exercise in pure mathematics (a game not entirely, since psychology also plays a part), and everyone who calls a problem "beautiful" is applauding mathematical beauty, even if it is beauty of a comparatively lowly kind. Chess problems are the hymn-tunes of mathematics.

We may learn the same lesson, at a lower level but for a wider public, from bridge, or descending further, from the puzzle columns of the popular newspapers. Nearly all their immense popularity is a tribute to the drawing power of rudimentary mathematics, and the better makers of puzzles, such as Dudeney or "Caliban," use very little else. They know their business; what the public wants is a little intellectual "kick," and nothing else has quite the kick of mathematics.

Questions and Topics for Discussion and Writing

1. Do you agree with Hardy's esthetic standard for the value of mathematics: "Beauty is the first test: there is no permanent place in the world for ugly mathematics"? What do you think characterizes beauty in mathematical endeavors? Discuss in a short paper.
2. The concept of power, for various reasons—cultural, historical, religious—has a bad name, especially in America. But Hardy says that power may be one of the "highly respectable motives which may lead men to prosecute research." Do you agree with Hardy that the desire for power may be one of the motives that lead to research? Do you agree that in it "there is nothing of which any decent man need be ashamed"? Why or why not?
3. Hardy speaks of mathematics as a "language." Discuss.
4. What do you understand to be the meaning of the metaphor that Greek mathematicians are, for modern mathematicians, "Fellows of another college"?
5. Do you agree with Hardy that "a mathematician has the best chance of whatever [immortality] may mean"? Write a paragraph discussing this idea.
6. What do you think Hardy means by the metaphor "Chess problems are the hymn-tunes of mathematics"?

Norbert Wiener

The American mathematician Norbert Wiener (1894–1964) entered academic life at what he calls "a rather unusual angle." Born in Missouri of Jewish parentage, Wiener suffered the "stresses and strains" of having been an infant prodigy. Brought up in a disciplined, intellectual atmosphere dominated by his father, Leo Wiener, a professor of Slavic languages and literature at Harvard University, Wiener early in life developed behavior to please his autocratic father.

Wiener later wrote that this striving tended to "isolate me from the world and to give me a certain aggressive, unlovable naiveté." He learned to read at three, and almost immediately "plunged into scientific reading of the most varied character." At eighteen, with his doctorate and a traveling fellowship from Harvard in hand, Wiener visited Europe's mathematical centers. At Cambridge, Bertrand Russell became the young Wiener's mentor. Calling his student "the infant phenomenon," Russell later confided, "The youth has been flattered . . . there is a perpetual contest between him and me." Another influence on Wiener's work was G. H. Hardy, whom he would later recall as "perhaps the greatest figure of his mathematical generation in England."

Wiener began his forty-one-year tenure at Massachusetts Institute of Technology in 1919, when mathematics departments were merely service departments to the "main center of life," engineering. He made occasional return visits to Cambridge. In 1939, Wiener put his mathematical talents at the disposal of a group of military mathematicians working on artillery weapons.

Wiener's thinking and the resulting vocabulary made a great impact on twentieth-century mathematics: He coined the term *cybernetics* (from the Greek word for steersman, *kubernetes*) to express "the art and science of control and communication in the animal and the machine." From the beginning an interdisciplinary concept, cybernetics was seen by its originators as "*trans*disciplinary." This unifying science brought into common parlance such terms as "feedback," "input," "filtering," "output," and "information control." In his book *Cybernetics*, Wiener synthesized his ideas: "These new concepts of communication and control involved a new interpretation of man, of man's knowledge of the universe, and of society." Sociology, anthropology, economics—all of these fields, he theorized, share in the general ideology of cybernetics. Further, he believed that cybernetics would affect the philosophy of science itself, particularly scientific method and epistemology, or the theory of knowledge. The gaps in his own knowledge, he believed, he filled by the act of writing *Cybernetics*—an immediate popular success.

Wiener's other writings include *God and Golem, Inc.; Nonlinear Problems in Random Theory; Extrapolation, Interpolation, and Smoothing of Stationary Time Series;* and two autobiographies.

"On the Nature of Mathematical Thinking" is an exposition of matters of mathematics and mathematical creation. Addressing a world that conceived of mathematics as "truth arrived at by deduction," Wiener concludes that "mathematics is an experimental science."

On the Nature of Mathematical Thinking

Norbert Wiener

If you divide the various sciences and learned disciplines in accordance with their subject matter, you will find the first and deepest line of cleavage between mathematics on the one side and the whole remaining body of human knowledge on the other. In what concerns its proved results, mathematics stands alone in its qualities of rigour, logical concatenation, precision, and conclusiveness. When to these marks of the subject matter of mathematics is conjoined the marvellous perfection of form possessed by that most familiar of all mathematical sciences, the geometry of the Greeks, it is entirely natural that the technique involved in obtaining such results and the mental processes of the mathematician should seem to the layman awful and mysterious. The latter attributes to the order of invention the characteristics of the order of presentation, and assigns to the nascent thoughts of the investigator something of the logical accuracy and sequence which appear in his published memoirs.

Now, there is perhaps no place where the order of being and the order of thinking need to be differentiated with such care as in mathematics. It needs but little reflection to see that any account of mathematics which makes logic not only the norm of the validity of its processes but also its chief heuristic tool is absurd on the face of it. The theory of the syllogism will tell you that when you have the propositions, "All A is B," and, "All B is C," you can derive the further proposition, "All A is C," but you will not find among all the works of Aristotle and Bertrand Russell combined, with the Schoolmen thrown in for good measure, one iota of information which will, without any further act of thought on your part, tell you when to use the syllogistic method, or what particular propositions to employ as major and minor premises. Logic will never answer a question for you until you have put it a definite question. Even then it will never volunteer any information. It has but two words in its language, and those are "yes" and "no." Logic is a critic, not a creator, even as regards its own laws of criticism. While a man endowed with logic alone would assuredly never do any bad mathematics, he would just as assuredly never do any good mathematics.

Mathematics is every bit as much an imaginative art as a logical science. As has just been said, if you wish to know the answer to a question, you must first ask it, and the art of mathematics is the art of asking the right questions. From any set of postulates or premises or assumptions there may be derived an infinite set of lemmata and theorems and conclusions, every one as sound in its logical deduction as any other. Some of these will be recognised by any mathematician as of transcendent importance, more will constitute the ordinary stock in trade of the mathematical journals, but by far the greatest part will be by common acceptance nugatory and trivial. This charge is entirely beyond the jurisdiction of logic, but the ability to discriminate between such trivial theorems and the really vital conclusions of a mathematical science is precisely that quality which the competent mathematician has and the incompetent mathematician lacks.

What is an important theorem? Some theorems are important because of their direct physical and technical applications, others because of their position in the development of a further theory which is of interest, and yet others because of the beauty, symmetry, and richness of the theory of which they form a part. These latter qualities are of a nature essentially aesthetic, and are of course bound up with individual and personal judgment after the fashion of all aesthetic qualities. In the general recognition of varying degrees of beauty and importance, together with the lack of any permanent and universally accepted norm of these characters, in the existence of fashions, of local and national standards and of individual eccentricity, mathematical taste shows its essential kinship with taste in the arts.

In order to do good mathematical work, then, and in fact to do *any* mathematical work, it is not enough to grind out mechanically the conclusions to be derived from a given set of axioms, as by some super-Babbage computing machine. We must select. The postulates with which we start contain our conclusions only in the sense in which the keyboard of the pianoforte contains a sonata, in the sense in which a yard of canvas and tubes of paint contain a painting, or a block of marble a statue.

The imagination is the mainspring of mathematical work, while logic is its balance-wheel. As in a watch, it is not until the mainspring has been wound up to a certain extent that the balance-wheel starts to move. It is not until after we have put ourselves a mathematical question, and have propounded at least a tentative answer to this question that there is any possibility of logical reasoning. Our tentative answer may be vague to begin with—very vague, and of a nature totally repugnant to logical thinking, for it may not even be in a form determinate enough to put down in black and white on paper. There is nothing more surprising than the power of the mind to formulate these vague yet useful hypotheses concerning a subject matter abstract and logical in character. What is it, I wonder, that forms the real content of our consciousness at one of the moments of reflective reverie which constitute so large a part of our periods of research? What we have can scarcely be a dim and confused image of the theorem at the end of our investigation, for the dim light of intuition may be a will-of-the-wisp, and our investigation may end frustrate. Those psychologies of meaning which see the psychological counterpart of the binomial theorem in an obscure strain at the back of the eyeballs are surely not very helpful in their analysis of the mental state of the mathematician.

This mathematical day-dreaming (in the midst of a difficult research, not a little of it is ordinary dreaming by night) is perhaps easiest to understand in the case of geometry, where it is largely dependent on a carefully cultivated power of spatial imagination. Even here it is remarkable how a crude two- or three-dimensional image can do service as the vehicle of a notion in four or five dimensions, or even in space of infinite dimensionality. In the highly rarefied regions of modern analysis, however, such aid as the spatial imagination can furnish, though it is of undeniable value, is fitful and occasional. No picture of an everywhere dense denumerable set of points, or of a continuous curve lacking a tangent at every point, is in the least adequate to the complexity of the situation which it represents. Throughout function theory, postulate theory, and the theory of assemblages, the whole mass of habits of thought which makes possible any imagination whatever is as much a new acquisition of the human mind as the body and organs of the butterfly are new acquisitions of the caterpillar.

Habits of thought—it is these rather than the sensory and imaginational content of the mind which constitute what is vital in mathematical imagining. Inasmuch as the mathematical imagination must sooner or later submit to the criticism of logic, it is essential that these habits should accord with logic. First and foremost among these habits is the habit that the mathematician should continually subject his ideas to trial by logic. He must incessantly try to draw the consequences inherent in his notions, and must instantly recognise when he is proving too much, and is drawing a conclusion which is manifestly false. He must arrange the steps of his proposed theory in a tentative logical order, narrowing the unproved gaps until his results cohere from beginning to end. He must revolve his system in his mind, trying it by all the examples his ingenuity can muster. When he finds a flaw, he must consider whether it is inherent in the very nature of his ideas, or merely adventitious and to be circumvented by a more ingenious approach. Whatever he builds up he must try to tear down, and whatever he tears down he must strive to build up again.

Not only must the mathematician employ his imagination in the invention of new problems and the discovery of *experimenta crucis* to test his answers to these problems, but he must ever be on the alert to see the widest consequences of the methods which lead to his conclusions. Many a theory is encumbered by restrictions which are either altogether unnecessary or are easily replaced by others of a more fundamental character. Many a research answers half a problem when it might just as readily answer the whole. In every branch of mathematics there is one plane of generality on which the theorems are easiest to prove, and needless complication arises as quickly by falling short of this as by exceeding it. It is a mark of the great mathematician to have taken a number of separate theories, fragmentary, intricate and tortuous, and by a profound perception of the true bearing and weight of their methods to have welded them into a single whole, clear, luminous, and simple.

Mathematics is an experimental science. The formulation and testing of hypotheses play in mathematics a part not other than in chemistry, physics, astronomy, or botany. Just as in the science of nature, old ways of regarding things are compared, tried against the facts, worn down by mutual attrition, until they take on a new and unfamiliar aspect. It matters little in which concerns scientific method and the mental processes of the investigator that the mathematician experiments with pencil and paper while the chemist uses test-tube and retort, or the biologist stains and the microscope. An experiment is the confronting of preconceived notions with hard facts, and the notions of the scientist are just as much the result of preconception, the facts just as hard, in mathematics as anywhere else. The only great point of divergence between mathematics and the other sciences lies in the far greater permanence of mathematical knowledge, in the circumstance that experience only whispers "yes" or "no" in reply to our questions, while logic shouts.

Since, however, pencil and paper are cheaper than retorts and microscopes, and since there are no long periods of waiting in mathematical research such as are incurred in the other sciences by the construction of apparatus, or the time-consuming propensities of chemical reactions, or any of the thousand and one petty worries which make the hair of the laboratory worker turn gray before its time, there is one great advantage with the mathematician: he may blunder to his heart's content, waste time in asking questions which he cannot answer, fumble and bungle and muddle, and if he can salvage one or two good ideas from this wreckage, neither he nor anyone else is a penny the worse off. So long as he keeps his published writings clear of error, if he welcomes every ghost of a shadow of an idea that comes his way, and tries it before casting it aside, he suffers no harm but great good; for it is just these waifs of notions that may furnish the new point of view which will found a new discipline or reanimate an old. He who lets his sense of the mathematically decorous inhibit the free flow of his imagination cuts off his own right hand.

Questions and Topics for Discussion and Writing

1. In his autobiography *I Am a Mathematician*, Norbert Wiener asserts, "Mathematics is very largely a young man's game. It is the athleticism of the intellect, making demands which can be satisfied to the full only when there is youth and strength." Write an essay that points out the value of this argument or one in which you develop this statement.
2. More than in any other discipline, according to Wiener, the "order of being" and the "order of thinking" need to be differentiated in mathematics. Why does he see this confusion as a crisis today?
3. According to Wiener, what is the role of logic in mathematical thinking?
4. Thinkers of the eighteenth and nineteenth centuries discussed faculties of the mind in terms of a hierarchy. The eighteenth century placed logic and reason in a superior position. The nineteenth placed imagination in the superior position. How does Wiener's categorization compare or contrast with these concepts?
5. Of the qualities of a theorem, Wiener includes beauty, symmetry, and richness, and says, "These latter qualities are of a nature essentially aesthetic, and are of course bound up with individual and personal judgment, after the fashion of all aesthetic qualities." Do these statements relating math to esthetics sound strange when applied to mathematical thinking? Are they in accord with other writings that you have read?
6. What is the role of chance in mathematical work? Compare Wiener's ideas to those of Bronowski in "The Idea of Chance" (Unit 5).
7. How does Wiener relate mathematical thinking to the theory of evolution? Explain.
8. According to Wiener, what is the most important "habit of thought" for the scientist? What is the mark of a great mathematician?
9. What is Wiener's definition of an experiment? Does his definition differ from your definition of an experiment?
10. What value does Wiener place on the imagination when he associates it with the mathematician's right hand?

Eugene P. Wigner

Unlike the names of Einstein, Teller, or Oppenheimer, the name Eugene Wigner may not excite the imagination, and yet that is the sort of company he keeps in the history of nuclear physics. Wigner was one of several European scientists to emigrate to the United States during the 1920s and 30s, where they later helped establish the core of this country's nuclear physics research program. He joined the staff of Princeton on a half-time basis in 1930, an appointment that became full-time in 1933, after President Hindenburg of the Weimar Republic established Adolf Hitler as his Chancellor.

Wigner (b. 1902), a native of Budapest, Hungary, was trained at the Technische Hochschule in Berlin as a chemical engineer.

Along with physicist Leo Szilard, Wigner approached Einstein with the information that Germany was engaging in uranium research, thus spurring Einstein to write President Roosevelt concerning the implications of such research. This action, in turn, triggered a chain-reaction of events that culminated in the production of the atomic bomb.

Although instrumental in initiating and procuring funding for the Los Alamos Project, Wigner has always championed the cause of the peaceful use of nuclear energy. During the war, while Los Alamos was working its way toward Hiroshima, Wigner headed the theoretical physics division at the Metallurgical Laboratory of the University of Chicago while it was engaged in the Manhattan Project, which led to the first, sustained fission chain-reaction.

After the war, Wigner joined the laboratories at Oak Ridge, Tennessee, where he served as co-director for a year, researching the practicality of nuclear power reactors, returning later to his research at Princeton. Much of this Princeton research—for which he shared the Nobel Prize in physics in 1963—was compiled in the volume *Group Theory and Its Application to the Quantum Mechanics of Atomic Spectra*. In "The Unreasonable Effectiveness of Mathematics," Wigner turns his mind to "the miracle of the appropriateness of the language of mathematics for the formulation of the laws of physics."

The Unreasonable Effectiveness of Mathematics

Eugene P. Wigner

"and it is probable that there is some secret here which remains to be discovered."

C. S. Peirce

There is a story about two friends, who were classmates in high school, talking about their jobs. One of them became a statistician and was working on population trends. He showed a reprint to his former classmate. The reprint started, as usual, with the Gaussian distribution and the statistician explained to his former classmate the meaning of the symbols for the actual population, for the average population, and so on. His classmate was a bit incredulous and was not quite sure whether the statistician was pulling his leg. "How can you know that?" was his query. "And what is this symbol here?" "Oh," said the statistician, "this is π." "What is that?" "The ratio of the circumference of the circle to its diameter." "Well, now you are pushing your joke too far," said the classmate, "surely the population has nothing to do with the circumference of the circle."

Naturally, we are inclined to smile about the simplicity of the classmate's approach. Nevertheless, when I heard this story, I had to admit to an eerie feeling because, surely, the reaction of the classmate betrayed only plain common sense. I was even more confused when, not many days later, someone came to me and expressed his bewilderment[1] with the fact that we make a rather narrow selection when choosing the data on which we test our theories. "How do we know that, if we made a theory which focuses its attention on phenomena we disregard and disregards some of the phenomena now commanding our attention, that we could not build another theory which has little in common with the present one but which, nevertheless, explains just as many phenomena as the present theory?" It has to be admitted that we have no definite evidence that there is no such theory.

The preceding two stories illustrate the two main points which are the subjects of the present discourse. The first point is that mathematical concepts turn up in entirely unexpected connections. Moreover, they often permit an unexpectedly close and accurate description of the phenomena in these connections. Secondly, just because of this circumstance, and because we do not understand the reasons of their usefulness, we cannot know whether a theory formulated in terms of mathematical concepts is uniquely appropriate. We are in a position similar to that of a man who was provided with a bunch of keys and who, having to open several doors in succession, always hit on the right key on the first or second trial. He became skeptical concerning the uniqueness of the coordination between keys and doors.

Most of what will be said on these questions will not be new; it has probably occurred to most scientists in one form or another. My principal aim is to illuminate it from several sides. The first point is that the enormous usefulness of mathematics in the natural sciences is something bordering on the mysterious and that there is no rational explanation for it. Second, it is just this uncanny usefulness of mathematical concepts that raises the question of the uniqueness of our physical theories. In order to establish the first point, that mathematics plays an unreasonably important role in physics, it will be useful to say a few words on the question, "What is mathematics?", then, "What is physics?", then, how mathematics enters physical theories, and last, why the success of mathematics in its role in physics appears so baffling. Much less will be said on the second point: the uniqueness of the theories of physics. A proper answer to this question would require elaborate experimental and theoretical work which has not been undertaken to date.

---

[1] The remark to be quoted was made by F. Werner when he was a student in Princeton.

WHAT IS MATHEMATICS?

Somebody once said that philosophy is the misuse of a terminology which was invented just for this purpose.[2] In the same vein, I would say that mathematics is the science of skillful operations with concepts and rules invented just for this purpose. The principal emphasis is on the invention of concepts. Mathematics would soon run out of interesting theorems if these had to be formulated in terms of the concepts which already appear in the axioms. Furthermore, whereas it is unquestionably true that the concepts of elementary mathematics and particularly elementary geometry were formulated to describe entities which are directly suggested by the actual world, the same does not seem to be true of the more advanced concepts, in particular the concepts which play such an important role in physics. Thus, the rules for operations with pairs of numbers are obviously designed to give the same results as the operations with fractions which we first learned without reference to "pairs of numbers." The rules for the operations with sequences, that is, with irrational numbers, still belong to the category of rules which were determined so as to reproduce rules for the operations with quantities which were already known to us. Most more advanced mathematical concepts, such as complex numbers, algebras, linear operators, Borel sets—and this list could be continued almost indefinitely— were so devised that they are apt subjects on which the mathematician can demonstrate his ingenuity and sense of formal beauty. In fact, the definition of these concepts, with a realization that interesting and ingenious considerations could be applied to them, is the first demonstration of the ingeniousness of the mathematician who defines them. The depth of thought which goes into the formulation of the mathematical concepts is later justified by the skill with which these concepts are used. The great mathematician fully, almost ruthlessly, exploits the domain of permissible reasoning and skirts the impermissible. That his recklessness does not lead him into a morass of contradictions is a miracle in itself: certainly it is hard to believe that our reasoning power was brought, by Darwin's process of natural selection, to the perfection which it seems to possess. However, this is not our present subject. The principal point which will have to be recalled later is that the mathematician could formulate only a handful of interesting theorems without defining concepts beyond those contained in the axioms and that the concepts outside those contained in the axioms are defined with a view of permitting ingenious logical operations which appeal to our aesthetic sense both as operations and also in their results of great generality and simplicity.[3]

The complex numbers provide a particularly striking example for the foregoing. Certainly, nothing in our experience suggests the introduction of these

---

[2] This statement is quoted here from W. Dubislav's *Die Philosophie der Mathematik in der Gegenwart* (Berlin: Junker and Dünnhaupt Verlag, 1932), p. 1.

[3] M. Polanyi, in his *Personal Knowledge* (Chicago: University of Chicago Press, 1958), says: "All these difficulties are but consequences of our refusal to see that mathematics cannot be defined without acknowledging its most obvious feature: namely, that it is interesting" (page 188).

quantities. Indeed, if a mathematician is asked to justify his interest in complex numbers, he will point, with some indignation, to the many beautiful theorems in the theory of equations, of power series, and of analytic functions in general, which owe their origin to the introduction of complex numbers. The mathematician is not willing to give up his interest in these most beautiful accomplishments of his genius.[4]

WHAT IS PHYSICS?

The physicist is interested in discovering the laws of inanimate nature. In order to understand this statement, it is necessary to analyze the concept, "law of nature."

The world around us is of baffling complexity and the most obvious fact about it is that we cannot predict the future. Although the joke attributes only to the optimist the view that the future is uncertain, the optimist is right in this case: the future is unpredictable. It is, as Schrödinger has remarked, a miracle that in spite of the baffling complexity of the world, certain regularities in the events could be discovered (1).[*] One such regularity, discovered by Galileo, is that two rocks, dropped at the same time from the same height, reach the ground at the same time. The laws of nature are concerned with such regularities. Galileo's regularity is a prototype of a large class of regularities. It is a surprising regularity for three reasons.

The first reason that it is surprising is that it is true not only in Pisa, and in Galileo's time, it is true everywhere on the Earth, was always true, and will always be true. This property of the regularity is a recognized invariance property and, as I had occasion to point out some time ago (2), without invariance principles similar to those implied in the preceding generalization of Galileo's observation, physics would not be possible. The second surprising feature is that the regularity which we are discussing is independent of so many conditions which could have an effect on it. It is valid no matter whether it rains or not, whether the experiment is carried out in a room or from the Leaning Tower, no matter whether the person who drops the rocks is a man or a woman. It is valid even if the two rocks are dropped, simultaneously and from the same height, by two different people. There are, obviously, innumerable other conditions which are all immaterial from the point of view of the validity of Galileo's regularity. The irrelevancy of so many circumstances which *could* play a role in the phenomenon observed has also been called an invariance (2). However, this invariance is of a different character from the preceding one since it cannot be formulated as a general principle. The exploration of the conditions

[4]The reader may be interested, in this connection, in Hilbert's rather testy remarks about intuitionism which "seeks to break up and to disfigure mathematics," *Abh. Math. Sem.*, Univ. Hamburg, 157 (1922), or *Gesammelte Werke* (Berlin: Springer, 1935), p. 188.
[*]The numbers in parentheses refer to the References at the end of the article.

which do, and which do not, influence a phenomenon is part of the early experimental exploration of a field. It is the skill and ingenuity of the experimenter
which show him phenomena which depend on a relatively narrow set of relatively easily realizable and reproducible conditions.[5] In the present case, Galileo's restriction of his observations to relatively heavy bodies was the most
important step in this regard. Again, it is true that if there were no phenomena
which are independent of all but a manageably small set of conditions, physics
would be impossible.

The preceding two points, though highly significant from the point of view of
the philosopher, are not the ones which surprised Galileo most, nor do they
contain a specific law of nature. The law of nature is contained in the statement
that the length of time which it takes for a heavy object to fall from a given
height is independent of the size, material, and shape of the body which drops.
In the framework of Newton's second "law," this amounts to the statement that
the gravitational force which acts on the falling body is proportional to its mass
but independent of the size, material, and shape of the body which falls.

The preceding discussion is intended to remind us, first, that it is not at all
natural that "laws of nature" exist, much less that man is able to discover them.[6]
The present writer had occasion, some time ago, to call attention to the succession of layers of "laws of nature," each layer containing more general and more
encompassing laws than the previous one and its discovery constituting a
deeper penetration into the structure of the universe than the layers recognized before (3). However, the point which is most significant in the present
context is that all these laws of nature contain, in even their remotest consequences, only a small part of our knowledge of the inanimate world. All the
laws of nature are conditional statements which permit a prediction of some
future events on the basis of the knowledge of the present, except that some
aspects of the present state of the world, in practice the overwhelming majority
of the determinants of the present state of the world, are irrelevant from the
point of view of the prediction. The irrelevancy is meant in the sense of the
second point in the discussion of Galileo's theorem.[7]

As regards the present state of the world, such as the existence of the earth
on which we live and on which Galileo's experiments were performed, the
existence of the sun and of all our surroundings, the laws of nature are entirely
silent. It is in consonance with this, first, that the laws of nature can be used to
predict future events only under exceptional circumstances—when all the rele-

---

[5] See, in this connection, the graphic essay of M. Deutsch, *Daedalus*, 87, 86 (1958). A. Shimony has
called my attention to a similar passage in C. S. Peirce's *Essays in the Philosophy of Science* (New
York: The Liberal Arts Press, 1957), p. 237.

[6] E. Schrödinger, in his *What Is Life* (Cambridge: Cambridge University Press, 1945), p. 31, says
that this second miracle may well be beyond human understanding.

[7] The writer feels sure that it is unnecessary to mention that Galileo's theorem, as given in the text,
does not exhaust the content of Galileo's observations in connection with the laws of freely falling
bodies.

vant determinants of the present state of the world are known. It is also in consonance with this that the construction of machines, the functioning of which he can foresee, constitutes the most spectacular accomplishment of the physicist. In these machines, the physicist creates a situation in which all the relevant coordinates are known so that the behavior of the machine can be predicted. Radars and nuclear reactors are examples of such machines.

The principal purpose of the preceding discussion is to point out that the laws of nature are all conditional statements and they relate only to a very small part of our knowledge of the world. Thus, classical mechanics, which is the best known prototype of a physical theory, gives the second derivatives of the positional coordinates of all bodies, on the basis of the knowledge of the positions, etc., of these bodies. It gives no information on the existence, the present positions, or velocities of these bodies. It should be mentioned, for the sake of accuracy, that we discovered about thirty years ago that even the conditional statements cannot be entirely precise: that the conditional statements are probability laws which enable us only to place intelligent bets on future properties of the inanimate world, based on the knowledge of the present state. They do not allow us to make categorical statements, not even categorical statements conditional on the present state of the world. The probabilistic nature of the "laws of nature" manifests itself in the case of machines also, and can be verified, at least in the case of nuclear reactors, if one runs them at very low power. However, the additional limitation of the scope of the laws of nature[8] which follows from their probabilistic nature will play no role in the rest of the discussion.

THE ROLE OF MATHEMATICS IN PHYSICAL THEORIES

Having refreshed our minds as to the essence of mathematics and physics, we should be in a better position to review the role of mathematics in physical theories.

Naturally, we do use mathematics in everyday physics to evaluate the results of the laws of nature, to apply the conditional statements to the particular conditions which happen to prevail or happen to interest us. In order that this be possible, the laws of nature must already be formulated in mathematical language. However, the role of evaluating the consequences of already established theories is not the most important role of mathematics in physics. Mathematics, or, rather, applied mathematics, is not so much the master of the situation in this function: it is merely serving as a tool.

Mathematics does play, however, also a more sovereign role in physics. This was already implied in the statement, made when discussing the role of applied

[8]See, for instance, E. Schrödinger, reference (1).

mathematics, that the laws of nature must have been formulated in the language of mathematics to be an object for the use of applied mathematics. The statement that the laws of nature are written in the language of mathematics was properly made three hundred years ago;[9] it is now more true than ever before. In order to show the importance which mathematical concepts possess in the formulation of the laws of physics, let us recall, as an example, the axioms of quantum mechanics as formulated, explicitly, by the great mathematician, von Neumann, or, implicitly, by the great physicist, Dirac (4, 5). There are two basic concepts in quantum mechanics: states and observables. The states are vectors in Hilbert space, the observables self-adjoint operators on these vectors. The possible values of the observations are the characteristic values of the operators—but we had better stop here lest we engage in a listing of the mathematical concepts developed in the theory of linear operators.

It is true, of course, that physics chooses certain mathematical concepts for the formulation of the laws of nature, and surely only a fraction of all mathematical concepts is used in physics. It is true also that the concepts which were chosen were not selected arbitrarily from a listing of mathematical terms but were developed, in many if not most cases, independently by the physicist and recognized then as having been conceived before by the mathematician. It is not true, however, as is so often stated, that this had to happen because mathematics uses the simplest possible concepts and these were bound to occur in any formalism. As we saw before, the concepts of mathematics are not chosen for their conceptual simplicity—even sequences of pairs of numbers are far from being the simplest concepts—but for their amenability to clever manipulations and to striking, brilliant arguments. Let us not forget that the Hilbert space of quantum mechanics is the complex Hilbert space, with a Hermitean scalar product. Surely to the unpreoccupied mind, complex numbers are far from natural or simple and they cannot be suggested by physical observations. Furthermore, the use of complex numbers is in this case not a calculational trick of applied mathematics but comes close to being a necessity in the formulation of the laws of quantum mechanics. Finally, it now begins to appear that not only complex numbers but so-called analytic functions are destined to play a decisive role in the formulation of quantum theory. I am referring to the rapidly developing theory of dispersion relations.

It is difficult to avoid the impression that a miracle confronts us here, quite comparable in its striking nature to the miracle that the human mind can string a thousand arguments together without getting itself into contradictions, or to the two miracles of the existence of laws of nature and of the human mind's capacity to divine them. The observation which comes closest to an explanation for the mathematical concepts' cropping up in physics which I know is Einstein's statement that the only physical theories which we are willing to accept

[9] It is attributed to Galileo.

are the beautiful ones. It stands to argue that the concepts of mathematics, which invite the exercise of so much wit, have the quality of beauty. However, Einstein's observation can at best explain properties of theories which we are willing to believe and has no reference to the intrinsic accuracy of the theory. We shall, therefore, turn to this latter question.

IS THE SUCCESS OF PHYSICAL THEORIES TRULY SURPRISING?

A possible explanation of the physicist's use of mathematics to formulate his laws of nature is that he is a somewhat irresponsible person. As a result, when he finds a connection between two quantities which resembles a connection well-known from mathematics, he will jump at the conclusion that the connection *is* that discussed in mathematics simply because he does not know of any other similar connection. It is not the intention of the present discussion to refute the charge that the physicist is a somewhat irresponsible person. Perhaps he is. However, it is important to point out that the mathematical formulation of the physicist's often crude experience leads in an uncanny number of cases to an amazingly accurate description of a large class of phenomena. This shows that the mathematical language has more to commend it than being the only language which we can speak; it shows that it is, in a very real sense, the correct language. . . . The miracle of the appropriateness of the language of mathematics for the formulation of the laws of physics is a wonderful gift which we neither understand nor deserve. We should be grateful for it and hope that it will remain valid in future research and that it will extend, for better or for worse, to our pleasure, even though perhaps also to our bafflement, to wide branches of learning. . . .

REFERENCES

1. Schrödinger, E., *Über Indeterminismus in der Physik* (Leipzig: J. A. Barth, 1932); also Dubislav, W., *Naturphilosophie* (Berlin: Junker und Dünnhaupt, 1933), Chap. 4.
2. Wigner, E. P., "Invariance in Physical Theory," *Proc. Am. Phil. Soc.*, 93, 521–526 (1949) . . .
3. Wigner, E. P., "The Limits of Science," *Proc. Am. Phil. Soc.*, 94, 422 (1950); also Margenau, H., *The Nature of Physical Reality* (New York: McGraw-Hill, 1950), Ch. 8 . . .
4. Dirac, P. A. M., *Quantum Mechanics*, 3rd ed. (Oxford: Clarendon Press, 1947).
5. von Neumann, J., *Mathematische Grundlagen der Quantenmechanik* (Berlin: Springer, 1932). English translation (Princeton, N.J.: Princeton Univ. Press, 1955).

Questions and Topics for Discussion and Writing

1. How does Wigner's opening narrative introduce his main points? Is it any more than an amusing story?
2. The author makes use of many scientific terms and examples that he does not define. What can you infer about his intended audience from this fact?
3. The third paragraph on page 178 is an excellent example of a forecast paragraph. Cite reasons why such a judgment can be made.
4. How many times does the author state his main point? Is his use of repetition effective?
5. According to Wigner, mathematics plays a crucial role in physics. In one paragraph, explain this crucial role.
6. Does Wigner's phrase "math is an end in itself" contradict his thesis that math is a "tool" of science?
7. How would you define the "sense of formal beauty" that Wigner attributes to the mind of the mathematician? Compare this statement to Hardy's assertion of the esthetic nature of mathematics.
8. How do you interpret Wigner's statement: "The great mathematician fully, almost ruthlessly, exploits the domain of permissible reasoning and skirts the impermissible"?
9. What does Wigner imply about his belief concerning human evolution by his digression: "Certainly it is hard to believe that our reasoning power was brought, by Darwin's process of natural selection, to the perfection which it seems to possess"?
10. Explain what Wigner means by the phrases "empirical law of epistemology" and "invariance of physical theories." Is his meaning clear?
11. Cite the three examples given by the author to illustrate the use of pure mathematical concepts to model physical reality accurately.
12. Can you attempt to explain the effectiveness of mathematics? In a short paper, write your explanation. (Consult Hamming's article in *American Mathematical Monthly* [February 1980].)

J. V. Cunningham

The American poet J. V. Cunningham is a native of Maryland (b. 1911) who became a professor of English at Brandeis University. Among his many books are *The Helmsman, The Judge's Fury, Doctor Drink, To What Strangers, What Welcome,* and *Some Salt.*

"Meditation on Statistical Method," from *Exclusions of a Rhyme,* is a reflection on the new patterns of thought in twentieth-century mathematical theory. While increasing our capabilities and revolutionizing our view of the world, these patterns have also emphasized probabilistic, statistical structure as a method of explanation and reduced the notion of individual certainty. By directly admonishing Plato to despair, the poem suggests a retreat from Platonic thought. Yet, from the paradoxical title forward, the poem is concerned with the human need for ideals rather than "empiric forms." It invites comparison with the prose writings on the consequences of twentieth-century mathematical theory.

Meditation on Statistical Method

J. V. Cunningham

Plato, despair!
We prove by norms
How numbers bear
Empiric forms,

How random wrong
Will average right
If time be long
And error slight,

But in our hearts
Hyperbole
Curves and departs
To infinity.

Error is boundless.
Nor hope nor doubt,
Though both be groundless,
Will average out.

Sheila Tobias

For people who are uncomfortable with mathematics, Sheila Tobias alleviates anxiety. One of the founders of the National Organization of Women (NOW), Sheila Tobias was educated at Radcliffe and Columbia University. Having worked as a teacher, administrator, and consultant, she is a partner in a Washington, D.C., consulting firm, Overcoming Math Anxiety, and co-director of the Institute for the Study of Anxiety in Learning at the Washington School of Psychiatry.

Interested in explaining technical subjects for the nonspecialist, Tobias is currently writing about military policy for nonmilitary people. With writer Peter Goudinoff, she co-authored "What Kinds of Guns Are They Buying for Your Butter?," a variation of her math-phobia pieces. Her essays appear in such widely circulated periodicals as *Ms., Today's Educator, The Center Magazine,* and *Psychology Today.*

In *Overcoming Math Anxiety* (1978), from which the following selection is taken, Tobias challenges the myths surrounding the "mathematical mind." Speaking from the nonmathematician's perspective, she provides advice and consolation for "math avoiders."

The Nature of Math Anxiety

Sheila Tobias

"The day they introduced fractions, I had the measles." Or the teacher was out for a month, the family moved, there were more snow days that year than ever before (or since). People who use events like these to account for their failure at math did, nevertheless, learn how to spell. True, math is especially cumulative. A missing link can damage understanding much as a dropped stitch ruins a knitted sleeve. But being sick or in transit or just too far behind to learn the next new idea is not reason enough for doing poorly at math forever after. It is unlikely that one missing link can abort the whole process of learning elementary arithmetic.

In fact, mathematical ideas that are rather difficult to learn at age seven or eight are much easier to comprehend one, two, or five years later if we try again. As we grow older, our facility with language improves; we have many more mathematical concepts in our minds, developed from everyday living; we can ask more and better questions. Why, then, do we let ourselves remain permanently ignorant of fractions or decimals or graphs? Something more is at work than a missed class.

It is of course comforting to have an excuse for doing poorly at math, better than having to concede that one does not have a mathematical mind. Still, the dropped stitch concept is often used by math anxious people to excuse their failure. It does not explain, however, why in later years they did not take the trouble to unravel the sweater and pick up where they left off.

Say they did try a review book. Chances are it would not be helpful. Few texts on arithmetic are written for adults. How insulting to go back to a "Run, Sport, run!" level of elementary arithmetic, when arithmetic can be infinitely clearer and more interesting if it is discussed at an adult level.

Moreover, when most of us learned math we learned dependence as well. We needed the teacher to explain, the textbook to drill us, the back of the book to tell us the right answers. Many people say that they never mastered the multiplication table, but I have encountered only one person so far who carries a multiplication table in his wallet. He may have no more skills than the others, but at least he is trying to make himself autonomous. The greatest value of using simple calculators in elementary school may, in the end, be to free pupils from dependence on something or someone beyond their control.

Adults can easily pick up those dropped stitches once they decide to do something about them. In one math counselling session for educators and psychologists, the following arithmetic bugbears were exposed:

How do you get a percentage out of a fraction like $\frac{7}{16}$?

Where does "pi" come from?

How do you do a problem like: Two men are painters. Each paints a room in a different time. How long does it take them to paint the room together?

The issues were taken care of within half an hour.

This leads me to believe that people are anxious not because they dropped a stitch long ago but rather because they accepted an ideology that we must reject: *that if we haven't learned something so far it is probably because we can't.*

FEAR OF BEING TOO DUMB OR TOO SMART

One of the reasons we did not ask enough questions when we were younger is that many of us were caught in a double bind between a fear of appearing too dumb in class and a fear of being too smart. Why anyone should be afraid of being too smart in math is hard to understand except for the prevailing notion that math whizzes are not normal. Boys who want to be popular can be hurt by this label. But it is even more difficult for girls to be smart in math. Matina Horner, in her survey of high-achieving college women's attitudes toward academic success, found that such women are especially nervous about competing with men on what they think of as men's turf. Since many people perceive ability in mathematics as unfeminine, fear of success may well interfere with ability to learn math.

The young woman who is frightened of seeming too smart in math must be very careful about asking questions in class because she never knows when a

question is a really good one. "My nightmare," one woman remembers, "was that one day in math class I would innocently ask a question and the teacher would say, 'Now that's a fascinating issue, one that mathematicians spent years trying to figure out.' And if that happened, I would surely have had to leave town, because my social life would have been ruined." This is an extreme case, probably exaggerated, but the feeling is typical. Mathematical precocity, asking interesting questions, meant risking exposure as someone unlike the rest of the gang.

It is not even so difficult to ask questions that gave the ancients trouble. When we remember that the Greeks had no notation for multi-digit numbers and that even Newton, the inventor of the calculus, would have been hard pressed to solve some of the equations given to beginning calculus students today, we can appreciate that young woman's trauma.

At the same time, a student who is too inhibited to ask questions may never get the clarification needed to go on. We will never know how many students developed fear of math and loss of self-confidence because they could not ask questions in class. But the math anxious often refer to this kind of inhibition. In one case, a counsellor in a math clinic spent almost a semester persuading a student to ask her math teacher a question *after* class. She was a middling math student, with a B in linear algebra. She asked questions in her other courses, but could not or would not ask them in math. She did not entirely understand her inhibition, but with the aid of the counsellor, she came to believe it had something to do with a fear of appearing too smart.

There is much more to be said about women and mathematics. At this point it is enough to note that some teachers and most pupils of both sexes believe that boys naturally do better in math than girls. Even bright girls believe this. When boys fail a math quiz their excuse is that they did not work hard enough. Girls who fail are three times more likely to attribute their lack of success to the belief that they "simply cannot do math." Ironically, fear of being too smart may lead to such passivity in math class that eventually these girls also develop a feeling that they are dumb. It may also be that these women are not as low in self-esteem as they seem, but by failing at mathematics they resolve a conflict between the need to be competent and the need to be liked. The important thing is that until young women are encouraged to believe that they have the right to be smart in mathematics, no amount of supportive, nurturant teaching is likely to make much difference.

Questions and Topics for Discussion and Writing

1. How would you characterize your attitude toward mathematics? Write an essay recalling the experiences that you think were significant in establishing that attitude.
2. In the first paragraph, Tobias uses the term *dropped stitch* to represent the loss of a link in the process of learning arithmetic. Does this simile reveal anything about the writer? What effect does it have on you as a reader?
3. For what audience is this essay written? Choose particular words and phrases that explain your analysis.
4. How does the author define "mathematical precocity"?
5. What is the ideology that many people have accepted as an excuse for not learning math? What is the "double bind" that prevents young students from learning math skills?
6. What does this essay suggest about the role of women's rights in increasing the likelihood of women becoming mathematical?
7. If there is a shortage of mathematicians today, how can such a shortage be alleviated? Prepare a controlling statement that presents your answer to this problem and write an essay developing your solution.

SUGGESTIONS FOR FURTHER READING

Beckmann, Peter. *The History of Pi.* Boulder: Golem Press, 1971.
Bell, Eric Temple. *Men of Mathematics.* New York: Simon & Schuster, 1965.
Davis, P. J., and R. Hersh. *The Mathematical Experience.* Boston: Birkhäuser, 1981.
Fermi, Laura. *Illustrious Immigrants.* Chicago: University of Chicago Press, 1971.
Gardner, Martin. *Mathematics, Magic and Mystery.* New York: Dover Publications, 1956.

Hofstader, Douglas. *Gödel, Bach, Escher: An Eternal Golden Braid*. New York: Basic Books, 1979.

Hogben, Lancelot. *Mathematics for the Millions*. New York: W. W. Norton, 1937.

Huntley, H. E. *The Divine Proportion: A Study in Mathematical Beauty*. New York: Dover Publications, 1970.

Kline, Morris. *Mathematics and the Search for Knowledge*. New York: Oxford University Press, 1985.

———. *Mathematics: The Loss of Certainty*. New York: Oxford University Press, 1980.

Osen, L. M. *Women in Mathematics*. Cambridge: MIT Press, 1974.

Spitznagel, Edward L., Jr. *Selected Topics in Mathematics*. New York: Holt, Rinehart and Winston, 1971.

Unit 5

PHYSICS: THE CLASSICAL AND THE NEW

Nicolaus Copernicus

"Sta, sol ne moveare"—"Stand, Sun, do not move" reads the monument to the founder of modern astronomy. The epoch-making revolution in conventional thought regarding the earth's position in the universe and the movements of heavenly bodies originated with the Polish astronomer Nicolaus Copernicus (1473–1543) in 1530. In a bold return to the forgotten theories of a few Greek philosophers, he "shattered" the traditional geocentric conception of the universe with his hypothesis of a heliocentric universe.

Much is made of the fact that Copernicus began his studies in 1492, the year of the discovery of America. One biographer observes that he started to think in the "atmosphere of anticipation created by the discovery of a new continent, and the virtual doubling of the known world. The discovery of America gave practical proof that the earth was a sphere." Educated in the fields of mathematics, medicine, and astronomy at universities in Cracow, Bologna, and Padua, he was awarded his doctorate in canon law at Ferrara. Through a powerful uncle, he became a canon at the Cathedral in Frombork, a position he held until his death.

From 1505 on, while residing in various cities in Prussia, Copernicus devoted his life to the study of astronomy. Since his student days, he had been troubled by what he considered serious errors in Ptolemy's astronomical system. Copernicus doubted that the motions of the planets and stars were as

192

complicated as his predecessor had made them. As he studied that subject more thoroughly, Copernicus came to the following conclusions: The sun is the center of the system; the earth, too, is a planet, like Mars and Venus; and all the planets revolve around the sun in an orderly way. He knew that the orbits these planets took around the sun were not circular, but was unable to determine their exact shape chiefly because of the existing technology. He had only crude wooden instruments, and was working one hundred years before the invention of telescopes.

Copernicus' revolutionary thinking and observations took him past the boundaries of the astronomical thinking of his day. His statement that the earth was a moving planet disturbed powerful people able to suppress his writings and even punish him, but some historians attribute his reluctance to publish less to his fear of death than to his fear of ridicule. Copernicus, at any rate, showed his handwritten thoughts and conclusions to only a few friends and scientists. Despite such precautions, however, the news of his discoveries spread. While admirers urged him to print his writings, thirty-six years passed before they appeared under the title *The Revolutions of the Heavenly Spheres*. The finished book reached him just hours before his death. His doctrine was developed by Galileo, Tycho Brahe, and Johannes Kepler.

In 1619, John Donne noted that Copernican ideas, which "may very well be true," were "creeping into every man's mind," and he gave classic expression to the modern malaise (see "The New Philosophy Calls All in Doubt," later in this unit).

In "The Motion of the Spheres," the universe according to Copernicus is presented.

The Motion of the Spheres

Copernicus

The first and highest of all is the sphere of the fixed stars, which contains itself and all things, and is therefore motionless. It is the location of the universe, to which the motion and position of all the remaining stars is referred. For though some consider that it also changes in some respect, we shall assign another cause for its appearing to do so in our deduction of the Earth's motion. There follows Saturn, the first of the wandering stars, which completes its circuit in thirty years. After it comes Jupiter which moves in a twelve-year long revolution. Next is Mars, which goes round biennially. An annual revolution holds the fourth place, in which as we have said is contained the Earth along with the lunar sphere which is like an epicycle. In fifth place Venus returns every nine months. Lastly, Mercury holds the sixth place, making a circuit in the space of eighty days. In the middle of all is the seat of the Sun. For who in this most beautiful of temples would put this lamp in any other or better place than the one from which it can illuminate everything at the same time? Aptly indeed is he named by some the lantern of the universe, by others the mind, by others the ruler. Trismegistus[1] . . . called him the visible God, Sophocles' Electra, the watcher over all things. Thus indeed the Sun as if seated on a royal throne governs his household of Stars as they circle round him. Earth also is by no means cheated of the Moon's attendance, but as Aristotle says in his book *On Animals* the Moon has the closest affinity with the Earth. Meanwhile the Earth conceives from the Sun, and is made pregnant with annual offspring. We find, then, in this arrangement the marvellous symmetry of the universe, and a sure linking together in harmony of the motion and size of the spheres, such as could be perceived in no other way. . . .

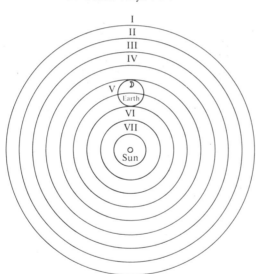

The universe according to Copernicus

    I. The unmoving sphere of the Fixed Stars.

    II. Saturn revolves in 30 years.

    III. The revolution of Jupiter is 12 years.

    IV. The two-yearly revolution of Mars.

    V. The annual revolution of the Earth with the Moon's orbit.

    VI. Venus nine-months.

    VII. Mercury in 80 days.

Questions and Topics for Discussion and Writing

1. Obtain a sketch or draw a sketch of the universe conceived by those preceding this new world-view.
2. Read an article that discusses the impact of Copernicus' cosmology on thinkers of

[1] Hermes Trismegistus, a legendary figure derived from the ancient Egyptian god Thoth, was supposed to be the author of a number of works in Greek on Platonic and Stoic philosophy, astrology, and alchemy.

his time. Write a paper describing how Copernicus' statement must have roused the interest or antagonism of religionists.

3. Some biographers note resemblances between Copernicus and Newton. Based on reading in the library, write a short report comparing the lives of the two.

4. In "The Rise of Science" (Unit 3), Bertrand Russell pays tribute to Copernicus, stating the qualities that make Copernicus a "true scientist." What are those qualities? What two great merits of Copernican astronomy does Russell cite?

5. In this excerpt, Copernicus explains his conception of the universe. Choose a natural or scientific phenomenon in astronomy, geology, physics, or some other field for research. Present your findings in a short explanatory paper.

6. In "Tradition in Science" (Unit 3), a talk given to celebrate the five-hundredth birthday of Copernicus, Heisenberg states that "the direction which [Copernicus] had chosen for his research in astronomy still determines to some extent the scientific work of our time." In a paper of evidence, explain how Copernicus' work is still a force in research.

Galileo Galilei

Calling Galileo "the father of modern times," Bertrand Russell writes: "Whatever we may like or dislike about the age in which we live, its increase of population, its improvement in health, its trains, motor-cars, radio, politics, and advertisements of soap—all emanate from Galileo."

Galileo Galilei (1564–1642), mathematician–astronomer, was born in Pisa. While his father, an impoverished nobleman, urged lessons in Latin, Greek, drawing, and music, Galileo preferred things mechanical. In 1581, he entered the University of Pisa to study medicine and Aristotelian philosophy, but achieved a dubious reputation for questioning established theories.

At twenty, Galileo made his first important scientific contribution: Watching the great lamp swing from the cathedral ceiling, he timed its motions with his pulse and suggested use of a pendulum to time the pulse rate of medical patients.

Five years later, having become a mathematics professor at the University of Pisa, Galileo discovered a law of falling bodies, which states that such bodies, regardless of their weight, are pulled by gravity to earth at the same speed. A story, possibly fictitious, relates that Galileo dropped a ten-pound and a one-pound weight from the Leaning Tower as fellow professors, students, and priests watched. His conclusion directly opposed the long-accepted theory of Aristotle that a heavier weight would fall faster. This challenge to tradition aroused such furor that Galileo was forced to resign his professorship. However, his experiments led to two "new" sciences, dynamics and ballistics.

Although not its inventor, Galileo made the telescope practical. Indeed, his improvements secured him tenure at the university. With the perfected instrument, he discovered that the moon, like the earth, displayed an uneven surface and that the sun had spots. His most remarkable discovery, however, was that of the four bright satellites of Jupiter. Moreover, Galileo inferred that the sun rotated on an inclined axis and became convinced of the Copernican theory that the earth does, indeed, move around the sun.

But fame earned him denunciation from churchmen and the label of heretic from enemies. His *Two Principal Systems of the World* took him before the Inquisition, where a lengthy trial forced him to recant his support for the Copernican theory. Papal order confined him to his villa in Florence.

Although in pain and partially blind during his last years, he directed his thought to the study of mechanics, the results of which are two laws of motion on which present-day physics and astronomy are founded. During this time, the blind poet John Milton visited Galileo and found inspiration for his characterization of the protagonist of *Samson Agonistes*.

The following selection, "Astronomical Message," communicates Galileo's sense of excitement and discovery in the use of the instruments he designed.

Astronomical Message

Which contains and explains recent observations
made with the aid of a new spyglass[1]
concerning the surface of the moon,
the Milky Way, nebulous stars, and
innumerable fixed stars,
as well as four planets never before seen, and
now named
THE MEDICEAN STARS

Galileo Galilei

Great indeed are the things which in this brief treatise I propose for observation and consideration by all students of nature. I say great, because of the excellence of the subject itself, the entirely unexpected and novel character of these things, and finally because of the instrument by means of which they have been revealed to our senses.

[1] The word "telescope" was not coined until 1611. A detailed account of its origin is given by Edward Rosen in *The Naming of the Telescope* (New York, 1947). In the present translation the modern term has been introduced for the sake of dignity and ease of reading, but only after the passage in which Galileo describes the circumstances which led him to construct the instrument. . . .

Surely it is a great thing to increase the numerous host of fixed stars previously visible to the unaided vision, adding countless more which have never before been seen, exposing these plainly to the eye in numbers ten times exceeding the old and familiar stars.

It is a very beautiful thing, and most gratifying to the sight, to behold the body of the moon, distant from us almost sixty earthly radii,[2] as if it were no farther away than two such measures—so that its diameter appears almost thirty times larger, its surface nearly nine hundred times, and its volume twenty-seven thousand times as large as when viewed with the naked eye. In this way one may learn with all the certainty of sense evidence that the moon is not robed in a smooth and polished surface but is in fact rough and uneven, covered everywhere, just like the earth's surface, with huge prominences, deep valleys, and chasms.

Again, it seems to me a matter of no small importance to have ended the dispute about the Milky Way by making its nature manifest to the very senses as well as to the intellect. Similarly it will be a pleasant and elegant thing to demonstrate that the nature of those stars which astronomers have previously called "nebulous" is far different from what has been believed hitherto. But what surpasses all wonders by far, and what particularly moves us to seek the attention of all astronomers and philosophers, is the discovery of four wandering stars not known or observed by any man before us. Like Venus and Mercury, which have their own periods about the sun, these have theirs about a certain star that is conspicuous among those already known, which they sometimes precede and sometimes follow, without ever departing from it beyond certain limits. All these facts were discovered and observed by me not many days ago with the aid of a spyglass which I devised, after first being illuminated by divine grace. Perhaps other things, still more remarkable, will in time be discovered by me or by other observers with the aid of such an instrument, the form and construction of which I shall first briefly explain, as well as the occasion of its having been devised. Afterwards I shall relate the story of the observations I have made.

About ten months ago a report reached my ears that a certain Fleming[3] had constructed a spyglass by means of which visible objects, though very distant from the eye of the observer, were distinctly seen as if nearby. Of this truly remarkable effect several experiences were related, to which some persons gave credence while others denied them. A few days later the report was con-

[2] The original text reads "diameters" here and in another place. That this error was Galileo's and not the printer's has been convincingly shown by Edward Rosen (*Isis*, 1952, pp. 344 ff.). The slip was a curious one, as astronomers of all schools had long agreed that the maximum distance of the moon was approximately sixty terrestrial radii. Still more curious is the fact that neither Kepler nor any other correspondent appears to have called Galileo's attention to this error; not even a friend who ventured to criticize the calculations in this very passage.

[3] Credit for the original invention is generally assigned to Hans Lipperhey, a lens grinder in Holland who chanced upon this property of combined lenses and applied for a patent on it in 1608.

firmed to me in a letter from a noble Frenchman at Paris, Jacques Badovere,[4] which caused me to apply myself wholeheartedly to inquire into the means by which I might arrive at the invention of a similar instrument. This I did shortly afterwards, my basis being the theory of refraction. First I prepared a tube of lead, at the ends of which I fitted two glass lenses, both plane on one side while on the other side one was spherically convex and the other concave. Then placing my eye near the concave lens I perceived objects satisfactorily large and near, for they appeared three times closer and nine times larger than when seen with the naked eye alone. Next I constructed another one, more accurate, which represented objects as enlarged more than sixty times. Finally, sparing neither labor nor expense, I succeeded in constructing for myself so excellent an instrument that objects seen by means of it appeared nearly one thousand times larger and over thirty times closer than when regarded with our natural vision.

It would be superfluous to enumerate the number and importance of the advantages of such an instrument at sea as well as on land. But forsaking terrestrial observations, I turned to celestial ones, and first I saw the moon from as near at hand as if it were scarcely two terrestrial radii away. After that I observed often with wondering delight both the planets and the fixed stars, and since I saw these latter to be very crowded, I began to seek (and eventually found) a method by which I might measure their distances apart.

Here it is appropriate to convey certain cautions to all who intend to undertake observations of this sort, for in the first place it is necessary to prepare quite a perfect telescope, which will show all objects bright, distinct, and free from any haziness, while magnifying them at least four hundred times and thus showing them twenty times closer. Unless the instrument is of this kind it will be vain to attempt to observe all the things which I have seen in the heavens, and which will presently be set forth. Now in order to determine without much trouble the magnifying power of an instrument, trace on paper the contour of two circles or two squares of which one is four hundred times as large as the other, as it will be when the diameter of one is twenty times that of the other. Then, with both these figures attached to the same wall, observe them simultaneously from a distance, looking at the smaller one through the telescope and at the larger one with the other eye unaided. This may be done without inconvenience while holding both eyes open at the same time; the two figures will appear to be of the same size if the instrument magnifies objects in the desired proportion.

Such an instrument having been prepared, we seek a method of measuring distances apart. This we shall accomplish by the following contrivance.

Let ABCD be the tube and E be the eye of the observer. Then if there were

---

[4] Badovere studied in Italy toward the close of the sixteenth century and is said to have been a pupil of Galileo's about 1598. When he wrote concerning the new instrument in 1609 he was in the French diplomatic service at Paris, where he died in 1620.

no lenses in the tube, the rays would reach the object FG along the straight lines ECF and EDG. But when the lenses have been inserted, the rays go along the refracted lines ECH and EDI; thus they are bought closer together, and those which were previously directed freely to the object FG now include only the portion of it III. The ratio of the distance EH to the line HI then being found, one may by means of a table of sines determine the size of the angle formed at the eye by the object HI, which we shall find to be but a few minutes of arc. Now, if to the lens CD we fit thin plates, some pierced with larger and some with smaller apertures, putting now one plate and now another over the lens as required, we may form at pleasure different angles subtending more or fewer minutes of arc, and by this means we may easily measure the intervals between stars which are but a few minutes apart, with no greater error than one or two minutes. And for the present let it suffice that we have touched lightly on these matters and scarcely more than mentioned them, as on some other occasion we shall explain the entire theory of this instrument. . . .

Questions and Topics for Discussion and Writing

1. This treatise may be classified as an "apology." Find words connoting defense and justification.
2. In what way does the language here differ from that of a modern scientific report? Paraphrase this essay in modern language.
3. Rethink the essays in Unit 2 on the scientific method. Bertrand Russell says that the scientific method "comes into the world full-fledged" with the work of Galileo. Discuss.
4. In a paper, discuss how Galileo's approach to dynamics was a "radical departure" from Aristotle's theories of dynamics.
5. In this treatise, Galileo calls his telescope a "spyglass." The name *telescope* was given by a Greek poet–theologian; thus began our custom of naming modern scientific instruments with terms of Greek origin. List several other inventions that bear such names.
6. Read the parable play *The Life of Galileo* by twentieth-century German playwright Bertolt Brecht. It is an example of Brecht's "epic theatre" and his *historicizing* approach, which uses great events and personages of the past to provoke audiences to think critically about their own cultures. Discuss how the events of Galileo's life and his ideas provide such a stimulus.
7. Like Galileo, other scientists faced conflict between science and religious orthodoxy. In the library, look up the problems encountered by Copernicus, Kepler, or Descartes. In a short paper, explain the way the scientist dealt with his problem.

John Donne

An English poet and divine, John Donne (1572–1631) was a Janus-figure. Like many of his seventeenth-century contemporaries, he looked in opposite directions: on one hand, toward the cosmic hierarchy and unified world-view of the Christian Middle Ages; on the other, toward the new patterns of the universe forming in the physical and natural sciences.

Raised a Catholic, Donne became an Anglican minister, then chaplain to King James I and Dean of St. Paul's Cathedral, still referred to as "John Donne's church" by poetry-loving visitors to London.

Donne's time was an age that gave authority to rhetoric, that respected art as a parallel to God's creation and emphasized wit. Ever mindful of the duality of human life, Donne displayed his wit in elaborate writing that incorporated Christian paradox together with elements from the science of the day. The poets who practiced this style of writing were called "the Metaphysicals." One of Ernest Hemingway's novels took its title from a line in one of Donne's *Devotions:* "send not to know for whom the bell tolls: it tolls for thee."

The selection below, from a poem of many parts (474 lines), "An Anatomie of the World," illustrates the cast of Donne's thought as he reflects on the effects of Copernican theory on the spirit of his age.

The New Philosophy Calls All in Doubt

John Donne

And new Philosophy calls all in doubt,
The element of fire is quite put out;
The Sun is lost, and th'earth, and no man's wit
Can well direct where to looke for it.
And freely men confesse that this world's spent,
When in the Planets, and the Firmament
They seeke so many new; they see that this
Is crumbled out againe to his Atomies.
'Tis all in peeces, all cohaerence gone;
All just supply, and all Relation:
Prince, Subject, Father, Sonne, are things forgot,
For every man alone thinkes he hath got
To be a Phoenix, and that there can bee
None of that kinde, of which he is, but hee. . . .

Man hath weav'd out a net, and this net throwne
Upon the Heavens, and now they are his owne.
Loth to goe up the hill, or labour thus
To goe to heaven, we make heaven come to us.
We spur, we reine the starres, and in their race
They're diversely content t'obey our pace.
But keepes the earth her round proportion still?

Questions and Topics for Discussion and Writing

1. This selection is taken from "An Anatomie of the World: The First Anniversary" (1611). Because language is dynamic, words change with time in spelling and even in meaning. Look in the OED (Oxford English Dictionary) to see what "anatomy" meant in Donne's time and what specialized meanings it holds in ours.
2. This passage has been selected to illustrate Donne's effort to reconcile the theory of Copernicus with theology. Find specific evidence in the poem of Donne's concern over the shift from a geocentric to a heliocentric world-view.
3. In Donne's time, most people felt that the social and religious order reflected the cosmic order. Do you see evidence in the poem that Donne is writing about loss of order in social and religious structure?
4. Write a paragraph paraphrasing the Donne selection.

5.   On the walls of the Greenwich Observatory in Sussex, England, are painted these two lines from Donne's poem:

> Man hath weav'd out a net, and this net throwne
> Upon the Heavens, and now they are his owne.

Look up the Greenwich Observatory in a reference book and comment on the appropriateness of this inscription. Be sure to check the definition of "prime meridian."

6.   Donne is sometimes described as a "Janus-figure" (see the figure on page 201). Consult a reference and write a short essay on how this term may apply to Donne or to one of the scientists cited in the text.

Isaac Newton

Sir Isaac Newton (1642–1727), English mathematician, astronomer, and physicist, produced grandiose scientific accomplishments and received remarkable public acclaim for them. Truly he was at the center of the scientific stage of his time. His summation of the laws of motion and of gravity (1687) acted like a Rosetta stone, deciphering many diverse natural phenomena in one stroke of synthesis. In the words of Daniel Boorstin, this unifying feat made Newton "the first popular hero of modern science." Other historians see him as an "Olympian figure" and label his contribution the "Newtonian Revolution." Newton's findings changed the contemporary world-view and brought him knighthood. Newton exercised influence not just by firing public imagination but through using organizational power: He was president of the Royal Society for twenty-four years, a post that allowed him to dominate the direction of scientific development.

It is no surprise to find that such an extraordinary man was deeply complex. A bachelor, Newton is described in contradictory terms: "modest and retiring," "secretive but quarrelsome," "understanding, gentle, and generous," "tortured and obsessive." Other biographers emphasize his insecurity and detail his dismay and even fury when he felt that other scientists did not respond fairly to his reports. He is known to have suffered spells of depression. Such complexity perhaps had its origins in difficult early years.

Born prematurely, on Christmas day the year that Galileo died, the infant Newton was small enough to fit into a "quart mug." His father, an illiterate yeoman, had died three months before. When the boy was two, his mother remarried and left him in the care of her own mother until he was ten. Widowed, she returned and took Newton from school in order that he might manage her farm. But he showed himself not adept at this task and, fortunately, was sent back to school. Progressing to Cambridge University, he found an environment in which he could flourish.

Newton's time at Cambridge was interrupted, however, for 1665 and 1666 were the dreadful years when the bubonic plague terrified Europe. The university closed and Newton went home. This time, the country years were astonishingly productive. At what he called "the prime of my age for invention," he conceived the binomial theorem and his method of "fluxions" (the infinitesimal calculus) and laid the foundations for the theory of colors and the epochal law of gravity. As reported of a conversation in his old age, the fall of an apple during these country years triggered his first reflections on gravity.

Newton wrote on chronology, chemistry, and—especially surprising to modern minds—alchemy, a hobby he practiced with dedication. He left lengthy writings on theology, where his mystical bent became apparent as he tried to fit the works of the Prophets into his rationally conceived universe. Newton invented (constructing it himself and hand-polishing it) the first reflective telescope. His greatest written work, *Philosophiae Naturalis Principia Mathematica* (1687), is generally considered the single most important work in modern science. The anxious Newton, reluctant to submit his ideas, might never have completed the manuscript except for the insistence of Halley—the astronomer of the famed comet—who urged his friend on and even paid for publication of the book.

Newton's own description of his prodigious works is much quoted: "I do not know what I may appear to the world, but, to myself, I seem to have been only like a boy playing on the seashore, and diverting myself in now and then finding a smoother pebble or a prettier shell than ordinary, whilst the great ocean of truth lay all undiscovered before me."

Together with his principles themselves, a second significant scientific contribution of Newton's is his methodology: rigorous mathematical reasoning supported by experiment. In "Demonstration of White Light," an excerpt from a letter to Henry Oldenburg of the Royal Society, Newton reports on his experiment with a prism, an account often considered the most lucid presentation ever written of an experiment.

Demonstration of White Light (Letter 40)

Newton to Oldenburg

6 February 1671/2[1]

Sir,

To perform my late promise to you . . . I procured me a Triangular glass-Prisme, to try therewith the celebrated *Phænomena of Colours*.[2] And in order thereto having darkened my chamber, and made a small hole in my window-shuts, to let in a convenient quantity of the Suns light, I placed my Prisme at its entrance, that it might be thereby refracted to the opposite wall. It was at first a very pleasing divertisement, to view the vivid and intense colours produced thereby; but after a while applying my self to consider them more circumspectly, I became surprised to see them in an *oblong* form; which, according to the received laws of Refraction, I expected should have been *circular*.

They were terminated at the sides with streight lines, but at the ends, the decay of light was so gradual, that it was difficult to determine justly, what was their figure; yet they seemed *semicircular*.

Comparing the length of this coloured *Spectrum* with its breadth, I found it about five times greater; a disproportion so extravagant, that it excited me to a more then ordinary curiosity of examining, from whence it might proceed. I could scarce think, that the various *Thickness* of the glass, or the termination with shadow or darkness, could have any Influence on light to produce such an effect; yet I thought it not amiss to examine first these circumstances, and so tryed, what would happen by transmitting light through parts of the glass of divers thicknesses, or through holes in the window of divers bignesses, or by setting the Prisme without, so that the light might pass through it, and be refracted before it was terminated by the hole: But I found none of those circumstances material. The fashion of the colours was in all these cases the same.

Then I suspected, whether by any *unevenness* in the glass, or other contingent irregularity, these colours might be thus dilated. And to try this, I took another Prisme like the former, and so placed it, that the light, passing through them both, might be refracted contrary ways, and so by the latter returned into that course, from which the former had diverted it. For, by this means I thought, the *regular* effects of the first Prisme would be destroyed by the second Prisme, but the *irregular* ones more augmented, by the multiplicity of refractions. The event was, that the light, which by the first Prisme was diffused into an *oblong* form, was by the second reduced into an *orbicular* one with as much regularity, as when it did not at all pass through them. So that, what ever was the cause of that length, 'twas not any contingent irregularity.

I then proceeded to examine more critically, what might be effected by the difference of the incidence of Rays coming from divers parts of the Sun; and to that end, measured the several lines and angles, belonging to the Image. . . .

Having made these observations, I first computed from them the refractive power of that glass, and found it measured by the *ratio* of the sines, 20 to 31.[3] And then, by that *ratio*, I computed the Refractions of two Rays flowing from opposite parts of the Sun's *discus*, so as to differ 31' in their obliquity of Incidence, and found, that the emergent Rays should have comprehended an angle of about 31', as they did, before they were incident.

But because this computation was founded on the Hypothesis of the proportionality of the *sines* of Incidence, and Refraction, which though by my own and others[4] Experience I could not imagine to be so erroneous, as to make that

Angle but 31′, which in reality was 2 deg. 49′; yet my curiosity caused me again to take my Prisme. And having placed it at my window, as before, I observed, that by turning it a little about its *axis* to and fro, so as to vary its obliquity to the light, more then by an angle of 4 or 5 degrees, the Colours were not thereby sensibly translated from their place on the wall, and consequently by that variation of Incidence, the quantity of Refraction was not sensibly varied. By this Experiment therefore, as well as by the former computation, it was evident, that the difference of the Incidence of Rays, flowing from divers parts of the Sun, could not make them after decussation diverge at a sensibly greater angle, than that at which they before converged; which being, at most, but about 31 or 32 minutes, there still remained some other cause to be found out, from whence it could be 2 deg. 49′.

Then I began to suspect, whether the Rays, after their trajection through the Prisme, did not move in curve lines, and according to their more or less curvity tend to divers parts of the wall. And it increased my suspition, when I remembred that I had often seen a Tennis-ball, struck with an oblique Racket, describe such a curve line. For, a circular as well as a progressive motion being communicated to it by that stroak, its parts on that side, where the motions conspire, must press and beat the contiguous Air more violently than on the other, and there excite a reluctancy and reaction of the Air proportionably greater.[5] And for the same reason, if the Rays of light should possibly be globular bodies, and by their oblique passage out of one medium into another acquire a circulating motion, they ought to feel the greater resistance from the ambient Æther, on that side, where the motions conspire, and thence be continually bowed to the other.[6] But notwithstanding this plausible ground of suspition, when I came to examine it, I could observe no such curvity in them. And besides (which was enough for my purpose) I observed, that the difference betwixt the length of the Image, and diameter of the hole, through which the light was transmitted, was proportionable to their distance.

The gradual removal of these suspitions at length led me to the *Experimentum Crucis*,[7] which was this: I took two boards, and placed one of them close behind the Prisme at the window, so that the light might pass through a small hole, made in it for that purpose, and fall on the other board, which I placed at about 12 foot distance, having first made a small hole in it also, for some of that Incident light to pass through. Then I placed another Prisme behind this second board, so that the light, trajected through both the boards, might pass through that also, and be again refracted before it arrived at the wall. This done, I took the first Prisme in my hand, and turned it to and fro slowly about its *Axis*, so much as to make the several parts of the Image, cast on the second board, successively pass through the hole in it, that I might observe to what places on the wall the second Prisme would refract them. And I saw by the variation of those places, that the light, tending to that end of the Image, towards which the refraction of the first Prisme was made, did in the second Prisme suffer a Refraction considerably greater then the light tending to the other end. And so the true cause . . . of the length of that Image was detected to be no other, then that *Light* consists of *Rays differently refrangible*, which, without any respect to a difference in their incidence, were, according to their degrees of refrangibility, transmitted towards divers parts of the wall.

When I understood this, I left off my aforesaid Glass-works; for I saw, that the perfection of Telescopes was hitherto limited, not so much for want of glasses truly figured according to the prescriptions of Optick Authors, . . . as because that Light it self is a *Heterogeneous mixture of differently refrangible Rays*. So that, were a glass so exactly figured, as to collect any one sort of rays into one point, it could not collect those also into the same point, which having the same Incidence upon the same Medium are apt to suffer a different refraction. Nay, I wondered, that seeing the difference of refrangibility was so great, as I found it, Telescopes should arrive to that perfection they are now at. . . .

Amidst these thoughts I was forced from *Cambridge* by the Intervening Plague, . . . and it was more then two years, before I proceeded further. . . .

Your humble Servt

Isaac Newton

NOTES

1. The original letter in Newton's handwriting has not been found, but a transcript, written by his copyist Wickins and bearing a few verbal corrections in Newton's own hand, is preserved in the Portsmouth Collection (U.L.C. Add. 3970. 3, fos. 460–6). With a few verbal changes the letter was printed in the *Philosophical Transactions* in the form here given, except that (i) the passages mentioned below in notes (7) and (9), and (ii) the signature, were omitted: these have been restored, so that the present version agrees verbally with the transcript. For many years Newton and Wickins shared rooms in college: and there exist copies of several of Newton's early manuscripts made by the latter.

    The letter brought the first notice of Newton's theory of colours to the Royal Society, being read on 8 February 1671/2 and published in *Phil. Trans.* 7 (1672), 3075; yet he had dealt with this in his lectures upon optics in the University of Cambridge (1669–71), starting probably in January 1669/70 (cf. a remark of Collins, p. 53), and late in 1671 he was preparing a series of twenty of the lectures for the press. The Royal Society invited the Bishop of Salisbury, Boyle and Hooke to peruse the letter and bring a report of it.

2. This may be a reference to the experiment with a prism described by Descartes in *Les Météores;* it was designed to account for the formation and the colours of a rainbow. Sunlight, falling almost perpendicularly upon the slanting face of a prism and passing through it, is refracted out of the further lowest and horizontal face at a slit made in an opaque support, and causes all the colours of the rainbow to appear below on a vertical screen, from red to blue. See Scott, *Descartes* (London, 1952), p. 76.

3. The ratio 20 to 31 roughly agrees with the corresponding figures 42 to 65 that occur in the subjoined table, which is taken from a copy of an early manuscript of Newton, "Of Refractions" (U.L.C. Add. 4000, fo. 33 v).

The sines measuring Re-fractions are in	Aire 42	Water 56	Glass 65	Christall 70
The proportions of ye motions of the Extreamely Heterogeneous Rays are in	39,4.40,4	$70\frac{2}{3}.71\frac{2}{3}$	$95\frac{1}{10}.96\frac{1}{10}$	$110\frac{1}{3}.111\frac{1}{3}$
The proportion of ye sines of refraction of ye Extreamely Heterogeneous Rays into Air out of		$90\frac{2}{3}.91\frac{2}{3}$	68.69	$61\frac{4}{5}.62\frac{4}{5}$
Their common sines of incidence		$68\frac{1}{3}$	$44\frac{1}{4}$	$36\frac{4}{5}$
Which subtracted, the Difference is		$22\frac{1}{3}.23\frac{1}{3}$	$23\frac{3}{4}.24\frac{3}{4}$	24.25
The like proportion for Refractions made into Water out of			$275\frac{3}{5},276\frac{3}{5}$ $238\frac{2}{5}$ $37\frac{2}{5},38\frac{2}{5}$	$196\frac{1}{3},197\frac{1}{3}$ $157\frac{4}{9}$ $39\frac{4}{9}.40\frac{1}{9}$

4. The words "& others" are omitted in the *Philosophical Transactions.*
5. The accuracy of Newton's explanation of a ball's swerve is noteworthy. Sir Graham Sutton, in his *Science of Flight*, 1949, p. 82, remarking on the transverse force of an object moving through a liquid if there is a circulation round the object, says that though the effect is associated with the name of Magnus, a German professor, in 1853, it was in fact recorded by Newton in 1672.
6. This remark closely follows the ideas of Descartes.
7. See Bacon, *Novum Organum,* II, xxxv, *instantia crucis;* Hooke, *Micrographia* (1665), p. 54, and *O.E.D., s.v.* crucial: "The passage in the *Novum Organum* shows clearly that Bacon is coining a phrase which he feels bound to explain by referring to the sign at a cross-roads." Hooke mentions Bacon, but writing from memory gets the phrase wrong, *experimentum crucis.* Newton is reminiscing from his reading of Hooke.

    See More's *Newton,* pp. 98–9, for the passage from Hooke: it deals with Descartes's hypothesis of the conversion of light ("whitenesse") into colour by adding a rotatory to the linear motion of globules.

Questions and Topics for Discussion and Writing

1. As implied in this letter, what does Newton judge to be the role of chance or accident in scientific discovery? Can it be creative? Look at the essays "The Creative Process" (Unit 3) and "The Idea of Chance" (below) by Jacob Bronowski and "The Lure of the Hunt" (Unit 6) for reflections on this subject.

2. Consider Newton's scientific methodology in this famous experiment. Compare Doyle's "The Science of Deduction" (Unit 2) or Roueché's "The Orange Man" (Unit 8) for techniques in the scientific method of ruling out possibilities; write a paper of comparison.
3. In a paragraph, summarize Newton's *Experimentum Crucis*.
4. Consider the critical judgment that this letter–report is the best presentation of an experiment ever written, the clarity of its prose perfectly transmitting the logic of an experiment. Do you agree with such a high evaluation? Does the seventeenth-century language create a barrier between Newton and you so that this clarity is obscured?
5. Read Robert Day's definition of a scientific paper, "What Is a Scientific Paper?" (Unit 9). In what ways do you think Newton's report conforms to Day's criteria? Do you see differences?

The Idea of Chance

Jacob Bronowski

You will find the biography for Jacob Bronowski before the essay "The Creative Process" in Unit 3.

There is of course nothing sacred about the causal form of natural laws. We are accustomed to this form, until it has become our standard of what every natural law ought to look like. If you halve the space which a gas fills, and keep other things constant, then you will double the pressure, we say. If you do such and such, the result will be so and so; and it will always be so and so. And we feel by long habit that it is this "always" which turns the prediction into a law. But of course there is no reason why laws should have this always, all-or-nothing form. If you self-cross the offspring of a pure white and a pure pink sweet pea, said Mendel, then on an average one quarter of these grandchildren will be white, and three quarters will be pink. This is as good a law as any other; it says what will happen, in good quantitative terms, and what it says turns out to be true. It is not any less respectable for not making that parade of every-time certainty which the law of gases makes. And indeed, the gas law takes its air of finality only from the accumulation of just such chances as Mendel's law makes explicit.

It is important to seize this point. If I say that after a fine week, it *always* rains on Sunday, then this is recognised and respected as a law. But if I say that after a fine week, it rains on Sunday more often than not, then this somehow is felt to be an unsatisfactory statement; and it is taken for granted that I have not really got down to some underlying law which would chime with our habit of wanting science to say decisively either "always" or "never." Even if I say that after a fine week, it rains on seven Sundays out of ten, you may accept this as a statistic, but it does not satisfy you as a law. Somehow it seems to lack the force of law.

Yet this is a mere prejudice. It is nice to have laws which say, This configuration of facts will always be followed by event A, ten times out of ten. But neither taste nor convenience really make this a more essential form of law than one which says, This configuration of facts will be followed by event A seven times out of ten, and by event B three times out of ten. In form the first is a causal law and the second a statistical law. But in content and in application, there is no reason to prefer one to the other. The laws of science have two functions, to be true and to be helpful; probably each of these functions includes the other. If the statistical law does both, that is all that can be asked of it. We may persuade ourselves that it is intellectually less satisfying than a causal law, and fails somehow to give us the same feeling of understanding the process of nature. But this is an illusion of habit. No law ever gave wider satisfaction than the law of gravitation. Yet we have seen that the explanation it gave of the workings of nature was false, and the understanding we got from it mistaken. What it really did, and did superbly, was to predict the movements of the heavenly bodies to an excellent approximation.

There is, however, a limitation within every law which does not contain the word "always." Bluntly, when I say that a configuration of facts will be followed sometimes by event A and at other times by B, I cannot be certain whether at the next trial A or B will turn up. I may know that A is to turn up seven times and B three times out of ten; but that brings me no nearer at all to knowing which is to turn up on the one occasion I have my eye on next time. Mendel's law is all very fine when you grow sweet peas by the acre; but it does not tell you, and cannot, whether the single second generation seed in your windowbox will flower white or pink. Mendel himself ran into this trouble because he had to do his experimental work in a rather small monastery garden.

So far, this is obvious enough. It is obvious that if we did know what is to happen precisely next time, then we would at once have not a statistical law, but a law of certainty into which we could write the word "always." But this limitation carries with it a less obvious one. If we are not sure whether A or B will turn up next time, then neither can we be sure which will turn up the time after, or the time after that. We know that A is to turn up seven times and B three; but this can never mean that every set of ten trials will give us exactly seven A's and three B's. In fact, it is not possible to write down an irregular string of A's and B's in such a way that every set of ten successive letters which we pick out from it, beginning where we like, is made up precisely of seven of one and three of the other. And of course it is quite impossible to write them down so that any choice of ten letters picked here and there will contain just seven A's.

Then what do I mean by saying that we expect A to turn up seven times to every three times which B turns up? I mean that among all the sets of ten trials which we can choose from an extended series, picking as we like, the greatest number will contain seven A's and three B's. This is the same thing as saying that if we have enough trials, the proportion of A's to B's will tend to the ratio of seven to three. But of course, no run of trials, however extended, is necessarily long enough. In no run of trials can we be sure of reaching precisely the balance of seven to three.

Then how do I know that the law is in fact seven A's and three B's? What do I mean by saying that the ratio tends to this in a long trial, when I never know if the trial is long enough? And more, when I know that at the very moment when we have reached precisely this ratio, the next single trial must upset it—because it must add either a whole A or a whole B, and cannot add seven tenths of one and three tenths of the other. I mean this. After ten trials, we may have eight A's and only two B's; it is not at all improbable. It is not very improbable that we may have nine A's, and it is not even excessively improbable that we may have ten. But it is very improbable that, after a hundred trials, we shall have as many as eighty A's. It is excessively improbable that after a thousand trials we shall have as many as eight hundred A's; indeed it is highly improbable that at this stage the ratio of A's and B's departs from seven to three by as much as five per cent. And if after a hundred thousand trials we should get a ratio which differs from our law by as much as one per cent, then we should have to face the fact that the law itself is almost certainly in error.

Let me quote a practical example. One of the French encyclopédists of the eighteenth century, the great naturalist Buffon, was a man of wide interests. His interest in geology and evolution got him into trouble with the Sorbonne, which made him formally recant his belief that the earth has changed since Genesis. His interest in the laws of chance was less perilous, but it prompted him to ask an interesting question. If a needle is thrown at random on a sheet of paper ruled with lines whose distance apart is exactly equal to the length of the needle, how often can it be expected to fall on a line and how often into a blank space? The answer is rather odd: it should fall on a line a little less than two times out of three—precisely, it should fall on a line two times out of π, where π is the familiar ratio of the circumference of a circle to its diameter, which has the value 3.14159265. . . . How near can we get to this answer in actual trials? This depends of course on the care with which we rule the lines and do the throwing; but, after that, it depends only on our patience. In 1901 an Italian mathematician, having taken due care, demonstrated his patience by making well over 3,000 throws. The value he got for π was right to the sixth place of decimals, which is an error of only a hundred thousandth part of one per cent.

This is the method to which modern science is moving. It uses no principle but that of forecasting with as much assurance as possible, but with no more than is possible. That is, it idealises the future from the outset, not as completely determined, but as determined within a defined area of uncertainty.

Let me illustrate the kind of uncertainty. We know that the children of two blue-eyed parents will certainly have blue eyes; at least, no exception has ever been found. By contrast, we cannot be certain that all the children of two brown-eyed parents will have brown eyes. And we cannot be certain of it even if they have already had ten children with brown eyes. The reason is that we can never discount a run of luck of the kind which Dr. Johnson once observed when a friend of his was breeding horses. "He has had," said Dr. Johnson, "sixteen fillies without one colt, which is an accident beyond all computation of chances." But what we can do is to compute the *odds* against such a run; this is not as hard as Johnson supposed. And from this we can compute the likelihood that the next child will have brown eyes. That is, we can make a forecast which states our degree of uncertainty in a precise form. Oddly enough, it is just here that Mendel's own account of his work is at fault. He assumed in effect that once a couple has had ten brown-eyed children, the chance that they may yet have blue-eyed children is negligible. But it was not.

This area of uncertainty shrinks very quickly in its proportion if we make our forecasts not about one family but about many. I do not know whether this or that couple will have a child next year; I do not even know whether I shall. But it is easy to estimate the number of children who will be born to the whole population, and to give limits of uncertainty to our estimate. The motives which lead to marriage, the trifles which cause a car to crash, the chanciness of today's sunshine or tomorrow's egg, are local, private and incalculable. Yet, as Kant saw long ago, their totals over the country in a year are remarkably steady; and even their ranges of uncertainty can be predicted.

This is the revolutionary thought in modern science. It replaces the concept of the *inevitable effect* by that of the *probable trend*. Its technique is to separate so far as possible the steady trend from local fluctuations. The less the trend has been overlaid by fluctuations in the past, the greater is the confidence with which we look along the trend into the future. We are not isolating a cause. We are tracing a pattern of nature in its whole setting. We are aware of the uncertainties which that large, flexible setting induces in our pattern. But the world cannot be isolated from itself: the uncertainty *is* the world. The future does not already exist; it can only be predicted. We must be content to map the places into which it may move, and to assign a greater or less likelihood to this or that of its areas of uncertainty.

These are the ideas of chance in science today. They are new ideas: they give chance a kind of order; they re-create it as the life within reality. These ideas have come to science from many sources. Some were invented by Renaissance brokers; some by seventeenth century gamblers; some by mathematicians who were interested in aiming-errors and in the flow of gases and more recently in radio-activity. The most fruitful have come from biology within little more than the last fifty years. I need not stress again how successful they have been in the last few years, for example in physics: Nagasaki is a monument to that. But we have not yet begun to feel their importance outside science altogether. For example, they make it plain that problems like Free Will or Determinism are simply misunderstandings of history. History is neither determined nor random. At any moment, it moves forward into an area whose general shape is known but whose boundaries are uncertain in a calculable way. A society moves under material pressure like a stream of gas; and on the average, its individuals obey the pressure; but at any instant, any individual may, like an atom of the gas, be moving across or against the stream. The will on the one hand and the compulsion on the other exist and play within these boundaries. In these ideas, the concept of chance has lost its old dry pointlessness and has taken on a new depth and power; it has come to life. Some of these ideas have begun to influence the arts: they can be met vaguely in the novels of the young French writers. In time they will liberate our literature from the pessimism which comes from our divided loyalties: our reverence for machines and, at odds with it, our nostalgia for personality. I am young enough to believe that this union, the union as it were of chance with fate, will give us all a new optimism.

Questions and Topics for Discussion and Writing

1. In the first paragraph, Bronowski speaks of Mendel. Who was Mendel and what was his contribution to science? Why does Bronowski use Mendel's law in his opening discussion?
2. What two kinds of laws does Bronowski discuss?
3. What rhetorical methods of development does Bronowski use?
4. What does Bronowski say is the method of modern science?
5. What is revolutionary about modern science?
6. Bronowski says that the ideas of chance come to us from many sources: Renaissance brokers, seventeenth-century gamblers, mathematicians, biologists. Research some of these sources. As an alternative, you might want to explore ideas of chance in literature. For example, see Dostoevsky's *The Gambler* for a presentation of psychological response to the laws of chance.
7. The concept of *Will* is one of the strong themes of nineteenth-century literature. The philosopher Schopenhauer, for example, developed this concept in his work *The World as Will and Idea* and states: "To Hume's question—What is causality?—we shall answer, Will." Pursue this idea through library research.
8. In an essay, explore the role of "chance" in your life. Provide details and examples to support your thesis.

Albert Einstein

Born in Ulm, Germany, Albert Einstein (1879–1955) did most of his studies in Switzerland and as a youth became a Swiss citizen. While studying for his doctorate at the University of Zurich and working for the Patent Office, he

evolved his "special" or "restricted" theory of relativity, which required modi-
fying the traditional laws of classical Newtonian mechanics. This first formulation was largely untestable, and Einstein's ideas did not gain much acceptance.

In 1913 he returned to Germany and resumed his German citizenship to
become director of the prestigious Kaiser Wilhelm Physical Institute. Then, in
1915, ten years after his first formulation, Einstein announced his general theory of relativity. Its astronomical implications were tested by British astronomers and lent some validity to his theory. Laboratory work by other scientists
measuring the mass of high-speed electrons offered further corroboration. Over
the years, the work of other colleagues inevitably validated Einstein's theory,
from Rutherford's verification of the relationship between mass and energy to
Fermi's initiation of the first controlled nuclear reaction.

In the meantime, Einstein engaged in other scientific work: extending
Planck's quantum theory, explaining Brownian movement, and discovering the
law of photoelectric effect (the principle of the "electric eye"). He was awarded
the 1921 Nobel Prize in physics. With the rise of the Nazi menace, Einstein
accepted a position at the Institute for Advanced Study in Princeton and later
became an American citizen. His eminent international stature helped foreign
scientists seeking asylum in the United States to penetrate bureaucracy and
alert the government to the implications of atomic fission and weapons.

Einstein effected a phenomenal revolution in scientific theory and was recognized to be one of the greatest thinkers of modern times, yet his simple and
lovable character has become legendary. An ardent Zionist, he also worked
tirelessly for world peace and humanitarian goals.

In the following selection, Einstein explains for the readers of the *London
Times* both his special and general theories of relativity, theories that have
become the "working tools" of modern physicists.

What Is the Theory of Relativity?

Albert Einstein

I gladly accede to the request of your colleague to write something for *The
Times* on relativity. After the lamentable breakdown of the old active inter-
course between men of learning, I welcome this opportunity of expressing my
feelings of joy and gratitude toward the astronomers and physicists of England.
It is thoroughly in keeping with the great and proud traditions of scientific work
in your country that eminent scientists should have spent much time and trouble, and your scientific institutions have spared no expense, to test the implications of a theory which was perfected and published during the war in the land

of your enemies. Even though the investigation of the influence of the gravitational field of the sun on light rays is a purely objective matter, I cannot forbear to express my personal thanks to my English colleagues for their work; for without it I could hardly have lived to see the most important implication of my theory tested.

We can distinguish various kinds of theories in physics. Most of them are constructive. They attempt to build up a picture of the more complex phenomena out of the materials of a relatively simple formal scheme from which they start out. Thus the kinetic theory of gases seeks to reduce mechanical, thermal, and diffusional processes to movements of molecules—i.e., to build them up out of the hypothesis of molecular motion. When we say that we have succeeded in understanding a group of natural processes, we invariably mean that a constructive theory has been found which covers the processes in question.

Along with this most important class of theories there exists a second, which I will call "principle-theories." These employ the analytic, not the synthetic, method. The elements which form their basis and starting-point are not hypothetically constructed but empirically discovered ones, general characteristics of natural processes, principles that give rise to mathematically formulated criteria which the separate processes or the theoretical representations of them have to satisfy. Thus the science of thermodynamics seeks by analytical means to deduce necessary conditions, which separate events have to satisfy, from the universally experienced fact that perpetual motion is impossible.

The advantages of the constructive theory are completeness, adaptability, and clearness, those of the principle theory are logical perfection and security of the foundations.

The theory of relativity belongs to the latter class. In order to grasp its nature, one needs first of all to become acquainted with the principles on which it is based. Before I go into these, however, I must observe that the theory of relativity resembles a building consisting of two separate stories, the special theory and the general theory. The special theory, on which the general theory rests, applies to all physical phenomena with the exception of gravitation; the general theory provides the law of gravitation and its relations to the other forces of nature.

It has, of course, been known since the days of the ancient Greeks that in order to describe the movement of a body, a second body is needed to which the movement of the first is referred. The movement of a vehicle is considered in reference to the earth's surface, that of a planet to the totality of the visible fixed stars. In physics the body to which events are spatially referred is called the coordinate system. The laws of the mechanics of Galileo and Newton, for instance, can only be formulated with the aid of a coordinate system.

The state of motion of the coordinate system may not, however, be arbitrarily chosen, if the laws of mechanics are to be valid (it must be free from rotation and acceleration). A coordinate system which is admitted in mechanics is called an "inertial system." The state of motion of an inertial system is according to mechanics not one that is determined uniquely by nature. On the contrary, the following definition holds good: a coordinate system that is moved uniformly and in a straight line relative to an inertial system is likewise an inertial system. By the "special principle of relativity" is meant the generalization of this definition to include any natural event whatever: thus, every universal law of nature which is valid in relation to a coordinate system C, must also be valid, as it stands, in relation to a coordinate system C', which is in uniform translatory motion relatively to C.

The second principle, on which the special theory of relativity rests, is the "principle of the constant velocity of light in vacuo." This principle asserts that light in vacuo always has a definite velocity of propagation (independent of the state of motion of the observer or of the source of the light). The confidence which physicists place in this principle springs from the successes achieved by the electrodynamics of Maxwell and Lorentz.

Both the above-mentioned principles are powerfully supported by experience, but appear not to be logically reconcilable. The special theory of relativity finally succeeded in reconciling them logically by a modification of kinematics— i.e., of the doctrine of the laws relating to space and time (from the point of view of physics). It became clear that to speak of the simultaneity of two events had no meaning except in relation to a given coordinate system, and that the shape of measuring devices and the speed at which clocks move depend on their state of motion with respect to the coordinate system.

But the old physics, including the laws of motion of Galileo and Newton, did not fit in with the suggested relativist kinematics. From the latter, general mathematical conditions issued, to which natural laws had to conform, if the above-mentioned two principles were really to apply. To these, physics had to be adapted. In particular, scientists arrived at a new law of motion for (rapidly moving) mass points, which was admirably confirmed in the case of electrically charged particles. The most important upshot of the special theory of relativity concerned the inert masses of corporeal systems. It turned out that the inertia of a system necessarily depends on its energy-content, and this led straight to the notion that inert mass is simply latent energy. The principle of the conservation of mass lost its independence and became fused with that of the conservation of energy.

The special theory of relativity, which was simply a systematic development of the electrodynamics of Maxwell and Lorentz, pointed beyond itself, however. Should the independence of physical laws of the state of motion of the coordinate system by restricted to the uniform translatory motion of coordinate systems in respect to each other? What has nature to do with our coordinate systems and their state of motion? If it is necessary for the purpose of describing nature, to make use of a coordinate system arbitrarily introduced by us, then the choice of its state of motion ought to be subject to no restriction; the laws ought to be entirely independent of this choice (general principle of relativity).

The establishment of this general principle of relativity is made easier by a fact of experience that has long been known, namely, that the weight and the inertia of a body are controlled by the same constant (equality of inertial and gravitational mass). Imagine a coordinate system which is rotating uniformly with respect to an inertial system in the Newtonian manner. The centrifugal forces which manifest themselves in relation to this system must, according to Newton's teaching, be regarded as effects of inertia. But these centrifugal forces are, exactly like the forces of gravity, proportional to the masses of the bodies. Ought it not to be possible in this case to regard the coordinate system as stationary and the centrifugal forces as gravitational forces? This seems the obvious view, but classical mechanics forbid it.

This hasty consideration suggests that a general theory of relativity must supply the laws of gravitation, and the consistent following up of the idea has justified our hopes.

But the path was thornier than one might suppose, because it demanded the abandonment of Euclidean geometry. This is to say, the laws according to which solid bodies may be arranged in space do not completely accord with the spatial laws attributed to bodies by Euclidean geometry. This is what we mean when we talk of the "curvature of space." The fundamental concepts of the "straight line," the "plane," etc., thereby lose their precise significance in physics.

In the general theory of relativity the doctrine of space and time, or kinematics, no longer figures as a fundamental independent of the rest of physics. The geometrical behavior of bodies and the motion of clocks rather depend on gravitational fields, which in their turn are produced by matter.

The new theory of gravitation diverges considerably, as regards principles, from Newton's theory. But its practical results agree so nearly with those of Newton's theory that it is difficult to find criteria for distinguishing them which are accessible to experience. Such have been discovered so far:

1. In the revolution of the ellipses of the planetary orbits round the sun (confirmed in the case of Mercury).
2. In the curving of light rays by the action of gravitational fields (confirmed by the English photographs of eclipses).
3. In a displacement of the spectral lines toward the red end of the spectrum in the case of light transmitted to us from stars of considerable magnitude (unconfirmed so far).[1]

The chief attraction of the theory lies in its logical completeness. If a single one of the conclusions drawn from it proves wrong, it must be given up; to modify it without destroying the whole structure seems to be impossible.

Let no one suppose, however, that the mighty work of Newton can really be superseded by this or any other theory. His great and lucid ideas will retain their unique significance for all time as the foundation of our whole modern conceptual structure in the sphere of natural philosophy.

[1] This criterion has since been confirmed.

Note: Some of the statements in your paper concerning my life and person owe their origin to the lively imagination of the writer. Here is yet another application of the principle of relativity for the delectation of the reader: today I am described in Germany as a "German savant," and in England as a "Swiss Jew." Should it ever be my fate to be represented as a *bête noire*, I should, on the contrary, become a "Swiss Jew" for the Germans and a "German savant" for the English.

Questions and Topics for Discussion and Writing

1. How does Einstein's language reflect an attitude of humility?
2. In the third paragraph, Einstein says that "principle-theories . . . employ the analytic, not the synthetic, method." Distinguish between these two terms.
3. Einstein uses an analogy, that of a two-story building, to explain the theory of relativity. How effective do you find the analogy?
4. Write a statement in your own words explaining the coordinate system.
5. Copy sentences wherein Einstein indicates that his theory is the outgrowth of that of someone else.
6. Write a brief paragraph describing Einstein's attitude toward traditional theories.
7. Comment on Einstein's use of listing.
8. What is the great tribute Einstein accords to Newton?
9. Look up the words "savant" and *"bête noire."* What is the tone and effect of Einstein's "Note"?

N. David Mermin

N. David Mermin is a professor of Physics at Cornell University and Director of the Laboratory of Atomic and Solid State Physics. He is a theorist, with interests in low-temperature, statistical, and solid-state physics. Scientific writing has occupied his attention for some time: In addition to about a hundred technical papers, he is co-author of an important textbook, *Solid State Physics;* has written a nontechnical book for the general reader on Einstein's special theory of relativity, *Space and Time in Special Relativity;* and from time to time publishes articles on various aspects of physics pedagogy.

His article on the "boojum" caused a minor sensation when it appeared in *Physics Today,* a monthly generally devoted to some of the more technical scientific and professional issues of concern to physicists. Abridged for this text by the author, the article explains how the word "boojum" became an internationally accepted scientific term, recognized in Webster and found in the most reputable journals of physics. Ostensibly about scientific nomenclature, "E Pluribus Boojum" also touches on matters such as professional competitiveness, scientific priority, and the manner in which research papers are (or are not) accepted by scientific journals, matters that are well known to all practicing scientists, but not always so vividly and openly discussed.

E Pluribus Boojum: The Physicist As Neologist

N. David Mermin

I know the exact moment when I decided to make the word "boojum" an internationally accepted scientific term. I was just back from a symposium at the University of Sussex near Brighton, honoring the discoverers of the superfluid phases of liquid helium-3.

One of the peculiar things about superfluid helium-3 is that unlike almost all liquids it has a grain, like a piece of wood. This grain is particularly pronounced in the form of the liquid called He^3-A. A network of lines weaves through the liquid He^3-A as if it were a sea of striped toothpaste, but unlike those stripes, the lines in He^3-A are built into the atomic structure of the liquid. They can be twisted, bent, or splayed, but never obliterated by stirring or otherwise disturbing the liquid.

Several of us at the Sussex Symposium had been thinking about how the grain of the He^3-A would arrange itself in a spherical drop of liquid. Nature often favors the most symmetrical pattern, and this would suggest that the grain should radiate outward from the center of the drop, like rays from the sun. There is an elegant mathematical argument, however, that such a pattern cannot be produced in He^3-A without at the same time producing a pair of little whirlpools, whose funnels connect the point of convergence of the lines of grain to points on the surface of the drop.

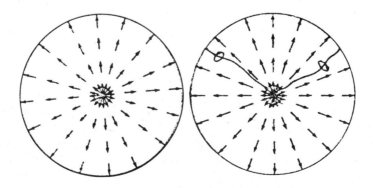

It appeared that if one did try to establish the symmetric pattern of radiating lines, then the accompanying whirlpools would draw the point of convergence of the lines to the surface of the drop, resulting in a final pattern that looked like this:

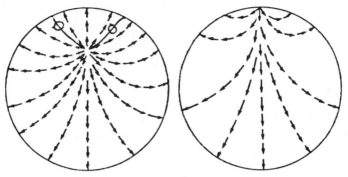

When I returned to Ithaca I began to prepare for the proceedings the final text of the talk I had given which examined, among other things, the question of the spherical drop. Although no remarks about the spherical drop were made after my talk, I decided to use the format of the discussion remark to describe the opinion that developed during the week: that the symmetric pattern would collapse to one in which the lines radiated from a point on the surface. I found myself describing this as the pattern that remained after the symmetric one had "softly and suddenly vanished away." Having said that, I could not avoid proposing that the new pattern should be called a boojum.

The term "boojum" is from Lewis Carroll's "Hunting of the Snark," which ends

He had softly and suddenly vanished away.
For the Snark was a Boojum you see.

Goodness knows why "boojum" suggested softly and suddenly vanishing away to Carroll, but the connection having been made, it was inevitable that softly and suddenly vanishing away should suggest "boojum" to me. I resolved then and there to get the word into the literature.

There would be competition. Other people at the symposium had proposed calling my boojum a flower or a bouquet. Philip W. Anderson, who was to win the Nobel Prize the next year, and who appears at several critical moments in my tale, was not at the Sussex Symposium, but he was also thinking about spherical He^3-A, and wanted to call the stable pattern a fountain.

The first step toward assuring the adoption of "boojum" was easily accomplished. The discussion remarks, including my nomenclatural proposal in the guise of a discussion remark, received the usual minimal editorial attention. Shortly thereafter the word "boojum" made its maiden appearance in the literature of modern physics.

My next step was clearly to publish something which put my nomenclatural proposal to use, calling a boojum a boojum without fanfare or quotation marks. I wasn't ready to fight with editors of journals, but I was to deliver a paper on superfluid helium-3 at a conference to be held on Sanibel Island in January 1977, and the conference proceedings were to be published as a book. The form of my contribution to the Sanibel proceedings as well as the intensity of my interest in the boojum was considerably influenced by a series of letters I exchanged with Phil Anderson that fall.

Our correspondence was somewhat constrained by the fact that although I knew I was writing to Anderson, he—at least for a while—did not know he was writing to me. I had been sent the text of a paper by Anderson and Gerard Toulouse to referee for *Physical Review Letters*. Anderson and Toulouse argued that He^3-A might not be as good a superfluid as people had expected, producing an ingenious reason why what might appear to be a conventional permanent supercurrent could in fact lose its flow.

I saw a possible flaw in their argument. The surface of a container has a rather peculiar effect on the anisotropy axis of He^3-A. The local axes are forced to line up perpendicular to the surface at the boundaries of the liquid (as they are in the pictures of the spherical drops shown above). Although this seemed to be of no relevance to the argument of Anderson and Toulouse, I worried that it might, in fact, invalidate the mechanism they proposed for the disappearance of the supercurrent. I suggested that such questions should be cleared up before the paper was published. In the only mildly acrimonious correspondence that ensued, the authors and I both started to realize that the ability of the surface to stabilize the supercurrent was indeed relevant, but that this stabilizing power could be lost if there was even a single boojum (none of us called it that) on the surface.

This was interesting enough for their paper to appear (though without the word boojum), and I found myself more committed than ever to establishing the term. It was now evident that boojums were more than an inert feature of the structure of He^3-A drops; they had a vital role to play in the most fundamental property of the liquid, its superfluid flow. Furthermore it was no longer a bad idea about the pattern in spherical drops that was softly and suddenly vanishing away, but the supercurrent itself, whose soft and sudden vanishing could be triggered by a well placed boojum. My nomenclatural impulse had acquired the character of a prophetic vision.

In "The Hunting of the Snark," a boojum is a singular variety of snark with the alarming ability to bring about the soft and sudden vanishing away of anyone encountering it. The boojum in He^3-A, being a point at which anisotropy axes in different directions all meet, is a mathematical singularity. A singularity in He^3-A responsible for the vanishing of a supercurrent *had* to be called a boojum.

Accordingly, at the appropriate point in my paper for the Sanibel Symposium, I let loose a flock of boojums:

> This twisted boojum is shown at the bottom of the torus in the cross section in which it resembles the hyperbolic twistless boojum. . . . If either of the booja encircles its part of the torus (see inset) then two quanta are subtracted from the circulation about the entire torus.

Inspection of this specimen reveals that I had adopted "booja" as the plural form. As we shall see this turned out to be a serious error. I believe it was the one false step I made in an otherwise impeccable campaign.

A few months later I talked about boojums at a conference in New Hampshire. No proceedings were published, but several Russians attended the meeting and it seemed important to get them thinking boojum too. I had hoped that the first person plural future form of the Russian verb to be ("budyem," more or less pronounced "BOODyim") would make them receptive to the term, but I was never then or thereafter able to convince any Russian that the two words resembled each other in any way. No matter. The Russians took to boojums at once, and one even said a boojum or two in his own talk.

I returned from New Hampshire convinced that the boojum was ready to make its debut in an established scientific journal, but before I could consider how to bring this about there were two alarming developments.

Chia-Ren Hu, whom I had met at Sanibel, became interested in boojums, and wrote a paper, "Exact Solution of a 'Boojum' Texture in He^3-A." He sent it to the *Journal of Low Temperature Physics* and in due course received a letter from the editor, which read in part

> I have just received the comments of our referee on your paper and I enclose a copy of them. As you will see, he considers that the paper should be published provided the word "Boojum" be replaced with a suitable scientific word or phrase in the title, abstract, and text (p. 4, lines 8 & 17; p. 6, line 6). I too as General Editor concur unreservedly in this requirement. If you are willing to make such changes, we shall be happy to publish the paper.

Dr. Hu sent this communication on to me, and informed me that it was his plan to substitute for the word "boojum" the acronym "SOSO" (for "singular on surface only"). Appalled at the imminent possibility of my boojum turning into a SOSO, I wrote immediately to the General Editor, dissenting unreservedly from his conclusions:

Professor John G. Daunt
Editor, Journal of Low Temperature Physics

Dear Professor Daunt:

The term "boojum" is now in wide use in the field of superfluid helium-3. It first appeared at the Sussex Symposium (August 1976) and can be found in print in *Physica*, 90B + C, 1 (1977). The term appears again in the proceedings of the Sanibel Symposium (January 1977) to be published by Plenum within a month or two. There its literary origins are explained and its use justified. Briefly, a boojum is a singular beast, the appearance of which causes the observer to "softly and suddenly vanish away." This is precisely the role played by the boojum in He^3-A. If a boojum is in the container it can catalyze the decay of the supercurrent. Such a process is unique to He^3-A and it requires a new nomenclature. The word "boojum" is sanctified by Webster's unabridged (2nd ed.) where it is defined essentially as I have done above. It is therefore as respectable a term as the currently fashionable "hedgehog" (and rather more respectable than "quark").

In addition to the two citations mentioned above, the term "boojum" will appear in print in the proceedings of the June 1977 Erice Summer School on Quantum Liquids; it appears in a recent preprint of a *review* article by Brinkman and Cross; it appears in a recent preprint from the Landau Institute; and it was widely used and understood in discussions at last month's Gordon Conference on non-equilibrium phenomena in quantum liquids, attended by most of the world's experts on superflow in He^3-A.

To ask that Dr. Hu resort to circumlocutions in the text of his article serves no linguistic or esthetic purpose, and obscures the physical significance of the point he wishes to make. Now that I have told you a little more about the meaning and widespread use of the term, may I urge you to let him reintroduce it in his text.

> Sincerely yours,
> N. *David Mermin*
> Professor of Physics

The reference to the dictionary is important. If Hu's paper had mentioned a "flower texture" or a "fountain texture" chances are it would not have sounded alarms in the editorial office. It had occurred to me that if "boojum" were in the dictionary the character of the dispute would change. Not surprisingly, dictionaries readily at hand did not contain "boojum," but I did find it listed as an ordinary common noun in a copy I have at home of Webster's Unabridged, which I had won as an American history prize in high school in 1951. This was a few years before the appearance of the notorious 3rd edition, a fact of central importance in what was to follow. Had I not won the American history prize thirty years ago, "boojum" would not today be an internationally accepted scientific term.

My argument that "boojum" was a harmless common noun did not persuade Professor Daunt:

> I myself was well aware of the meaning that you have attached to the word "Boojum" since, amongst other occasions, I was in your audience at Sanibel last year. I am, of course, aware of its origin. However, at the moment it is not only my opinion, but also that of the reviewers of Dr. Hu's paper, that the physico-technical meaning of the word is insufficiently known to the international audience of our Journal to warrant its use as Dr. Hu wished to use it in his paper.

Clearly the gauntlet had been thrown down. To his credit, Daunt was as ruthlessly impermissive with the odious SOSO as he had been with my gentle boojum, but it was now essential that "boojum" appear in print in the most authoritative and widely circulated of all the international journals of physics: *Physical Review Letters.*

Of central importance to the success of this enterprise was the second alarming development. I received the text of a review of recent developments in the theory of superfluid helium-3 by W. F. Brinkman and M. C. Cross, both of Bell Labs. Leafing through the section of greatest interest to me, I was mortified to read

the expected configuration for a sphere is the "boogum" shown in Fig. 3.

The careful reader of my letter to Daunt will notice that I tried to make the best of even a bad business like this, but here was clearly a new kind of trouble. The manuscript was being circulated prior to publication, but would assume an authoritative position when it did appear. It was essential to set the spelling straight. I thought I would simply cite Webster to Brinkman and, being in my office rather than at home, went to the nearby Physical Science Library to look up the exact citation. I was appalled to find:

boogum or boojum (perh. fr. boojum, an imaginary creature in *The Hunting of the Snark* by Lewis Carroll (C. S. Dodgson) 1898 Eng. mathematician & writer; fr its grotesque appearance): a spiny tree (*Idria columnaris*) of the family Fouquieriacea chiefly of Lower California, sometimes arching over and rooting at its tips.

The library edition of Webster's was the third. Back at home I read again the lucid and concise second edition:

boojum, n. In Lewis Carroll's *Hunting of the Snark,* a species of snark the hunters of which "softly and suddenly vanish away."

I compared the two editions on snark:
2nd edition:

snark, n. (A blend of *snake* and *shark*). A nonsense creature invented by Lewis Carroll (Charles L. Dodgson), in his poem, *The Hunting of the Snark* (1876). One variety is known as the *boojum* (which see).

3rd edition:

snark (prob alter of snork) dial Brit: snore, snort.

In some ways this was absolutely uncanny. One has to understand, to begin with, that physicists from Bell Labs have a celebrated and annoying habit of disagreeing with you and being right, or agreeing with you but getting there first. More than once had Brinkman casually corrected public slips of mine, or pointed out politely that some of my more beautiful thoughts had already been thought by him—as I would have realized had I troubled to read the proceedings of last year's (or even the year before's Scottish Summer School), for example. If you look again at the picture of a boojum in He^3-A and think of a tree, "arching over and rooting at its tips," you do indeed begin to wonder whether it shouldn't, in fact, be called a boogum instead. This was not to be the last time that the telephone company threatened to snatch my boojum away from me.

As a result of this unfortunate episode I not only felt it essential to get "boojum" into the most distinguished of journals with the greatest possible speed, but also realized it would be necessary to ward off the upstart "boogum" while doing so.

There are two problems facing one wanting to get "boojum" into *Physical Review Letters*. The first is getting the article into *Physical Review Letters;* the second is getting the boojum into the article. I had been thinking about some puzzling aspects of supercurrents in He^3-A and while pondering the problem with some associates, the resolution became apparent. While not an earthshaking advance, it was a likely candidate for one of the major discoveries of the week, and *Physical Review Letters* seemed an entirely appropriate vehicle.

Without any distortion of our central point, I was able to introduce a remark about boojums. Anticipating resistance to the boojums even if the article were accepted, I carefully supplied footnotes documenting the scientific and literary origins of the word. I even added a reference to Webster's second edition. I had no doubt this would be excised by the editor, but I hoped it would convey to him the fact that "boojum" was, after all, no more than an ordinary English common noun, and therefore not a candidate for rigorous editorial scrutiny. Mindful, however, of the mess I would land in if they happened to check my citation with the third edition, I added the phrase

In this, as in many matters, the views of the 3rd edition should be spurned.

Then I sent the manuscript off.

Surprisingly soon after an associate editor of *Physical Review Letters*, Gene Wells, phoned me in my office. Our article had been accepted, he told me. Instantly I readied myself for the central battle of the campaign. *Physical Review Letters* does not announce the acceptance of papers by telephone, unless something is up, and I knew what was up: boojums!

There was a small problem he explained; in the second paragraph we used a word. . . . I gave him the essence of my letter to Daunt. To my gratification he acknowledged that this might be a case where a judiciously selected exception could fortify the general rule against neologisms. But the term needed to have a rigorous justification. For example *Physical Review Letters* was even taking a strong stand against "instanton." I congratulated him. But "boojum" was something else. Was it? he wanted to know, and then he put me through a cross-examination such as I have not had since my PhD qualifying exam. What aspect of the Boojum was pertinent? What was it that vanished away? Could the metaphor be construed as mixed? And, perhaps most importantly, if they let me get away with "boojum" would I be back to them with "snark"?

I swore then and there a solemn oath never to try to make the word "snark" an internationally accepted scientific term. I promised never again to try to introduce any new word at all. "Boojum" would be quite enough for me; indeed, the way things were going it might turn out to be altogether too much.

Wells said they would have to discuss the matter. Some time later he called again. "Boojum" had been approved. But there was the important question of the plural. I had used the form "booja" in our manuscript, not because I favor Latin plurals, but because I had always thought that Boojums was a common name for a rather unpleasantly fluffy kind of beribboned cat. We considered the possibilities. I had already discussed the question that summer in New Hampshire with A. J. Leggett of the University of Sussex, who has made profound contributions to the theory of superfluid helium-3 and who did an undergraduate degree in classics at Oxford. Leggett's position was simple: It would be evident to any ancient Roman that boojum was a word of foreign origin, and words of foreign origin are indeclinable. Therefore if one did want to form a latinate plural that plural should be not "booja" but "boojum." One boojum, two boojum.

I liked Leggett's logic, but had persisted in using "booja" in the belief that superfluid He^3-A was complicated enough without the added problem of distinguishing singular from plural. Now, however, *Physical Review Letters* and I were setting a standard for the generations to come. I thought (incorrectly, as we shall see) that Leggett's argument against "booja" was unassailable. Wells frowned on "boojum" for the same reasons that I did, and that left us only with "boojums," so "boojums" the plural became.

The debut of "boojum" as a fully authorized scientific term was quiet and dignified:

> . . . Surface current can be reduced (with an accompanying reduction to bulk current) by the motion of a special type of surface point singularity (a "boojum").[5,6] The relative importance of these mechanisms depends on details of the pinning, nucleation, and equilibrium populations of boojums, as well as on the energetics of vortex texture formation. . . .

I was surprised to find that my lexicographic footnote had not been deleted:

> [4]N. Webster, in *New International Dictionary* (Merriam, Springfield, Mass., 1934), 2nd ed., p. 308. See also p. 2379. (In this, as in many other matters, the views of the third edition should be spurned.)

Almost a year later I happened to meet Gene Wells, who confided to me that my lexicographic footnote had won the day for the boojum. It seems that the weight of opinion among the editors was against the term. However the editor-in-chief had loathed the third edition of Webster's for many years.

The unprecedented opportunity I had handed him to print a brisk attack on it in his own journal was more than he could resist; he forsook one set of linguistic principles for a higher one, and let the boojum in.

The final decisive moment of the campaign came when I again received for review from *Physical Review Letters* a paper of which P. W. Anderson was an author. And it had boojums. The word, in fact, appeared in title, abstract, text, *and* a figure caption. Anderson had just won his Nobel Prize. If the paper appeared nothing could sink the boojum. Eagerly I read it, and realized with dismay that it was wrong. I thus faced an unusual moral dilemma.

Relations between authors and referees are almost always strained. Authors are convinced that the malicious stupidity of the referee is alone preventing them from laying their discoveries before an admiring world. Referees are convinced that authors are too arrogant and obtuse to recognize blatant fallacies in their own reasoning, even when these have been called to their attention with crystalline lucidity. All physicists know this, because all physicists are both authors and referees, but it does no good. The ability of one person to hold both views is an example of what Bohr called complementarity.

In this case, however, the referee wanted the paper to appear more than the authors could have imagined. I nevertheless did the honorable thing. Believing that at best I would be rewarded with invective and abuse and at worst, if I was truly persuasive, I would prevent the culminating moment of my own hard fought campaign, I wrote a long thoughtful report, listing all of my objections.

I received a most courteous reply, thanking me for my help. Many of my suggestions were adopted and many of my objections were deftly and effectively dealt with. The central one was not, though the resubmittal letter politely but firmly insisted that it was.

What to do? The harmony could not survive another exchange of letters. I had been wrong before, particularly with authors associated with Bell Laboratories. And the paper, even though—or, I should say, if—wrong, was undoubtedly thought provoking.

I let it through. And it looked glorious. There were boojums all the way through.

As I was learning, however, things have a curious way of softly and suddenly going awry, when boojums are concerned. I was not to go unpunished for my breach of professional ethics, though retribution was another two years coming.

Meanwhile there were promising developments, of which I mention only one: the first Russian boojum. It occurred in a preprint I received last year from the Landau Institute. *Mermin nazval "budzhum,"* it declares: Mermin called it "boojum." It goes on to explain that the word is taken from *Lyuis Kerrol's* "*Okhota na Snarka.*" The Russian boojum was a counterexample to Leggett's theory that highly inflected languages would treat the word as indeclinable, and demonstrated that *booja* could have been an acceptable plural. In one page I found the nominative plural (*budzhumi*), the genitive plural (*budzhumov*), and, my favorite, the instrumental singular (*budzhumom*).

On Tuesday, 3 June 1980, my long delayed comeuppance arrived through the unexpected medium of *The New York Times.* An article appeared on whimsy in scientific nomenclature. It talked about quarks for a while, turned to Lewis Carroll, and then finally, at the end, it said

> Some snarks are dangerous to hunt, of course, because they may actually be boojums—beings that annihilate their hunters by making them disappear forever. Boojums found their way into science thanks to Philip W. Anderson, a 1977 Nobel Prize winner, who needed them as personae in a difficult notion about the broken symmetries of nature.

This terrible thing was brought to my attention, with a smirk, by my own graduate student. "Bell Labs always wins out in the end" he cheerily opined, and danced off.

There was only one thing to do. I wrote a short but firm letter to the *Times*, and gave it to my trusted friend and colleague, N. W. Ashcroft, to send under his own name.

The day after Ashcroft's letter went off I received a nice note from Anderson:

> Dear David:
>
> I note a depressingly typical example of the Matthew effect in today's *Science Times*. Do you want me to try to correct it? He didn't talk to me or anybody who knew anything.
>
> Regards, and sorry
> *Phil*

I sent back a cheery reply, passing it off as a case of *sic transit gloria boojorum*. I did, however, ask what the Matthew effect was. I got an immediate reply: "Matthew effect: R. Merton: 'to him that hath shall be given, etc.'" I knew R. Merton hadn't said that, and turning to my Bible found that "etc." stood for "but from him that hath not shall be taken away even that which he hath." There I was, once again at the wrong end of the Matthew effect, and not even knowing it until I was told.

But he who deals in boojums does not stay down for long. The *Times* printed Ashcroft's letter; it has yet to print Anderson's reply to Ashcroft's letter. And in that vast land beyond the Bell System where *The New York Times* cannot be had for love or money, they will all shortly be learning that *Mermin nazval budzhum*.

Questions and Topics for Discussion and Writing

1. Part of the title of this essay is an allusion to the motto of the United States. After reading Mermin's essay, discuss how this allusion, combined with the word *boojum*, influences your expectations for the essay.
2. Look up the word *neologism*. Does a physicist seem an appropriate person to be the creator of one? Write a paragraph in which you defend Mermin as a neologist.

3. Mermin mentions three other words and, later, one acronym that physicists suggested for what became the boojum. Do any of these terms seem more appealing or suitable to you? If so, why? If not, why do your prefer "boojum"?

4. Mermin writes, "My nomenclatural impulse had acquired the character of a prophetic vision." Explain this idea in your own words.

5. At first, Mermin makes the plural of *boojum* the word *booja*. Why do you think he chose this form? Explain why he came to consider this form a "serious error."

6. List the steps Mermin used to introduce the term *boojum* into scientific literature. How is Mermin's tale about his first attempts to publish something that would include the word an assertion that scientists can also be linguists?

7. What does Mermin say is the vital role of boojums in superfluid flow of He^3-A?

8. Mermin compares his term to the accepted *hedgehog* and *quark*. Look up these terms as used in physics today and write a brief definition of each.

9. Read *The Hunting of the Snark*, written by Charles Lutwidge Dodgson (1832–1898), an Oxford University mathematics don who is better known as Lewis Carroll. Under this pseudonym, he wrote nonsense verse and a number of children's books, including *Alice's Adventures in Wonderland*. "Snark" is a poem evidently attractive to mathematicians. Martin Gardner, for example, edited *The Annotated Snark* in 1962. What qualities and characteristics do you find in "Snark" that may explain this attraction? Write a short paper explaining the basis for physicist Mermin's attraction to Carroll.

10. Read a brief biography of Lewis Carroll (Charles Lutwidge Dodgson), and read the poem "Jabberwocky." Determine, if you can, what qualities of mind would cause a serious mathematician to create nonsense verse.

11. Why does Mermin speak of "the notorious 3rd edition" of Webster's Dictionary?

12. Why does Mermin use battle imagery? Explain the metaphor "the gauntlet had been thrown down." Does such a metaphor—in high scientific circles—surprise you? Does Mermin continue the use of this kind of imagery throughout? If so, list other examples and explain their purpose.

13. Describe Mermin's view of the physicists at Bell Laboratories. Does he imply that the "telephone company" has a "personality"? Explain.

14. What is a "mixed metaphor"? Why would Mermin wish to avoid such a form?

15. Mermin refers to Danish physicist Niels Bohr's theory of *complementarity*. Look up this theory and, in a short paper, discuss some of its consequences. You may, for example, want to pursue the debate between Bohr and Einstein on this subject. Does Mermin's reference to the theory add support to his account? How?

16. In a short paper, discuss the role of creativity and playfulness in science as demonstrated in this essay.

Fritjof Capra

Fritjof Capra works in theoretical high-energy physics at the Lawrence Berkeley Laboratory and lectures at the University of California at Berkeley. A doctor of physics from the University of Vienna (1966), he came to public attention through his controversial book *The Tao of Physics* (1975). In it, he tried to synthesize modern physics and the *yin-yang* of Eastern religious thought. He followed with *The Turning Point: Science, Society and Rising Culture* (1982), which expands upon the interconnectedness of animate and inanimate matter, systems theory, and a new way of perceiving the nature of things.

Capra's central theme is that today's crises result from our tendency to perceive the world in Newtonian terms. Although Capra readily acknowledges that the digital/analytical Newtonian approach led to impressive technological advances, he points out that this view is mechanistic. Held by politicians, social scientists, economists, and industrialists, it has led to the competitive, isolationist, war mentality that pollutes the planet and continually threatens us with nuclear self-destruction. What the world needs now, Capra holds, is a perspective shift from a mechanistic to a holistic, systems view. While many modern physicists criticize his views as too sweeping, Capra sees the world as an integrated series of systems within systems. The concept of static structure, too, must be abandoned, in favor of the more dynamic notion of matter as the result of rhythmic patterns of organization of energy, thus completing the shift from a Newtonian world-view to a relativistic, Einsteinian one.

In the Epilogue to *The Tao of Physics*, Capra reveals that his parallels do not present a "rigorous demonstration" but, rather, the opportunity to enlarge one's paradigm of the cosmos, or world-view.

Epilogue,
The Tao of Physics
Fritjof Capra

The Eastern religious philosophies are concerned with timeless mystical knowledge which lies beyond reasoning and cannot be adequately expressed in words. The relation of this knowledge to modern physics is but one of its many aspects and, like all the others, it cannot be demonstrated conclusively but has to be experienced in a direct intuitive way. What I hope to have achieved, to some extent, therefore, is not a rigorous demonstration, but rather to have given the reader an opportunity to relive, every now and then, an experience which has become for me a source of continuing joy and inspiration; that the principal theories and models of modern physics lead to a view of the world which is internally consistent and in perfect harmony with the views of Eastern mysticism.

For those who have experienced this harmony, the significance of the parallels between the world views of physicists and mystics is beyond any doubt. The interesting question, then, is not *whether* these parallels exist, but *why;* and, furthermore, what their existence implies.

In trying to understand the mystery of Life, man has followed many different approaches. Among them, there are the ways of the scientist and mystic, but there are many more; the ways of poets, children, clowns, shamans, to name but a few. These ways have resulted in different descriptions of the world, both verbal and non-verbal, which emphasize different aspects. All are valid and useful in the context in which they arose. All of them, however, are only descriptions, or representations, of reality and are therefore limited. None can give a complete picture of the world.

The mechanistic world view of classical physics is useful for the description of the kind of physical phenomena we encounter in our everyday life and thus appropriate for dealing with our daily environment, and it has also proved extremely successful as a basis for technology. It is inadequate, however, for the description of physical phenomena in the submicroscopic realm. Opposed to the mechanistic conception of the world is the view of the mystics which may be epitomized by the word "organic," as it regards all phenomena in the universe as integral parts of an inseparable harmonious whole. This world view emerges in the mystical traditions from meditative states of consciousness. In their description of the world, the mystics use concepts which are derived from these non-ordinary experiences and are, in general, inappropriate for a scientific description of macroscopic phenomena. The organic world view is not advantageous for constructing machines, nor for coping with the technical problems in an overpopulated world.

In everyday life, then, both the mechanistic and the organic views of the universe are valid and useful; the one for science and technology, the other for a balanced and fulfilled spiritual life. Beyond the dimensions of our everyday environment, however, the mechanistic concepts lose their validity and have to be replaced by organic concepts which are very similar to those used by the mystics. This is the essential experience of modern physics which has been the subject of our discussion. Physics in the twentieth century has shown that the concepts of the organic world view, although of little value for science and technology on the human scale, become extremely useful at the atomic and subatomic level. The organic view, therefore, seems to be more fundamental than the mechanistic. Classical physics, which is based on the latter, can be derived from quantum theory, which implies the former, whereas the reverse is not possible. This seems to give a first indication why we might expect the world views of modern physics and Eastern mysticism to be similar. Both emerge when man enquires into the essential nature of things—into the deeper realms of matter in physics; into the deeper realms of consciousness in mysticism—when he discovers a different reality behind the superficial mechanistic appearance of everyday life.

The parallels between the views of physicists and mystics become even more plausible when we recall the other similarities which exist in spite of their different approaches. To begin with, their method is thoroughly empirical. Physicists derive their knowledge from experiments; mystics from meditative insights. Both are observations, and in both fields these observations are acknowledged as the only source of knowledge. The object of observation is of course very different in the two cases. The mystic looks within and explores his or her consciousness at its various levels, which include the body as the physical manifestation of the mind. The experience of one's body is, in fact, emphasized in many Eastern traditions and is often seen as the key to the mystical experience of the world. When we are healthy, we do not feel any separate parts in our body but are aware of it as an integrated whole, and this awareness generates a feeling of well-being and happiness. In a similar way, the mystic is aware of the wholeness of the entire cosmos which is experienced as an extension of the body. In the words of Lama Govinda,

> To the enlightened man . . . whose consciousness embraces the universe, to him the universe becomes his "body," while his physical body becomes a manifestation of the Universal Mind, his inner vision an expression of the highest reality, and his speech an expression of eternal truth and mantric power. . . .

In contrast to the mystic, the physicist begins his enquiry into the essential nature of things by studying the material world. Penetrating into ever deeper realms of matter, he has become aware of the essential unity of all things and events. More than that, he has also learnt that he himself and his consciousness are an integral part of this unity. Thus the mystic and the physicist arrive at the same conclusion; one starting from the inner realm, the other from the outer world. The harmony between their views confirms the ancient Indian wisdom that *Brahman*, the ultimate reality without, is identical to *Atman*, the reality within.

A further similarity between the ways of the physicist and mystic is the fact that their observations take place in realms which are inaccessible to the ordinary senses. In modern physics, these are the realms of the atomic and subatomic world; in mysticism they are non-ordinary states of consciousness in which the sense world is transcended. Mystics often talk about experiencing higher dimensions in which impressions of different centres of consciousness are integrated into a harmonious whole. A similar situation exists in modern physics where a four-dimensional "space-time" formalism has been developed which unifies concepts and observations belonging to different categories in the ordinary three-dimensional world. In both fields, the multi-dimensional experiences transcend the sensory world and are therefore almost impossible to express in ordinary language.

We see that the ways of the modern physicist and the Eastern mystic, which seem at first totally unrelated, have, in fact, much in common. It should not be too surprising, therefore, that there are striking parallels in their descriptions of the world. Once these parallels between Western science and Eastern mysticism are accepted, a number of questions will arise concerning their implications. Is modern science, with all its sophisticated machinery, merely rediscovering ancient wisdom, known to the Eastern sages for thousands of years? Should physicists, therefore, abandon the scientific method and begin to meditate? Or can there be a mutual influence between science and mysticism; perhaps even a synthesis?

I think all these questions have to be answered in the negative. I see science and mysticism as two complementary manifestations of the human mind; of its rational and intuitive faculties. The modern physicist experiences the world through an extreme specialization of the rational mind; the mystic through an extreme specialization of the intuitive mind. The two approaches are entirely different and involve far more than a certain view of the physical world. However, they are complementary, as we have learned to say in physics. Neither is comprehended in the other, nor can either of them be reduced to the other, but both of them are necessary, supplementing one another for a fuller understanding of the world. To paraphrase an old Chinese saying, mystics understand the roots of the *Tao* but not its branches; scientists understand its branches but not its roots. Science does not need mysticism and mysticism does not need science; but man needs both. Mystical experience is necessary to understand the deepest nature of things, and science is essential for modern life. What we need, therefore, is not a synthesis but a dynamic interplay between mystical intuition and scientific analysis.

So far, this has not been achieved in our society. At present, our attitude is too *yang*—to use again Chinese phraseology—too rational, male and aggressive. Scientists themselves are a typical example. Although their theories are leading to a world view which is similar to that of the mystics, it is striking how little this has affected the attitudes of most scientists. In mysticism, knowledge cannot be separated from a certain way of life which becomes its living manifestation. To acquire mystical knowledge means to undergo a transformation; one could even say that the knowledge *is* the transformation. Scientific knowledge, on the other hand, can often stay abstract and theoretical. Thus most of today's physicists do not seem to realize the philosophical, cultural and spiritual implications of their theories. Many of them actively support a society which is still based on the mechanistic, fragmented world view, without seeing that science points beyond such a view, towards a oneness of the universe which includes not only our natural environment but also our fellow human beings. I believe that the world view implied by modern physics is inconsistent with our present society, which does not reflect the harmonious interrelatedness we observe in nature. To achieve such a state of dynamic balance, a radically different social and economic structure will be needed: a cultural revolution in the true sense of the word. The survival of our whole civilization may depend on whether we can bring about such a change. It will depend, ultimately, on our ability to adopt some of the *yin* attitudes of Eastern mysticism; to experience the wholeness of nature and the art of living with it in harmony.

Questions and Topics for Discussion and Writing

1. List the terms Capra uses to define mysticism. Define the Oriental terms *Tao*, *Brahman*, *Atman*, *yang*, and *yin*.
2. What does Capra mean by a "macroscopic world view"? Provide examples of macroscopic phenomena for which, according to Capra, the mystical approach is not advantageous.

3. What does Capra offer as the limits of the world-view of classical physics?
4. What evidence do you find for Capra's claim to personal mystical experience?
5. Artists are often said to offer a mystical world-view. How, then, in the light of Capra's argument, can art serve to make life more meaningful for modern individuals?
6. What does Capra mean by "reality"? Does he imply that the physicist, like the Indian mystic, experiences a *divine* reality in his awareness of the essential unity of all things?
7. How do "objects of observation" of the physicist and of the mystic differ?
8. What does Capra mean by "specialization" when discussing "two complementary manifestations of the human mind"?
9. Capra concludes that "man needs both" science and mysticism. In an essay explain his rationale for that conclusion.
10. *The Tao of Physics* generated controversy. Look up book reviews describing its reception, and pinpoint the ideas considered to be most controversial. Examine carefully the immediate reactions. Classify and summarize them in a short paper.

Alexander Calandra

Alexander Calandra (b. 1911) is an American chemist and physicist. A native New Yorker, he received his bachelor's degree in science from Brooklyn College and his doctorate in statistics from New York University. As an educator, Calandra has been a consultant to the American Council for Education. In 1950, he became professor of physics at Washington University in St. Louis. He has been a leader in educational television.

The following short story reveals its author as a student of the history of science, a keen observer of educational methods, and a teacher who recognizes the potential of young people. With wit and humor, "The Barometer Story: Angels on a Pin" offers a challenge to modern teachers of the talented and gifted.

The Barometer Story: Angels on a Pin

Alexander Calandra

Some time ago, I received a call from a colleague who asked if I would be the referee on the grading of an examination question. He was about to give a student a zero for his answer to a physics question, while the student claimed he should receive a perfect score and would if the system were not set up against the student. The instructor and the student agreed to submit this to an impartial arbiter, and I was selected.

I went to my colleague's office and read the examination question: "Show how it is possible to determine the height of a tall building with the aid of a barometer."

The student had answered: "Take the barometer to the top of the building, attach a long rope to it, lower the barometer to the street, and then bring it up, measuring the length of the rope. The length of the rope is the height of the building."

I pointed out that the student really had a strong case for full credit, since he had answered the question completely and correctly. On the other hand, if full credit were given, it could well contribute to a high grade for the student in his physics course. A high grade is supposed to certify competence in physics, but the answer did not confirm this. I suggested that the student have another try at answering the question. I was not surprised that my colleague agreed, but I was surprised that the student did.

I gave the student six minutes to answer the question, with the warning that his answer should show some knowledge of physics. At the end of five minutes, he had not written anything. I asked if he wished to give up, but he said no. He had many answers to this problem; he was just thinking of the best one. I excused myself for interrupting him, and asked him to please go on. In the next minute, he dashed off his answer which read:

"Take the barometer to the top of the building and lean over the edge of the roof. Drop the barometer, timing its fall with a stopwatch. Then, using the formula $S = \frac{1}{2} at^2$, calculate the height of the building."

At this point, I asked my colleague if *he* would give up. He conceded, and I gave the student almost full credit.

In leaving my colleague's office, I recalled that the student had said he had other answers to the problem, so I asked him what they were. "Oh, yes," said the student. "There are many ways of getting the height of a tall building with the aid of a barometer. For example, you could take the barometer out on a sunny day and measure the height of the barometer, the length of its shadow, and the length of the shadow of the building, and by the use of a simple proportion, determine the height of the building."

"Fine," I said. "And the others?"

"Yes," said the student. "There is a very basic measurement method that you will like. In this method, you take the barometer and begin to walk up the stairs. As you climb the stairs, you mark off the length of the barometer along the wall. You then count the number of marks, and this will give you the height of the building in barometer units. A very direct method.

"Of course, if you want a more sophisticated method, you can tie the barometer to the end of a string, swing it as a pendulum, and determine the value of 'g' at the street level and at the top of the building. From the difference between the two values of 'g,' the height of the building can, in principle, be calculated."

Finally he concluded, there are many other ways of solving the problem. "Probably the best," he said, "is to take the barometer to the basement and knock on the superintendent's door. When the superintendent answers, you speak to him as follows: 'Mr. Superintendent, here I have a fine barometer. If you will tell me the height of this building, I will give you this barometer.'"

At this point, I asked the student if he really did not know the conventional answer to this question. He admitted that he did, but said that he was fed up with high school and college instructors trying to teach him how to think, to use the "scientific method," and to explore the deep inner logic of the subject in a pedantic way, as is often done in the new mathematics, rather than teaching him the structure of the subject. With this in mind, he decided to revive scholasticism as an academic lark to challenge the Sputnik-panicked classrooms of America.

Questions and Topics for Discussion and Writing

1. Look up *scholasticism*, a medieval school of thought. In what way is the allusion of the subtitle, "Angels on a Pin," related to this school? Explain.
2. The Latin poet and critic Horace said that literature exists to *delight* and to *instruct*. You might apply this notion to this story or a story of your choice. Write a short paper in which you agree or disagree.
3. In the light of your readings in Unit 2, comment on the student's application of the scientific method.
4. Consider the "real-life" student in "The A-Bomb Kid" (Unit 7). In an essay, compare and contrast the two pieces in terms of characterization, method, and tone.

SUGGESTIONS FOR FURTHER READING

Bergmann, P. G. *Basic Theories of Physics, Volumes I and II*. New York: Dover Publications, 1949.

Bernal, John Desmond. *The Extension of Man: A History of Physics Before the Quantum*. Cambridge: MIT Press, 1972.

Cole, K. C. *Sympathetic Vibrations: Reflections on Physics as a Way of Life*. New York: William Morrow, 1985.

Einstein, Albert, and Leopold Infeld. *The Evolution of Physics*. New York: Simon & Schuster, 1961.

Fritzsch, H. *Quarks: The Stuff of Matter*. New York: Basic Books, 1983.

Gamow, George. *Biography of Physics*. New York: Harper & Row, 1961.

———. *Mr. Tompkins in Paperback* (Composite of *Mr. Tompkins Is Wonderful* and *Mr. Tompkins Explores the Atom*). London: Cambridge University Press, 1965.

Graetzer, H. G., and D. Anderson. *The Discovery of Nuclear Fission.* New York: Arno Press, 1981.

Jones, Roger S. *Physics as Metaphor.* New York: New American Library, 1983.

Kelves, D. *The Physicists.* New York: Alfred Knopf, 1978.

Pagels, H. R. *The Cosmic Code: Quantum Physics as the Language of Nature.* New York: Bantam, 1983.

Snow, C. P. *The Physicists.* Boston: Little, Brown. 1981. (Novel).

Wolf, Abraham. *A History of Science, Technology and Philosophy in the 18th Century.* London: Allen and Unwin, 1952.

Zukav, Gary. *The Dancing Wu Li Masters: An Overview of the New Physics.* New York: Morrow, 1979.

Unit 6

CHEMISTRY AND GEOLOGY: SEARCHING OUT MATTER

Lucretius

Lucretius adds a voice from the ancient world to discussion of "the nature of things." He expresses a surprisingly "modern scientific outlook," according to many readers—or, at least, a vision of the universe with which many scientific readers say they feel a kinship. A Roman citizen whose dates of birth and death are not known with exactness, Titus Lucretius Carus was a younger contemporary of Julius Caesar's. At the time of Lucretius, Rome was at the height of its power, and one of its "imports" from subjugated peoples was ideas. His acquaintance with the ideas of the Greek philosopher Epicurus seems to have been the most significant happening of his life. With an imagination captured by the Epicurean vision of the physical universe, Lucretius devoted his creative energies to its expression.

Epicurean philosophy was actually founded on the physics of previous thinkers, Leucippus and Democritus, the first atomists. They tried to explain the world in purely rational terms, contending that basic matter consists of minute, moving, unchanging particles. The existence of the "void" allows matter to be divided into these particles and allows the particles to move. Epicurus took up their physics and based his ethical theory on their system.

Lucretius, in turn, was impressed by the Epicurean notion that the world of human beings is ruled by the movement of atoms. The message of his famous philosophical poem *De Rerum Natura (On the Nature of Things)* is that the

237

atomic explanation should banish human fear of death and fear of the gods. Lucretius believed that assimilating his view would allow the reader "to be able to regard all that is with a mind at peace." While many readers find the vision of such utter materialism in the poem to be bleak and terrible in one way, they also praise the vivid imagery of life understood through the senses as bringing the poem an extraordinary tension. In an analysis of *De Rerum Natura* entitled *The Lyre of Science,* literary critic Richard Minadeo praises the poem for its successful fusion of scientific philosophy and poetry.

In the following selection from Book I, Lucretius uses imaginative deduction to demonstrate that invisible bodies must exist and that Nature acts by means of these invisible particles, as we can ascertain from observation.

On the Nature of Things

Lucretius

You know I have said creation out of nothing
Is nonsense. So is destruction of things to nothingness.
But since you may doubt the validity of a doctrine
Requiring the existence of invisible elements,
I should like to draw your attention to certain bodies
Which must be allowed to exist, although we can't see them.

Think of the winds, which beat up the sea with their blows,
Wrecking the largest vessels, scattering the clouds;
And sometimes, driving a hurricane over the plains,
Strewing great trees on the ground, and with shattering blasts
Lashing the mountaintops: a roaring, a fury,
There is rage to come in their smallest menacing murmur.
No doubt at all, the winds are invisible bodies
Which sweep across the sea, the earth and the sky
And toss the clouds and carry them off in a storm.
You may compare them and the damage they do
To what is done by water, whose nature is gentle,
Yet when the rivers are swollen by terrible downpours
Collected on mountain slopes and sent hurtling down,
They carry before them branches and even whole trees;
No bridges are strong enough for the sudden onrush,
They crumple; the river, carrying the rains in its arms,

Crashes against the piers and pushes them forward;
They fall with a roar, and they are under the water,
Immense blocks: nothing could stand against the river.
So with the winds; it must be, their action is similar
For like a river they lash wherever they choose,
Pushing whatever impedes them, overturning
In one or several assaults; sometimes they lift
And carry things upwards in an eddying swirl.
It proves, it must prove, that winds are invisible bodies,
For by their action and habit they rival the rivers
Which no one denies are made of a visible substance.

Or again, take smell. We perceive all manner of odours
But never observe one on its way to our noses.
Nor does sight communicate blazing heat or cold weather
Or enable us to detect or distinguish a sound;
Yet the nature of all these things must of course be physical
Since otherwise they could not impress our senses
—For impression means touch, and touch means the touch of bodies.

Then observe, if you hang clothes out where the waves are breaking,
They get wet, just as they dry if they're spread in the sun.
Yet nobody ever saw how the damp gets into them
Or how it gets out when the weather is hot.
It follows that moisture must be composed of particles
So small it is not possible they should be seen.
In the same way, if you wear a ring on your finger,
After many years it will wear perceptibly thin;
A drip will hollow a stone; the blade of a plough
In time will secretly wear away in the fields;
And paving-stones grow smooth and thin with crowds
Who tread on them year by year: by a city gate
You may see a statue of bronze with the right hand worn
Where travellers have kissed it as they went on their way.
These things diminish, we see, by little and little,
But what is lost at any particular time
Is something that nature does not allow us to see
Any more than she allows us to see what is added
To bodies in the course of their natural growth.
The same is true of what is taken away
From bodies when they are wasted by time and age;
And there are half-eaten cliffs overhanging the sea,
But who ever saw the salt removing a mouthful?
Nature does all these things with invisible substances.

Not everywhere, however, is crowded with matter,
For nature is such that everything has its emptiness.
This is a necessary part of the lesson,
Without which nature would continue to mystify you
And my theories of it would in fact be incomplete.
There is the emptiness of unoccupied space,
Without which, clearly, nothing could ever move.
The function of matter is to get in the way;
If there were no space nothing could ever move
But everything get in the way of everything else.
Nothing would ever give, and nothing would budge.
But in fact we see the seas move, the earth, the clouds,
The stars sweep by, and everything has its movement.
If there were no emptiness none of this could happen,
Nothing indeed could ever change or begin;
There would be close-packed matter and that would be all.

The fact is, things which appear to us to be solid
Are really made of somewhat rarified stuff.
That is why water drips through the roof of a cave
And it looks as if thick slabs of rock had burst into tears;
That is why food distributes itself through the body:
Trees grow, and manage in time to produce their fruit
Because what they feed on is carried from roots to the trunk
And so in the end to the very tip of the branches.
Noises don't stop at a wall but are carried right through,
It makes no difference that the house is shut up;
The cold gets into our bones: and none of these things
Could happen, unless there were spaces matter could get through.

And why is it some things weigh a lot more than others
Although the volume is exactly the same?
A lump of bread and a lump of wool, for example?
The difference must be in the proportion of matter.
The nature of matter is to press everything down
While the nature of empty space is to be without weight.
It follows that, with objects of equal volume,
The lighter must be the one which contains more space
And the heavier must be the one which contains more matter
While the space it contains must be accordingly smaller.
This demonstrates that the composition of things
Includes, as well as matter, some empty space.

*   *   *

First of all, since it is clear that nature is twofold,
Consisting of elements of quite different kinds,
Body, and space in which all events take place,
These two must be quite separate from each other.
For where there is space with nothing in it but emptiness,
There can be no body there; and where there is body,
It clearly will not do to talk about emptiness.
So particles are quite solid and have no space in them.

Since there is emptiness in created things,
It must be surrounded by something solid:
For could things hide such emptiness in their interior
If there were no material around to hide it?
And what could this be except a collection of particles
Arranged to form a sort of screen for the void?
Matter, consisting entirely of solid particles,
Can be eternal, though everything made of it dies.
Then, if there were no such thing as empty space,
Everything would be solid; on the other hand,
If there were not bodies which filled up all the space
They occupied, so that nothing else could intrude,
The universe would be nothing but emptiness.
But matter and space, in fact, are alternatives,
They cannot be both in one place; the world is neither
Made up entirely of the one nor of the other
And this mixed nature of things would only be possible
If there were bodies which did not give way to the void.

These bodies—the particles—cannot break up at a blow,
Nor can anything get past their outer defences,
Nor can they yield or give way whatever may come
—All these points that I have already made.

                    *   *   *

The original particles, although themselves invisible,
Must have limits, which means, a series of points
And these must be the smallest bodies in nature,
Without parts; points moreover which never existed
In isolation, or could so ever exist
Since they are only parts of another body
—Units which, joined together with others like them,
Make up the bodies of the original particles.
And since these points have no existence apart,
They must remain eternally glued together.

So the particles are of solid and simple nature,
Made up of crowded irreducible points
And not the product of any act of assembly,
But such that they have always existed in that conjunction:
No kind of separation or any subtraction
From the particles which are the seeds of everything.

Questions and Topics for Discussion and Writing

1. Lucretius uses analogy as his method of argument. In a paragraph, comment on his use of the action of visible things as an analogy to argue for the existence of invisible things.
2. Lucretius is not a narrative poet, but a didactic one. He sees his purpose as a moral one: to teach the truth about "the nature of things"—which for him is a thoroughly materialistic vision—in order to free humanity from its fears. To accomplish his purpose, he is sometimes what may be called "technical." Cite words or phrases that evince this technical quality.
3. What importance does Lucretius give to the evidence of the senses? Do you find that he employs careful or intense observation? If so, list examples.
4. Although his observations of the world do not have the support of controlled experiment, Lucretius is often called a "scientific poet." Do you consider his outlook "scientific"? Explain.
5. Lucretius's vision has been described by some as "bleak" or "gray." Yet others have praised the poem as lifted "beyond logic into lyricism." Do you find any evidence of lyricism—perhaps of delight in nature?
6. Research the theories of Leucippus and Democritus, and, in a short paper, relate your findings to modern atomic theory.
7. Write a paragraph explaining how you think Lucretius (who considered the atom an irreducible particle) might react to the modern achievement of splitting the atom.

Lewis Thomas

The self-styled "biology watcher" Lewis Thomas (b. 1913) is a medical doctor, a cancer researcher, and an illuminating writer on human physical and philosophical concerns. His eminent medical career can be said in one way to be the result of family expectations: When he was three or four, his physician father began taking him on house calls to prepare the boy to be a doctor also. Perhaps early memories of his father's "having a good time in this work" gave Thomas the outlook in his popular medical writing that blends the serious and the lighthearted.

According to Thomas, the worst thing that has happened to science education is that students no longer see the "fun" in science. They don't see science as the "high adventure it really is, the wildest of all explorations . . . the chance to catch close views of things never seen before." The merging of the imagination and insight of a creative writer with the scientist's passion for detail goes far to explain the strength of his style. To biomedical research, Thomas has brought his own individuality, interpreting his specialized field in essays that have won a wide reading public. About his musings, Stephen Jay Gould writes, "He has a vision—a kind and humane vision—and visions are in dreadfully short supply in an intellectual world of intense specialization and cynicism."

A prolific writer, Thomas has written more than 200 technical papers on virology, immunology, experimental pathology, and infectious disease, and has received twenty honorary degrees in science, law and letters, and music. In 1971, he began writing a monthly column in the weekly *New England Journal of Medicine* under the general heading "Notes of a Biology Watcher." Many of these essays later appeared in *The Lives of a Cell* (1974), which was a tremendous success in both literary and scientific circles. In 1975, *Lives* earned the National Book Award. Thomas has also produced *The Medusa and the Snail* (1979), which won the American Book Award for 1981, *The Youngest Science* (1983), and *Late Night Thoughts On Listening to Mahler's Ninth Symphony* (1983).

Thomas graduated from Princeton in 1933 and from Harvard Medical School, *cum laude*, in 1937. He began reading poetry while a student at Princeton, where he was particularly inspired by the poetry of Ezra Pound and T. S. Eliot. During World War II, he worked in a United States Navy laboratory of virology. Since 1973 he has been professor of medicine and pathology at the Cornell University Medical College and, since 1975, an adjunct professor at The Rockefeller University. Thomas is currently President Emeritus of Memorial Sloan-Kettering Cancer Center and University Professor at the State University of New York and trustee of the Guggenheim Foundation.

"Alchemy," the following selection, examines the origins of chemistry in the "magical" experiments and lore of medieval practitioners.

Alchemy

Lewis Thomas

Alchemy began long ago as an expression of the deepest and oldest of human wishes: to discover that the world makes sense. The working assumption—that everything on earth must be made up from a single, primal sort of matter—led to centuries of hard work aimed at isolating the original stuff and rearranging it to the alchemists' liking. If it could be found, nothing would lie beyond human grasp. The transmutation of base metals to gold was only a modest part of the prospect. If you knew about the fundamental substance, you could do much more than make simple money: you could boil up a cure-all for every disease affecting humankind, you could rid the world of evil, and, while doing this, you could make a universal solvent capable of dissolving anything you might want to dissolve. These were heady ideas, and generations of alchemists worked all their lives trying to reduce matter to its ultimate origin.

To be an alchemist was to be a serious professional, requiring long periods of apprenticeship and a great deal of late-night study. From the earliest years of the profession, there was a lot to read. The documents can be traced back to Arabic, Latin, and Greek scholars of the ancient world, and beyond them to Indian Vedic texts as far back as the tenth century B.C. All the old papers contain a formidable array of information, mostly expressed in incantations, which were required learning for every young alchemist and, by design, incomprehensible to everyone else. The word "gibberish" is thought by some to refer back to Jabir ibn Hayyan, an eighth-century alchemist, who lived in fear of being executed for black magic and worded his doctrines so obscurely that almost no one knew what he was talking about.

Indeed, black magic was what most people thought the alchemists were up to in their laboratories, filled with the fumes of arsenic, mercury, and sulphur and the bubbling infusions of all sorts of obscure plants. We tend to look back at them from today's pinnacle of science as figures of fun, eccentric solitary men wearing comical conical hats, engaged in meaningless explorations down one blind alley after another. It was not necessarily so: the work they were doing was hard and frustrating, but it was the start-up of experimental chemistry and physics. The central idea they were obsessed with—that there is a fundamental, elementary particle out of which everything in the universe is made—continues to obsess today's physicists.

They never succeeded in making gold from base metals, nor did they find a universal elixir in their plant extracts; they certainly didn't rid the world of evil. What they did accomplish, however, was no small thing: they got the work going. They fiddled around in their laboratories, talked at one another incessantly, set up one crazy experiment after another, wrote endless reams of notes, which were then translated from Arabic to Greek to Latin and back again, and the work got under way. More workers became interested and then involved in the work, and, as has been happening ever since in science, one thing led to another. As time went on and the work progressed, error after error, new and accurate things began to turn up. Hard facts were learned about the behavior of metals and their alloys, the properties of acids, bases, and salts were recognized, the mathematics of thermodynamics were worked out, and, with just a few jumps through the centuries, the helical molecule of DNA was revealed in all its mystery.

The current anxieties over what science may be doing to human society, including the worries about technology, are no new thing. The third-century Roman emperor Diocletian decreed that all manuscripts dealing with alchemy were to be destroyed, on grounds that such enterprises were against nature. The work went on in secrecy, and, although some of the material was lost, a great deal was translated into other languages, passed around, and preserved.

The association of alchemy with black magic has persisted in the public mind throughout the long history of the endeavor, partly because the objective—the transmutation of one sort of substance to another—seemed magical by definition. Partly also because of the hybrid term: *al* was simply the Arabic article, but *chemy* came from a word meaning "the black land," *Khemia*, the Greek name for Egypt. Another, similar-sounding word, *khumeia*, meant an infusion or elixir, and this was incorporated as part of the meaning. The Egyptian origin is very old, extending back to Thoth, the god of magic (who later reappeared as Hermes Trismegistus, master of the hermetic seal required by alchemists for the vacuums they believed were needed in their work). The notion of alchemy may be as old as language, and the idea that language and magic are somehow related is also old. "Grammar," after all, was a word used in the Middle Ages to denote high learning, but it also implied a practicing familiarity with alchemy. *Gramarye*, an older term for grammar, signified occult learning and necromancy. "Glamour," of all words, was the Scottish word for grammar, and it meant, precisely, a spell, casting enchantment.

Medicine, from its dark origins in old shamanism millennia ago, became closely linked in the Middle Ages with alchemy. The preoccupation of alchemists with metals and their properties led to experiments—mostly feckless ones, looking back—with the therapeutic use of all sorts of metals. Paracelsus, a prominent physician of the sixteenth century, achieved fame from his enthusiastic use of mercury and arsenic, based on what now seems a wholly mystical commitment to alchemical philosophy as the key to understanding the universe and the human body simultaneously. Under his influence, three centuries of patients with all varieties of illness were treated with strong potions of metals, chiefly mercury, and vigorous purgation became standard medical practice.

Physics and chemistry have grown to scientific maturity, medicine is on its way to growing up, and it is hard to find traces anywhere of the earlier fumblings toward a genuine scientific method. Alchemy exists only as a museum piece, an intellectual fossil, so antique that we no longer need be embarrassed by the memory, but the memory is there. Science began by fumbling. It works because the people involved in it work, and *work together*. They become excited and exasperated, they exchange their bits of information at a full shout, and, the most wonderful thing of all, they keep *at* one another.

Something rather like this may be going on now, without realizing it, in the latest and grandest of all fields of science. People in my field, and some of my colleagues in the real "hard" sciences such as physics and chemistry, have a tendency to take lightly and often disparagingly the efforts of workers in the so-called social sciences. We like to refer to their data as soft. We do not acknowledge as we should the differences between the various disciplines within behavioral research—we speak of analytical psychiatry, sociology, linguistics, economics, and computer intelligence as though these inquiries were all of a piece, with all parties wearing the same old comical conical hats. It is of course not so. The principal feature that the social sciences share these days is the attraction they exert on considerable numbers of students, who see the prospect of exploring human behavior as irresistible and hope fervently that a powerful scientific method for doing the exploring can be worked out. All of the matters on the social-science agenda seem more urgent to these young people than they did at any other time in human memory. It may turn out, years hence, that a solid discipline of human science will have come into existence, hard as quantum physics, filled with deep insights, plagued as physics still is by ambiguities but with new rules and new ways of getting things done. Like, for instance, getting rid of thermonuclear weapons, patriotic rhetoric, and nationalism all at once. If anything like this does turn up we will be looking back at today's social scientists, and their close colleagues the humanists, as having launched the new science in a way not all that different from the accomplishment of the old alchemists, by simply working on the problem—this time, the fundamental, primal universality of the human mind.

Questions and Topics for Discussion and Writing

1. Thomas has titled his work simply "Alchemy." Would a better title be "A Defense of Alchemy"? Why or why not?
2. Thomas suggests we owe much to the alchemists for their contribution to the development of language. Does his tribute seem a digression?
3. According to Thomas, what sciences were precipitated by alchemy?
4. Thomas says, "The central idea they [the alchemists] were obsessed with—that there is a fundamental, elementary particle out of which everything in the universe is made—continues to obsess today's physicists." Does this statement add understanding to your reading of Capra's essay, the Epilogue to *The Tao of Physics* (Unit 5)? To your reading of Lucretius's poem (earlier in this unit)?
5. What is implied about scientific discovery and progress in Thomas's writing of the alchemists' chaotic approach to problems?
6. The word *enthusiasm* has not always meant what it denotes today. Look up this word in the OED (*Oxford English Dictionary*), and write a sentence in which you state what Thomas means by saying that Paracelsus "achieved fame from his enthusiastic use of mercury and arsenic. . . ."
7. Thomas predicts future generations' judgment of today's scientists. How will the future judge the activity and discoveries of today's scientists, according to Thomas?

8. Aylmer, the hero of "The Birthmark" by Nathaniel Hawthorne, is depicted as an alchemist. Read this short story and write an essay in which you illustrate the ways that Hawthorne's character fits Thomas's description.

Paracelsus

Paracelsus (1493–1534) gained a wide, but troubled, reputation as an alchemist, a medical reformer, and a practitioner of miraculous cures. A native of Switzerland, he entered the University of Basel at sixteen, ostensibly to follow in the footsteps of his father, a well-regarded physician. Although a promising student, he opposed the authorities and was forced to leave, thus beginning a nomadic trek throughout Europe. After years about which little is known, he appeared in Italy as an army surgeon, where his remarkable cures began. Returning to Germany and Switzerland with mineral medicines, he performed cures on notable patients. In Basel, he was offered a post at his old university, even without his medical diploma.

Paracelsus, however, outraged the orthodox faculty by such belligerent actions as writing his own manifesto to replace the Hippocratic oath and even burning respected texts. At length, he was again driven from the university and began a career as a wandering "gadfly" to the medical profession. He traveled widely, trying to "sting" the opposition into listening to his ideas as he studied the occupational diseases of miners and metal workers.

His vision of a new, unorthodox medicine was passionate, founded in his belief that divine order is reflected in the human body. Disease is caused by agents outside the body. This concept went against the prevailing opinion that an individual imbalance in the four bodily "humors"—blood, phlegm, choler, and melancholy—caused disease. Although Paracelsus often saw these "outside agents" as effects from the stars and expressed his ideas in the occult language of astrology, he anticipated the modern medical insight that the causes of disease are specific. In seeking cures through alchemy and knowledge of metallurgy, he established a basis for the application of chemistry to medicine today.

As much honored by mystics—for example, the Rosicrucians—as by medical historians, Paracelsus left a story wrapped in legend. His death was mysterious: Some say that he was thrown from a window, others that he was murdered by assassins hired by medical faculty. Even his name manifests elements of legend, for his real name was Theophrastus Philippus Aureolus Bombastus von Hohenheim! Some authorities say that other alchemists conferred the name "Paracelsus" on him to show that he was greater than Celsus, a physician famous in the ancient world; others emphasize that the prefix *para-* may refer to his use of paradox. While the modern reader may be tempted to see in "Bombastus" the origin of "bombast" (a term meaning inflated, extravagant language), the notion that the term comes from his name is not based in fact. His rhetoric, if somewhat bombastic, is developed to challenge, combat, overwhelm, and sometimes confound by prodigious intuition.

The following brief selection from Paracelsus's "Book of Vexations" provides a sense of his disposition and illustrates his rhetoric and style.

Matter and Magic: Alchemy

Paracelsus

THE SCIENCE AND NATURE OF ALCHEMY, AND WHAT OPINION SHOULD BE FORMED THEREOF.
Regulated by the Seven Rules or Fundamental Canons according to the seven commonly known Metals; and containing a Preface with certain Treatises and Appendices.

THE PREFACE
of Theophrastus Paracelsus to All
Alchemists and Readers
of This Book

You who are skilled in Alchemy, and as many others as promise yourselves great riches or chiefly desire to make gold and silver, which Alchemy in different ways promises and teaches; equally, too, you who willingly undergo toil and vexations, and wish not to be freed from them, until you have attained your rewards, and the fulfilment of the promises made to you; experience teaches this every day, that out of thousands of you not even one accomplishes his desire. Is this a failure of Nature or of Art? I say, no; but it is rather the fault of fate, or of the unskilfulness of the operator.

Since, therefore, the characters of the signs, of the stars and planets of heaven, together with the other names, inverted words, receipts, materials, and instruments are thoroughly well known to such as are acquainted with this art, it would be altogether superfluous to recur to these same subjects in the present book, although the use of such signs, names, and characters at the proper time is by no means without advantage.

But herein will be noticed another way of treating Alchemy different from the previous method, and deduced by Seven Canons from the sevenfold series of the metals. This, indeed, will not give scope for a pompous parade of words, but, nevertheless, in the consideration of those Canons everything which should be separated from Alchemy will be treated at sufficient length, and, moreover, many secrets of other things are herein contained. . . . It is tedious to read long descriptions, and everybody wishes to be advised in straightforward words. Do this, then; proceed as follows, and you will have Sol and Luna, by help whereof you will turn out a very rich man. Wait awhile, I beg, while this process is described to you in few words, and keep these words well digested, so that out of Saturn, Mercury, and Jupiter you may make Sol and Luna. There is not, nor ever will be, any art so easy to find out and practise, and so effective in itself. The method of making Sol and Luna by Alchemy is so prompt that there is no more need of books, or of elaborate instruction, than there would be if one wished to write about last year's snow.

Concerning the Receipts of Alchemy

What, then, shall we say about the receipts of Alchemy, and about the diversity of its vessels and instruments? These are furnaces, glasses, jars, waters, oils, limes, sulphurs, salts, saltpetres, alums, vitriols, chrysocollæ, copper-greens, atraments, auri-pigments, fel vitri, ceruse, red earth, thucia, wax, lutum sapientiæ, pounded glass, verdigris, soot, testæ ovorum, crocus of Mars, soap, crystal, chalk, arsenic, antimony, minimum, elixir, lazurium, gold-leaf, salt-nitre, sal ammoniac, calamine stone, magnesia, bolus armenus, and many other things. Moreover, concerning preparations, putrefactions, digestions, probations, solutions, cementings, filtrations, reverberations, calcinations, graduations, rectifications, amalgamations, purgations, etc., with these alchemical books are crammed. Then, again, concerning herbs, roots, seeds, woods, stones, animals, worms, bone dust, snail shells, other shells, and pitch. These and the like, whereof there are some very far-fetched in Alchemy, are mere incumbrances of work; since even if Sol and Luna could be made by them they rather hinder and delay than further one's purpose. But it is not from these—to say the truth—that the Art of making Sol and Luna is to be learnt. So, then, all these things should be passed by, because they have no effect with the five metals, so far as Sol and Luna are concerned. Someone may ask, What, then, is this short and easy way, which involves no difficulty, and yet whereby Sol and Luna can be made? Our answer is, this has been fully and openly explained in the Seven Canons. It would be lost labour should one seek further to instruct one who does not understand these. It would be impossible to convince such a person that these matters could be so easily understood, but in an occult rather than in an open sense.

THE ART IS THIS: After you have made heaven, or the sphere of Saturn, with its life to run over the earth, place on it all the planets, or such, one or more, as you wish, so that the portion of Luna may be the smallest. Let all run, until heaven, or Saturn, has entirely disappeared. Then all those planets will remain dead with their old corruptible bodies, having meanwhile obtained another new, perfect, and incorruptible body.

That body is the spirit of heaven. From it these planets again receive a body and life, and live as before. Take this body from the life and the earth. Keep it. It is Sol and Luna. Here you have the Art altogether, clear and entire. If you do not yet understand it, or are not practised therein, it is well. It is better that it should be kept concealed, and not made public.

Questions and Topics for Discussion and Writing

1. In what way does Lewis Thomas's essay prepare readers for an understanding or appreciation of Paracelsus's work?
2. What are the two reasons cited by Paracelsus for alchemists' failure in the transmutation of metals?
3. What is meant by the term *occult?* Write an extended definition of the word in a paragraph.
4. Alchemy seems to modern readers hopelessly medieval. What, then, besides the hope for riches, could have attracted a rigorous mind like that of Sir Isaac Newton, for example, to alchemical reading, writing, and experiments for over three decades while he was simultaneously contributing to the "new physics" of his age?
5. What does Paracelsus see as the practical outcome of alchemy?
6. Paracelsus's stance may be called elitist. He states that if one does not understand the method of alchemy, then one probably lacks intelligence. How do you judge your mental capability after reading this essay?
7. Does the personality of the writer come through? Note words and phrases that characterize his tone.
8. Write a paragraph in which you relate Paracelsus's communication purpose to his style.
9. The English poet Robert Browning wrote a poem in five scenes, each depicting a critical moment in the life of Paracelsus. The poem portrays both Paracelsus's failures and his enthusiastic quest for knowledge. Locate the poem in your library and paraphrase it, scene by scene. What do you think it reveals about Paracelsus?

Marie S. Curie

Marie Curie (1867–1934) was born Marya Sklodovska in Warsaw, Poland, the youngest of five gifted children. While her father, a teacher of physics, guided her initial study of science, she was barred from formal study at the University of Warsaw because she was a woman. Moreover, to be a Pole seeking education was in itself difficult: Poland at that time was partitioned among Germany, Austria, and Russia. To hold the citizens of "Russian Poland" under control, the representatives and spies of the Czar were everywhere, inspecting the performances of pupils in the schools, imprisoning or executing revolutionary Polish patriots, and even attempting to stamp out nationalism by suppressing the use of the Polish language. All official activities—even school lessons—had to be presented in Russian. Like other Poles, Sklodovska dreamed of liberty.

France stood for liberty in her mind and that of her older sister Bronya. Their dream focused on going to Paris to study at the Sorbonne, but poverty made the goal remote. Then Sklodovska formed a plan of mutual aid: She would find employment to help support her sister's medical studies and Bronya would, in turn, send for her. Working as a governess and sending half her salary to Bronya, Sklodovska also conducted secret classes in Polish for poor children and working women, taught herself physics and chemistry from textbooks, and attended a "floating university" of secret classes held by other patriotic young Poles who held a passion for science. Eventually, her sister was able to make a home for her, and Sklodovska made her way to Paris.

At the Sorbonne, she completed her master's degree in physics, undertook her master's degree in mathematics, and began a study of the magnetic properties of steels. While engaged in this work, she met Pierre Curie, a distinguished physicist who, with his brother, discovered piezoelectricity, the phenomenon in which pressure creates electrical charges in certain crystals. With their marriage in 1895 Pierre and Marie Curie became research partners. Not long before, Wilhelm Roentgen (1845–1923) had discovered that invisible rays, or X-rays, exist and Henri Becquerel (1852–1908) had discovered that uranium soils produce invisible rays. The phenomenon of radiation appealed to Curie as the ideal subject for her doctoral investigation, and her husband soon left his own line of research to collaborate on her project. They devoted themselves to investigating what she termed *radioactivity*, not only from the theoretical standpoint but also from the practical side.

The Curies set themselves the task of analyzing pitchblende. Soon they announced the probable existence of the element polonium, named after "the homeland of one of us," then of radium. Determined to prove the existence of unknown elements and to isolate them, the Curies first had to struggle to obtain a supply of pitchblende. In refining a ton of the material, Marie took (in her daughter's words) the "man's job," laboring in an ill-equipped, ill-ventilated "laboratory" to cart buckets of ore and to stir the heated ore with an iron pole "almost as big as she was." After four years of strenuous work, she succeeded in isolating pure radium salt and determining its atomic weight.

After Pierre Curie's untimely death in a traffic accident, Marie took over his teaching post, becoming the first woman professor in the 650-year history of the Sorbonne. She continued their important research on radiation, combining these efforts with her role as a parent of their two daughters, Irène and Eve. In tribute to the pioneering nature of their work, both Curies are lastingly honored in scientific nomenclature: A *curie* (named after Marie) is a unit of radioactivity; the *curie point* (named after Pierre) is a significant temperature in physiochemical research.

Pierre Curie, Marie Curie, and Henri Becquerel all shared the Nobel Prize in physics in 1903. For her further work on radiation after her husband's death, Marie Curie in 1911 received a second Nobel Prize, that for chemistry; this time she was the sole recipient, making her the first person to receive the prize twice. Her friend Albert Einstein attributed her accomplishments "not merely to bold intuition but to a devotion and tenacity in execution under the most extreme hardships imaginable, such as the history of experimental science has not often witnessed."

Within a short time after the discovery of radioactive elements, Marie Curie realized that the substances promised remarkable medical applications. This recognition engendered a burst of patriotic activity during World War I, when she distinguished herself in X-ray technology, assisted by her daughter Irène. The medical possibilities of radium caught the imagination of the world, and women, in particular, chose to honor Marie Curie by collecting funds to permit further research. During two triumphal tours of America in 1921 and 1929, Madame Curie received gifts to promote the use of radiology. Ironically, leukemia—induced by exposure to radioactivity—caused her death.

The following excerpt is from an address Madame Curie delivered to an audience at Vassar College in gratitude for the gift from American women of a gram of radium.

The Discovery of Radium

Marie Curie

I could tell you many things about radium and radioactivity and it would take a long time. But as we cannot do that, I shall give you only a short account of my early work about radium. Radium is no more a baby; it is more than twenty years old, but the conditions of the discovery were somewhat peculiar, and so it is always of interest to remember them and to explain them.

We must go back to the year 1897. Professor Curie and I worked at that time in the laboratory of the School of Physics and Chemistry where Professor Curie held his lectures. I was engaged in some work on uranium rays which had been discovered two years before by Professor Becquerel. I shall tell you how these uranium rays may be detected. If you take a photographic plate and wrap it in black paper and then on this plate, protected from ordinary light, put some uranium salt and leave it a day, and the next day the plate is developed, you notice on the plate a black spot at the place where the uranium salt was. This spot has been made by special rays which are given out by the uranium and are able to make an impression on the plate in the same way as ordinary light. You can also test those rays in another way, by placing them on an electroscope. You know what an electroscope is. If you charge it, you can keep it charged several hours and more, unless uranium salts are placed near to it. But if this is the case the electroscope loses its charge and the gold or aluminum leaf falls gradually in a progressive way. The speed with which the leaf moves may be used as a measure of the intensity of the rays; the greater the speed, the greater the intensity.

I spent some time in studying the way of making good measurements of the uranium rays, and then I wanted to know if there were other elements, giving out rays of the same kind. So I took up a work about all known elements and their compounds and found that uranium compounds are active and also all thorium compounds, but other elements were not found active, nor were their compounds. As for the uranium and thorium compounds, I found that they were active in proportion to their uranium or thorium content. The more uranium or thorium, the greater the activity, the activity being an atomic property of the elements, uranium and thorium.

Then I took up measurements of minerals and I found that several of those which contain uranium or thorium or both were active. But then the activity was not what I could expect; it was greater than for uranium or thorium compounds, like the oxides which are almost entirely composed of these elements. Then I thought that there should be in the minerals some unknown element having a much greater radioactivity than uranium or thorium. And I wanted to find and to separate that element, and I settled to that work with Professor Curie. We thought it would be done in several weeks or months, but it was not so. It took many years of hard work to finish that task. There was not *one* new element; there were several of them. But the most important is radium, which could be separated in a pure state.

All the tests for the separation were done by the method of electrical measurements with some kind of electroscope. We just had to make chemical separations and to examine all products obtained, with respect to their activity. The product which retained the radioactivity was considered as that one which had kept the new element; and, as the radioactivity was more strong in some products, we knew that we had succeeded in concentrating the new element. The radioactivity was used in the same way as a spectroscopical test.

The difficulty was that there is not much radium in a mineral; this we did not know at the beginning. But we now know that there is not even one part of radium in a million parts of good ore. And, too, to get a small quantity of pure radium salt, one is obliged to work up a huge quantity of ore. And that was very hard in a laboratory.

We had not even a good laboratory at that time. We worked in a hangar where there were no improvements, no good chemical arrangements. We had no help, no money. And because of that, the work could not go on as it would have done under better conditions. I did myself the numerous crystallizations which were wanted to get the radium salt separated from the barium salt, with which it is obtained, out of the ore. And in 1902 I finally succeeded in getting pure radium chloride and determining the atomic weight of the new element, radium, which is 226, while that of barium is only 137.

Later I could also separate the metal radium, but that was a very difficult work; and, as it is not necessary for the use of radium to have it in this state, it is not generally prepared that way.

Now, the special interest of radium is in the intensity of its rays, which is several million times greater than the uranium rays. And the effects of the rays make the radium so important. If we take a practical point of view, then the most important property of the rays is the production of physiological effects on the cells of the human organism. These effects may be used for the cure of several diseases. Good results have been obtained in many cases. What is considered particularly important is the treatment of cancer. The medical utilization of radium makes it necessary to get that element in sufficient quantities. And so a factory of radium was started, to begin with, in France, and later in America, where a big quantity of ore named carnotite is available. America does not produce many grams of radium every year but the price is still very high because the quantity of radium contained in the ore is so small. The radium is more than a hundred thousand times dearer than gold.

But we must not forget that when radium was discovered no one knew that it would prove useful in hospitals. The work was one of pure science. And this is a proof that scientific work must not be considered from the point of view of the direct usefulness of it. It must be done for itself, for the beauty of science, and then there is always the chance that a scientific discovery may become, like the radium, a benefit for humanity.

But science is not rich; it does not dispose of important means; it does not generally meet recognition before the material usefulness of it has been proved. The factories produce many grams of radium every year, but the laboratories have very small quantities. It is the same for my laboratory, and I am very grateful to the American women who wish me to have more of radium, and give me the opportunity of doing more work with it.

The scientific history of radium is beautiful. The properties of the rays have been studied very closely. We know that particles are expelled from radium with a very great velocity, near to that of light. We know that the atoms of radium are destroyed by expulsion of these particles, some of which are atoms of helium. And in that way it has been proved that the radioactive elements are constantly disintegrating, and that they produce, at the end, ordinary elements, principally helium and lead. That is, as you see, a theory of transformation of atoms, which are not stable, as was believed before, but may undergo spontaneous changes.

Radium is not alone in having these properties. Many having other radioelements are known already: the polonium, the mesothorium, the radiothorium, the actinium. We know also radioactive gases, named emanations. There is a great variety of substances and effects in radioactivity. There is always a vast field left to experimentation and I hope that we may have some beautiful progress in the following years. It is my earnest desire that some of you should carry on this scientific work, and keep for your ambition the determination to make a permanent contribution to science.

Eve Curie

Eve Curie, the younger daughter of the physicists Pierre and Marie Curie, turned her intellectual energy to the humanities. An accomplished pianist, Eve gave her first concert in Paris in 1925. A career as music and theater critic followed. During World War II, she worked both in France and abroad as a lecturer and organizer for the Free French cause, and she worked as a correspondent covering the Eastern fronts.

In contrast, her older sister Irène followed their parents into the field of physics, assisting Marie Curie in the laboratory. Later, Irène worked with her husband, Frédéric Joliot-Curie, to develop the atomic bomb, and they later shared a Nobel Prize for physics. After World War II, the two served on the French Atomic Commission until their outspoken Communist sympathies forced their dismissal.

Eve Curie was sixteen when she and Irène accompanied their mother in 1921 on a lecture series in the United States; Marie Curie made a second such trip in 1929. Both times she received financial aid from American women for her research. Eve's recollections portray her mother as viewing the honors given her as awards to science, not to herself.

After Marie Curie's death, Eve organized her mother's manuscripts and wrote the biography *Madame Curie*, a work translated into several languages. The following excerpt, "Four Years in a Shed," portrays the dedication of mind and body that led to a momentous event in the history of science.

Four Years in a Shed
Eve Curie

The shed in the Rue Lhomond surpassed the most pessimistic expectations of discomfort. In summer, because of its skylights, it was as stifling as a hothouse. In winter one did not know whether to wish for rain or frost; if it rained, the water fell drop by drop, with a soft, nerve-racking noise, on the ground or on the worktables, in places which the physicists had to mark in order to avoid putting apparatus there. If it froze, one froze. There was no recourse. The stove, even when it was stoked white, was a complete disappointment. If one went near enough to touch it one received a little heat, but two steps away and one was back in the zone of ice.

It was almost better for Marie and Pierre to get used to the cruelty of the outside temperature, since their technical installation—hardly existent—possessed no chimneys to carry off noxious gases, and the greater part of their treatment had to be made in the open air, in the courtyard. When a shower came the physicists hastily moved their apparatus inside: to keep on working without being suffocated they set up draughts between the opened door and windows. . . .

In such conditions M. and Mme Curie worked for four years from 1898 to 1902.

During the first year they busied themselves with the chemical separation of radium and polonium and they studied the radiation of the products (more and more active) thus obtained. Before long they considered it more practical to separate their efforts. Pierre Curie tried to determine the properties of radium, and to know the new metal better. Marie continued those chemical treatments which would permit her to obtain salts of pure radium.

In this division of labor Marie had chosen the "man's job." She accomplished the toil of a day laborer. Inside the shed her husband was absorbed by delicate experiments. In the courtyard, dressed in her old dust-covered and acid-stained smock, her hair blown by the wind, surrounded by smoke which stung her eyes and throat, Marie was a sort of factory all by herself.

At this period we were entirely absorbed by the new realm that was, thanks to an unhoped-for discovery, opening before us [Marie was to write]. In spite of the difficulties of our working conditions, we felt very happy. Our days were spent at the laboratory. In our poor shed there reigned a great tranquillity: sometimes, as we watched over some operation, we would walk up and down, talking about work in the present and in the future; when we were cold a cup of hot tea taken near the stove comforted us. We lived in our single preoccupation as if in a dream.

. . . We saw only very few persons at the laboratory; among the physicists and chemists there were a few who came from time to time, either to see our experiments or to ask for advice from Pierre Curie, whose competence in several branches of physics was well-known. Then took place some conversations before the blackboard—the sort of conversation one remembers well because it acts as a stimulant for scientific interest and the ardor for work without interrupting the course of reflection and without troubling that atmosphere of peace and meditation which is the true atmosphere of a laboratory.

Whenever Pierre and Marie, alone in this poor place, left their apparatus for a moment and quietly let their tongues run on, their talk about their beloved radium passed from the transcendent to the childish.

"I wonder what *It* will be like, what *It* will look like," Marie said one day with the feverish curiosity of a child who has been promised a toy. "Pierre, what form do you imagine *It* will take?"

"I don't know," the physicist answered gently. "I should like it to have a very beautiful color. . . ."

In the course of the years 1899 and 1900 Pierre and Marie Curie published a report on the discovery of "induced radioactivity" due to radium, another on the effects of radioactivity, and another on the electric charge carried by the rays. And at last they drew up, for the Congress of Physics of 1900, a general report on the radioactive substances, which aroused immense interest among the scientists of Europe.

The development of the new science of radioactivity was rapid, overwhelming—the Curies needed fellow workers. . . . They now required technicians of the first order. Their discovery had important extensions in the domain of chemistry, which demanded attentive study. . . . Thus, even before radium and polonium were isolated, a French scientist, André Debierne, had discovered a "brother," *actinium*.

Marie continued to treat, kilogram by kilogram, the tons of pitchblende residue which were sent her on several occasions from St Joachimsthal. With her terrible patience, she was able to be, every day for four years, a physicist, a chemist, a specialized worker, an engineer and a laboring man all at once. Thanks to her brain and muscle, the old tables in the shed held more and more concentrated products—products more and more rich in radium. Mme Curie was approaching the end: she no longer stood in the courtyard, enveloped in bitter smoke, to watch the heavy basins of material in fusion. She was now at the stage of purification and of the "fractional crystallization" of strongly radioactive solutions. But the poverty of her haphazard equipment hindered her work more than ever. It was now that she needed a spotlessly clean workroom and apparatus perfectly protected against cold, heat and dirt. In this shed, open to every wind, iron and coal dust was afloat which, to Marie's despair, mixed itself into the products purified with so much care. Her heart sometimes constricted before these little daily accidents, which took so much of her time and her strength.

Pierre was so tired of the interminable struggle that he would have been quite ready to abandon it. Of course, he did not dream of dropping the study of radium and of radioactivity. But he would willingly have renounced, for the time being, the special operation of preparing pure radium. The obstacles seemed insurmountable. Could they not resume this work later on, under better conditions? More attached to the meaning of natural phenomena than to their material reality, Pierre Curie was exasperated to see the paltry results to which Marie's exhausting effort had led. He advised an armistice.

He counted without his wife's character. Marie wanted to isolate radium and she would isolate it. She scorned fatigue and difficulties, and even the gaps in her own knowledge which complicated her task. After all, she was only a very young scientist: she still had not the certainty and great culture Pierre had acquired by twenty years' work, and sometimes she stumbled across phenomena or methods of calculation which she knew very little, and for which she had to make hasty studies.

So much the worse! With stubborn eyes under her great brow, she clung to her apparatus and her test tubes.

In 1902, forty-five months after the day on which the Curies announced the probable existence of radium, Marie finally carried off the victory in this war of attrition: she succeeded in preparing a decigram of pure radium, and made a first determination of the atomic weight of the new substance, which was 225.

The incredulous chemists—of whom there were still a few—could only bow before the facts, before the superhuman obstinacy of a woman.

Radium officially existed.

It was nine o'clock at night. Pierre and Marie Curie were in their little house at 108 Boulevard Kellermann, where they had been living since 1900. . . . The day's work had been hard, and it would have been more reasonable for the couple to rest. But Pierre and Marie were not always reasonable. As soon as they had put on their coats and told Dr Curie of their flight, they were in the street. They went on foot, arm in arm, exchanging few words. After the crowded streets of this queer district, with its factory buildings, wastelands and poor tenements, they arrived in the Rue Lhomond and crossed the little courtyard. Pierre put the key in the lock. The door squeaked, as it had squeaked thousands of times, and admitted them to their realm, to their dream.

"Don't light the lamps!" Marie said in the darkness. Then she added with a little laugh:

"Do you remember the day when you said to me 'I should like radium to have a beautiful color'?"

The reality was more entrancing than the simple wish of long ago. Radium had something better than "a beautiful color": it was spontaneously luminous. And in the somber shed where, in the absence of cupboards, the precious particles in their tiny glass receivers were placed on tables or on shelves nailed to the wall, their phosphorescent bluish outlines gleamed, suspended in the night.

"Look . . . Look!" the young woman murmured.

She went forward cautiously, looked for and found a straw-bottomed chair. She sat down in the darkness and silence. Their two faces turned toward the pale glimmering, the mysterious sources of radiation, toward radium—their radium. Her body leaning forward, her head eager, Marie took up again the attitude which had been hers an hour earlier at the bedside of her sleeping child.

Her companion's hand lightly touched her hair.

She was to remember forever this evening of glowworms, this magic.

Questions and Topics for Discussion and Writing

1. Both Marie Curie's speech and Eve Curie's account use the metaphor of radium as a baby. Writings are frequently unified and developed by an extended metaphor. In Eve Curie's account, trace and list the words and phrases that create such a metaphor comparing the discovery of radium with the birth of a child. How effective is this device in conveying the significance of the Curies' work as scientists?

2. Eve Curie says that Pierre and Marie "let their tongues run on, their talk about their beloved radium passed from the transcendent to the childish." What are some of the "transcendent" ideas they may have discussed?

3. What are the characteristics that made Marie Curie a "scientist of genius"? What dimensions does reading the daughter's account add to your perceptions of Marie Curie's endeavors?

4. How does the phrase "terrible patience" characterize Marie Curie as a scientist? What are the differences in character and attitude as scientists evidenced by Pierre and Marie?

5. Can you find evidence that Eve Curie had greater admiration for her mother's than for her father's scientific accomplishments?

6. What are "important extensions" in science that resulted from the Curies' work?

7. In what way was Marie Curie isolated from those around her?

8. Contrast the description of the Curies' laboratory that is placed at the beginning of Eve Curie's account with that at the end.

9. George Bernard Shaw once concluded that "the reasonable man adapts himself to the world while the unreasonable man insists on adapting the world to himself. Therefore, all progress depends on the unreasonable man." How does the comment "Pierre and Marie were not always reasonable" serve to characterize their contributions to the advancement of science?

Miroslav Holub

Born in Czechoslovakia in 1923, Miroslav Holub early exhibited the dual involvement with the arts and sciences that distinguishes his career. By age three, he was composing and reciting poems, which his mother dutifully wrote down. He complemented the esthetic attraction any youngster feels in collecting brightly colored butterflies with his concern over their precise, scientific cataloguing.

Holub's medical studies, begun in 1947, led to a career as an internationally respected immunologist. In conjunction with Herbert Fischer of Germany, Holub discovered the role of the omentum (a flap of tissue in a mouse's belly) in the immunological system of mice. He has also been Writer in Residence at Oberlin College, Ohio. He continues his scientific research at the Institute of Clinical and Experimental Medicine in Prague and contributes to the cultural life of his country, amassing an impressive array of literary publications (prose and poetry), including a popular travel book on America and a scholarly study of Edgar Allan Poe. When he works in the laboratory, Holub writes in his off-hours; when he teaches writing, he manages to work scientific experiments into the schedule. Enriching each discipline with a special interplay of perspectives, while his poetic wonder at life permeates his research, Holub informs his poetry with a scientific concern for orderliness and structural precision.

Although poetry and science have obvious differences, Holub sees important similarities. Both science and poetry deal, not with "objective" reality as such, but with the human being's relation to reality, to nature. Defending the sciences against the accusation of "stark, cold reality," he stresses that both scientists and poets dissect the world and experience, but only in order to construct an internally consistent world-view.

As a teacher of poetry, Holub takes a structuralist approach, which has been likened to an archeologist's digging: He poses deep and demanding questions. Leaping beyond concerns over an image or a single line, Holub probes further, as in the following poem where he reflects on the role of the human as scientific observer.

Brief Thoughts on a Test-tube

Miroslav Holub

You take
 a bit of fire, a bit of water,
 a bit of rabbit or tree
 or any little piece of man,
 you mix it, shake well, cork it up,
 put it in a warm place, in darkness, in light, in frost,
 leave it alone for a while—though things don't leave you alone—
 and that's the whole point.

And then
 you have a look—and it grows,
 a little sea, a little volcano,
 a little tree, a little heart, a little brain,
 so small you don't hear it pleads to be let out,
 and that's the whole point, not to hear.

Then you go
 and record it, all the minuses or
 all the pluses, some with an exclamation-mark,

all the zeros, or all the numbers, some with an
  exclamation-mark,
and the point is that the test-tube
is an instrument for changing question-
into exclamation-marks,

And the point is
that for the moment you forget
you yourselves are

In the test-tube.

Questions and Topics for Discussion and Writing

1. Considering the overall movement of the poem, state how the structure follows laboratory procedure.
2. Why does Holub separate *man* from other test-tube ingredients by the use of the conjunction *or* and by the use of the word *piece* instead of *bit*?
3. Look up the word *point*. Note the several uses of the word in the poem. How many different meanings does Holub convey by his use of the word?
4. Punctuation and line structure as well as spacing suggest meaning in a poem. What does Holub's use of dashes add? How do line structure and spacing in the last three lines indicate the scientist's absorption in his work?
5. Why, after the singular *you* throughout, does Holub use the reflexive plural *yourselves* in the next-to-last line?
6. Holub uses specific phrases related to time twice in the poem: "for a while" and "for the moment." How do these phrases offer different concepts of time? How does "for the moment" suggest the scientist's immersion in his work?
7. Is there significance to the order of listing in Stanza 2?
8. Why is the word *it* so ambiguous? What is its antecedent?
9. What does Holub mean by saying that the scientist records both with exclamation marks and with question marks?
10. Explain Holub's meaning in these lines:

> the test-tube
> is an instrument for changing question-
> into exclamation-marks. . . .

Linus Pauling

No More War! exclaims the title of the urgent book by Linus Pauling, the internationally known American physicist and chemist. In his Preface, Pauling writes: "We of the mid-twentieth century live in a most extraordinary time. We are living through that unique epoch in the history of civilization when war will cease to be the means of settling great world problems." This book against nuclear weapons and war underscores the fact that Pauling has had two careers: scientist and crusader for peace. Remarkably, he has been honored with the Nobel Prize in each—the only person to win two unshared prizes. Based on his research into the nature of the chemical bond, he holds the 1954 award for chemistry; for his work toward a nuclear test-ban treaty, he was awarded the 1962 prize for peace.

Although he was born in Oregon in 1901, the son of a pharmacist, and received his undergraduate degree from Oregon Agricultural College, Pauling's long career in teaching and research is associated with California. Awarded his PhD in 1925 from the California Institute of Technology, he went to Europe on a Guggenheim Fellowship to study quantum mechanics, but returned to CIT, where, in 1931, he became head of the Division of Chemistry and Chemical Engineering. In 1967, Pauling moved to the San Diego campus of the University of California as research professor of chemistry, then to Stanford University in 1969. In 1973, he founded the Linus Pauling Institute of Science and Medicine, also in Palo Alto, which has become the center for his activities.

The main subject of Pauling's research has been molecular structure. His knowledge of quantum mechanics gave him a theoretical tool for original work on X-ray and electron diffraction. In the 1930s, he expanded his study to biological molecules, research that led to the general theory of protein structure. Pauling's particular thrust was in studies of molecules of medical interest; in 1940, he published his first paper on the structure of antibodies. During World War II, he worked on immunological problems and developed an interest in sickle-cell anemia, which he showed to be a molecular disease caused by abnormal hemoglobin molecules. Pauling has since extended his study to the molecular basis of memory.

For his profoundly significant research and his role as an educator, Pauling has received many awards in addition to his two Nobel prizes. Along with numerous honorary degrees from universities in the United States and abroad, he received the Royal Society's Sir Humphrey Davy Award in 1947 and the Presidential Medal for Merit in 1948.

In the 1950s, Pauling used his reputation as a scientist to become a humanitarian voice in issues connected with science. Ever since, he has continued to use his energy to speak out against nuclear arms. He is quick to disagree with established policy when armaments are the issue: "Genuine world peace cannot be achieved by stockpiling ingenious and horrifying weapons for mass destruction. . . ."

In scientific matters as well, Pauling generates and welcomes controversy. An outstanding instance is his work on Vitamin C. Pauling has always held that Vitamin C (ascorbic acid), if taken in large enough doses, will be beneficial in aiding the body to fight off many diseases. Practicing what he preaches, Pauling continues to take megadoses of the vitamin himself, in defiance of the several studies that report little of the effect he claims. Indeed, Pauling has carried his theory a step further and believes that ascorbic acid can also fight cancer, as he asserts in his book *Cancer and Vitamin C* (1979), written with Scottish physician Ewan Cameron.

"Vitamin C and Evolution," from *Vitamin C and the Common Cold* (1970), presents Pauling's explanation of how human dependence on Vitamin C arose.

Vitamin C and Evolution

Linus Pauling

A human being requires many different foods in order to be in good health. In addition to carbohydrates, proteins, essential fats, and minerals, he requires ascorbic acid and a number of other vitamins.

The protein in our diet is the principal source of the nitrogen required for the nitrogenous substances in our body, proteins and nucleic acids.

The proteins in the human body, and in other living organisms, are linear chains of residues of about twenty different amino acids—glycine, alanine, serine, lysine, phenylalanine, and fifteen others. It is not necessary that all of the amino acids be present in the diet. Some of them can be synthesized in the human body. But eight amino acids, called the essential amino acids, cannot be synthesized in the human body, and must be present in the food that is ingested. The eight essential amino acids are threonine, valine, methionine, lysine, histidine, phenylalanine, tryptophan, and leucine. The disease kwashiorkor (protein starvation) results from an inadequate intake of the essential amino acids.

We are accustomed to thinking of man as the highest of all species of living organisms. In one sense he is: he has achieved effective control over a large part of the earth, and has even begun to extend his realm as far as the moon. But in his biochemical capabilities he is inferior to many other organisms, including even unicellular organisms, such as bacteria, yeasts, and molds.

The red bread mold *(Neurospora)*, for example, is able to carry out in its cells a great many chemical reactions that human beings are unable to carry out. The red bread mold can live on a very simple medium, consisting of water, inorganic salts, an inorganic source of nitrogen, such as ammonium nitrate, a suitable source of carbon, such as sucrose, and a single "vitamin," biotin. All other substances required by the red bread mold are synthesized by it, with use of its internal mechanisms. The red bread mold does not need to have any amino acids in its diet, because it is able to synthesize all of them, and also to synthesize all of the vitamins except biotin.

The red bread mold owes its survival, over hundreds of millions of years, to its great biochemical capabilities. If, like man, it were unable to synthesize the various amino acids and vitamins it would not have survived, because it could not have solved the problem of getting an adequate diet.

From time to time a gene in the red bread mold undergoes a mutation, such as to cause the cell to lose the ability to manufacture one of the amino acids or vitamin-like substances essential to its life. This mutated spore gives rise to a deficient strain of red bread mold, which could stay in good health only with an addition to the diet that suffices for the original type of the mold. The scientists G. W. Beadle and E. L. Tatum carried on extensive studies of mutated strains of the red bread mold, when they were working in Stanford University, beginning about 1938. They were able to keep the mutant strains alive in the laboratory by providing each strain with the additional food that it needed for good health, as shown by a normal rate of growth.

The substance thiamine (vitamin B_1) is needed by human beings to keep them from dying of the disease beriberi. It has been found, in fact, that thiamine is needed as an essential food for all other animal species that have been studied, including the domestic pigeon, the laboratory rat, the guinea pig, the pig, the cow, the domestic cat, and the monkey, and that chickens fed on a diet that contains none of this food also die of a neurological disease resembling beriberi.

We may surmise that the need of all of these animal species for thiamine as an essential food, which they must ingest in order not to develop a disease resembling beriberi in human beings, resulted from an event that took place over 500 million years ago. Let us consider the epoch, early in the history of life on earth, when the early animal species from which present-day birds and mammals have evolved populated a part of the earth. We assume that the animals of this species nourished themselves by eating plants, possibly together with other food. Many plants contain thiamine. Accordingly the animals would have in their bodies thiamine that they had ingested with the foodstuffs that they had eaten, as well as the thiamine that they themselves synthesized by use of their own synthetic mechanism. Now let us assume that a mutant animal appeared in the population, an animal that, as the result of impact of a cosmic ray on a gene or of the action of some other mutagenic agent, had lost the biochemical machinery that still permitted the other members of the species to manufacture thimaine from other substances. The amount of thiamine provided by the ingestion of food would suffice to keep the mutant well nourished, essentially as well nourished as the unmutated animals; and the mutant would have an advantage over the unmutated animals, in that it would be liberated of the burden of the machinery for manufacturing its own thiamine. As a result the mutant would be able to have more offspring than the other animals in the population.

By reproduction the mutated animal would pass its advantageously mutated gene along to some of its offspring, and they too would have more than the average number of offspring. Thus in the course of time this advantage, the advantage of not having to do the work of manufacturing thiamine or to carry within itself the machinery for this manufacture, could permit the mutant type to replace the original type.

Many different kinds of molecules must be present in the body of an animal in order that the animal be in good health. Some of these molecules can be synthesized by the animal; others must be ingested as foods. If the substance is available as a food, it is advantageous to the animal species to rid itself of the burden of the machinery for synthesizing it.

It is believed that, over the millennia, the ancestors of human beings were enabled, over and over again, by the availability of certain substances as foods, including the essential amino acids and the vitamins, to simplify their own biochemical lives by shuffling off the machinery that had been needed by their ancestors for synthesizing these substances. It is evolutionary processes of this sort that gradually, over periods of millions of years, led to the appearance of new species, including man.

Some very interesting experiments have been carried out on competition between strains of organisms that require a certain substance as food and those that do not require the substance, because of the ability to synthesize it themselves. These experiments were carried out in the University of California, Los Angeles, by Zamenhof and Eichhorn, who published their findings in 1967. They studied a bacterium, *Bacillus subtilis*, by comparing a strain that had the power of manufacturing the amino acid tryptophan and a mutant strain that had lost the ability to manufacture this amino acid. If the same numbers of cells of the two strains were put in a medium that did not contain any tryptophan, the strain that could manufacture tryptophan survived, whereas the other strain died out. If, however, some cells of the two strains were put together in a medium containing a good supply of tryptophan the scales were turned: the mutant strain, which had lost the ability to manufacture the amino acid, survived, and the original strain, with the ability to manufacture the amino acid, died out. The two strains of bacteria differed only in a single mutation, the loss of the ability to manufacture the amino acid tryptophan. We are hence led to conclude that the burden of using the machinery for tryptophan synthesis was disadvantageous to the strain possessing this ability, and hampered it, in its competition with the mutant strain, to such an extent as to cause it to fail in this competition. The number of generations (cell divisions) required for take-over in this series of experiments (starting with an equal number of cells, to a million times as many cells of the victorious strain) was about fifty, which would correspond to only about 1500 years for man (30 years per generation).

We may say that Zamenhof and Eichhorn carried out a small-scale experiment about the process of the evolution of species. This experiment, and several others that they also carried out, showed that it can be advantageous to be free of the internal machinery for synthesizing a vital substance, if the vital substance can be obtained instead as a food from the immediate environment.

Most of the vitamins required by man for good health are also required by animals of other species. Vitamin A is an essential nutrient for all vertebrates for vision, maintenance of skin tissue, and normal development of bones. Riboflavin (vitamin B_2), pantothenic acid, pyridoxine (vitamin B_6), nicotinic acid (niacin), and cyanocobalamin (vitamin B_{12}) are required for good health by the cow, pig, rat, chicken, and other animals. It is likely that the loss of the ability to synthesize these essential substances, like the loss of the ability to synthesize thiamine, occurred rather early in the history of life on earth, when the primitive animals began living largely on plants, which contain a supply of these nutrients.

Dr. Irwin Stone pointed out in 1965 that, whereas most species of animals can synthesize ascorbic acid, man and other primates that have been tested, including the rhesus monkey, the Formosan long tail monkey, and the ringtail or brown capuchin monkey, are unable to synthesize the substance, and require it as a vitamin. He concluded that the loss of the ability to synthesize ascorbic acid probably occurred in the common ancestor of the primates. A rough estimate of the time at which this mutational change occurred is twenty-five million years ago. . . .

The guinea pig and an Indian fruit-eating bat are the only other mammals known to require ascorbic acid as a vitamin. The red-vented bulbul and some other Indian birds (of the order Passeriformes) also require ascorbic acid. The overwhelming majority of mammals, birds, amphibians, and reptiles have the ability to synthesize the substance in their tissues, usually in the liver or the kidney. The loss of the ability by the guinea pig, the fruit-eating bat, and the red-vented bulbul and some other species of passeriform birds probably resulted from independent mutations in populations of these species of animals living in an environment that provided an ample supply of ascorbic acid in the available foodstuffs.

We may ask why ascorbic acid is not required as a vitamin in the food of the cow, pig, horse, rat, chicken, and many other species of animals that do require the other vitamins required by man. Ascorbic acid is present in green plants, along with these other vitamins. When green plants became the steady diet of the common ancestor of man and other mammals, hundreds of millions of years ago, why did not this ancestor undergo the mutation of eliminating the mechanism for synthesizing ascorbic acid, as well as the mechanisms for synthesizing thiamine, pantothenic acid, pyridoxine, and other vitamins?

I think that the answer to this question is that for optimum health more ascorbic acid was needed than could be provided under ordinary conditions by the usually available green plants.

Let us consider the common precursor of the primates, at a time about twenty-five million years ago. This animal and his ancestors had for hundreds of millions of years continued to synthesize ascorbic acid from other substances that they had ingested. Let us assume that a population of this species of animals was living, at that time, in an area that provided an ample supply of food with an unusually large content of ascorbic acid, permitting the animals to obtain from their diet approximately the amount of ascorbic acid needed for optimum health. A cosmic ray or some other mutagenic agent then caused a mutation to occur, such that the enzyme in the liver that catalyzes the conversion of L-gulonolactone to ascorbic acid was no longer present in the liver. Some of the progeny of this mutant animal would have inherited the loss of the ability to synthesize ascorbic acid. These mutant animals would, in the environment that provided an ample supply of ascorbic acid, have an advantage over the ascorbic-acid-producing animals, in that they had been relieved of the burden of constructing and operating the machinery for producing ascorbic acid. Under these conditions the mutant would gradually replace the earlier strain.

A mutation that involves the loss of the ability to synthesize an enzyme occurs often. Such a mutation requires only that the gene be damaged in some way or be deleted. (The reverse mutation, leading to the ability to produce the enzyme, is difficult, and would occur only extremely rarely.) Once the ability to synthesize ascorbic acid has been lost by a species of animals, that species depends for its existence on the availability of ascorbic acid as a food.

The fact that most species of animals have not lost the ability to manufacture their own ascorbic acid shows that the supply of ascorbic acid available generally in foodstuffs is not sufficient to provide the optimum amount of this substance. Only in an unusual environment, in which the available food provided unusually large amounts of ascorbic acid, have circumstances permitted a species of animal to abandon its own powers of synthesis of this important substance. These unusual circumstances occurred for the precursor of man and other primates, for the guinea pig, for the Indian fruit-eating bat, and for the precursor of the red-vented bulbul and some other species of passeriform birds, but have not occurred, through the hundreds of millions of years of evolution, for the precursors of the cow, the horse, the pig, the rat, and hundreds of other animals. Thus the consideration of evolutionary processes, as presented in the foregoing analysis, indicates that the ordinarily available foodstuffs might well provide essentially the optimum amounts of thiamine, riboflavin, niacin, vitamin A, and other vitamins that are required as essential nutrients by all mammalian species, but be deficient in ascorbic acid. For this food, essential for man but synthesized by many other species of animals, the optimum rate of intake is indicated to be larger than the rate associated with the ingestion of the ordinarily available diet.

I have checked the amounts of various vitamins present in 110 raw, natural plant foods, as given in the tables in the metabolism handbook published by the Federation of American Societies for Experimental Biology. . . . When the amounts of vitamins corresponding to one day's food for an adult (the amount that provides 2500 kilocalories of energy) are calculated, it is found that for most vitamins this amount is about three times the daily allowance recommended by the Food and Nutrition Board. For ascorbic acid, however, the average amount in the daily ration of the 110 plant foodstuffs is 2.3 g, about forty-two times the amount recommended as the daily allowance for a person with a caloric requirement of 2500 kilocalories per day (see Table 1 . . .).

TABLE 1. Vitamin Content of 110 Raw Natural Plant Foods Referred to Amount Giving 2500 Kilocalories of Food Energy

Foods	Thiamine	Riboflavin	Niacin	Ascorbic acid
Nuts and grains (11)	3.2 mg	1.5 mg	27 mg	0 mg
Fruit, low C (21)	1.9	2.0	19	600
Beans and peas (15)	7.5	4.7	34	1000
Berries, low C (8)	1.7	2.0	15	1200
Vegetables, low C (25)	5.0	5.9	39	1200
Intermediate-C foods (16)	7.8	9.8	77	3400
High-C foods (6)	8.1	19.6	58	6000
Very high-C foods (8)	6.1	9.0	68	12000
Averages for 110 foods	5.0	5.4	41	2300
Recommended daily allowance for adult	1.0 to 1.6 mg	1.3 to 1.7 mg	13 to 20 mg	50 to 60 mg
Ratio of plant food average to average recommended allowance	3.8	3.6	2.5	42

Nuts and grains: almonds, filberts, macadamia nuts, peanuts, barley, brown rice, whole grain rice, sesame seeds, sunflower seeds, wheat, wild rice.

Fruit (low in vitamin C, less than 2500 mg): apples, apricots, avocadoes, bananas, cherries (sour red, sweet), coconut, dates, figs, grapefruit, grapes, kumquats, mangoes, nectarines, peaches, pears, pineapple, plums, crabapples, honeydew melon, watermelon.

Beans and peas: broad beans (immature seeds, mature seeds), cowpeas (immature seeds, mature seeds), lima beans (immature seeds, mature seeds), mung beans (seeds, sprouts), peas (edible pod, green mature seeds), snapbeans (green, yellow), soybeans (immature seeds, mature seeds, sprouts).

Berries (low C, less than 2500 mg): blackberries, blueberries, cranberries, loganberries, raspberries, currants (red), gooseberries, tangerines.

Vegetables (low C, less than 2500 mg): bamboo shoots, beets, carrots, celeriac root, celery, corn, cucumber, dandelion greens, egg-plant, garlic cloves, horseradish, lettuce, okra, onions (young, mature), parsnips, potatoes, pumpkins, rhubarb, rutabagas, squash (summer, winter), sweet potatoes, green tomatoes, yams.

Intermediate-C foods (2500 to 4900 mg): artichokes, asparagus, beet greens, cantaloupe, chicory greens, chinese cabbage, fennel, lemons, limes, oranges, radishes, spinach, zucchini, strawberries, swiss chard, ripe tomatoes.

High-C foods (5000 to 7900 mg): brussels sprouts, cabbage, cauliflower, chives, collards, mustard greens.

Very high-C foods (8000 to 16500 mg): broccoli spears, black currants, kale, parsley, hot chili peppers (green, red), sweet peppers (green, red).

If the need for ascorbic acid were really as small as the daily allowance rec-ommended by the Food and Nutrition Board the mutation would surely have occurred 500 million years ago, and dogs, cows, pigs, horses, and other animals would be obtaining ascorbic acid from their food, instead of manufacturing it in their own liver cells.

Therefore, I conclude that 2.3 g per day is less than the optimum rate of intake of ascorbic acid for an adult human being.

The average ascorbic-acid content of the fourteen plant foodstuffs richest in this vitamin is 9.4 g per 2500 kilocalories. Peppers (hot or sweet, green or red) and black currants are richest of all, with 15 g per 2500 kilocalories. These amounts indicate an upper limit for the optimum daily intake for man.

I conclude that the optimum daily intake of ascorbic acid for most adult human beings lies in the range 2.3 g to 9 g. The amount of individual biochemi-cal variability . . . is such that for a large population the range may be as great as from 250 mg to 10 g or more per day.

The foregoing argument represents an extension and refinement of Bourne's gorilla argument, and it leads to a similar conclusion. The conclusion about the optimum intake is also nearly the same as that from Stone's rat argument (1.8 g to 4.1 g per day).

It is, of course, almost certain that some evolutionary effective mutations have occurred in man and his immediate predecessors rather recently (within the last few million years) such as to permit life to continue on an intake of ascorbic acid less than that provided by high-ascorbic-acid raw plant foods. These mutations might involve an increased ability of the kidney tubules to pump ascorbic acid back into the blood from the glomerular filtrate (dilute urine, being concentrated on passage along the tubules) and an increased abil-ity of certain cells to extract ascorbic acid from the blood plasma. It is likely that the adrenal glands act as a storehouse of ascorbic acid, extracting it from the blood when green plant foods are available, in the summer, and releasing it slowly when the supply is depleted. On general principles we can conclude, however, that these mechanisms require energy and are a burden to the orga-nism. The optimum rate of intake of ascorbic acid might still be within the range given above, 2.3 g per day or more, or might be somewhat less; and, of course, there is always the factor of biochemical individuality.

Questions and Topics for Discussion and Writing

1. What are "essential amino acids"?
2. In what way does Pauling's specialized perspective as a biochemist appear in his view of the human race?
3. Explain in your own words the dependence of many animal species on thiamine.
4. In what way did Zamenhof and Eichhorn carry out "a small-scale experiment about the process of evolution of species"?

5. How does Pauling relate ascorbic acid to the subject of human evolution?
6. Review Albert Szent-Györgyi's essay "Lost in the Twentieth Century" (Unit 1). What links do you see between his work and that of Pauling? In the library, research the history of ascorbic acid (Vitamin C) and write a short summary.
7. What is the value of the graphic aid Pauling supplies? Why do you suppose that he used the tabular form?
8. What is the "Vitamin C controversy"? Search your library for the most up-to-date findings. In an argument paper, weigh carefully both sides of the controversy. Present your position, being certain to address the opposition.

George Washington Carver

How does one characterize a man who developed more than 300 by-products from the peanut and 118 from the sweet potato? George Washington Carver (1861–1943), botanist, chemist, and educator, instigated significant advances in agriculture in an era when it was the largest single occupation of Americans. He helped to revolutionize the economy of the South by freeing it from its dependence on cotton. Born the son of slaves on a Missouri farm, he was kidnapped with his mother from her owner, Moses Carver, who bought back the infant for a horse valued at $300. Although the Emancipation Proclamation freed him, he remained on the plantation until he was ten or twelve. Then, retaining the Carver surname, he left to seek an education.

A youth with a keen and inquiring mind, he wandered about Kansas doing odd jobs to earn money for schooling. Despite seeing Carver's obvious talent in art and music, his mentors encouraged him, because of his skill with plants, to study agriculture. Earning a degree in 1894 from Iowa State Agricultural College, he joined the Ames Experiment Station. In 1896, Carver earned his master's degree in horticulture, and Booker T. Washington quickly offered Carver the position as head of the new Department of Agriculture at Tuskegee Institute.

At this time, the overproduction of tobacco and cotton had depleted the nutrients of much of the land in the South. Using the poor land around the Institute as a laboratory for experiments, Carver began a "school on wheels" program, a traveling classroom that taught Alabama farmers the basics of soil conservation and crop diversification. His concept of the moving school was soon adopted by the United States Department of Agriculture and by many foreign countries. When farmers heeded his advice on planting, but found no markets for their new products, Carver began experimenting with the nitrogen cycle and discovered that many plants had uses other than just as "foodstuffs." These by-products included plastics, dyes, medicines, oils, cereals, flour, soaps, powdered milk, wood stains, and fertilizer.

During the wartime food shortage of 1918, he demonstrated in Washington how food substitutes might be produced in abundance. Collaborating with the United States Department of Agriculture and the Smithsonian Institution, Carver catalogued medicinal flora and fungi. His fame as a scientist and educator spread through the world; many governments requested his assistance with their agricultural programs. Joseph Stalin invited him to Russia to assist in the ambitious cotton-growing experiment in the Ural Mountains, but Carver refused to go, sending instead his students.

Many persons, including Thomas Edison and Henry Ford, offered him employment at enormous salaries, but he refused to leave Tuskegee, unwilling even to accept a raise there. Carver did not care to commercialize the results of his experiments. What funds accumulated from uncashed salary checks, he willed to the Carver Foundation, to provide scholarships for young people with the potential to aid farmers through science.

Carver's pamphlet *Feeding Acorns to Livestock* (1898) was the first in a series of forty-five publications resulting from his research. His many honors included election to the Royal Society, the Spingarn Medal, the Roosevelt Medal, and the Thomas A. Edison Foundation Award. In "The Undiscovered Sweet Potato," Carver extols the virtues of the ignored "aristocrat of the vegetable kingdom" and explores his vision of its "potentialities."

The Undiscovered Sweet Potato

George Washington Carver

Of all the crops that have passed, as it were, in panoramic form before the critical eyes of the chemist, the careful and painstaking work of the dietitian and the stern but just judge of the palate, none have stood the test so well as the sweet potato.

As a conserver of wheat flour it has stood the most rigid tests, and is pronounced, almost without reserve, as the best substitute yet discovered. This, of course, makes it of vital importance as a war measure.

There are six kinds of flour ("meal," as millers insist upon calling them); two unusually choice breakfast foods sufficiently sweet to serve without sugar; a starch just as good if not superior to corn starch for puddings, blanc mange, custards, gravies, etc. The pulp, after the starch has been removed, makes pies, puddings, etc., similar to shredded cocoanut. It is easily dried and can be put away for future use the same as any dried fruit or vegetable.

In Alabama alone many thousands of bushels of potatoes were lost outright or rendered unfit for human food by the severe winter. The various fungus trou-

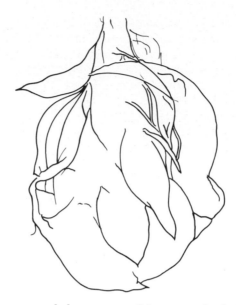

The sweet potato is recommended on account of the ease with which it can be dried and the unusual quality of the dried product. In this state it has unlimited possibilities and may become just as important an article of food as apples, peaches, prunes or fruits of any kind.

bles make the problem of storage a matter of considerable doubt, and the loss in the aggregate is altogether too much. So, therefore, it seems to me that drying has the following things to recommend it:

(a) It puts the potato into an imperishable form.
(b) The ease with which it can be dried.
(c) The superior quality of the dried product.
(d) The ease with which it can be marketed.
(e) The simplicity and cheapness puts the drying process within the reach of any farmer.
(f) It enables him to work up very small, cut, bruised and broken potatoes, so that there need be no loss or waste at all.
(g) With just a little preparation they are ready to serve (without waste).
(h) The peelings, strings, etc., can be dried . . . so that there need be no loss or waste.

On February 18, 1918, I selected four pounds of the Triumph variety, steamed them until done, peeled and dried them. They yielded two pounds of dried potato and four ounces of stock food (the peelings). This yield of dried product may be a little high, as the percentage of water in the potatoes was low when taken.

In connection with cornmeal or wheat flour I do not think of the sweet potato or any of its products as wholly replacing either, but intimately and admirably blending with and conserving both to a remarkable degree, without sacrificing the taste or attractiveness of the product, but indeed adding a delightful piquancy to it, and without very materially changing the nutritive properties of the product (varying, of course, with the percentage of admixture).

For breads, cakes, pies, custards, etc., it is wholly unnecessary to make the sweet potato flour. If the potatoes are steamed or boiled until done, peeled, sliced thin, granulated or run through a ricer or food grinder and dried, they may be soaked in a little water or milk until soft, which requires only an hour or so, according to the thickness of the pieces. They may now be made into a paste and used in the same manner as the fresh potato.

Those who do not want to take the time to soak the nibs of dried potatoes will find the flour attractive, and indeed it has considerable commercial possibilities.

It is probable that the sweet potato is going to do more towards solving the problem of grain shortage than anything else yet found. It will furnish us much sugar and starch as well.

Peanuts, soy beans, cow peas and field beans in variety can supply the proteids and fats cheaply and abundantly.

The drying of the sweet potato is not simply a war measure destined to die as soon as the war is over, but a commercial necessity, and it will become one of the staple articles of commerce, on account of its superior quality, imperishable form and ease of transportation. It seems not only possible but probable that the dried sweet potato will become as important an article of food as dried apples, peaches, prunes or fruits of any kind.

I have mentioned only a few of the eighty or more different products made from the sweet potato, but trust enough has been said to convince the most skeptical that the sweet potato is truly an aristocrat of the vegetable kingdom with almost unlimited possibilities, and that God has bequeathed to the South a wonderful heritage in its fine climate and soils peculiarly adapted to large yields of almost every variety of this splendid vegetable.

Questions and Topics for Discussion and Writing

1. In a short paper, explain the significance of Carver's work with the sweet potato.
2. During Carver's time, other leading Black scientists included Elmer S. Imes, Ernest E. Just, Julian Lewis, William Hinton, Percy Julian, Charles Drew, and Daniel Hale Williams. After research in the library on the work of at least two of these figures, write a report summarizing their contributions.
3. Examine the language in this report. In a short paper, characterize the tone.
4. Carver calls the sweet potato the "aristocrat of the vegetable kingdom." How does

this use of a metaphor affect your reading of the report? What criteria does Carver employ to determine this ranking?

5.   List topics in current agricultural research. Compile an annotated bibliography of sources that record research in one field (for example, the development of special strains of wheat or corn by genetic selection).

John W. Harrington

"I'm a minstrel, a teacher. I sing of ancient stories told by rocks, rivers, mountains, and plains. Follow me. . . ." opens John W. Harrington's *Dance of the Continents.* As a "Pied Piper" of geology taking us on an extended geological field trip, Harrington is well worth following. He gives his subject appeal, always illustrating his belief that "Geologists are hunters of *things* in order to be hunters of the *mind.*"

Harrington (b. 1918) grew up in Richmond, Virginia, where he early discovered the fascination of rocks. Recalling a first-grade incident of a failed spelling test, he says one thought guided him: to "bury the evidence." He writes, "As I knelt in a lonely spot to commit this crime against integrity, I saw the most beautiful rock imaginable. It can only be described in purple prose: my rock was tinted by the setting sun a glistening mass of golden, glassy grains that sparkled in a wondrous way. It made me think of a picture I had seen of the Hall of Mirrors in Versailles. The spelling test was forgotten as a new fantasy world spread before me. . . ."

When his engineer father dismissed as impractical a career in geology, Harrington settled for a degree in mining engineering as an alternative way to live with rocks. He says, "The engineer's required Sin–Cos–Theta brand of thinking, however, proved too austere to be taken seriously. Fortunately, the G.I. Education Bill offered a way to take a PhD in geology, and a future life of open-ended romance centered first on the lure of the hunt in petroleum exploration."

By 1949, Harrington had established three concurrent careers. A consulting practice searching for oil in Texas, Canada, and other areas "paid the bills." Teaching geology at Southern Methodist University in Dallas and later at Wofford College in South Carolina "supplied endless chains of new ideas" and surrounded him with "the infectious excitement of students." Scientific research based on "information bred in the oil hunt" added the "thrill of seeking frontiers." But time passed and "a mellowing occurred."

"Life at Wofford College" he writes, "was encased in the classical liberal arts tradition of Cardinal Newman's mid–nineteenth century thesis, *The Idea of a University.* Faculty conversations were humanistic and rarely concerned with thoughts of rocks. The only way I could be a part of it all was to join in and imitate the faculty value system. This led to an entirely new career in the history and philosophy of science." This life also honed his rhetorical skills.

Harrington's flair for anecdote, for vivid examples, and for the dramatic relationships of ideas makes him one of the most readable geologists. To date, three books have chronicled his adventures, *To See a World, Discovering Science,* and *Dance of the Continents.* "The most creative part of this work," according to Harrington, "consisted of hunting out the original data and outcrops used by pioneering geologists as they first learned to read the story of the earth. These men and women were great hunters whose exploits can be shared by anyone willing to visit the historic localities, look at the same rocks, and read the published accounts in this context."

"The Lure of the Hunt" is a sample of Harrington's observations.

The Lure of the Hunt

John Harrington

"We were hunters before we were farmers."
                    Henry S. Johnson, Jr., Consulting Geologist
                    Personal Communication

We were hunters and gatherers for more than five million years before learning to farm successfully. A gene-imprinted lust to follow the old game trails yet may stir within us all. If so, field geologists are certainly members of the breed for they are a curiously driven set of explorers, trackers, hunters, and gatherers. We can learn a great deal about the science of geology by watching a few of them at work. Anticipation, derring-do, and the intellectual and physical fun of it all: traits that belong to sportsmen and artists characterize scientists as well. It's sometimes difficult to distinguish their work from their play, or hardship from adventure. Geologists tend to romanticize things. A pocketful of peanuts and raisins for lunch becomes a picnic, and tavern talk in the evening, a university. Geologists seem to live in the hunter's state of expectant hope: who knows what surprise may turn up at any moment?

On the day of Queen Elizabeth's Silver Jubilee, I was at Lyme Regis, a small town on the southern coast of England, the setting for John Fowles's best-selling novel, *The French Lieutenant's Woman.* Every amateur fossil collector in Great Britain appeared to be on the cliffs that day, searching in the crumbling, dark limestones and shales for 140-million-year-old marine fossils of Jurassic age. It was hard for me to imagine that anything of value could have been overlooked by this mob of hunters and gatherers. Nevertheless, I had come to Lyme Regis to sample its treasures and was determined not to leave empty-handed.

The beach was a great mass of broken pieces of chert and flint, worn from the cliffs along various parts of the coastline. These are ideal materials from which to make stone tools and weapons. Primitive man used them extensively. Here was an unexpected opportunity to follow the trail of the ancient Britons.

The search began among the boulders clustered on the shore. Perhaps through the years they had offered enough protection from harassing storm waves and high tides to have preserved something for me. In less than three minutes I had found half of an ancient, but recently broken, stone axhead wedged between the rocks. Its edges were a little dulled by wave wear; yet the delicate chipping that defined the ax as the work of a human was undeniable. Weathering had changed the outside texture and covered the older surfaces with a soft, porous patina. Newly made breaks were distinguished by exposures of fresh, unweathered rock.

I was less impressed by the technical factors of the ax's survival in the violent beach environment than I was by the human factors of its creation. Survival on the beach was simply a matter of sufficient protection from the mechanical and chemical agents that can destroy rocks; explaining that sort of thing is routine for a geologist. The more exciting questions dealt with people: who made the ax and when was it done? The work was obviously that of a craftsperson. Did he or she share the same kinds of aesthetic pleasures that I was experiencing millennia later?

Heraclitus, a Greek philosopher of the fifth century B.C., said that no man could step into the same river twice. That's true in one way, yet my Stone Age friend had seen, heard, and smelled the same part of the English Channel that I was experiencing. Our hands had touched the same piece of stone. We were both hunters who thrilled with the anticipation of enriching our lives by looking in the best places for the best game. Separation in time was all that kept us from sharing my peanuts and raisins as well as some of our thoughts about the world and its ways. For me, Jubilee Day was spent in the outer fringes of reality somewhere between science and romance. Surely, this is the way William Blake felt when he wrote of holding infinity in the palm of his hand.

Experiences of this sort are quite common in field-oriented sciences, where a practitioner is often alone with his or her thoughts. Years ago when I was still an undergraduate studying mining engineering, I spent a happy Sunday afternoon in the Black Heath of Chesterfield County near Richmond, Virginia. I had gone there to find plant fossils in the shales of the Richmond Triassic Basin. The rocks are about 180 million years old and contain a few low-grade coal seams of submarginal economic value. Optimistic entrepreneurs had initiated sporadic attempts to mine the area since colonial times. After poking my head into three or four eighteenth-century mine tunnels exposed in the walls of a more recently cut open pit, I decided to split open a block of black shale with the chisel edge of my geologist's pick. I chipped carefully along the line of a bedding plane and watched a crack open completely around the block. Suddenly, the halves fell apart revealing a magnificent, bladelike leaf about fifteen inches long and an inch and a half wide. Fossils are usually seen as casts or molds of stony material that follow the shape of some plant or animal that had been entombed while the rock was being deposited. This one was different. The actual leaf lay before me as a thin, mechanically free film of carbon. Every vein and pore was there. For a moment the leaf was perfect. Suddenly the breeze caught the edge of the leaf and began to break it into brittle, jagged fragments. I thought of the original green leaf that had waved so easily in the winds of Triassic age before it was shut off from sunlight by death and burial. What to do? For a moment I was paralyzed by the enormity of my responsibility. Suddenly, I had an unusual flash of genius, and a sense of proportion: the world's finest Triassic salad was going to waste. I couldn't let that happen . . . so I ate it!

Eating the data may be all right for students, but working geologists must be pragmatists. They are paid to deal with tangible things in a logical manner. Yet even in commercial ventures there is an element of romance. Exploration for phosphate deposits beneath concealing sands of the coastal plain of the southeastern states may lead to better nutrition and a better life for millions of people in Asia and Africa, since phosphate fertilizers improve soil productivity in areas of marginal agricultural potential. The chain of relationships linking geological prospecting for minerals and the quality of life available to all people who live in an industrial society is unbroken.

Back in the 1970s, Henry Johnson, a farsighted, consulting geologist based in Charleston, South Carolina, had this experience while working at the base of the food chain.

His assignment was to create new wealth by discovering phosphate deposits that could be mined and sold at profit. Johnson's method consisted of drilling holes in strategic places and studying core samples taken at various depths. The plan was comparable to hunting for apples in an unknown country by searching the wastelands to find an orchard, then a tree, and eventually the apple. This is a method of homing in on success by eliminating places where failure has been experienced. On the last day of the scheduled field season Johnson reacted intuitively as a gambler.

Years earlier a driller had reported encountering about fifty feet of phosphatic sand in a water well many miles away from Johnson's prospecting area. The report indicated that the deposit, even if properly identified, was too deep to be mined successfully. Johnson's idea was, first, to drill a new hole beside the water well to verify the report and, then, to try to locate a place where mining could be done profitably.

Things went badly from the start. Technical difficulties forced Johnson's driller to give up the first test hole and move the rig to a new location. Wild thunderstorms, encircling the rig, threatened the work with glowering, black clouds, heavy rains, and spectacular displays of lightning. Late in the afternoon the driller finally reached the proper depth and brought up a core from the critical section. Johnson and his assistant waited at the core barrel as the rock slid from its steel case. Their pulses raced in suspense.

Twenty feet of blackish-brown fish-roe-sized pellets of concentrated phosphate emerged. It seemed as pure as fine caviar! In order to see the interior of the core, Johnson split the soft, uncemented rock with his heavy butcher knife. There was no error: it was solid phosphate. The two geologists looked up simultaneously, then their heads turned in unison as they stared around the full circle of the horizon. Johnson broke the silence.

"By God, we've found the scent of the bloody meat! There is a phosphate factory out there somewhere! We're going to shut this rig down until we can figure out where to find it."

That's exactly what they did: paused, evaluated their information, and eventually discovered many millions of tons of mineable phosphate ore.

Field geologists function as detectives. The acts of nature usually have been completed long before they arrive on the scene. Their task becomes one of reconstructing the events that produced the earth as they found it. Their technique is to start with the most recent event and work back in time. Sherlock Holmes called this method, thinking backward. Henry Johnson's phosphate hunt offers a beautiful example.

He began with the knowledge that phosphate pebbles are round because they have been rolled and worn by ocean currents drifting across the sea floor. This meant that a linear current system must have been in operation between the site of the natural phosphate "factory" and the drill hole where Henry "picked up the scent." Johnson also knew that phosphate rocks are marine sediments that were laid down as part of the submerged apron of sand geologists call the continental shelf. Using the pattern of contemporary ocean currents on the continental shelf as a guide, it was an easy matter to predict the best direction in which to explore.

The natural phosphate factory site was located under a thin cover of younger sediments. Originally it had been a submerged hill where the physical conditions were just right for the chemical precipitation of phosphate pellets from seawater. The ore body was waiting for him just below the crest of the hill. It's easy to imagine the triumphant joy Johnson felt when the shape, thickness, and value of the ore body were defined by further drilling.

Geological fieldwork is a hunt for meaning. Rocks, minerals, fossils, and structures are just intermediate steps, missing links that must be identified before ignorance gives way to understanding. . . .

Sometimes a little imagination allows us to enjoy reliving a particularly interesting field trip that someone else has taken. One of my favorites is a "loaf of bread and a jug of wine" situation that occurred about 1730 in Somerset in the southwest of England. The heroine was Mrs. Mary Chandler, a lively milliner and poetess who published this astounding verse in 1734.

"Description of Bath"

The shatter'd Rocks and Strata seem to say,
"Nature is old, and tends to her Decay":

Yet, lovely in Decay and green in Age,
Her Beauty lasts to her latest Stage.

The striking thing about this verse is the way that Mary Chandler has combined an accurate description with an awareness of the way rocks indicate the
passage of time. When I first read the poem I couldn't imagine how its author
was able to learn enough geology to think this way. Who was her teacher?

Several years after discovering Mary's poem, I was probing some broader
questions in the history of geology and by accident found what I imagine to be
the answer. A modern geologist, John G. C. M. Fuller, published a paper in
1969 with the formidable title "The Industrial Basis of Stratigraphy: John
Strachey, 1671–1743, and William Smith, 1769–1839." Fuller's sources included two very old papers written by a coal mining engineer named John
Strachey in 1719 and 1725 and published in the *Philosophical Transactions of
the Royal Society of London.* Strachey proved to be a geological genius with
broad field experience.

Strachey's engineering assignments took him to a number of coal mines in
the northern part of Somerset near the beautiful old Roman town of Bath,
where Mary lived. The hills in this area are capped by flat-lying rocks that
partially cover a set of steeply dipping coal seams visible only in the valleys.
Strachey made exhaustive notes on the distribution and attitudes of all these
rocks as they were exposed both above ground and in the mine shafts and
tunnels. His data were presented in two magnificent vertical cross-sections . . .
which form a three-dimensional model of this area. Strachey even tried to
explain the origin of the tilted strata by proposing that the beds were wrapped
over one another in response to the eastward rotation of the earth. This idea of
a two-stage history in which deposition was followed by tilting (i.e., Mary's
"The shatter'd Rocks and Strata . . .") offers a tenuous thread connecting geologic hero and heroine.

Apparently, no one else in that part of Somerset had this much technical and
poetic interest in geology. Therefore it is highly probably that John and Mary
knew one another as teacher and student. Could it also have been that the
fifty-nine-year-old engineer-geologist and the vivacious forty-three-year-old
poetess had a romantic attachment? John Strachey, an ancestor of Lytton
Strachey, was a well-known rake in his day. He was a busy man with nineteen
children by two wives: perhaps he needed to get away from time to time. If so,
would either he or Mary have suspected that they could be caught 250 years
after the fact? I have wandered like a musing Omar Khayyam through the lush
meadows, shady forests, and quiet lanes of Somerset hoping to be able to tip
my hat at their passing carriage. Field trips are great fun!

A good field geologist must learn to practice the art that Shakespeare called
"jumping o'er times." It helps establish the validity of rock history. That's how
we can learn to escape from the culture-bound time trap into which we were
born. Those of us who excel at the art of jumping o'er times establish logical,

yet emotionally exciting bonds with both past and future. At the very best it is even possible to imagine living in all times. . . .

Science is the progressive discovery of the nature of nature. This definition implies that science is open-ended. The people who work at it are hunters, gatherers, and travelers who try to find and understand things they have never known before. That's appealing. Geology is a perfect example of a progressive science. Its ancient Greek roots, *geo* ("earth") and *logy* ("discourse"), fix the limits of this adventure to the rocks, minerals, fossils, structures, land forms, oceans, atmosphere, and processes of our planet. All of this was known long ago.

The Reverend John Walker who taught the first systematic course in geology at the University of Edinburgh from 1781 to 1803 told his students exactly what they had to do: "The objects of nature sedulously [i.e., zealously] examined in their native state, the fields and mountains must be traversed, the woods and waters explored, the ocean must be fathomed and its shores scrutinized by everyone that would become proficient in natural knowledge. The way to knowledge of natural history is to go to the fields, the mountains, the oceans, and observe, collect, identify, experiment and study."

That takes a good deal of physical and intellectual energy. John Walker's point was demonstrated to me very clearly years ago when I was serving as an unpaid peon with my first field party. We were working in the Blue Ridge Mountains of Virginia and a famous geologist dropped by to help us. Her name was Anna Jonas Stose, and she was agile as a mountain goat. Our first stop that morning was at an exposure of weathered rock in a deep road cut being made for the then-unfinished Skyline Drive. Anna amazed me by scrambling thirty feet up the embankment to stand on a loose boulder and squint at a piece of rock through her hand lens. Just as she was pronouncing the specimen to be, "Wissahickon Schist . . . " the rock she was standing on began to turn and fall. I shouted a warning but Anna didn't even look up. She just stepped nimbly over to the next foothold and finished her sentence, " . . . there's no doubt about it." I've loved her ever since.

Anna showed me another endearing quality that every hunter must possess. It had been a long, sweaty day, and I had been riding on the runningboard of our field car, holding onto the open doorframe for balance. The windows were down and the dust of the unpaved mountain roads was as bad inside the car as it was out. We all resembled sweat-stained gingerbread cookies. About 4:30 in the afternoon we stopped at a cool spot where the James River had cut a spec-tacular gorge through the nearly 600-million-year-old Cambrian quartzites. Most of the group began to look for trilobite fossils, but Anna would have none of it.

"John, by this time of day I don't care if I never see another rock. However, I'm quitting for good whenever the excitement doesn't return again by 8:00 A.M. the next day." There we have it: the lure of the hunt is in the mind of the hunter.

Questions and Topics for Discussion and Writing

1. Why does Harrington stress the practical aspects of geology?
2. How is the scientific method of the geologist like that of a detective? Compare Harrington's statement about the function of the geologist with the dramatic presentation of the deductive method in Doyle's "The Science of Deduction" (Unit 2).
3. Discuss the use of transitional phrases from one episode to the next.
4. What is the effect of including the examples in this essay?
5. What was the role played by intuition in the great discovery of phosphate deposits described in this essay? Explain.
6. What is the significance of Harrington's use of the term "picked up the scent" in describing geological fieldwork?
7. William Blake, in describing his mystical experience, wrote that he "held infinity in the palm of his hand." By alluding to Blake's poem, what does Harrington imply about his own experience on Jubilee Day?
8. What about his discovery gave him the most pleasure?
9. Geology is usually given prominence as a precursor to biological science. How does this essay prove that geology in itself is a continuing and vital science? What is suggested as the drawing force leading a geologist to his research?
10. Emulating Harrington's style, describe an adventure that had special meaning for you. Keep the description as exact as possible, maintaining Harrington's proportion of examples and details.
11. Look for passages that show the writer to be a man of wit and humor.

SUGGESTIONS FOR FURTHER READING

Asimov, Isaac. *The Chemicals of Life*. New York: Signet Science Library, 1954.

Attenborough, David. *Life on Earth*. Boston: Little, Brown & Company, 1979.

Boys, V. *Soap Bubbles*. New York: Dover Publications, 1959.

Calder, Nigel. *Time Scale: An Atlas of the Fourth Dimension*. New York: Viking Press, 1983.

Keller, Edward A. *Environmental Geology*. Columbus: Merrill, 1985.

Leicester, Henry M., and Herbert S. Klickstein. *Source Book in Chemistry*. Cambridge: Harvard University Press, 1968.

Levi, Primo. *The Periodic Table [Autobiography]*. Trans. Raymond Rosenthal. New York: Schocken Books, 1984.

Mahon, Thomas. *Principles of American Nuclear Chemistry: A Novel*. Boston: Little, Brown & Company, 1970.

Partington, James R. *History of Chemistry*. New York: Harper & Row, 1960.

Rosenfield, Israel, Edward Ziff, and Borin van Loon. *DNA for Beginners*. New York: W. W. Norton, 1983.

Sayre, A. *Rosalind Franklin and DNA*. New York: W. W. Norton, 1975.

Stryer, Lubert. *Biochemistry*. San Francisco: W. H. Freeman and Company, 1981.

Watson, James. *The Double Helix: A Personal Account of the Discovery of the Structure of DNA*. Ed. Gunther S. Stent. New York: W. W. Norton, 1980.

Unit 7

LIFE SCIENCES: SIZE, SCALE, AND DESIGN

We believe that it requires great enthusiasm to deal accurately with little things; and that it is, consequently, impossible to meet with a reasonable or sober entomologist.

Edinburgh Review, 1822

Anton Van Leeuwenhoek

Anton Van Leeuwenhoek (1632–1723) was an amateur microscopist and anatomist whose scientific studies are among the seventeenth century's most dramatic discoveries. Born in Delft, Holland, he received limited schooling. At sixteen, he was apprenticed to a cloth merchant in Amsterdam, returning four years later to establish himself as a draper, then as chamberlain to the sheriffs of Delft. Thus, little in his background would indicate that Leeuwenhoek would come to be considered the Father of Microbiology.

Leeuwenhoek began grinding lenses as a hobby and used them to study small organisms, the "animalcules" that he began to discover all around him. This avocation became a major activity. Through tedious hand-grinding techniques that he developed himself, Leeuwenhoek created his own microscopes, preferring single, double-convex lenses of short focus to the compound microscope. His lenses were mounted between flat brass plates fitted with handles, enabling the viewer to hold them close to the eye. With these lenses, he achieved magnification of 50 to 300 times. It is said that he never sold any, but did give some away. Leeuwenhoek bequeathed a large number of his lenses to the Royal Society of London; examination of them indicates that he must have developed a special method of illumination to increase their power. However, he never revealed it.

Despite keeping this technique a secret, Leeuwenhoek shared the meticulous records of his careful observations. When he first began his research, Leeuwenhoek studied such diverse materials as water from various sources, teeth scrapings, spermatozoa, and yeast cells. A most important advance was his description of red blood cells, extending knowledge of capillary circulation, which first had been demonstrated by Malpighi in 1660. Less known for theory than for observations, Leeuwenhoek opposed the idea of spontaneous generation of lower forms of life, proving that the sea mussel, the eel, and even the flea—"this minute and despised creature"—bred by the regular course of generation. He thus made a significant contribution to the theory of biology.

Although his observations seem, in a sense, playful or even childlike attempts at scientific categorization, they set a high standard for biologists who were to follow. They are so exact that they can be easily repeated and confirmed. Many observations, however, such as his discoveries of what were obviously bacteria, were not reported again for over 100 years.

In the selection below, part of a letter (1675) to the Royal Society of London, Leeuwenhoek describes his "animalcules" and reveals the ingenuity of his experimental technique.

Animalcules

Anton Van Leeuwenhoek

[1st Observation on Rain-water.]

In the year 1675, about half-way through September[1] (being busy with studying air, when I had much compressed it by means of water[2]), I discovered living creatures in rain, which had stood but a few days[3] in a new tub, that was painted blue within.[4] This observation provoked me to investigate this water more narrowly; and especially because these little animals were, to my eye, more than ten thousand times smaller[5] than the animalcule which Swammerdam[6] has portrayed, and called by the name of Water-flea, or Water-louse,[7] which you can see alive and moving in water with the bare eye.

Of the first sort[8] that I discovered in the said water, I saw, after divers observations, that the bodies consisted of 5, 6, 7, or 8 very clear globules, but without being able to discern any membrane or skin that held these globules together, or in which they were inclosed. When these animalcules bestirred 'emselves, they sometimes stuck out two little horns,[9] which were continually moved, after the fashion of a horse's ears. The part between these little horns was flat, their body else being roundish, save only that it ran somewhat to a point at the hind end; at which pointed end it had a tail, near four times as long as the whole body, and looking as thick, when viewed through my microscope, as a spider's web.[10] At the end of this tail there was a pellet, of the bigness of one of the globules of the body; and this tail I could not perceive to be used by them for their movements in very clear water. These little animals were the most wretched creatures that I have ever seen; for when, with the pellet, they did but hit on any particles or little filaments[11] (of which there are many in water, especially if it hath but stood some days), they stuck intangled in them; and then pulled their body out into an oval, and did struggle, by strongly stretching themselves, to get their tail loose; whereby their whole body then sprang back towards the pellet of the tail, and their tails then coiled up serpentwise, after the fashion of a copper or iron wire that, having been wound close about a round stick, and then taken off, kept all its windings.[12] This motion, of stretching out and pulling together the tail, continued; and I have seen several hundred animalcules, caught fast by one another in a few filaments, lying within the compass of a coarse grain of sand.[13]

I also discovered a second sort[14] of animalcules, whose figure was an oval; and I imagined that their head was placed at the pointed end. These were a little bit bigger than the animalcules first mentioned. Their belly is flat, provided with divers incredibly thin little feet, or little legs,[15] which were moved very nimbly, and which I was able to discover only after sundry great efforts, and wherewith they brought off incredibly quick motions. The upper part of their body was round, and furnished inside with 8, 10, or 12 globules: otherwise these animalcules were very clear. These little animals would change their body into a perfect round, but mostly when they came to lie high and dry. Their body was also very yielding: for if they so much as brushed against a tiny filament, their body bent in, which bend also presently sprang out again; just as if you stuck your finger into a bladder full of water, and then, on removing the finger, the inpitting went away. Yet the greatest marvel was when I brought any of the animalcules on a dry place, for I then saw them change themselves at last into a round, and then the upper part of the body rose up pyramid-like, with a point jutting out in the middle; and after having thus lain moving with their feet for a little while, they burst asunder, and the globules and a watery humour flowed away on all sides, without my being able to discern even the least sign of any skin wherein these globules and the liquid had, to all appearance, been inclosed; and at such times I could discern more globules than when they were alive. This bursting asunder I figure to myself to happen thus: imagine, for example, that you have a sheep's bladder filled with shot, peas, and water; then, if you were to dash it apieces on the ground, the shot, peas, and water would scatter themselves all over the place.[16]

Furthermore, I discovered a third sort[17] of little animals, that were about twice as long as broad, and to my eye quite eight times smaller[18] than the animalcules first mentioned: and I imagined, although they were so small, that I could yet make out their little legs, or little fins. Their motion was very quick, both roundabout and in a straight line.

The fourth sort[19] of animalcules, which I also saw a-moving, were so small, that for my part I can't assign any figure to 'em. These little animals were more than a thousand times less than the eye of a full-grown louse[20] (for I judge the diameter of the louse's eye to be more than ten times as long as that of the said creature), and they surpassed in quickness the animalcules already spoken of. I have divers times seen them standing still, as 'twere, in one spot, and twirling themselves round with a swiftness such as you see in a whip-top a-spinning before your eye;[21] and then again they had a circular motion, the circumference whereof was no bigger than that of a small sand-grain; and anon they would go straight ahead, or their course would be crooked.[22]

Furthermore, I also discovered sundry other sorts of little animals; but these were very big, some as large as the little mites on the rind of cheese, others bigger and very monstrous.[23] But I intend not to specify them; and will only say, that they were for the most part made up of such soft parts, they they burst asunder whenever the water happened to run off them. . . .

In the summer, when I feel disposed to look at all manner of little animals, I just take the water that has been standing a few days in the leaden gutter up on my roof, or the water out of stagnant shallow ditches: and in this I discover marvellous creatures.

And whether I put in the water whole white pepper, black pepper, coarse pounded pepper, or pepper pounded as fine as flour, animalcules always turn up in it, even on the coldest days in winter, provided only that the water doesn't get frozen.

This day [26 December 1678] there are in my pepper-water some animalcules . . . on which I can make out the paws too, which are also pleasant to behold, because of their swift motions. The paws of these animalcules are very big, in proportion to their bodies.[24] . . . And I am persuaded that thirty million of these animalcules together wouldn't take up as much room, or be as big, as a coarse grain of sand.

NOTES

1. *In den jare 1675 ontrent half September* MS. "In the year 1675" *Phil. Trans.* By neglecting to translate the latter part of the original statement, Oldenburg left the precise date of the discovery in doubt; and a controversy arose later between

Ehrenberg and Haaxman in consequence—the former alleging that the observations were made in April, while the latter, on evidence furnished by the MS. letter from L. to Const. Huygens, believed the correct date to be mid-September. Cf. Haaxman (1875). It is now obvious that Haaxman was right.

2.   Some account of L.'s experiments "on the compression of the air" had already appeared in *Phil. Trans.* (1673), Vol. IX, p. 21. (*Letter 2*, 15 August 1673. MS. Roy. Soc.)

3.   "four days" Saville Kent (Vol. I, p. 3). This is merely due to careless copying, and is not in the originals.

4.   *in een nieuwe ton, die van binnen blauw geverft was* MS. "in a new earthen pot, glased blew within" *Phil. Trans.* Oldenburg here mistranslated L.'s words, which were quite plain and were rendered concordantly by Chr. Huygens "*dans un tonneau peint en huile par dedans*". The vessel was obviously not of Delft porcelain.

5.   *i.e.*, in bulk—not in diameter. This expression means, with L., that he judged the animalcules to have roughly one twenty-fifth of the diameter of the bigger creatures.

6.   Jan Swammerdam (1637–1680). For his life see especially his *Biblia Naturae* (1737) and Sinia (1878).

7.   Swammerdam's "*watervlooy*" was *Daphnia*—as all students of the *Biblia Naturae* are well aware (cf. *B.N.* Vol. I, p. 86, Pl. XXXI). But as this work was not published until 1737—long after his death—it is clear that L. here alludes to his earlier Dutch publication (Swammerdam, 1669), in which the water-flea is shown on Pl. I. Swammerdam himself called it "the branched water-flea", and attributed the name "water-louse" to Goedaert.

8.   *Vorticella* sp. The following admirable description makes the identification certain.

9.   The optical section of the wreath of cilia round the peristome—so interpreted by most of the early observers.

10.   *i.e.*, as thick as a spider's web looks to the naked eye.

11.   *maer quamen aen eenige deeltgens of veseltgens* MS. "if they chanced to light upon the least filament or string, or other such particle" *Phil. Trans.* These words of Oldenburg are amusingly mistranslated by Nägler (1918, p. 7) "*Wenn man diese kleinen Kreaturen zufällig belichtete.*" (He apparently supposes that "to light upon" means "to illuminate"!)

12.   Apparently it never occurred to L., at this time, that the contraction and extension of the stalk ("tail") of *Vorticella* could have any other significance than that here attributed to them. The idea of a *stalked* and normally *sessile animal* probably never entered his head; and consequently he jumped to the incorrect conclusion that the animals were endeavouring to "get their tails loose"—which, of course, was a mistake, though a very natural one. L. published pictures of Vorticellids later, in *Phil. Trans.*, Vol. XXIII (Letter dated 25 Dec. 1702): and still later he arrived at a more correct interpretation of the function of the "tail," and of the organization of these remarkable animals (*Send-brief* VII, dated 28 June 1713).

13.   *inde spatie van een grof sant* MS. "within the space of a grain of gross sand" *Phil. Trans.* This is a very common expression with L., and Oldenburg fully understood its meaning; but Nägler (1918, p. 9) mistranslates his words "*in der Höhlung eines grossen Sandkorns*"—as though L. had seen the animalcules lying in a cavity in an actual grain of sand!

14. Not identifiable with certainty, but undoubtedly a ciliate.
15. *i.e.*, cilia.
16. The foregoing graphic account of the bursting of the "little animals" is of great interest, as it shows clearly that L. was really observing protozoa. An animal whose body consisted entirely of soft "protoplasm"—without any skeletal parts or obvious skin—was, of course, a considerable novelty at this date.
17. Not identifiable. Probably a small ciliate.
18. *i.e.*, having a diameter equal to about half that of the *Vorticella*.
19. Probably—from the ensuing description—a species of *Monas*. Certainly not bacteria of any kind.
20. This makes the diameter of the protozoon here described about 6–8 μ, and is agreeable with its interpretation as *Monas vulgaris*.
21. If the description applies to *Monas*—as I strongly suspect—then the "spinning" here described was an illusion. I fancy L. saw a *Monas* attached by its caudal filament, and mistook the swirl of the water at its anterior end (occasioned by the movements of the small accessory flagellum) for a motion caused by the rotation of the body as a whole.
22. *en dan weder soo regt uijt, als crom gebogen* MS. These words are hard to understand. The above seems to me to be L.'s meaning: but Oldenburg translates "and then extending themselves streight forward, and by and by lying in a bending posture" *(Phil. Trans.)*. It is hardly likely that L. could have observed "a bending posture" in an organism so small that he could discern "no figure" in it: and as the "circular motion" just mentioned evidently refers to the orbit described by the organism—not to the animalcule itself—I imagine that *"regt uijt"* and *"crom gebogen"* likewise refer to the path traversed. I should point out, however, that L. elsewhere *(Letter 38)* applies precisely the same words to the *shape* of the spermatozoa of a frog.
23. Some of these were doubtless protozoa, but the "monsters" were perhaps rotifers. Much later, when describing these animals, L. mentions that he had previously discovered them in rain-water, in which he had steeped pepper and ginger. See *Letter 144*, 9 Feb. 1702.
24. L. probably here refers to the cirrhi on a small hypotrichous ciliate such as *Euplotes*.

Questions and Topics for Discussion and Writing

1. The quotation from the *Edinburgh Review* that opens this unit declares that "it requires great enthusiasm to deal accurately with little things. . . ." How would you apply this statement to Leeuwenhoek?
2. Go to the OED *(Oxford English Dictionary)* and trace the meanings of *enthusiasm*. Does the original meaning affect your understanding of the quotation?
3. Do you consider Leeuwenhoek's method "scientific"? Explain.
4. In what way does Leeuwenhoek personify the subjects of his investigation? Does this approach interfere with scientific objectivity? Cite specific words and phrases to explain your evaluation.

5.   By using the rhetorical method of historical tracing and after consulting the OED, write a short definition paper on one of the following terms: *sophistication, apology, Hobson's choice, etiquette, anatomy.*

Robert Frost

Robert Frost (1874–1963) observed in much of his poetry that the realms of nature and of science are mysterious, exciting, sometimes frightening, but always worth contemplating. In a lecture given at Harvard in 1938, he said, "Science puts it into our heads that there must be new ways to be new." He was referring to the then-new, experimental, but often linguistically obscure poetry that had come into vogue after World War I. Frost argued for a return to the literary conservatism, which held that the old, clear, and simple ways of making poems were the best ones. While his remark was taken by some to be criticism of the scientific approach for seeming too complicated, Frost was speaking here about the communication of ideas.

Frost was, by no means, criticizing science itself or the scientific mode; for good poets, like good scientists, are fascinated to find out, by whatever methods available, *why* things are. In fact, "Design," the second poem presented here, poses an epistemological question: Can we find a rational pattern in nature's random phenomena? Surely this question is one that Darwin, for example, asked himself, again and again.

Frost asked such questions from a personal experience filled with disruption and difficulty. Born in California, Frost moved to Massachusetts after his father's early death in 1885; he remained in the East to become our most important exponent of the New England literary tradition. After unsuccessful attempts to earn a living for his family at both farming and teaching, he removed them to England in 1912 in order to concentrate seriously on writing as a career. There, he published two books of poetry, *A Boy's Will* and *North of Boston.* These books were so well received that he returned home as a poet of recognized stature. He was then in a position to lecture, teach, and write, from time to time, at Amherst, Harvard, Dartmouth, and the University of Michigan.

Between 1924 and 1943, Frost received four Pulitzer Prizes for his poetry; and in 1950, on his seventy-fifth birthday, the United States Senate passed a resolution to honor him as an American poet. In 1961, he read his poem "The Gift Outright" at the presidential inauguration of New Englander John F. Kennedy.

The first poem presented here, "A Considerable Speck," raises a time-worn but still unanswered question about how much intelligence is verifiable in the natural but nonhuman world. The poet uses metaphor, symbol, and image to engage the mind, but we are aware that his curiosity about life and death is no less fervent than that of the scientist in the laboratory—each one is seeking answers. Frost himself was to note that a poem "always ends in wisdom."

A Considerable Speck: Microscopic

Robert Frost

A speck that would have been beneath my sight
On any but a paper sheet so white
Set off across what I had written there.
And I had idly poised my pen in air
To stop it with a period of ink,
When something strange about it made me think.
This was no dust speck by my breathing blown,
But unmistakably a living mite
With inclinations it could call its own.
It paused as with suspicion of my pen,
And then came racing wildly on again
To where my manuscript was not yet dry;
Then paused again and either drank or smelt—
With loathing, for again it turned to fly.
Plainly with an intelligence I dealt.
It seemed too tiny to have room for feet,
Yet must have had a set of them complete
To express how much it didn't want to die.
It ran with terror and with cunning crept.
It faltered: I could see it hesitate;
Then in the middle of the open sheet
Cower down in desperation to accept
Whatever I accorded it of fate.
I have none of the tenderer-than-thou
Collectivistic regimenting love
With which the modern world is being swept.
But this poor microscopic item now!
Since it was nothing I knew evil of
I let it lie there till I hope it slept.

I have a mind myself and recognize
Mind when I meet with it in any guise.
No one can know how glad I am to find
On any sheet the least display of mind.

Questions and Topics for Discussion and Writing

1. Consider the title of this poem. Why does Frost include the subtitle "Microscopic"?
2. An *oxymoron* is a figure of speech. Look up the term. Is the pairing in the term *considerable speck* an oxymoron?
3. Frost deals with the relations between human beings and animals. How does the poem serve as a statement of the human responsibility for nature? Support your judgment.
4. In Line 14, why does Frost attribute "loathing" to "this microscopic item"? Comment on Frost's phrase "it turned to fly."
5. Because couplets allow for small units of thought, they are often used in satire. Are couplets well used here?
6. One of the vital functions that rhyme serves in poetry is to unite thought expressed in separate lines. With the functions of couplets and of rhyme in mind, examine lines 14 and 18. Discuss the way these lines work.
7. Find words that attribute sensitivity and intelligence to minute animals.
8. Reed Whittemore's poem "The Tarantula" later in this unit is rendered in the first person, with the tarantula as speaker. In "Speck," Frost speaks from a human perspective. In a paragraph, compare the difference in tone created by this difference in point of view.
9. The conclusion of the poem is relevant to Frost's attempts at writing the satiric. What comment is made on his writing? What is the object of the satire?

Design

Robert Frost

I found a dimpled spider, fat and white,
On a white heal-all, holding up a moth
Like a white piece of rigid satin cloth—
Assorted characters of death and blight
Mixed ready to begin the morning right,
Like the ingredients of a witches' broth—

A snow-drop spider, a flower like a froth,
And dead wings carried like a paper kite.

What had that flower to do with being white,
The wayside blue and innocent heal-all?
What brought the kindred spider to that height,
Then steered the white moth thither in the night?
What but design of darkness to appall?—
If design govern in a thing so small.

Alexander Petrunkevitch

Most people would not include spiders or scorpions on their list of favorite anythings, but for Alexander Petrunkevitch (1875–1964), the foremost arachnologist of his time, those small creatures represented not a phobia embodied, but rather life-on-eight-legs, clinging tenaciously to the highest and lowest, hottest and coldest ecological niches on the planet. Petrunkevitch himself was in many ways as tenacious and far-ranging as the creatures he studied, keeping regular hours at Yale's Osborn Zoological Laboratory to continue his research years after his official "retirement" in 1944. He wrote articles on topics in other fields, such as philosophy, poetry, history, and photography. His arachnological writings include the article on spiders for the 1950 edition of the *Encyclopaedia Britannica* (an article triple the length of its predecessor), as well as the 790-page "Bible" on American arachnids, *The Index Catalogue of Spiders of North, Central and South America* (1911).

The son of a Russian nobleman, Petrunkevitch left the University of Moscow to continue his studies at the University of Freiburg in Germany, as well as to avoid political imprisonment. While in Germany, he met his wife, an American student of classical philosophy and the niece of the inventor of roll-up window shades. Himself the inventor of several widely used scientific instruments and aids, Petrunkevitch never received a single royalty, however. Scientific investigation should be free, he believed, and patents on scientific devices needlessly interfere with the pursuit of knowledge.

Ironically, Petrunkevitch became interested in black widows and other eight-legged creatures as a result of the poor health of his wife. When his wife became an invalid after a bout with influenza, Petrunkevitch left his position at Harvard to care for her. He found spiders to be convenient research subjects, being easily studied and stored in the makeshift laboratory he established in the house. This research continued part time, during his wife's periods of relapse, until she was finally diagnosed as tubercular, at which point Petrunkevitch joined the staff of Yale to earn money for her treatment.

Although his subject may often be evoked as an example of the efficient cruelty to be found in nature, with tales of the spider's penchant for cannibalism being the focus of more than one natural history film, Petrunkevitch himself has been characterized as an extremely gentle man. He found reprimanding even his most delinquent students almost impossible. "On a walk with him," one former student has written, "you have a feeling of being with an immense, benign intelligence, sweetly in key with the humbler part of the universe."

"The Spider and the Wasp," a much-anthologized essay, is a good introduction to Petrunkevitch's world of arachnids.

The Spider and the Wasp

Alexander Petrunkevitch

To hold its own in the struggle for existence, every species of animal must have a regular source of food, and if it happens to live on other animals, its survival may be very delicately balanced. The hunter cannot exist without the hunted; if the latter should perish from the earth, the former would, too. When the hunted also prey on some of the hunters, the matter may become complicated.

This is nowhere better illustrated than in the insect world. Think of the complexity of a situation such as the following: There is a certain wasp, *Pimpla inquisitor*, whose larvae feed on the larvae of the tussock moth. *Pimpla* larvae in turn serve as food for the larvae of a second wasp, and the latter in their turn nourish still a third wasp. What subtle balance between fertility and mortality must exist in the case of each of these four species to prevent the extinction of all of them! An excess of mortality over fertility in a single member of the group would ultimately wipe out all four.

This is not a unique case. The two great orders of insects, Hymenoptera and Diptera, are full of such examples of interrelationship. And the spiders (which are not insects but members of a separate order of arthropods) also are killers and victims of insects.

The picture is complicated by the fact that those species which are carnivorous in the larval stage have to be provided with animal food by a vegetarian mother. The survival of the young depends on the mother's correct choice of a food which she does not eat herself.

In the feeding and safeguarding of their progeny the insects and spiders exhibit some interesting analogies to reasoning and some crass examples of blind instinct. The case I propose to describe here is that of the tarantula spiders and their arch-enemy, the digger wasps of the genus Pepsis. It is a classic example of what looks like intelligence pitted against instinct—a strange situation in which the victim, though fully able to defend itself, submits unwittingly to its destruction.

Most tarantulas live in the Tropics, but several species occur in the temperate zone and a few are common in the southern U.S. Some varieties are large and have powerful fangs with which they can inflict a deep wound. These formidable looking spiders do not, however, attack man; you can hold one in your hand, if you are gentle, without being bitten. Their bite is dangerous only to insects and small mammals such as mice; for a man it is no worse than a hornet's sting.

Tarantulas customarily live in deep cylindrical burrows, from which they emerge at dusk and into which they retire at dawn. Mature males wander about after dark in search of females and occasionally stray into houses. After mating, the male dies in a few weeks, but a female lives much longer and can mate several years in succession. In a Paris museum is a tropical specimen which is said to have been living in captivity for 25 years.

A fertilized female tarantula lays from 200 to 400 eggs at a time; thus it is possible for a single tarantula to produce several thousand young. She takes no care of them beyond weaving a cocoon of silk to enclose the eggs. After they hatch, the young walk away, find convenient places in which to dig their burrows and spend the rest of their lives in solitude. Tarantulas feed mostly on insects and millepedes. Once their appetite is appeased, they digest the food for several days before eating again. Their sight is poor, being limited to sensing a change in the intensity of light and to the perception of moving objects. They apparently have little or no sense of hearing, for a hungry tarantula will pay no attention to a loudly chirping cricket placed in its cage unless the insect happens to touch one of its legs.

But all spiders, and especially hairy ones, have an extremely delicate sense of touch. Laboratory experiments prove that tarantulas can distinguish three types of touch: pressure against the body wall, stroking of the body hair and riffling of certain very fine hairs on the legs called trichobothria. Pressure against the body, by a finger or the end of a pencil, causes the tarantula to move off slowly for a short distance. The touch excites no defensive response unless the approach is from above where the spider can see the motion, in which case it rises on its hind legs, lifts its front legs, opens its fangs and holds this threatening posture as long as the object continues to move. When the motion stops, the spider drops back to the ground, remains quiet for a few seconds and then moves slowly away.

The entire body of a tarantula, especially its legs, is thickly clothed with hair. Some of it is short and woolly, some long and stiff. Touching this body hair produces one of two distinct reactions. When the spider is hungry, it responds with an immediate and swift attack. At the touch of a cricket's antennae the tarantula seizes the insect so swiftly that a motion picture taken at the rate of 64 frames per second shows only the result and not the process of capture. But when the spider is not hungry, the stimulation of its hairs merely causes it to shake the touched limb. An insect can walk under its hairy belly unharmed.

The trichobothria, very fine hairs growing from disklike membranes on the legs, were once thought to be the spider's hearing organs, but we now know that they have nothing to do with sound. They are sensitive only to air movement. A light breeze makes them vibrate slowly without disturbing the common hair. When one blows gently on the trichobothria, the tarantula reacts with a quick jerk of its four front legs. If the front and hind legs are stimulated at the same time, the spider makes a sudden jump. This reaction is quite independent of the state of its appetite.

These three tactile responses—to pressure on the body wall, to moving of the common hair and to flexing of the trichobothria—are so different from one another that there is no possibility of confusing them. They serve the tarantula adequately for most of its needs and enable it to avoid most annoyances and dangers. But they fail the spider completely when it meets its deadly enemy, the digger wasp Pepsis.

These solitary wasps are beautiful and formidable creatures. Most species are either a deep shiny blue all over, or deep blue with rusty wings. The largest have a wing span of about four inches. They live on nectar. When excited, they give off a pungent odor—a warning that they are ready to attack. The sting is much worse than that of a bee or common wasp, and the pain and swelling last longer. In the adult stage the wasp lives only a few months. The female produces but a few eggs, one at a time at intervals of two or three days. For each egg the mother must provide one adult tarantula, alive but paralyzed. The tarantula must be of the correct species to nourish the larva. The mother wasp attaches the egg to the paralyzed spider's abdomen. Upon hatching from the egg, the larva is many hundreds of times smaller than its living but helpless victim. It eats no other food and drinks no water. By the time it has finished its single gargantuan meal and become ready for wasphood, nothing remains of the tarantula but its indigestible chitinous skeleton.

The mother wasp goes tarantula-hunting when the egg in her ovary is almost ready to be laid. Flying low over the ground late on a sunny afternoon, the wasp looks for its victim or for the mouth of a tarantula burrow, a round hole edged by a bit of silk. The sex of the spider makes no difference, but the mother is highly discriminating as to species. Each species of Pepsis requires a certain species of tarantula, and the wasp will not attack the wrong species. In a cage with a tarantula which is not its normal prey the wasp avoids the spider, and is usually killed by it in the night.

Yet when a wasp finds the correct species, it is the other way about. To identify the species the wasp apparently must explore the spider with her antennae. The tarantula shows an amazing tolerance to this exploration. The wasp crawls under it and walks over it without evoking any hostile response. The molestation is so great and so persistent that the tarantula often rises on all eight legs, as if it were on stilts. It may stand this way for several minutes. Meanwhile the wasp, having satisfied itself that the victim is of the right species, moves off a few inches to dig the spider's grave. Working vigorously with legs and jaws, it excavates a hole 8 to 10 inches deep with a diameter slightly larger than the spider's girth. Now and again the wasp pops out of the hole to make sure that the spider is still there.

When the grave is finished, the wasp returns to the tarantula to complete her ghastly enterprise. First she feels it all over once more with her antennae. Then her behavior becomes more aggressive. She bends her abdomen, protruding her sting, and searches for the soft membrane at the point where the spider's leg joins its body—the only spot where she can penetrate the horny skeleton. From time to time, as the exasperated spider slowly shifts ground, the wasp turns on her back and slides along with the aid of her wings, trying to get under the tarantula for a shot at the vital spot. During all this maneuvering, which can last for several minutes, the tarantula makes no move to save itself. Finally the wasp corners it against some obstruction and grasps one of its legs in her powerful jaws. Now at last the harassed spider tries a desperate but vain defense. The two contestants roll over and over on the ground. It is a terrifying sight and the outcome is always the same. The wasp finally manages to thrust her sting into the soft spot and holds it there for a few seconds while she pumps in the poison. Almost immediately the tarantula falls paralyzed on its back. Its legs stop twitching; its heart stops beating. Yet it is not dead, as is shown by the fact that if taken from the wasp it can be restored to some sensitivity by being kept in a moist chamber for several months.

After paralyzing the tarantula, the wasp cleans herself by dragging her body along the ground and rubbing her feet, sucks the drop of blood oozing from the wound in the spider's abdomen, then grabs a leg of the flabby, helpless animal in her jaws and drags it down to the bottom of the grave. She stays there for many minutes, sometimes for several hours, and what she does all that time in the dark we do not know. Eventually she lays her egg and attaches it to the side of the spider's abdomen with a sticky secretion. Then she emerges, fills the grave with soil carried bit by bit in her jaws, and finally tramples the ground all around to hide any trace of the grave from prowlers. Then she flies away, leaving her descendant safely started in life.

In all this the behavior of the wasp evidently is qualitatively different from that of the spider. The wasp acts like an intelligent animal. This is not to say that instinct plays no part or that she reasons as man does. But her actions are to the point; they are not automatic and can be modified to fit the situation. We do not know for certain how she identifies the tarantula—probably it is by some olfactory or chemo-tactile sense—but she does it purposefully and does not blindly tackle a wrong species.

On the other hand, the tarantula's behavior shows only confusion. Evidently the wasp's pawing gives it no pleasure, for it tries to move away. That the wasp is not simulating sexual stimulation is certain, because male and female tarantulas react in the same way to its advances. That the spider is not anesthetized by some odorless secretion is easily shown by blowing lightly at the tarantula and making it jump suddenly. What, then, makes the tarantula behave as stupidly as it does?

No clear, simple answer is available. Possibly the stimulation by the wasp's antennae is masked by a heavier pressure on the spider's body, so that it reacts as when prodded by a pencil. But the explanation may be much more complex. Initiative in attack is not in the nature of tarantulas: most species fight only when cornered so that escape is impossible. Their inherited patterns of behavior apparently prompt them to avoid problems rather than attack them. For example, spiders always weave their webs in three dimensions, and when a spider finds that there is insufficient space to attach certain threads in the third dimension, it leaves the place and seeks another, instead of finishing the web in a single plane. This urge to escape seems to arise under all circumstances, in all phases of life and to take the place of reasoning. For a spider to change the pattern of its web is as impossible as for an inexperienced man to build a bridge across a chasm obstructing his way.

In a way the instinctive urge to escape is not only easier but often more efficient than reasoning. The tarantula does exactly what is most efficient in all cases except in an encounter with a ruthless and determined attacker dependent for the existence of her own species on killing as many tarantulas as she can lay eggs. Perhaps in this case the spider follows its usual pattern of trying to escape, instead of seizing and killing the wasp, because it is not aware of its danger. In any case, the survival of the tarantula species as a whole is protected by the fact that the spider is much more fertile than the wasp.

Questions and Topics for Discussion and Writing

1. This article is a discourse on the predatory in nature. Petrunkevitch makes clear that there is both a creative and a destructive order to nature. How does this essay serve as a defense of the predatory?
2. Scientists generally find the exclamation mark "out of place" in scientific writing. Yet Petrunkevitch uses one in the second paragraph. Do you find its use suitable?
3. Scientific writing is generally described as "objective." Yet the first five paragraphs contain a number of "value" words, modifiers that express a subjective view. List words that fit this category.

4.   Many statements in this essay elicit a subjective response from the reader. What is your response, for example, to the statement "The survival of the young depends on the mother's correct choice of a food which she does not eat herself"?

5.   In the first segment, what evidence do you find that Petrunkevitch is writing for an audience wider than fellow arachnologists?

6.   Read the seventh paragraph with care. Then read Reed Whittemore's poem "The Tarantula," the next selection in this unit. In a short paper, compare the impact of Petrunkevitch's paragraph with that of Whittemore's poem.

7.   What are the three types of touch that Petrunkevitch attributes to the tarantula? What is the rhetorical importance of discussion of these three types as preface to Petrunkevitch's distinction between the tarantula and the wasp?

8.   Study the concluding sentence of the twelfth paragraph. In what ways is it an effective "clincher" and sentence of transition?

9.   Read the thirteenth paragraph with special care. Is the language always "scientific"? Why, or why not? Look up *gargantuan*. Is this a literary term? What effect does it provide? What are the implications of the term *wasphood*?

10.   Examine the conclusion of the paragraph that opens, "When the grave is finished. . . ." Does the information in the biographical headnote add any implication to the objective, impersonal presentation of information?

11.   Petrunkevitch accepts the scientist's facing of insoluble problems, while being left with only a hypothesis. How does this acceptance reveal him as a twentieth-century scientist? Read, or reread, Planck's essay "Phantom Problems in Science" (Unit 3) and Szent-Györgyi's "Lost in the Twentieth Century" (Unit 1). Do the essays help your understanding of Petrunkevitch's perspective?

12.   Does Petrunkevitch's essay seem unfinished to you? Are you "unsatisfied" with the conclusion? Does such seeming suspension reflect the scientific method? If you could rewrite the conclusion, what would you modify or add?

Reed Whittemore

Reed Whittemore (b. 1919) received his bachelor's degree from Yale University and went on to become senior program adviser to the National Institute of Public Affairs. Whittemore taught English for twenty years at Carleton College and is now Professor Emeritus at the University of Maryland. He has twice been Consultant in Poetry at the Library of Congress. His books include *An American Takes a Walk*, *Self-Made Man*, *Boy from Iowa*, *Fascination of the Abomination*, *Poems: New and Selected*, *The Feel of Rock*, and a biography of William Carlos Williams.

   Whittemore's wide-ranging interests include scientific reading. In "The Tarantula," he builds on Petrunkevitch's essay, which precedes this poem, to present the life of a tarantula as seen through the eyes of a poet. In doing so, Whittemore moves from the universal—"Everyone"—to the particular—"little me. . . . William/Too."

The Tarantula

Reed Whittemore

Everyone thinks I am poisonous. I am not.
Look up and read the authorities on me, especially
One Alexander Petrunkevitch, of Yale, now retired,
Who has said of me (and I quote): my "bite is dangerous
Only
To insects and small mammals such as mice."
I would have you notice that "only"; that is important,
As you who are neither insect nor mouse can appreciate.
I have to live as you do,
And how would you like it if someone construed your rela-
    tions
With the chicken, say, as proof of your propensities?
Furthermore,
Petrunkevitch has observed, and I can vouch for it,
That I am myopic, lonely and retiring. When I am born
I dig a burrow for me, and me alone,
And live in it all my life except when I come
Up for food and love (in my case the latter
Is not really satisfactory: I
"Wander about after dark in search of females,
And occasionally stray into houses," after which I
Die). How does that sound?
Furthermore,
I have to cope with the digger wasp of the genus
Pepsis; and despite my renown as a killer (nonsense, of
    course),
I can't. Petrunkevitch says so.
Read him. He's good on the subject. He's helped *me*.

Which brings me to my point here. You carry
This image about of me that is at once libelous
And discouraging, all because you, who should know better,
Find me ugly. So I am ugly. Does that mean that you
Should persecute me as you do? Read William Blake.
Read William Wordsworth.
Read Williams in general, I'd say. There was a book
By a William Tarantula once, a work of some consequence

In my world on the subject of beauty,
Beauty that's skin deep only, beauty that some
Charles (note the "Charles") of the Ritz can apply and take
    off
At will, beauty that—
                        but I digress.
What I am getting at
Is that you who are blessed (I have read) with understand-
    ing
Should understand me, little me. My name is William
Too.

Questions and Topics for Discussion and Writing

1. Does your reading of this poem following the reading of Petrunkevitch's essay modify your interpretation of each work? If so, how?
2. How does Whittemore's presentation differ from the scientific one? What does the poet add?
3. How does the use of the first person—the presentation of the tarantula as speaker—affect your idea of the tarantula?
4. What effect does Whittemore obtain by quoting passages from Petrunkevitch's essay? By having the tarantula speak them?
5. Poets sometimes place a single word on a single line. Observe this practice in Line 5, "*Only.*" How does this placement emphasize the meaning of the poem?
6. How does the tarantula's question about our "relations/With the chicken" affect our reading? Does the poem attack our logic?
7. In a poem, words and phrases can serve to modify ambiguously. How can the parenthetical clause "(I have read)" serve two ways, placed where it is in the line?
8. Look up the allusions to the name *William* and the name *Charles*. Be ready to add your findings to class discussion of the connotations.
9. Do you find evidences of humor in the poem? Cite them.
10. How would you define the tone of the poem? Point out several passages that help to produce this tone.

Edward Osborne Wilson

Living the dual life of academic scientist and field biologist, Edward O. Wilson is an internationally known entomologist. Born in Alabama (1929), he received his B.S. and M.S. degrees from the University of Alabama and his PhD from Harvard University, where he is professor of science. Field trips have taken him to remote areas of the world, and each physically adventurous journey reflects a rich mental journey. For Wilson, the "journey to the interior" has resulted in a concept he calls *biophilia*. From his adventures in the mind and in the natural world, he developed a new way of looking at the phenomenon of life and helped form a new discipline, sociobiology.

Wilson defines *biophilia* as the "innate tendency to focus on life and lifelike processes." He explains: "From infancy we concentrate happily on ourselves and other organisms. We learn to distinguish life from the inanimate and move toward it like moths to a porch light. . . . To explore and affiliate with life is a deep and complicated process in mental development. To an extent still under-valued in philosophy and religion, our existence depends on this propensity, our spirit is woven from it, hope rises on its current."

In the following excerpt, Wilson recalls a memorable day in the rain forests he calls "South America's awesome ecological heartland." In the tiny village of Bernhardsdorp—its unlikely name the evidence of Dutch colonial heritage— Wilson applies his expertise to the study of ants. Physically, he carries the "standard equipment of a field biologist: camera; canvas satchel containing for-ceps, trowel, ax, mosquito repellent, jars, vials of alcohol, and notebook; a twenty-power hand lens swinging with a reassuring tug around the neck; partly fogged eyeglasses sliding down the nose and khaki shirt plastered to the back with sweat." Intellectually, he carries what he calls the "mental set—call it the naturalist's trance, the hunter's trance—by which biologists locate more elu-sive organisms."

Biophilia: A Day at Bernhardsdorp

Edward O. Wilson

Think of scooping up a handful of soil and leaf litter and spreading it out on a white ground cloth, in the manner of the field biologist, for close examination. This unprepossessing lump contains more order and richness of structure, and particularity of history, than the entire surfaces of all the other (lifeless) planets. It is a miniature wilderness that can take almost forever to explore.

Tease apart the adhesive grains with the aid of forceps, and you will expose the tangled rootlets of a flowering plant, curling around the rotting veins of humus, and perhaps some larger object such as the boat-shaped husk of a seed. Almost certainly among them will be a scattering of creatures that measure the world in millimeters and treat this soil sample as traversable: ants, spiders, springtails, armored oribatid mites, enchytraeid worms, millipedes. With the aid of a dissecting microscope, proceed on down the size scale to the round-worms, a world of scavengers and fanged predators feeding on them. In the hand-held microcosm all these creatures are still giants in a relative sense. The organisms of greatest diversity and numbers are invisible or nearly so. When the soil-and-litter clump is progressively magnified, first with a compound light microscope and then with scanning electron micrographs, specks of dead leaf expand into mountain ranges and canyons, soil particles become heaps of boulders. A droplet of moisture trapped between root hairs grows into an underground lake, surrounded by a three-dimensional swamp of moistened humus. The niches are defined by both topography and nuances in chemistry, light, and temperature shifting across fractions of a millimeter. Organisms now come into view for which the soil sample is a complete world. In certain places are found the fungi: cellular slime molds, the one-celled chitin-producing chytrids, minute gonapodyaceous and oomycete soil specialists, Kickxellales, Eccrinales, Endomycetales, and Zoopagales. Contrary to their popular reputation, the fungi are not formless blobs, but exquisitely structured organisms with elaborate life cycles. . . .

Still smaller than the parasitic fungi are the bacteria, including colony-forming polyangiaceous species, specialized predators that consume other bacteria. All around them live rich mixtures of rods, cocci, coryneforms, and slime azotobacteria. Together these microorganisms metabolize the entire spectrum of live and dead tissue. At the moment of discovery some are actively growing and fissioning, while others lie dormant in wait for the right combination of nutrient chemicals. Each species is kept at equilibrium by the harshness of the environment. Any one, if allowed to expand without restriction for a few weeks, would multiply exponentially, faster and faster, until it weighed more than the entire Earth. But in reality the individual organism simply dissolves and assimilates whatever appropriate fragments of plants and animals come to rest near it. If the newfound meal is large enough, it may succeed in growing and reproducing briefly before receding back into the more normal state of physiological quiescence.

Biologists, to put the matter as directly as possible, have begun a second reconnaissance into the land of magical names. In exploring life they have commenced a pioneering adventure with no imaginable end. The abundance of organisms increases downward by level, like layers in a pyramid. The handful of soil and litter is home for hundreds of insects, nematode worms, and other larger creatures, about a million fungi, and ten billion bacteria. Each of the species of these organisms has a distinct life cycle fitted, as in the case of the predatory fungus, to the portion of the microenvironment in which it thrives and reproduces. The particularity is due to the fact that it is programed by an exact sequence of nucleotides, the ultimate molecular units of the genes.

The amount of information in the sequence can be measured in bits. One bit is the information required to determine which of two equally likely alternatives is chosen, such as heads or tails in a coin toss. English words average two bits per letter. A single bacterium possesses about ten million bits of genetic information, a fungus one billion, and an insect from one to ten billion bits according to species. If the information in just one insect—say an ant or beetle—

were to be translated into a code of English words and printed in letters of standard size, the string would stretch over a thousand miles. Our lump of earth contains information that would just about fill all fifteen editions of the *Encyclopaedia Britannica.* . . .

With advanced techniques it has been possible to begin mapping insect nervous systems in sufficient detail to draw the equivalent of wiring diagrams. Each brain consists of somewhere between a hundred thousand and a million nerve cells, most of which send branches to a thousand or more of their neighbors. Depending on their location, individual cells appear to be programed to assume a particular shape and to transmit messages only when stimulated by coded discharges from neighbor units that feed into them. In the course of evolution, the entire system has been miniaturized to an extreme. The fatty sheaths surrounding the axon shafts of the kind found in larger animals have been largely stripped away, while the cell bodies are squeezed off to one side of the multitudinous nerve connections. Biologists understand in very general terms how the insect brain might work as a complete on-board computer, but they are a long way from explaining or duplicating such a device in any detail.

The great German zoologist Karl von Frisch once said of his favorite organism that the honeybee is like a magic well: the more you draw from it, the more there is to draw. But science is in no other way mystical. Its social structure is such that anyone can follow most enterprises composing it, as observer if not as participant, and soon you find yourself on the boundaries of knowledge. . . .

Every species is a magic well. Biologists have until recently been satisfied with the estimate that there are between three and ten million of them on Earth. Now many believe that ten million is too low. The upward revision has been encouraged by the increasingly successful penetration of the last great unexplored environment of the planet, the canopy of the tropical rain forest, and the discovery of an unexpected number of new species living there. This layer is a sea of branches, leaves, and flowers crisscrossed by lianas and suspended about one hundred feet above the ground. It is one of the easiest habitats to locate—from a distance at least—but next to the deep sea the most difficult to reach. The tree trunks are thick, arrow-straight, and either slippery smooth or covered with sharp tubercles. Anyone negotiating them safely to the top must then contend with swarms of stinging ants and wasps. A few athletic and adventurous younger biologists have begun to overcome the difficulties by constructing special pulleys, rope catwalks, and observation platforms from which they can watch high arboreal animals in an undisturbed state. Others have found a way to sample the insects, spiders, and other arthropods with insecticides and quick-acting knockdown agents. They first shoot lines up into the canopy, then hoist the chemicals up in canisters and spray them out into the surrounding vegetation by remote control devices. The falling insects and other organisms are caught in sheets spread over the ground. The creatures discovered by these two methods have proved to be highly specialized in their food habits, the part of the tree in which they live, and the time of the year when they are active. So an unexpectedly large number of different kinds are able to coexist. Hundreds can fit comfortably together in a single tree top. On the basis of a preliminary statistical projection from these data, Terry L. Erwin, an entomologist at the National Museum of Natural History, has estimated that there may be thirty million species of insects in the world, most limited to the upper vegetation of the tropical forests.

Although such rough approximations of the diversity of life are not too difficult to make, the exact number of species is beyond reach because — incredibly—the majority have yet to be discovered and specimens placed in museums. Furthermore, among those already classified no more than a dozen have been studied as well as the honeybee. Even *Homo sapiens*, the focus of billions of dollars of research annually, remains a seemingly intractable mystery. All of man's troubles may well arise, as Vercors suggested in *You Shall Know Them*, from the fact that we do not know what we are and do not agree on what we want to become. This crucial inadequacy is not likely to be remedied until we have a better grasp of the diversity of the life that created and sustains us. So why hold back? It is a frontier literally at our fingertips, and the one for which our spirit appears to have been explicitly designed.

I walked on through the woodland at Bernhardsdorp to see what the day had to offer. In a decaying log I found a species of ant previously known only from the midnight zone of a cave in Trinidad. With the aid of my hand lens I identified it from its unique combination of teeth, spines, and body sculpture. A month before I had hiked across five miles of foothills in central Trinidad to find it in the original underground habitat. Now suddenly here it was again, nesting and foraging in the open. Scratch from the list what had been considered the only "true" cave ant in the world—possessed of workers pale yellow, nearly eyeless, and sluggish in movement. Scratch the scientific name *Spelaeomyrmex*, meaning literally cave ant, as a separate taxonomic entity. I knew that it would have to be classified elsewhere, into a larger and more conventional genus called *Erebomyrma*, ant of Hades. A small quick victory, to be reported later in a technical journal that specializes on such topics and is read by perhaps a dozen fellow myrmecologists. I turned to watch some huge-eyed ants with the formidable name *Gigantiops destructor*. When I gave one of the foraging workers a freshly killed termite, it ran off in a straight line across the forest floor. Thirty feet away it vanished into a small hollow tree branch that was partly covered by decaying leaves. Inside the central cavity I found a dozen workers and their mother queen—one of the first colonies of this unusual insect ever recorded. All in all, the excursion had been more productive than average. Like a prospector obsessed with ore samples, hoping for gold, I gathered a few more promising specimens in vials of ethyl alcohol and headed home, through the village and out onto the paved road leading north to Paramaribo.

Later I set the day in my memory with its parts preserved for retrieval and closer inspection. Mundane events acquired the raiment of symbolism, and this is what I concluded from them: That the naturalist's journey has only begun and for all intents and purposes will go on forever. That it is possible to spend a lifetime in a magellanic voyage around the trunk of a single tree. That as the exploration is pressed, it will engage more of the things close to the human heart and spirit. And if this much is true, it seems possible that the naturalist's vision is only a specialized product of a biophilic instinct shared by all, that it can be elaborated to benefit more and more people. Humanity is exalted not because we are so far above other living creatures, but because knowing them well elevates the very concept of life.

Questions and Topics for Discussion and Writing

1.   In a passage before this excerpt, Wilson describes his own attitude as that of "a transient of no consequence in this familiar yet deeply alien world that I had come to love," yet also that of one who can focus on the scene and "will" animals to material-ize when he is ready to observe them. How do you think Wilson integrates two such points of view?
2.   Wilson speaks of the "naturalist's trance, the hunter's trance." What do you under-stand by these phrases? Read Harrington's "The Lure of the Hunt" (Unit 6). What similarities do you find between the two scientists? Write a paper comparing their attitudes and writing styles.
3.   At Bernhardsdorp, as Wilson describes just before this excerpt, he is strongly re-minded of the nineteenth-century landscape painters Albert Bierstadt, Frederick Edwin Church, Thomas Cole, and their contemporaries. Look at some of the work of these painters and explain what qualities you perceive that make their paintings relevant to the ideas that Wilson discusses. For example, look in particular for their treatment of light.
4.   Wilson cites the work of zoologist Karl von Frisch on the honeybee, referring to the "waggle dance" as a mode of communication. Look up Frisch's book *The Dancing Bees* and write a brief summary of the sequence of the waggle dance.
5.   Explain Wilson's figure of speech "to spend a lifetime in a magellanic voyage around the trunk of a single tree."
6.   What is your opinion of Wilson's concept of a "biophilic instinct"? Discuss.

Herman Melville

Read from a twentieth-century perspective, Herman Melville looms as a herald of the modern scientific attitude. A New Yorker, he was born in 1819, during the declining years of the American Enlightenment. When he died in 1891, his life had spanned a time of great transition in American and scientific thought. As evidenced in his works from *Typee* to *Billy Budd*, he looks to the American Romantic Movement and to the Realist, Naturalist, and literary Existentialism movements. Moreover, like his friend Nathaniel Hawthorne, he wrote from the fringes of the Transcendental Movement.

As a Romantic, Melville stressed the value of imagination, intuition, and mysticism, as well as the beauty of nature, meanwhile challenging the advancing science of his time. Simultaneously, as a Realist, he praised the use of reason and stressed the terrifying aspects of nature and the need for a scientific challenge to religious orthodoxy.

Melville has been commended by critics for his knowledge of history, philosophy, foreign literatures, and science. He is illuminated by his novels as one well read in astronomy, cetology, geology, and zoology. Several sea voyages as a member of the crew—his "Harvard and his Yale"—provided him with the topics for his writings. His voyage to the Galapagos Islands resulted in description that invites comparison with Darwin's travel account of the same locale. Melville sees the islands as desolate and changeless, surrounded by a rock-bound coast, with "little but reptile life"—the "chief sound" he heard, "a hiss."

As a critic himself, Melville chastised the Romantic and Transcendental Platonists who would ignore the reality of nature, and he challenged science for thinking it could master reality by classifying facts. He concluded that human beings cannot understand the mystery of nature simply by defining and classifying observable phenomena, often isolating themselves from the humanity that science should serve. He is unlike Ahab, the self-isolated captain of the *Pequod*, who would obtain knowledge of a first cause by penetrating the "mask" of natural phenomena, symbolized by the whale Moby Dick. Ishmael, the narrator–hero, in contrast, transcends his first attempt to classify whales and thus to understand the whole of nature, the universe, and the divine. Ishmael examines various parts of the whale, ending with the dead whale's skeleton; he accepts his inability to attain such all-encompassing knowledge.

The following chapter from *Moby Dick*, often acclaimed as the greatest American novel, must be understood as one part of Ishmael's analysis of other parts of the whale. Like other chapters in which Ishmael describes a part of the whale, it develops from a realistic level on which Ishmael uses logic and mathematics to describe the whale's skeleton. It concludes on a more metaphysical level on which Melville–Ishmael warns those who would attempt to comprehend the incomprehensible and, under the guise of humor, satirizes orthodox religions that would impose a supernatural structure on the natural.

Measurement of a Whale's Skeleton

Herman Melville

In the first place, I wish to lay before you a particular, plain statement, touching the living bulk of this leviathan, whose skeleton we are briefly to exhibit. Such a statement may prove useful here.

According to a careful calculation I have made, and which I partly base upon Captain Scoresby's estimate, of seventy tons for the largest-sized Greenland whale of sixty feet in length; according to my careful calculation, I say, a sperm whale of the largest magnitude, between eighty-five and ninety feet in length, and something less than forty feet in its fullest circumference, such a whale will weigh at least ninety tons; so that, reckoning thirteen men to a ton, he would considerably outweigh the combined population of a whole village of one thousand one hundred inhabitants.

Think you not then that brains, like yoked cattle, should be put to this leviathan, to make him at all budge to any landsman's imagination?

Having already in various ways put before you his skull, spout-hole, jaw, teeth, tail, forehead, fins, and divers other parts, I shall now simply point out what is most interesting in the general bulk of his unobstructed bones. But as the colossal skull embraces so very large a proportion of the entire extent of the skeleton; as it is by far the most complicated part; and as nothing is to be repeated concerning it in this chapter, you must not fail to carry it in your mind, or under your arm, as we proceed, otherwise you will not gain a complete notion of the general structure we are about to view.

In length, the sperm whale's skeleton at Tranque measured seventy-two feet; so that when fully invested and extended in life, he must have been ninety feet long; for in the whale, the skeleton loses about one-fifth in length compared with the living body. Of this seventy-two feet, his skull and jaw comprised some twenty feet, leaving some fifty feet of plain backbone. Attached to this backbone, for something less than a third of its length, was the mighty circular basket of ribs which once enclosed his vitals.

To me this vast ivory-ribbed chest, with the long, unrelieved spine, extending far away from it in a straight line, not a little resembled the hull of a great ship new-laid upon the stocks, when only some twenty of her naked bow-ribs are inserted, and the keel is otherwise, for the time, but a long, disconnected timber.

The ribs were ten on a side. The first, to begin from the neck, was nearly six feet long; the second, third, and fourth were each successively longer, till you came to the climax of the fifth, or one of the middle ribs, which measured eight feet and some inches. From that part, the remaining ribs diminished, till the tenth and last only spanned five feet and some inches. In general thickness, they all bore a seemly correspondence to their length. The middle ribs were the most arched. In some of the Arsacides they are used for beams whereon to lay footpath bridges over small streams.

In considering these ribs, I could not but be struck anew with the circumstance, so variously repeated in this book, that the skeleton of the whale is by no means the mould of his invested form. The largest of the Tranque ribs, one of the middle ones, occupied that part of the fish which, in life, is greatest in depth. Now, the greatest depth of the invested body of this particular whale must have been at least sixteen feet; whereas, the corresponding rib measured but little more than eight feet. So that this rib only conveyed half of the true notion of the living magnitude of that part. Besides, for some way, where I now saw but a naked spine, all that had been once wrapped round with tons of added bulk in flesh, muscle, blood, and bowels. Still more, for the ample fins, I here saw but a few disordered joints; and in place of the weighty and majestic, but boneless flukes, an utter blank!

How vain and foolish, then, thought I, for timid untravelled man to try to comprehend aright this wondrous whale, by merely poring over his dead attenuated skeleton, stretched in this peaceful wood. No. Only in the heart of quickest perils; only when within the eddyings of his angry flukes; only on the profound unbounded sea, can the fully invested whale be truly and livingly found out.

But the spine. For that, the best way we can consider it is, with a crane, to pile its bones high up on end. No speedy enterprise. But now it's done, it looks much like Pompey's Pillar.

There are forty and odd vertebrae in all, which in the skeleton are not locked together. They mostly lie like the great knobbed blocks on a Gothic spire, forming solid courses of heavy masonry. The largest, a middle one, is in width something less than three feet, and in depth more than four. The smallest, where the spine tapers away into the tail, is only two inches in width, and looks something like a white billiard ball. I was told that there were still smaller ones, but they had been lost by some little cannibal urchins, the priests' children, who had stolen them to play marbles with. Thus we see how that the spine of even the hugest of living things tapers off at last into simple child's play.

Questions and Topics for Discussion and Writing

1. Melville challenges attempts to categorize nature and the universe. Why do you think he compares the size of the whale to that of the human being?
2. Melville creates the *Pequod* as a microcosm. Its crew (with the exception of Ishmael) is annihilated because of the obsession of its captain, Ahab, who would penetrate the "mask" of nature to reach the underlying cause. What warning and foreshadowing are created by his comparison of the whale to a ship?
3. Why does Melville include a description of the skeleton of a whale in his observations of the various parts of the living whale?
4. Is there a part of Melville's categorization of the whale that you, with some education in modern science, would criticize?
5. One particular passage is an obvious warning to scientists. Identify the passage and summarize in your own words Melville's warning.
6. Melville's prose has been termed "masculine." Would you agree with this description? Write a paragraph in which you explain, by using examples from the chapter, why such judgment is or is not just.

7.   Melville describes the vertebrae of the whale's spine as a "Gothic spire" and describes lost parts of the spine as playthings for "priests' children." What does Melville imply by such descriptions? Does the humorous reference to "child's play" suggest a particular belief about religion, life, and death?

Stephen Jay Gould

At the age of five, after viewing the reconstructed skeleton of the *Tyrannosaurus Rex* at the American Museum of Natural History in Manhattan, Stephen Jay Gould (b. 1941) abandoned his earlier ambition of becoming a garbage collector and settled, instead, upon a career in paleontology. Not swaying from that early commitment, Gould has since come to be known as "a serious and gifted interpreter of biological theory, of the history of ideas and the cultural context of ideas." Both articulate and accessible, Gould is also known as an often witty popularizer and defender of evolutionary theory. In 1981, he served as an expert witness at a trial in Little Rock, Arkansas, where creationists were trying to redefine religion as science by claiming the right to equal time for "creation science" in biology textbooks.

Although popularly known as an advocate of Darwinism, within professional circles Gould has been accused of breaking ranks with classical Darwinists. According to the accepted version of the theory, evolution works incrementally, with new species gradually arising from the old through a steady build-up of minute mutations. Gould has contested this view, also known as gradualism, by proposing (with Niles Eldredge) the following: Instead of a smooth progression toward perfection, evolution is better typified by long stretches of relative stability punctuated by periods of rapid change in small groups, with nature selecting not perfect, but quirky, make-do paths with no discernible or predictable direction. This theory, one of the more controversial among modern, evolutionary biologists, is known as punctuated equilibrium.

The author of numerous scientific and "popular" books and articles, Gould teaches biology, geology, and the history of science at Harvard, where his classes are often "standing-room only." He is perhaps best known to the general public, however, through his monthly column "This View of Life," which he writes for *Natural History* magazine, and for the three volumes collected from that column, *Ever Since Darwin, The Panda's Thumb,* and *Hen's Teeth and Horse's Toes.* A fourth book, *The Mismeasure of Man,* in which Gould details the history of "scientific racism" by exposing the inconsistencies of scientists bent on validating racial suppression, won the 1982 National Book Critics' Circle Award.

"A Biological Homage to Mickey Mouse" is from *The Panda's Thumb,* which was the 1981 winner of the American Book Award for Science.

A Biological Homage to Mickey Mouse

Stephen Jay Gould

Age often turns fire to placidity. Lytton Strachey, in his incisive portrait of Florence Nightingale, writes of her declining years:

> Destiny, having waited very patiently, played a queer trick on Miss Nightingale. The benevolence and public spirit of that long life had only been equalled by its acerbity. Her virtue had dwelt in hardness. . . . And now the sarcastic years brought the proud woman her punishment. She was not to die as she had lived. The sting was to be taken out of her; she was to be made soft; she was to be reduced to compliance and complacency.

I was therefore not surprised—although the analogy may strike some people as sacrilegious—to discover that the creature who gave his name as a synonym for insipidity had a gutsier youth. Mickey Mouse turned a respectable fifty last year [1978]. To mark the occasion, many theaters replayed his debut performance in *Steamboat Willie* (1928). The original Mickey was a rambunctious, even slightly sadistic fellow. In a remarkable sequence, exploiting the exciting new development of sound, Mickey and Minnie pummel, squeeze, and twist the animals on board to produce a rousing chorus of "Turkey in the Straw." They honk a duck with a tight embrace, crank a goat's tail, tweak a pig's nipples, bang a cow's teeth as a stand-in xylophone, and play bagpipe on her udder.

Christopher Finch, in his semiofficial pictorial history of Disney's work, comments: "The Mickey Mouse who hit the movie houses in the late twenties was not quite the well-behaved character most of us are familiar with today. He was mischievous, to say the least, and even displayed a streak of cruelty." But Mickey soon cleaned up his act, leaving to gossip and speculation only his unresolved relationship with Minnie and the status of Morty and Ferdie. Finch continues: "Mickey . . . had become virtually a national symbol, and as such he was expected to behave properly at all times. If he occasionally stepped out of line, any number of letters would arrive at the Studio from citizens and organizations who felt that the nation's moral well-being was in their hands. . . . Eventually he would be pressured into the role of straight man."

As Mickey's personality softened, his appearance changed. Many Disney fans are aware of this transformation through time, but few (I suspect) have recognized the coordinating theme behind all the alterations—in fact, I am not sure that the Disney artists themselves explicitly realized what they were doing, since the changes appeared in such a halting and piecemeal fashion. In

Mickey's evolution during 50 years (left to right). As Mickey became increasingly well behaved over the years, his appearance became more youthful. Measurements of three stages in his development revealed a larger relative head size, larger eyes, and an enlarged cranium—all traits of juvenility. © Walt Disney Productions

short, the blander and inoffensive Mickey became progressively more juvenile in appearance. (Since Mickey's chronological age never altered—like most cartoon characters he stands impervious to the ravages of time—this change in appearance at a constant age is a true evolutionary transformation. Progressive juvenilization as an evolutionary phenomenon is called neoteny. More on this later.)

The characteristic changes of form during human growth have inspired a substantial biological literature. Since the head-end of an embryo differentiates first and grows more rapidly in utero than the foot-end (an antero-posterior gradient, in technical language), a newborn child possesses a relatively large head attached to a medium-sized body with diminutive legs and feet. This gradient is reversed through growth as legs and feet overtake the front end.

Heads continue to grow but so much more slowly than the rest of the body that relative head size decreases.

In addition, a suite of changes pervades the head itself during human growth. The brain grows very slowly after age three, and the bulbous cranium of a young child gives way to the more slanted, lower-browed configuration of adulthood. The eyes scarcely grow at all and relative eye size declines precipitously. But the jaw gets bigger and bigger. Children, compared with adults, have larger heads and eyes, smaller jaws, a more prominent, bulging cranium, and smaller, pudgier legs and feet. Adult heads are altogether more apish, I'm sorry to say.

Mickey, however, has traveled this ontogenetic pathway in reverse during his fifty years among us. He has assumed an ever more childlike appearance as the ratty character of *Steamboat Willie* became the cute and inoffensive host to a magic kingdom. By 1940, the former tweaker of pig's nipples gets a kick in the ass for insubordination (as the *Sorcerer's Apprentice* in *Fantasia*). By 1953, his last cartoon, he has gone fishing and cannot even subdue a squirting clam.

The Disney artists transformed Mickey in clever silence, often using suggestive devices that mimic nature's own changes by different routes. To give him the shorter and pudgier legs of youth, they lowered his pants line and covered his spindly legs with a baggy outfit. (His arms and legs also thickened substantially—and acquired joints for a floppier appearance.) His head grew relatively larger and its features more youthful. The length of Mickey's snout has not altered, but decreasing protrusion is more subtly suggested by a pronounced thickening. Mickey's eye has grown in two modes: first, by a major, discontinuous evolutionary shift as the entire eye of ancestral Mickey became the pupil of his descendants, and second, by gradual increase thereafter.

Mickey's improvement in cranial bulging followed an interesting path since his evolution has always been constrained by the unaltered convention of representing his head as a circle with appended ears and an oblong snout. The circle's form could not be altered to provide a bulging cranium directly. Instead, Mickey's ears moved back, increasing the distance between nose and ears, and giving him a rounded, rather than a sloping, forehead.

To give these observations the cachet of quantitative science, I applied my best pair of dial calipers to three stages of the official phylogeny—the thin-nosed, ears-forward figure of the early 1930s (stage 1), the latter-day Jack of Mickey and the Beanstalk (1947, stage 2), and the modern mouse (stage 3). I measured three signs of Mickey's creeping juvenility: increasing eye size (maximum height) as a percentage of head length (base of the nose to top of rear ear); increasing head length as a percentage of body length; and increasing cranial vault size measured by rearward displacement of the front ear (base of the nose to top of front ear as a percentage of base of the nose to top of rear ear).

All three percentages increased steadily—eye size from 27 to 42 percent of head length; head length from 42.7 to 48.1 percent of body length; and nose to front ear from 71.7 to a whopping 95.6 percent of nose to rear ear. For compari-

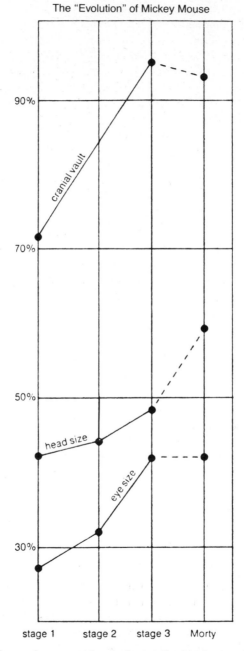

The "Evolution" of Mickey Mouse

At an early stage in his evolution, Mickey had a smaller head, cranial vault, and eyes. He evolved toward the characteristics of his young nephew Morty (connected to Mickey by a dotted line).

son, I measured Mickey's young "nephew" Morty Mouse. In each case, Mickey has clearly been evolving toward youthful stages of his stock, although he still has a way to go for head length.

You may, indeed, now ask what an at least marginally respectable scientist has been doing with a mouse like that. In part, fiddling around and having fun, of course. (I still prefer *Pinocchio* to *Citizen Kane*.) But I do have a serious point—two, in fact—to make. We must first ask why Disney chose to change his most famous character so gradually and persistently in the same direction? National symbols are not altered capriciously and market researchers (for the doll industry in particular) have spent a good deal of time and practical effort learning what features appeal to people as cute and friendly. Biologists also have spent a great deal of time studying a similar subject in a wide range of animals.

In one of his most famous articles, Konrad Lorenz argues that humans use the characteristic differences in form between babies and adults as important behavioral cues. He believes that features of juvenility trigger "innate releasing mechanisms" for affection and nurturing in adult humans. When we see a living creature with babyish features, we feel an automatic surge of disarming tenderness. The adaptive value of this response can scarcely be questioned, for we must nurture our babies. Lorenz, by the way, lists among his releasers the very features of babyhood that Disney affixed progressively to Mickey: "a relatively large head, predominance of the brain capsule, large and low-lying eyes, bulging cheek region, short and thick extremities, a springy elastic consistency, and clumsy movements." (I propose to leave aside for this article the contentious issue of whether or not our affectionate response to babyish features is truly innate and inherited directly from ancestral primates—as Lorenz argues—or whether it is simply learned from our immediate experience with babies and grafted upon an evolutionary predisposition for attaching ties of affection to certain learned signals. My argument works equally well in either case for I only claim that babyish features tend to elicit strong feelings of affection in adult humans, whether the biological basis be direct programming or the capacity to learn and fix upon signals. I also treat as collateral to my point the major thesis of Lorenz's article—that we respond not to the totality or *Gestalt*, but to a set of specific features acting as releasers. This argument is important to Lorenz because he wants to argue for evolutionary identity in modes of behavior between other vertebrates and humans, and we know that many birds, for example, often respond to abstract features rather than *Gestalten*. Lorenz's article, published in 1950, bears the title *Ganzheit und Teil in der tierischen und menschlichen Gemeinschaft*—"Entirety and part in animal and human society." Disney's piecemeal change of Mickey's appearance does make sense in this context—he operated in sequential fashion upon Lorenz's primary releasers.)

Lorenz emphasizes the power that juvenile features hold over us, and the abstract quality of their influence, by pointing out that we judge other animals by the same criteria—although the judgment may be utterly inappropriate in an evolutionary context. We are, in short, fooled by an evolved response to our own babies, and we transfer our reaction to the same set of features in other animals.

Many animals, for reasons having nothing to do with the inspiration of affection in humans, possess some features also shared by human babies but not by human adults—large eyes and a bulging forehead with retreating chin, in particular. We are drawn to them, we cultivate them as pets, we stop and admire them in the wild—while we reject their small-eyed, long-snouted relatives who might make more affectionate companions or objects of admiration. Lorenz points out that the German names of many animals with features mimicking human babies end in the diminutive suffix *chen*, even though the animals are often larger than close relatives without such features—*Rotkehlchen* (robin), *Eichhörnchen* (squirrel), and *Kaninchen* (rabbit), for example.

In a fascinating section, Lorenz then enlarges upon our capacity for biologically inappropriate response to other animals, or even to inanimate objects that mimic human features. "The most amazing objects can acquire remarkable, highly specific emotional values by 'experimental attachment' of human properties. . . . Steeply rising, somewhat overhanging cliff faces or dark storm-clouds piling up have the same, immediate display value as a human being who is standing at full height and leaning slightly forwards"—that is, threatening.

We cannot help regarding a camel as aloof and unfriendly because it mimics, quite unwittingly and for other reasons, the "gesture of haughty rejection" common to so many human cultures. In this gesture, we raise our heads, placing our nose above our eyes. We then half-close our eyes and blow out through our nose—the "harumph" of the stereotyped upperclass Englishman or his well-trained servant. "All this," Lorenz argues quite cogently, "symbolizes resistance against all sensory modalities emanating from the disdained counterpart." But the poor camel cannot help carrying its nose above its elongate eyes, with mouth drawn down. As Lorenz reminds us, if you wish to know whether a camel will eat out of your hand or spit, look at its ears, not the rest of its face.

In his important book *Expression of the Emotions in Man and Animals*, published in 1872, Charles Darwin traced the evolutionary basis of many common gestures to originally adaptive actions in animals later internalized as symbols in humans. Thus, he argued for evolutionary continuity of emotion, not only of form. We snarl and raise our upper lip in fierce anger—to expose our nonexistent fighting canine tooth. Our gesture of disgust repeats the facial actions associated with the highly adaptive act of vomiting in necessary circumstances. Darwin concluded much to the distress of many Victorian contemporaries: "With mankind some expressions, such as the bristling of the hair under the influence of extreme terror, or the uncovering of the teeth under that of furious rage, can hardly be understood, except on the belief that man once existed in a much lower and animal-like condition."

In any case, the abstract features of human childhood elicit powerful emotional responses in us, even when they occur in other animals. I submit that Mickey Mouse's evolutionary road down the course of his own growth in reverse reflects the unconscious discovery of this biological principle by Disney and his artists. In fact, the emotional status of most Disney characters rests on the same set of distinctions. To this extent, the magic kingdom trades on a biological illusion—our ability to abstract and our propensity to transfer inappropriately to other animals the fitting responses we make to changing form in the growth of our own bodies.

Donald Duck also adopts more juvenile features through time. His elongated beak recedes and his eyes enlarge; he converges on Huey, Louie, and Dewey as surely as Mickey approaches Morty. But Donald, having inherited the mantle of Mickey's original misbehavior, remains more adult in form with his projecting beak and more sloping forehead.

Mouse villains or sharpies, contrasted with Mickey, are always more adult in appearance, although they often share Mickey's chronological age. In 1936, for example, Disney made a short entitled *Mickey's Rival*. Mortimer, a dandy in a yellow sports car, intrudes upon Mickey and Minnie's quiet country picnic. The thoroughly disreputable Mortimer has a head only 29 percent of body length, to Mickey's 45, and a snout 80 percent of head length, compared with Mickey's 49. (Nonetheless, and was it ever different, Minnie transfers her affection until an obliging bull from a neighboring field dispatches Mickey's rival.) Consider also the exaggerated adult features of other Disney characters—the swaggering bully Peg-leg Pete or the simple, if lovable, dolt Goofy.

Dandified, disreputable Mortimer (here stealing Minnie's affections) has strikingly more adult features than Mickey. His head is smaller in proportion to body length; his nose is a full 80 percent of head length. © Walt Disney Productions

As a second, serious biological comment on Mickey's odyssey in form, I note that his path to eternal youth repeats, in epitome, our own evolutionary story. For humans are neotenic. We have evolved by retaining to adulthood the originally juvenile features of our ancestors. Our australopithecine forebears, like Mickey in *Steamboat Willie,* had projecting jaws and low vaulted craniums.

Our embryonic skulls scarcely differ from those of chimpanzees. And we follow the same path of changing form through growth: relative decrease of the cranial vault since brains grow so much more slowly than bodies after birth, and continuous relative increase of the jaw. But while chimps accentuate these changes, producing an adult strikingly different in form from a baby, we proceed much more slowly down the same path and never get nearly so far. Thus, as adults, we retain juvenile features. To be sure, we change enough to produce a notable difference between baby and adult, but our alteration is far smaller than that experienced by chimps and other primates.

Cartoon villains are not the only Disney characters with exaggerated adult features. Goofy, like Mortimer, has a small head relative to body length and a prominent snout. © Walt Disney Productions

A marked slowdown of developmental rates has triggered our neoteny. Primates are slow developers among mammals, but we have accentuated the trend to a degree matched by no other mammal. We have very long periods of gestation, markedly extended childhoods, and the longest life span of any mammal. The morphological features of eternal youth have served us well. Our enlarged brain is, at least in part, a result of extending rapid prenatal growth rates to later ages. (In all mammals, the brain grows rapidly in utero but often very little after birth. We have extended this fetal phase into postnatal life.)

But the changes in timing themselves have been just as important. We are preeminently learning animals, and our extended childhood permits the transference of culture by education. Many animals display flexibility and play in childhood but follow rigidly programmed patterns as adults. Lorenz writes, in the same article cited above: "The characteristic which is so vital for the human peculiarity of the true man—that of always remaining in a state of development—is quite certainly a gift which we owe to the neotenous nature of mankind."

In short, we, like Mickey, never grow up although we, alas, do grow old. Best wishes to you, Mickey, for your next half-century. May we stay as young as you, but grow a bit wiser.

Questions and Topics for Discussion and Writing

1. To many readers, that a scientist would take notice of a cartoon character seems strange. What does Gould's so doing suggest about the scientist's keen observations?
2. Although the title relates to biology, is Gould, as revealed in this essay, working in this area of thinking as a psychologist or anthropologist?
3. What does the sketch depicting Mickey Mouse's evolution add to the clarity of Gould's paper? What is suggested by the fact that each character sketches the next?
4. Gould writes in this essay: "Progressive juvenilization as an evolutionary phenomenon is called neoteny." Rewrite, in your own words, this sentence of definition.
5. Read the seventh paragraph again. In context, does the sentence beginning "By 1940 . . ." simply add humor to the essay, or does it intrude on the sensibility of the reader and thus disturb the tone of the essay?
6. Study the graph "The 'Evolution' of Mickey Mouse." Is this graph offered humorously or seriously?
7. At what point did you discover that Gould was "fiddling around and having fun" with his subject and his audience? Does he spend too much time and waste too much ink on this cartoon character before changing to a more serious tone? Does his prolonged discussion of Mickey's altered appearance prepare you for better understanding of his scientific thesis?
8. The phenomenon of attributing human characteristics to animals is called *anthropomorphizing*. Does Gould suggest such transfer?

9. Gould suggests that Disney and his artists created Mickey Mouse's evolution unconsciously. How might such artists benefit from reading Gould's paper?
10. Will you look at human features differently after reading Gould's article?
11. What does Gould mean by saying that "humans are neotenic"?
12. What has caused humans' neoteny, according to Gould?
13. For education, what are the implications of Gould's statement that humans enjoy an "extended childhood"?
14. To test your comprehension of Gould's argument, write a paragraph in which you discuss how the sketch of Goofy evidences exaggerated adult features.

J. B. S. Haldane

John Burdon Sanderson Haldane (1892–1964) was a British biologist who did distinguished work in genetics, physiology, and biometry. Like many other leading scientists of our century, he also became a writer of popular works in order to interpret his work for the public. While satisfied with a philosophy of scientific materialism both professionally and personally, Haldane held that treating the subject of the human being from one view only, the biological, was insufficient. In a talk given in 1947 at Princeton, "Human Evolution: Past and Future," Haldane stated: "Only evil can come from forgetting that man must be considered from many angles. You can think of him as a producer and consumer. This concept is fully justified provided that you do not think that the economic angle is the only angle. You can treat him as a thinker, as an individual, as a member of society, as a being capable of moral choice, as a creator and appreciator of beauty, and so on. Concentration on only one of these aspects is disastrous."

Haldane's many-angled view of humanity did not prevent his working in tightly specialized scientific fields, nor from holding a strong political orientation. His respect for scientific materialism made him sympathetic with Marxist thinking; in fact, he credited Friedrich Engels (co-author with Karl Marx of *The Communist Manifesto*), along with Darwin, for most influencing his thinking. This view, however, still allowed Haldane to identify himself with the thinking of Gandhi and Nehru.

Haldane is recognized for his work on *selection pressure*, a figurative term that expresses the magnitude of the force of natural selection. He worked by the principle that the pressure of selection during the course of evolution is measured by the rate at which one of the hereditary factors known as "alleles" replaces another. Appropriately, his work on evolution and genetics brought him the Darwin Medal of the Royal Society in 1952. Haldane also won recognition for his work during World War II on the physiology of diving.

Haldane's many books include subjects still highly appealing to today's readers: *Adventures of a Biologist, Everything Has a History*, and, with Julian

Huxley, *Animal Biology.* The essay that follows, "On Being the Right Size," from *Possible Worlds,* continues to be much anthologized for its skill in embodying abstract ideas in concrete examples.

On Being the Right Size

J. B. S. Haldane

The most obvious differences between different animals are differences of size, but for some reason the zoologists have paid singularly little attention to them. In a large textbook of zoology before me I find no indication that the eagle is larger than the sparrow, or the hippopotamus bigger than the hare, though some grudging admissions are made in the case of the mouse and the whale. But yet it is easy to show that a hare could not be as large as a hippopotamus, or a whale as small as a herring. For every type of animal there is a most convenient size, and a large change in size inevitably carries with it a change of form.

Let us take the most obvious of possible cases, and consider a giant man sixty feet high—about the height of Giant Pope and Giant Pagan in the illustrated *Pilgrim's Progress* of my childhood. These monsters were not only ten times as high as Christian, but ten times as wide and ten times as thick, so that their total weight was a thousand times his, or about eighty to ninety tons. Unfortunately the cross sections of their bones were only a hundred times those of Christian, so that every square inch of giant bone had to support ten times the weight borne by a square inch of human bone. As the human thigh-bone breaks under about ten times the human weight, Pope and Pagan would have broken their thighs every time they took a step. This was doubtless why they were sitting down in the picture I remember. But it lessens one's respect for Christian and Jack the Giant Killer.

To turn to zoology, suppose that a gazelle, a graceful little creature with long thin legs, is to become large, it will break its bones unless it does one of two things. It may make its legs short and thick, like the rhinoceros, so that every pound of weight has still about the same area of bone to support it. Or it can compress its body and stretch out its legs obliquely to gain stability, like the giraffe. I mention these two beasts because they happen to belong to the same order as the gazelle, and both are quite successful mechanically, being remarkably fast runners.

Gravity, a mere nuisance to Christian, was a terror to Pope, Pagan, and Despair. To the mouse and any smaller animal it presents practically no dangers. You can drop a mouse down a thousand-yard mine shaft; and, on arriving at the bottom, it gets a slight shock and walks away. A rat would probably be killed, though it can fall safely from the eleventh story of a building; a man is killed, a horse splashes. For the resistance presented to movement by the air is proportional to the surface of the moving object. Divide an animal's length, breadth, and height each by ten; its weight is reduced to a thousandth, but its surface only to a hundredth. So the resistance to falling in the case of the small animal is relatively ten times greater than the driving force.

An insect, therefore, is not afraid of gravity; it can fall without danger, and can cling to the ceiling with remarkably little trouble. It can go in for elegant and fantastic forms of support like that of the daddy-long-legs. But there is a force which is as formidable to an insect as gravitation to a mammal. This is surface tension. A man coming out of a bath carries with him a film of water of about one-fiftieth of an inch in thickness. This weighs roughly a pound. A wet mouse has to carry about its own weight of water. A wet fly has to lift many times its own weight and, as every one knows, a fly once wetted by water or any other liquid is in a very serious position indeed. An insect going for a drink is in as great danger as a man leaning out over a precipice in search of food. If it once falls into the grip of the surface tension of the water—that is to say, gets wet—it is likely to remain so until it drowns. A few insects, such as water-beetles, contrive to be unwettable, the majority keep well away from their drink by means of a long proboscis.

Of course tall land animals have other difficulties. They have to pump their blood to greater heights than a man and, therefore, require a larger blood pressure and tougher blood-vessels. A great many men die from burst arteries, especially in the brain, and this danger is presumably still greater for an elephant or a giraffe. But animals of all kinds find difficulties in size for the following reason. A typical small animal, say a microscopic worm or rotifer, has a smooth skin through which all the oxygen it requires can soak in, a straight gut with sufficient surface to absorb its food, and a simple kidney. Increase its dimensions tenfold in every direction, and its weight is increased a thousand times, so that if it is to use its muscles as efficiently as its miniature counterpart, it will need a thousand times as much food and oxygen per day and will excrete a thousand times as much of waste products.

Now if its shape is unaltered its surface will be increased only a hundredfold, and ten times as much oxygen must enter per minute through each square millimetre of skin, ten times as much food through each square millimetre of intestine. When a limit is reached to their absorptive powers their surface has to be increased by some special device. For example, a part of the skin may be drawn out into tufts to make gills or pushed in to make lungs, thus increasing the oxygen-absorbing surface in proportion to the animal's bulk. A man, for example, has a hundred square yards of lung. Similarly, the gut, instead of being smooth and straight, becomes coiled and develops a velvety surface, and other organs increase in complication. The higher animals are not larger than the lower because they are more complicated. They are more complicated because they are larger. Just the same is true of plants. The simplest plants, such as the green algae growing in stagnant water or on the bark of trees, are mere round cells. The higher plants increase their surface by putting out leaves and roots. Comparative anatomy is largely the story of the struggle to increase surface in proportion to volume.

Some of the methods of increasing the surface are useful up to a point, but not capable of a very wide adaptation. For example, while vertebrates carry the oxygen from the gills or lungs all over the body in the blood, insects take air directly to every part of their body by tiny blind tubes called tracheae which open to the surface at many different points. Now, although by their breathing movements they can renew the air in the outer part of the tracheal system, the oxygen has to penetrate the finer branches by means of diffusion. Gases can diffuse easily through very small distances, not many times larger than the average length travelled by a gas molecule between collisions with other molecules. But when such vast journeys—from the point of view of a molecule—as a quarter of an inch have to be made, the process becomes slow. So the portions of an insect's body more than a quarter of an inch from the air would always be short of oxygen. In consequence hardly any insects are much more than half an inch thick. Land crabs are built on the same general plan as insects, but are much clumsier. Yet like ourselves they carry oxygen around in their blood, and are therefore able to grow far larger than any insects. If the insects had hit on a plan for driving air through their tissues instead of letting it soak in, they might well have become as large as lobsters, though other considerations would have prevented them from becoming as large as man.

Exactly the same difficulties attach to flying. It is an elementary principle of aeronautics that the minimum speed needed to keep an aeroplane of a given shape in the air varies as the square root of its length. If its linear dimensions are increased four times, it must fly twice as fast. Now the power needed for the minimum speed increases more rapidly than the weight of the machine. So the larger aeroplane, which weighs sixty-four times as much as the smaller, needs one hundred and twenty-eight times its horsepower to keep up. Applying the same principles to the birds, we find that the limit to their size is soon reached. An angel whose muscles developed no more power weight for weight than those of an eagle or a pigeon would require a breast projecting for about four feet to house the muscles engaged in working its wings, while to economize in weight, its legs would have to be reduced to mere stilts. Actually a large bird such as an eagle or kite does not keep in the air mainly by moving its wings. It is generally to be seen soaring, that is to say balanced on a rising column of air. And even soaring becomes more and more difficult with increasing size. Were this not the case eagles might be as large as tigers and as formidable to man as hostile aeroplanes.

But it is time that we passed to some of the advantages of size. One of the most obvious is that it enables one to keep warm. All warm-blooded animals at rest lose the same amount of heat from a unit area of skin, for which purpose they need a food-supply proportional to their surface and not to their weight. Five thousand mice weigh as much as a man. Their combined surface and food or oxygen consumption are about seventeen times a man's. In fact a mouse eats about one quarter its own weight of food every day, which is mainly used in keeping it warm. For the same reason small animals cannot live in cold countries. In the arctic regions there are no reptiles or amphibians, and no small mammals. The smallest mammal in Spitzbergen is the fox. The small birds fly away in the winter, while the insects die, though their eggs can survive six months or more of frost. The most successful mammals are bears, seals, and walruses.

Similarly, the eye is a rather inefficient organ until it reaches a large size. The back of the human eye on which an image of the outside world is thrown, and which corresponds to the film of a camera, is composed of a mosaic of "rods and cones" whose diameter is little more than a length of an average light wave. Each eye has about a half a million, and for two objects to be distinguishable their images must fall on separate rods or cones. It is obvious that with fewer but larger rods and cones we should see less distinctly. If they were twice as broad two points would have to be twice as far apart before we could distinguish them at a given distance. But if their size were diminished and their number increased we should see no better. For it is impossible to form a definite image smaller than a wave-length of light. Hence a mouse's eye is not a small-scale model of a human eye. Its rods and cones are not much smaller than ours, and therefore there are far fewer of them. A mouse could not distinguish one human face from another six feet away. In order that they should be of any use at all the eyes of small animals have to be much larger in proportion to their bodies than our own. Large animals on the other hand only require relatively small eyes, and those of the whale and elephant are little larger than our own.

For rather more recondite reasons the same general principle holds true of the brain. If we compare the brain-weights of a set of very similar animals such as the cat, cheetah, leopard, and tiger, we find that as we quadruple the body-weight the brain-weight is only doubled. The larger animal with proportionately larger bones can economize on brain, eyes, and certain other organs.

Such are a very few of the considerations which show that for every type of animal there is an optimum size. Yet although Galileo demonstrated the contrary more than three hundred years ago, people still believe that if a flea were as large as a man it could jump a thousand feet into the air. As a matter of fact the height to which an animal can jump is more nearly independent of its size than proportional to it. A flea can jump about two feet, a man about five. To jump a given height, if we neglect the resistance of the air, requires an expenditure of energy proportional to the jumper's weight. But if the jumping muscles form a constant fraction of the animal's body, the energy developed per ounce of muscle is independent of the size, provided it can be developed quickly enough in the small animal. As a matter of fact an insect's muscles, although they can contract more quickly than our own, appear to be less efficient; as otherwise a flea or grasshopper could rise six feet into the air.

And just as there is a best size for every animal, so the same is true for every human institution. In the Greek type of democracy all the citizens could listen to a series of orators and vote directly on questions of legislation. Hence their philosophers held that a small city was the largest possible democratic state. The English invention of representative government made a democratic nation possible, and the possibility was first realized in the United States, and later elsewhere. With the development of broadcasting it has once more become possible for every citizen to listen to the political views of representative orators, and the future may perhaps see the return of the national state to the Greek form of democracy. Even the referendum has been made possible only by the institution of daily newspapers.

To the biologist the problem of socialism appears largely as a problem of size. The extreme socialists desire to run every nation as a single business concern. I do not suppose that Henry Ford would find much difficulty in running Andorra or Luxembourg on a socialistic basis. He has already more men on his pay-roll than their population. It is conceivable that a syndicate of Fords, if we could find them, would make Belgium Ltd. or Denmark Inc. pay their way. But while nationalization of certain industries is an obvious possibility in the largest of states, I find it no easier to picture a completely socialized British Empire or United States than an elephant turning somersaults or a hippopotamus jumping a hedge.

Questions and Topics for Discussion and Writing

1. A certain childlike tone characterizes the title and the opening paragraphs of this essay; a balance between a conversational tone and a more formal one characterizes the rest of the essay. Comment on the appropriateness of the tone to the subject.
2. What advantage does Haldane find in challenging the descriptions of fictional characters?
3. One difference between Darwin's and Lamarck's theories of evolution is that for Lamarck, the *will* plays an important part in mutation. Do you find any evidence of Haldane's use of Lamarck's perspective in his discussion of "convenient size"? Prepare a list of words and phrases. Do they seem "at home" in this work that assumes Darwinian evolution as fact?
4. Study the fourth paragraph. By what pattern or patterns of reasoning is it organized?
5. Write a paragraph of your own to illustrate Haldane's generalization, "Comparative anatomy is largely the story of the struggle to increase surface in proportion to volume."

SUGGESTIONS FOR FURTHER READING

Asimov, Isaac. *The Intelligent Man's Guide to the Biological Sciences.* New York: Basic Books, 1960.

Crick, Francis. *Life Itself: Its Origin and Nature.* New York: Simon & Schuster, 1981.

de Kruif, Paul. *The Microbe Hunters.* New York: Harcourt Brace Jovanovich, 1954.

Evans, Howard Ensign. *Life on a Little-Known Planet.* New York: Dell, 1970.

Huxley, Julian S. *Man Stands Alone.* New York: Harper & Row, 1941.

Medawar, P. B. *The Life Sciences: Current Ideas of Biology.* New York: Harper & Row, 1977.

————, and J. S. Medawar. *Aristotle to Zoos: A Philosophical Dictionary of Biology.* Cambridge: Harvard University Press, 1983.

Miller, Jonathan, and Borin van Loon. *Darwin for Beginners.* New York: W. W. Norton, 1982.

Miller, Stanley L., and Leslie E. Orgel. *The Origins of Life on Earth.* Englewood Cliffs: Prentice-Hall, 1974.

Orgel, Leslie E. *The Origins of Life.* New York: John Wiley and Sons, 1973.

Schmidt-Nielsen, Knut. *Scaling: Why Is Animal Size So Important?* New York: Cambridge University Press, 1984.

Thomas, Lewis. *The Lives of a Cell: Notes of a Biology Watcher.* New York: Viking Press, 1974.

von Frisch, Karl. *Bees: Their Vision, Chemical Senses and Languages.* Ithaca: Cornell University Press, 1971.

Wilson, Edward O. *The Insect Societies.* Cambridge: Harvard University Press, 1981.

Unit 8

SCIENCE AND MEDICINE

Berton Roueché

Anthrax bacillus, jimson weed, rabid bats, trichina worms—all these and more organisms are villains in Berton Roueché's medical detective stories. For four decades, science writer Roueché has been investigating unusual diseases through his case studies. He has made medical journalism a significant and compelling genre. Even doctors have profited by his reporting of medical investigations.

Born in Kansas City, Missouri, and educated at the University of Missouri, Roueché began as a reporter on the *Kansas City Star*. He became the St. Louis reporter for *The New Yorker* in 1942 and originated their "Annals of Medicine" department. As one reviewer writes, Roueché uses a "combination of intellectual deduction and detective legwork" to present "a near-perfect synthesis of entertainment and information." Charting the course of human medical problems, Roueché adds drama to medical insight through his "mystery stories."

Appropriately, his awards come from both sides, ranging from the Mystery Writers of America Award to the American Medical Association Annual Journalism Award. His prolific writings include *Curiosities of Medicine, Annals of Epidemiology*, several novels, and the recent *Sea to Shining Sea: People, Travels, Places*.

In "The Orange Man," from *The Medical Detectives*, the reader observes painstaking accumulation of data and logical reasoning that organizes the data into a pattern. Roueché summarizes Dr. Wooten's method as a "process of elimination."

The Orange Man

Berton Roueché

Around eleven-thirty on the morning of December 15, 1960, Dr. Richard L. Wooten, an internist and an assistant professor of internal medicine at the University of Tennessee College of Medicine, in Memphis, was informed by the receptionist in the office he shares with several associates that a patient named (I'll say) Elmo Turner was waiting to see him. Dr. Wooten remembered Turner, but not much about him. He asked the receptionist to fetch him Turner's folder, and then, when she had done so, to send Turner right on in. The folder refreshed his memory. Turner was fifty-three years old, married, and a plumber by trade, and over the past ten years Dr. Wooten had seen him through an attack of pneumonia and referred him along for treatment of a variety of troubles, including a fractured wrist and a hip-joint condition. There were footsteps in the hall. Dr. Wooten closed the folder. The door opened, and Turner—a short, thick, muscular man—came in. Dr. Wooten had risen to greet him, but for a moment he could only stand and stare. Turner's face was orange—a golden, pumpkin orange. So were both his hands.

Dr. Wooten found his voice. He gave Turner a friendly good morning, asked him to sit down, and remarked that it had been a couple of years since their last meeting. Turner agreed that it had. He had been away. He had been working up in Alaska—in Fairbanks. He and his wife were back in Memphis only on a matter of family business. But, being in town, he thought he ought to pay Dr. Wooten a visit. There was something that kind of bothered him. Dr. Wooten listened with half an ear. His mind was searching through the spectrum of pathological skin discolorations. There were many diseases with pigmentary manifestations. There was the paper pallor of pituitary disease. There was the cyanotic blue of congenital heart disease. There was the deep Florida tan of thyroid dysfunction. There was the jaundice yellow of liver damage. There was the bronze of hemochromatosis. As far as he knew, however, there was no disease that colored its victims orange. Turner's voice recalled him. In fact, he was saying, he was worried. Dr. Wooten nodded. Just what seemed to be the trouble? Turner touched his abdomen with a bright-orange hand. He had a pain down there. His abdomen had been sore off and on for over a year, but now it was more than sore. It hurt. Dr. Wooten gave an encouraging grunt, and waited. He waited for Turner to say something about his extraordinary color. But Turner had finished. He had said all he had to say. Apparently, it was only his abdomen that worried him.

Dr. Wooten stood up. He asked Turner to come along down the hall to the examining room. His color, however bizarre, could wait. A chronic abdominal pain came first. And not only that. The cause of Turner's pain was probably also the cause of his color. That seemed, at least, a reasonable assumption. They entered the examining room. Dr. Wooten switched on the light above the examination table and turned and looked at Turner. The light in his office had been an ordinary electric light, and ordinary electric light has a faintly yellow tinge. The examining room had a true-color daylight light. But Turner's color owed nothing to tricks of light. His skin was still an unearthly golden orange. Turner stripped to the waist and got up on the table and stretched out on his back. His torso was as orange as his face. Dr. Wooten began his examination. He found the painful abdominal area, and carefully pressed. There was something there. He could feel an abnormality—a deep-seated mass about the size of an apple. It was below and behind the stomach, and he thought it might be sited at the liver. He pressed again. It wasn't the liver. It was positioned too near the center of the stomach for that. It was the pancreas.

Dr. Wooten moved away from the table. He had learned all he could from manual exploration. He waited for Turner to dress, and then led the way back to his office. He told Turner what he had found. He said he couldn't identify the mass he had felt, and he wouldn't attempt to guess. Its nature could be determined only by a series of X-ray examinations. That, he was sorry to say, would require a couple of days in the hospital. The pancreas was seated too deep to be accessible to direct X-ray examination, and an indirect examination took time and special preparation. Turner listened, and shrugged. He was willing to do whatever had to be done. Dr. Wooten swiveled around in his chair and picked up the telephone. He put in a call to the admitting office of Baptist Memorial Hospital, an affiliate of the medical school, and had a few words with the reservations clerk. He swiveled back to Turner. It was all arranged. Turner would be expected at Baptist Memorial at three o'clock that afternoon. Turner nodded, and got up to go. Dr. Wooten waved him back into his chair. There was one more thing. It was about the color of his skin. How long had it been like that? Turner looked blank. Color? What color? What was wrong with the color of his skin? Dr. Wooten hesitated. He was startled. There was no mistaking Turner's reaction. He was genuinely confused. He didn't know about his color—he really didn't know. And that was an interesting thought. It was, in fact, instructive. It clearly meant that Turner's change of color was not a sudden development. It had come on slowly, insidiously, imperceptibly. He realized that Turner was waiting, that his question had to be answered. Dr. Wooten answered it. Turner looked even blanker. He gazed at his hands, and then at Dr. Wooten. He didn't see anything unusual about his color. His skin was naturally ruddy. It always looked this way.

Dr. Wooten let it go at that. There was no point in pressing the matter any further right then. It would only worry Turner, and he was worried enough already. The matter would keep until the afternoon, until the next day, until he had a little more information to work with. He leaned back and lighted a cigarette, and changed the subject. Or seemed to. Had Turner ever met the senior

associate here? That was Dr. Hughes—Dr. John D. Hughes. No? Well, in that
case . . . Dr. Wooten reached for the telephone. Dr. Hughes's office was just
next door, and he arrived a moment later. He walked into the room and
glanced at Turner, and stopped—and stared. Dr. Wooten introduced them. He
described the reason for Turner's visit and the mass he had found in the region
of the pancreas. Dr. Hughes subdued his stare to a look of polite attention.
They talked for several minutes. When Turner got up again to go, Dr. Wooten
saw him to the door. He came back to his desk and sat down. Well, what did
Dr. Hughes make of that? Had he ever seen or read or even heard of a man that
color before? Dr. Hughes said no. And he didn't know what to think. He was
completely flabbergasted. He was rather uneasy, too. That, Dr. Wooten said,
made two of them.

Turner was admitted to Baptist Memorial Hospital for observation that afternoon at a few minutes after three. He was given the usual admission examination and assigned a bed in a ward. An hour or two later, Dr. Wooten, in the
course of his regular hospital rounds, stopped by Turner's bed for the ritual visit
of welcome and reassurance. Turner appeared to be no more than reasonably
nervous, and Dr. Wooten found that satisfactory. He then turned his attention
to Turner's chart and the results of the admission examination. They were, as
expected, unrevealing. Turner's temperature was normal. So were his pulse
rate (seventy-eight beats a minute), his respiration rate (sixteen respirations a
minute), and his blood pressure (a hundred and ten systolic, eighty diastolic).
The results of the urinalysis and of an electrocardiographic examination were
also normal. Before resuming his rounds, Dr. Wooten satisfied himself that the
really important examinations had been scheduled. These were comprehensive
X-ray studies of the chest, upper gastrointestinal tract, and colon. The first two
examinations were down for the following morning.

They were made at about eight o'clock. When Dr. Wooten reached the hospital on a midmorning tour, the radiologist's report was in and waiting. It more
than confirmed Dr. Wooten's impression of the location of the mass. It defined
its nature as well. The report read, "Lung fields are clear. Heart is normal.
Barium readily traversed the esophagus and entered the stomach. In certain
positions, supine projections, an apparent defect was seen on the stomach.
However, this was extrinsic to the stomach. It may well represent a pseudocyst
of the pancreas. No lesions of the stomach itself were demonstrated. Duodenal
bulb and loop appeared normal. Stomach was emptying in a satisfactory manner." Dr. Wooten put down the report with a shiver of relief. A pancreatic
cyst—even a pseudocyst—is not a trifling affliction, but he welcomed that
diagnosis. The mass on Turner's pancreas just might have been a tumor. It
hadn't been a likely possibility—the mass was too large and the symptoms were
too mild—but it had been a possibility.

Dr. Wooten went up to Turner's ward. He told Turner what the X-ray examination had shown and what the findings meant. A cyst was a sac retaining a

liquid normally excreted by the body. A pseudocyst was an empty sac—a mere dilation of space. The only known treatment of a pancreatic cyst was surgical, and surgery involving the pancreas was difficult and dangerous. Surgery was difficult because of the remote location of the pancreas, and dangerous because of the delicacy of the organs surrounding the pancreas (the stomach, the spleen, the duodenum) and the delicacy of the functions of the pancreas (the production of enzymes essential to digestion and the secretion of insulin). Fortunately, however, treatment was seldom necessary. Most cysts—particularly pseudocysts—had a way of disappearing as mysteriously as they had come. It was his belief that this was such a cyst. In that case, there was nothing much to do but be patient. And careful. Turner was to guard his belly from sudden bumps or strains. A blow or a wrench could cause a lot of trouble.

Nevertheless, Dr. Wooten went on, he wanted Turner to remain in the hospital for at least another day. There was a final X-ray of the colon to be made, and several other tests. In view of this morning's findings, the examination was, he admitted, very largely a matter of form. The cause of Turner's abdominal pain was definitely a pseudocyst of the pancreas. But prudence required an X-ray, and it would probably be done the next day. It was usual, for technical reasons, to let a day elapse between an upper-gastrointestinal study and a colon examination. Two of the other tests were indicated by the X-ray findings. One was a test for diabetes—the glucose-tolerance test. Diabetes was a possible complication of a cyst of the pancreas. Pressure from the cyst could produce diabetes by disrupting the production of insulin in the pancreatic islets of Langerhans. Such pressure could also cause another complication—a blockage of the common bile duct. The diagnostic test for that was a chemical analysis of the blood serum for the presence of the bile pigment known as bilirubin. Dr. Wooten paused. The time had come to reopen the subject that he had tactfully dropped the day before. He reopened it. It was possible, he said, that the bilirubin test might help explain the unusual color of Turner's skin. And Turner's skin *was* a most unusual color. He held up an adamant hand. No. Turner was mistaken. His color *had* changed in the past year or two. It wasn't a natural ruddiness. It was a highly unnatural orange. It was a sign that something was wrong, and he intended to find out what. That was the reason for a third test he had ordered. It was a diagnostic blood test for a condition called hemochromatosis. Hemochromatosis was a disturbance of iron metabolism that deposited iron in the skin and stained it the color of bronze. To be frank, he didn't hope for much from either of the pigmentation tests. Turner's color wasn't the bronze of hemochromatosis, and it wasn't the yellow of jaundice. The possibility of jaundice was particularly remote. The whites of Turner's eyes were still white, and that was usually where jaundice made its first appearance. But he had to carry out the tests. He had to be sure. The process of elimination was always an instructive process. And they didn't have long to wait. The results of the tests would be ready sometime that afternoon. He would be back to see Turner then.

Dr. Wooten spent the next few hours at the hospital and his office. He had other patients to see, other problems to consider, other decisions to make. But Turner remained on his mind. His first impression, like so many first impressions, had been mistaken. It now seemed practically certain that Turner's color had no connection with Turner's pancreatic cyst. They were two quite different complaints. And that returned him to the question he had asked himself when Turner walked into his office. What did an orange skin signify? What disease had the power to turn its victims orange? The answer, as before, was none. But perhaps this wasn't in the usual sense a disease. Perhaps it was a drug-induced reaction. Many chemicals in common therapeutic and diagnostic use were capable of producing conspicuous skin discolorations. Or it might be related to diet.

The question hung in Dr. Wooten's mind all day. It was still hanging there when he headed back to Turner's ward. On the way, he picked up the results of the tests he had ordered that morning, and they did nothing to resolve it. Turner's total bilirubin level was 0.9 milligrams per hundred milliliters, or normal. The total iron-binding capacity was also normal—286 micrograms per hundred milliliters. And he didn't have diabetes. When Dr. Wooten came into the ward, he found Turner's wife at his bedside and Turner in a somewhat altered state of mind. He said he had begun to think that maybe Dr. Wooten was right about the color of his skin. There must be something peculiar about it. There had been a parade of doctors and nurses past his bed ever since early morning. Mrs. Turner looked bewildered. She hadn't noticed anything unusual about her husband's color. She hadn't thought about it—the question had never come up. But now that it had, she had to admit that he did look kind of different. He did look kind of orange. But what was the reason? What in the world could cause a thing like that? Dr. Wooten said he didn't know. The most he could say at the moment was that certain possibilities had been eliminated. He summarized the results of the three diagnostic tests. Another possible cause, he then went on to say, was drugs. Medicinal drugs. Certain medicines incorporated dyes or chemicals with pigmentary properties. Turner shook his head. Maybe so, he said, but that was out. It had been months since he had taken any kind of drug except aspirin.

Dr. Wooten was glad to believe him. Drugs had been a rather farfetched possibility. The color changes they produced were generally dramatically sudden and almost never lasting. He turned to another area—to diet. What did Turner like to eat? What, for example, did he usually have for breakfast? That was no problem, Turner said. His breakfast was almost always the same—orange juice, bacon and eggs, toast, coffee. And what about lunch? Well, that didn't change much, either. He ate a lot of vegetables—carrots, rutabagas, squash, beans, spinach, turnips, things like that. Mrs. Turner laughed. That, she said, was putting it mildly. He ate carrots the way some people eat candy. Dr. Wooten sat erect. Carrots, he was abruptly aware, were rich in carotene. So were eggs, oranges, rutabagas, squash, beans, spinach, and turnips. And carotene was a powerful yellow pigment. What, he asked Mrs. Turner, did she mean about the way her husband ate carrots? Mrs. Turner laughed again. She meant just what she said. Elmo was always eating carrots. Eating carrots and drinking tomato juice. Tomato juice was his favorite drink. And carrots were his favorite snack. He ate raw carrots all day long. He ate four or five of them a day. Why, driving down home from Alaska last week, he kept her busy just scraping and slicing and feeding him carrots. Turner gave an embarrassed grin. His wife was right. He reckoned he did eat a lot of carrots. But he had his reasons. You needed extra vitamins when you lived in Alaska. You had to make up for the long, dark winters—the lack of sunlight up there. Dr. Wooten stood up to go. What the Turners had told him was extremely interesting. He was sure, he said, that Turner's appetite for carrots was a clue to the cause of his color. It was also, as it happened, misguided. The so-called "sunshine vitamin" was Vitamin D. The vitamin with which carrots and other yellow vegetables were abundantly endowed was Vitamin A.

There was a telephone just down the hall from Turner's ward. Dr. Wooten stopped and put in a call to the hospital laboratory. He arranged with the technician who took the call for a sample of Turner's blood to be tested for an abnormal concentration of carotene. Then he left the hospital and cut across the campus to the Mooney Memorial Library. He asked the librarian to let him see what she could find in the way of clinical literature of carotenemia and any related nutritional skin discolorations. He was elated by what he had learned from the Turners, but he knew that it wasn't enough. He had seen several cases of carotenemia. An excessive intake of carotene was a not uncommon condition among health-bar habitués and other amateur nutritionists. But carotene didn't color people orange. It colored them yellow. Or such had been his experience.

The librarian reported that papers on carotenemia were scarce. She had, however, found three clinical studies that looked as though they might be useful. Here was one of them. She handed Dr. Wooten a bound volume of the *Journal of the American Medical Association* for 1919, and indicated the relevant article. It was a report by two New York City investigators—Alfred F. Hess and Victor C. Myers—entitled "Carotenemia: A New Clinical Picture." Dr. Wooten knew their report, at least by reputation. It was the original study in the field. The opening descriptive paragraphs refreshed his memory and confirmed his judgment. They read:

About a year ago one of us (A.F.H.) observed that two children in a ward containing about twenty-five infants, from a year to a year and a half in age, were developing a yellowish complexion. This coloration was not confined to the face, but involved, to a less extent, the entire body, being most evident on the palms of the hands. . . . For a time, we were at a loss to account for this peculiar phenomenon, when our attention was directed to the fact that these two children, and only these two, were receiving a daily ration of carrots in addition to their milk and cereal. For some time we had been testing the food value of dehydrated vegetables, and when the change in color was noted, had given these babies the equivalent of 2 tablespoonfuls of fresh carrots for a period of six weeks.

It seemed as if this mild jaundiced hue might well be the result of the introduction into the body of a pigment rather than the manifestation of a pathologic condition. Attention was accordingly directed to the carrots, and the same amount of this vegetable was added to the dietary of two other children of about the same age. In the one instance, after an interval of about five weeks, a yellowish tinge of the skin was noted, and about two weeks later the other baby had become somewhat yellow. There was a decided difference in the intensity of color of the four infants, indicating probably that the alteration was in part governed by individual idiosyncrasy. On omission of the carrots from the dietary, the skin gradually lost its yellow color, and in the course of some weeks regained its normal tint.

The librarian returned to Dr. Wooten's table with the other references. Both were contributions to the *New England Journal of Medicine*. One was entitled "Skin Changes of Nutritional Origin," and had been written by Harold Jeghers, an associate professor of medicine at the Boston University School of Medicine, in 1943. The other was the work of three faculty members of the Harvard Medical School—Peter Reich, Harry Shwachman, and John M. Craig—and was entitled "Lycopenemia: A Variant of Carotenemia." It had appeared in 1960. Dr. Wooten looked first at "Skin Changes of Nutritional Origin." It was a comprehensive survey, and it read, in part:

> The carotenoid group of pigments color the serum and fix themselves to the fat of the dermis and subcutaneous tissues, to which they impart the yellow tint. . . . Edwards and Duntley showed by means of spectrophotometric analysis of skin color in human beings that carotene is present in every normal skin and is one of the five basic pigments that determine the skin color of every living person. Clinically, therefore, carotenemia refers to the presence of an excess over normal of carotene in the skin and serum. . . . In most cases carotenemia results simply from excess use of foods rich in the carotenoid pigments. Individuals probably vary in the ease with which carotenemia develops, which is evidenced by the fact that many vegetarians do not develop it. It is said to develop more readily in those who sweat profusely. Except for the yellow color produced, it appears to be harmless, even though present for months. It eventually disappears over several weeks to months when the carotene consumption is reduced.

Dr. Wooten moved on to the third report. ("This investigation concerns a middle-aged woman whose prolonged and excessive consumption of tomato juice led to the discoloration of her skin.") He read it slowly through from beginning to end, and then turned back and reread certain passages:

> Although carotenemia due to the ingestion of foods containing a high concentration of beta carotene is a commonly described disorder, a similar condition secondary to the ingestion of tomatoes and associated with high serum levels of lycopene has not previously been reported. . . . Lycopene is a common carotenoid pigment widely

distributed through nature. It is most familiar as the red pigment of tomatoes, but has been detected in many animals and vegetables. . . . It is also frequently found in human serum and liver, especially when tomatoes are eaten. But lycopene is not well known medically because, unlike beta carotene, it is physiologically inert and has not been involved in any form of illness.

Dr. Wooten closed the volume. Turner was not only a heavy eater of carrots. He was also a heavy drinker of tomato juice. Carrots are rich in carotene and tomatoes are rich in lycopene. Carotene is a yellow pigment and lycopene is red. And yellow and red make orange.

Dr. Wooten completed his record of the case with a double diagnosis: pseudocyst of the pancreas and carotenemia–lycopenemia. The results of the X-ray examination of Turner's colon were normal ("Terminal ileum was visualized. No pathology was demonstrated in the colon"), and the carotene test showed a high concentration of serum carotenoids (495 micrometers per hundred milliliters, compared to a normal concentration of 50 to 350 micrometers per hundred milliliters). The diagnosis of lycopenemia was made from the clinical evidence. Turner was discharged from the hospital on December 17. His instructions were to avoid abdominal blows, carrots (and other yellow vegetables), and tomatoes in any form. Four months later, on April 16, 1961, he reported to Dr. Wooten that his skin had recovered its normal ruddiness. Two years later, in 1963, he returned again to Memphis and dropped in on Dr. Wooten for a visit. His abdominal symptoms had long since disappeared, and a comprehensive examination showed no sign of the pseudocyst.

Elmo Turner was the first recorded victim of the condition known as carotenemia–lycopenemia. He is not, however, the only one now on record. Another victim turned up in 1964. She was a woman of thirty-five, a resident of Memphis, and a patient of Dr. Wooten's. He had been treating her for a mild diabetes since 1962, and had put her at that time on the eighteen-hundred-calorie diet recommended by the American Diabetes Association. She had faithfully followed the diet, but in order to do so had eaten heavily of low-calorie vegetables, and (as she confirmed, with some surprise, when questioned) the vegetables she ate most heavily were carrots and tomatoes. She ate at least two cups of carrots and at least two whole tomatoes every day. Dr. Wooten was unaware of this until she walked into his office one October day in 1964 for her semi-annual consultation. He greeted her as calmly as he could, and asked her to sit down. He would be back in just a moment. He stepped along the hall to Dr. Hughes's office and looked in. Dr. Hughes was alone.

"Have you got a minute, John?" Dr. Wooten said.

"Sure," Dr. Hughes said. "What is it?"

"I'd like you to come into my office," Dr. Wooten said. "I'd like to show you something."

Questions and Topics for Discussion and Writing

1. A case study tells a story. Therefore, it has some of the elements of fiction. For example, foreshadowing is an important device to unify fiction. What evidence do you find in the opening paragraph that the cause of the problem is foreshadowed?
2. This case study may also be said to fit in the genre of the detective story, with a logical reading of clues. List the clues as they enter the doctor's train of reasoning.
3. In the diagnosis of Elmo Turner's disease, what role did chance play?
4. When did you realize that the story was really an account of the steps inherent in diagnosis?
5. Narration has been defined as a "moving picture," a sequence of events. How has Roueché been able to write a medical case study as though it is an eventful narrative?
6. Several definitions are stated in the story of this case, for example, the definition of the "really important examinations." Do they impede the flow of the narrative?
7. Study the eighth paragraph. What is the dual audience this paragraph serves?
8. Two fairly long parenthetical expressions, as well as two expressions set off by dashes, appear in the same paragraph. What function do these serve? What is their effect on readability?
9. In this "story" of Turner's case, what is the "climax"? Find what you consider the climactic sentence.
10. How does Turner's taking the "wrong vitamin" bear on the case?
11. How does this case underscore the importance of writing by scientists about their findings? What does it say about retrieval systems for scientific information?
12. Why does Roueché end the narrative with the beginning of a new case?

On Warts

Lewis Thomas

You will find the biography for Lewis Thomas before the essay "Alchemy" in Unit 6.

Warts are wonderful structures. They can appear overnight on any part of the skin, like mushrooms on a damp lawn, full grown and splendid in the complexity of their architecture. Viewed in stained sections under a microscope, they are the most specialized of cellular arrangements, constructed as though for a purpose. They sit there like turreted mounds of dense, impenetrable horn, impregnable, designed for defense against the world outside.

In a certain sense, warts are both useful and essential, but not for us. As it turns out, the exuberant cells of a wart are the elaborate reproductive apparatus of a virus.

You might have thought from the looks of it that the cells infected by the wart virus were using this response as a ponderous way of defending themselves against the virus, maybe even a way of becoming more distasteful, but it is not so. The wart is what the virus truly wants; it can flourish only in cells undergoing precisely this kind of overgrowth. It is not a defense at all; it is an overwhelming welcome, an enthusiastic accommodation meeting the needs of more and more virus.

The strangest thing about warts is that they tend to go away. Fully grown, nothing in the body has so much the look of toughness and permanence as a wart, and yet, inexplicably and often very abruptly, they come to the end of their lives and vanish without a trace.

And they can be made to go away by something that can only be called thinking, or something like thinking. This is a special property of warts which is absolutely astonishing, more of a surprise than cloning or recombinant DNA or endorphin or acupuncture or anything else currently attracting attention in the press. It is one of the great mystifications of science: warts can be ordered off the skin by hypnotic suggestion.

Not everyone believes this, but the evidence goes back a long way and is persuasive. Generations of internists and dermatologists, and their grandmothers for that matter, have been convinced of the phenomenon. I was once told by a distinguished old professor of medicine, one of Sir William Osler's original bright young men, that it was his practice to paint gentian violet over a wart and then assure the patient firmly that it would be gone in a week, and he never saw it fail. There have been several meticulous studies by good clinical investigators, with proper controls. In one of these, fourteen patients with seemingly intractable generalized warts on both sides of the body were hypnotized, and the suggestion was made that all the warts on one side of the body would begin to go away. Within several weeks the results were indisputably positive; in nine patients, all or nearly all of the warts on the suggested side had vanished, while the control side had just as many as ever.

It is interesting that most of the warts vanished precisely as they were instructed, but it is even more fascinating that mistakes were made. Just as you might expect in other affairs requiring a clear understanding of which is the right and which the left side, one of the subjects got mixed up and destroyed the warts on the wrong side. In a later study by a group at the Massachusetts General Hospital, the warts on both sides were rejected even though the instructions were to pay attention to just one side.

I have been trying to figure out the nature of the instructions issued by the unconscious mind, whatever that is, under hypnosis. It seems to me hardly enough for the mind to say, simply, get off, eliminate yourselves, without providing something in the way of specifications as to how to go about it.

I used to believe, thinking about this experiment when it was just published, that the instructions might be quite simple. Perhaps nothing more detailed than a command to shut down the flow through all the precapillary arterioles in and around the warts to the point of strangulation. Exactly how the mind would accomplish this with precision, cutting off the blood supply to one wart while leaving others intact, I couldn't figure out, but I was satisfied to leave it there anyhow. And I was glad to think that my unconscious mind would have to take the responsibility for this, for if I had been one of the subjects I would never have been able to do it myself.

But now the problem seems much more complicated by the information concerning the viral etiology of warts, and even more so by the currently plausible notion that immunologic mechanisms are very likely implicated in the rejection of warts.

If my unconscious can figure out how to manipulate the mechanisms needed for getting around that virus, and for deploying all the various cells in the correct order for tissue rejection, then all I have to say is that my unconscious is a lot further along than I am. I wish I had a wart right now, just to see if I am that talented.

There ought to be a better word than "Unconscious," even capitalized, for what I have, so to speak, in mind. I was brought up to regard this aspect of thinking as a sort of private sanitarium, walled off somewhere in a suburb of my brain, capable only of producing such garbled information as to keep my mind, my proper Mind, always a little off balance.

But any mental apparatus that can reject a wart is something else again. This is not the sort of confused, disordered process you'd expect at the hands of the kind of Unconscious you read about in books, out at the edge of things making up dreams or getting mixed up on words or having hysterics. Whatever, or whoever, is responsible for this has the accuracy and precision of a surgeon. There almost has to be a Person in charge, running matters of meticulous detail beyond anyone's comprehension, a skilled engineer and manager, a chief executive officer, the head of the whole place. I never thought before that I possessed such a tenant. Or perhaps more accurately, such a landlord, since I would be, if this is in fact the situation, nothing more than a lodger.

Among other accomplishments, he must be a cell biologist of world class, capable of sorting through the various classes of one's lymphocytes, all with quite different functions which I do not understand, in order to mobilize the right ones and exclude the wrong ones for the task of tissue rejection. If it were left to me, and I were somehow empowered to call up lymphocytes and direct them to the vicinity of my wart (assuming that I could learn to do such a thing), mine would come tumbling in all unsorted, B cells and T cells, suppressor cells and killer cells, and no doubt other cells whose names I have not learned, incapable of getting anything useful done.

Even if immunology is not involved, and all that needs doing is to shut off the blood supply locally, I haven't the faintest notion how to set that up. I assume that the selective turning off of arterioles can be done by one or another chemical mediator, and I know the names of some of them, but I wouldn't dare let things like these loose even if I knew how to do it.

Well, then, who does supervise this kind of operation? Someone's got to, you know. You can't sit there under hypnosis, taking suggestions in and having them acted on with such accuracy and precision, without assuming the existence of something very like a controller. It wouldn't do to fob off the whole intricate business on lower centers without sending along a quite detailed set of specifications, way over my head.

Some intelligence or other knows how to get rid of warts, and this is a disquieting thought.

It is also a wonderful problem, in need of solving. Just think what we would know, if we had anything like a clear understanding of what goes on when a wart is hypnotized away. We would know the identity of the cellular and chemical participants in tissue rejection, conceivably with some added information about the ways that viruses create foreignness in cells. We would know how the traffic of these reactants is directed, and perhaps then be able to understand the nature of certain diseases in which the traffic is being conducted in wrong directions, aimed at the wrong cells. Best of all, we would be finding out about a kind of superintelligence that exists in each of us, infinitely smarter and possessed of technical know-how far beyond our present understanding. It would be worth a War on Warts, a Conquest of Warts, a National Institute of Warts and All.

Questions and Topics for Discussion and Writing

1. How did you react to the opening line of Thomas's essay?
2. How would you characterize Thomas's prose?
3. How is this essay a confirmation of the concept of "mind over matter"?
4. Write a definition of the word *etiology*.
5. Thomas uses analogy to clarify the role of the "Unconscious": "I was brought up to regard this aspect of thinking as a sort of private sanitarium, walled off somewhere in a suburb of my brain. . . ." What analogy comes to your mind when you think of the parts of your own mind? You may profit from reading Freud's essay on psychoanalysis (Unit 10).
6. Thomas writes poetry and alludes to the major poets in his essays. List words and phrases from this essay that evidence his poetical nature.
7. Point out at least one example of Thomas's use of humor.
8. Thomas opens and closes his essay by using the word *wonderful*. What are the two meanings of the word that Thomas suggests in context?
9. Discuss Thomas's use of the term *wonderful problem*.
10. Beginning with a dictionary definition and extending it into a full essay, write on one of the following: mental depression, pornography, sin, or censorship. Use at least two methods of defining.

John Ciardi

John Ciardi, according to Peter Gorner of the *Chicago Tribune*, has long enjoyed the status of that "rare American who could walk into a bank, declare his occupation as 'poet,' and emerge with a mortgage." The son of Italian immigrants, Ciardi (1916–1986) received his master's degree from the University of Michigan in 1939 and went on to become an English instructor at several universities, including Harvard and Rutgers.

In 1961, after producing classic-verse translations of Dante's *Inferno* and *Purgatorio*, Ciardi broke away from academia to devote all his energies to his own literary pursuits, including not only "serious" poetry, but also children's poems, verse translations, and criticism (most popularly, as poetry critic for *Saturday Review*, 1956–1972). Although he considers the writing of poetry a "love affair, not a sales campaign," Ciardi himself has made a comfortable living from his pen.

Ciardi tends to be more intellectual than emotional, his intricate works reflecting his conviction that "a poet needs at least as much training as a concert pianist." Preferring precise language and form to the spontaneous overflow of emotional imagery and excitation for its own sake, Ciardi has been critical of—and criticized by—many modern poets, particularly the Beats. "We live in a cascade of imprecision of feeling," Ciardi has said. "Poetry tries to make for *precision* of feeling. . . ."

In "Bufo Vulgaris," study Ciardi's use of language to summon the reader's emotional responses to the experience of the laboratory.

Bufo Vulgaris

John Ciardi

From a day in the time of my own warts I recall
bufo vulgaris, the common toad, asprawl
in biology lab. First we dissected him dead,
and many parts of the toad were stuffed in my head
gray green and bitter with formaldehyde,
though all I remember now for ready reference
is the vast difference made by the *vas deferens*.

Next, we probed him living. Under the lens
I saw his blood cells pumped alive between
the albumen shadows of his webs outspread
on the light that pierced from below. On a
    radar screen.
electrons dance the way I remember his blood
drawn in its endless chain, they also tracing
the invisible land below, that country racing
in the space between our senses. So I stood
at a peephole into Genesis watching that blood
particled and bright in its great flow
and whirligig (but in order) down below.
In and out of focus it danced as I turned
the gnurled knob at my thumb like Captain God
observing from a space ship.

                        So the toad
pumped me with light all a long afternoon
when I was Godlike. From a star cocoon
my eye broke wingéd. High above Polaris
I looked into his blood—*bufo vulgaris*
meaning, as noted above, the common
toad, but suggesting also "vulgar clown."

Questions and Topics for Discussion and Writing

1. What is the importance of the first line of the poem?
2. What does Ciardi mean by "the time of my own warts"?
3. What is the role of the Latin terms in the first stanza?
4. Study the use of verbs in the poem and comment. What is the function, for example, of the *-ing* suffix in the second stanza?
5. What is the importance of the parenthetical expression in the second stanza?
6. What is conveyed by the religious allusions and imagery, such as the phrases "Captain God," "peephole into Genesis," and "I was Godlike"?
7. What is the implied comparison to the "gnurled knob"?
8. Explain how the toad "pumps" the speaker of the poem with "light."
9. Look up the word *vulgar* in the OED (*Oxford English Dictionary*). As you consider the various definitions, what changes in meaning do you see in *bufo vulgaris* from the first stanza to the last?

Florence Nightingale

The Englishwoman who "created" the modern profession of nursing and revolutionized health care for soldier and civilian alike began life as an unlikely candidate for the role of international authority on scientific nursing. Florence Nightingale (1820–1910) was born in Florence, Italy, where her parents were on an extended tour of the continent; they named her after that city. For many years she lived, at least externally, the life of ease and luxury associated with the upper class, focusing on poise, grace, and other attributes of a lady in society. She early felt a call, however, to do more with her life, even professing to hear God's voice on several occasions. At age six, she recorded such a call in her journal, the only true confidante of her lifetime.

In her era, nursing was considered a disreputable career, nurses usually being equated with prostitutes and drunkards. Society prescribed strict behavior codes for a woman of Nightingale's class and frowned upon involvement, however humanitarian, with the lower classes. Hence her family's opposition. But Nightingale felt strongly about her calling. The result was a nearly lifelong struggle with her family, the cause of bitterness and psychosomatic illness. The quarrel intensified especially when she refused marriage to the "man I adore" lest she compromise her dedication to her vocation.

From her father, however, Nightingale did receive some support and guidance. Although uncomfortable with her choice of nursing, he recognized his daughter's intellectual gifts and tutored her in the classical learning normally reserved for men, even arranging for her to learn mathematics. This knowledge later became her statistical tool to demonstrate the effect on the sick of poor sanitation.

Having been trained in Germany at Kaiserwerth Institute, which prepared deaconesses to nurse the sick poor, she was immediately sought by government officials when the Crimean War broke out. Nightingale considered the hapless British soldiers her children and labored incessantly to improve their condition in war, then in peace. For these efforts, she achieved her great fame.

After the war, she participated in the Royal Commission, using her influence, her connections, and her threats to reveal to the public the reports she had compiled of unsanitary and inefficient methods. In the process, she managed to interest Queen Victoria in her work, thus ensuring official response. A semi-invalid, she founded the Nightingale School of Nursing in London, where she guided and ruled the candidates with iron-willed resolve. Though she achieved her vocation with a tenacious, almost tyrannical control of those helping her, Nightingale was revered by the British Army and the public for her tireless work on their behalf. She became the first woman to receive the British Order of Merit.

From the perspective of the nineteenth century, in "What Is a Nurse?," she defines nursing as both a profession and a vocation.

What Is a Nurse?

Florence Nightingale

The very alphabet of a nurse is to be able to read every change which comes over a patient's countenance, without causing him the exertion of saying what he feels. What would many a nurse do otherwise than she does, if her patient were a valuable piece of furniture or a sick cow? I do not know. Yet a nurse must be something more than a lift or a broom. A patient is not merely a piece of furniture, to be kept clean and ranged against the wall, and saved from injury or breakage—though to judge from what many a nurse does and does not do you would say he was. But watch a good old-fashioned monthly nurse with the infant; she is firmly convinced, not only that she understands everything it "says," and that no one else can understand it, but also that it understands everything she says, and understands no one else.

Now a nurse *ought* to understand in the same way every change of her patient's face, every change of his attitude, every change of his voice. And she ought to study them till she feels sure that no one else understands them so well. She may make mistakes, but she is *on the way* to being a good nurse. Whereas the nurse who never observes her patient's countenance at all, and never expects to see any variation, any more than if she had the charge of delicate china, is on the way to nothing at all. She never will be a nurse.

"He hates to be watched," is the excuse of every careless nurse. Very true. All sick people and all children "hate to be watched." But find a nurse who really knows and understands her children and her patients, and see whether these are aware that they have been "watched." It is not the staring at a patient which tells the really observant nurse the little things she ought to know.

People often talk of a nurse who has been ten or fifteen years with the sick, as being an "experienced nurse." But it is observation only which makes experience; and a woman who does not observe might be fifty or sixty years with the sick and never be the wiser.

Nay, more, experience sometimes tells in the opposite direction. A farmer "who practises the blunders of his predecessors," is often said to be "a practical man"; and she who perpetuates the "blunders of her predecessors" is often called an experienced nurse. The friends of a patient have been known to recommend the lodging in which he fell ill, just for the very reason which made him ill. A nurse has alleged as her reason for doing the things by which her predecessor ruined her own and her patient's health, that her predecessor "had always done them." People have taken a house because it had been emptied by death of all its occupants. These are they whom *no* experience will teach—viz., those who cannot see or understand the practical results of what they and others do. Now it is *no* reason that A. did it for B. to do it. It would be a reason if the results of A's doing it had been proved to be good.

What strikes one most with many women, who call themselves nurses, is that they have not learnt this A B C of a nurse's education. The A of a nurse ought to be to know what a sick human being is. The B to know how to behave to a sick human being. The C to know that her patient is a sick human being and not an animal.

What is it to feel a *calling* for any thing? Is it not to do your work in it to satisfy your own high idea of what is the *right*, the *best*, and not because you will be "found out" if you don't do it? This is the "enthusiasm" which every one, from a shoemaker to a sculptor, must have, in order to follow his "calling" properly. Now the nurse has to do, not with shoes, or with chisel and marble, but with human beings; and if she, for her own satisfaction, does not look after her patients, no *telling* will make her capable of doing so.

A nurse who has such a "calling" will, for her own satisfaction and interest in her patient, inform herself as to the state of his pulse, which can be quite well done without disturbing him. She will have observed the state of the secretions, whether told to do so or not. Nay, the very appearance of them, a slight difference in colour, will betray to her observing eye that the utensil has not been emptied after each motion.

She will, in like manner, have observed the state of the skin, whether there is dryness or perspiration—the effect of the diet, of the medicines, the stimulants. And it is remarkable how often the doctor is deceived in private practice by not being told that the patient has just had his meal or his brandy. She will most carefully have watched any redness or soreness of the skin, always on her guard against bedsores. Any loss of flesh will never take place unknown to her. Nor will she ever mistake puffing or swelling for gaining in flesh. She will be well acquainted with the different eruptions of fevers, measles, etc., and premonitory symptoms. She will know the shiver which betrays that matter is forming—that which slows the unconscious patient's desire to pass water—that which precedes fever. She will observe the changes of animal heat in her patient, and whether periodical, and not consider him as a piece of wood or stone, in keeping him warm or cool.

A nurse who has such a "calling" will look at all the medicine-bottles delivered to her for her patients, smell each of them, and, if not satisfied, taste each. Nine hundred and ninety-times there will be no mistake, but the thousandth time there may be a serious mistake detected by her means. But if she does not do this for her own satisfaction, it is no use telling her, because you may be sure that she will use neither smell nor taste to any purpose.

A nurse who has *not* such a "calling," will never be able to learn the sound of her patient's bell from that of others.

She will, when called to for hot brandy-and-water for her fainting patient, offer the weekly "Punch" (fact). Or she will wait to bring the cordial till she brings his tea (fact).

Under such a nurse, the patient never gets a hot drink. She pours out his tea, then she makes a journey to the larder for the butter, then she remembers that she has forgotten the toast, and has another journey to the kitchen fire to make the toast, and then she fills a hot water bottle, and last of all she takes him his tea.

Such a nurse will never know whether her patient is awake or asleep. She will rouse him up to ask him "if he wants anything," and leave him uncared for when he *is* up.

She will make the room like an oven when he is feverish at night, and let out the fire when he is cold in the morning.

Such a nurse seems to have neither eyes, nor ears, nor hands.

She never touches anything without a crash or an upset.

She does not shut the door, but pulls it after her, so that it always bursts open again.

She cannot rub in an embrocation without making a sore, which, in too many cases, never heals during the patient's life.

She catches up a cup and saucer in one hand, and pokes the fire with the other. Both, of course, come to "grief." Or she carries in a tray in one hand, and a coal-scuttle in the other. Both of course tip out their contents. And she, in stooping to pick them up, knocks over the bedside table upon the patient with her head (fact).

Tables are made for things to stand upon—beds for patients to lie in.

But such a nurse puts down a heavy flower-pot upon the bed, or a large book or bolster which has rolled upon the floor.

Yet these things are not done by drinking old females, but by respectable women.

Yet we are often told that a nurse needs only to be "devoted and obedient."

This definition would do just as well for a porter. It might even do for a horse. It would not do for a policeman. Consider how many women there are who have nothing to devote—neither intelligence, nor eyes, nor ears, nor hands. They will sit up all night by the patient, it is true; but their attendance is worth nothing to him, nor their observations to the doctor.

Cases have been known where the patient was cold before the nurse had observed he was dead—and yet she was not asleep—many cases where she supposed him comfortably sleeping, and he was insensible—very many where she never knew he was dying, unless he told her so himself.

But let no woman suppose that obedience to the doctor is not absolutely necessary. Only, neither doctor nor nurse lay sufficient stress upon *intelligent* obedience, upon the fact that obedience *alone* is a very poor thing.

I have known an obedient nurse, told not to disturb a very sick patient as usual at ten o'clock, with some customary service which she used to perform for him then, actually leave him in the dark all night, alleging this order as her reason for not carrying in his night-light as usual.

Everybody has known the window left open in heavy fog or rain, or shut when the patient was fainting, by such obedient nurses.

There seems to be no medium for them between a furnace of a fire and no fire at all; and one is actually obliged in this variable climate to divide the year into two parts, and tell them—"Now no fire," "Now fire"; as if they were volunteer riflemen. You cannot trust them to make a *small* fire, although in England it is a question whether, except when the air without is hotter than the air within, patients are not always the better of some fire, if only to promote ventilation. But no; such nurses make it impossible.

The elements of a nurse's duty are to observe the state of the pulse; the effect of the diet,—of sleep, whether it has been disturbed; whether there have been startings up in bed—a common mark of fatal disease; whether it has been a heavy, dull sleep, with stertorous breathing; whether there has been twitching of the bed-clothes,—to observe the state of the expectoration, the rusty expectoration of pneumonia, the frothy expectoration of pleurisy, the viscid mucous expectoration of bronchitis, the blood-streaked, dense, heavy expectoration which often occurs in consumption,—the nature of the cough itself by which the expectoration is expelled,—to observe the state of the secretions (yet nine-tenths of all nurses know nothing about these), whether the motions are costive or relaxed, and what is their colour, or whether there are alternations every few days of diarrhœa, and of no action of the bowels at all; whether the urine is high-coloured or pale, excessive or scanty, muddy or clear, or whether it is high-coloured when the bowels do not act, and pale when there is diarrhœa; whether there is ever blood in the motions,—in children, whether there are worms. All these things most nurses do not appear to consider it their business to observe.

The condition of the breathing and the position in which the patient breathes most easily, is another thing essential for the nurse to observe. In heart complaints life is often extinguished by the patient "accidentally" falling into a position in which he cannot breathe—and life is preserved by an "accidental" change of position. Now, what a thing it is to have to say of a nurse that it was not through her means, but through an "accident" that her patient was able to breathe.

Another essential duty of the nurse is, to observe the action of medicine; as, for instance, that of quinine. The sore throat, the deafness, the tight feeling in the head, are well-known effects of quinine. But the loss of memory it often occasions, is seldom known except to a very observant nurse. Indeed, she has often not memory enough herself to remember that the patient has forgotten.

A good nurse scarcely ever asks a patient a question—neither as to what he feels nor as to what he wants. But she does not take for granted, either to herself or to others, that she knows what he feels and wants, without the most careful observation and testing of her own observations.

But why, for instance, should a nurse ask a patient every day, "Shall I bring your coffee?" or "your broth?" or whatever it is—when she has every day brought it to him at that hour. One would think she did it for the sake of making the patient speak. Now, what the patient most wants is, never to be called upon to speak about such things.

Remember, every nurse should be one who is to be depended upon; in other words, capable of being a "confidential" nurse. She does not know how soon she may find herself placed in such a situation; she must be no gossip, no vain talker; she should never answer questions about her sick except to those who have a right to ask them; she must, I need not say, be strictly sober and honest; but more than this, she must be a religious and devoted woman; she must have a respect for her own calling, because God's precious gift of life is often literally placed in her hands; she must be a sound, and close, and quick observer; and she must be a woman of delicate and decent feeling.

Questions and Topics for Discussion and Writing

1. List the characteristics of a good nurse, according to Florence Nightingale.
2. Look up the word *alphabet*. Why do you think Nightingale used it in her phrase "the very alphabet of a nurse"?
3. Carefully read the comparisons in the first paragraph. Do these comments expose the role of women in her time and reveal Nightingale's protest as a feminist?
4. In what way is the nurse more important than the doctor, according to Nightingale? How does she qualify the nurse's subordinate position to the doctor?
5. To Nightingale, nursing was to be equated with art. In what way is a nurse an artist?
6. What does Nightingale mean by saying that nursing is a "calling"?
7. One of the most often repeated words in this essay is "observation." Why do you think it occurs so frequently?
8. Compare this essay with either Roueché's "The Orange Man" (earlier in this unit) or Doyle's "The Science of Deduction" (Unit 2). Write a paper on the subject of the role of medical personnel in making observations.
9. Toward the end of the essay, Nightingale observes, "Now, what a thing it is to have to say of a nurse that it was not through her means, but through an 'accident' that her patient was able to breathe." This sentence seems to stand out as different in tone from the rest of the essay. How is it different in tone?

Jean D. Lockhart

As director of the Department of Maternal, Child and Adolescent Health at the American Academy of Pediatrics, Jean Douglas Lockhart heads an association of more than 27,000 Board-certified pediatricians in North and South America. Organized more than five decades ago, the association works to improve the physical, mental, and social health of children in the Americas. The very success of their work has increased the task of pediatricians, for better sanitation and disease control have allowed them to expand their definition of pediatric health problems to include in their focus child abuse, alcohol abuse, accidents, and even obesity.

Lockhart's work involves her with numerous committees (Bioethics, Fetus and Newborn, Nutrition, Infectious Diseases, Environmental Hazards, Genetics, and Drugs, for example) and necessitates frequent travel. Often, this work takes her to Washington, D.C., where, after obtaining her medical degree in 1940, she practiced pediatrics and worked for a decade for the Food and Drug Administration.

As editor-in-chief of *Current Problems in Pediatrics*, a monthly journal of pediatric monographs, Lockhart conducts what she calls her "avocation." The skill and concern with language required for this post are evident in "PPIs As an Art Form," whose deft and lively parodies convey a message.

PPIs As an Art Form

Jean D. Lockhart

The problem with the Patient-Package-Insert concept is not with the concept itself—everyone wants patients to be better informed, or says so. The problem is, Who is to write the inserts? As any publishing house knows, some writers "sell" better than others, which is to say, more people like to read them. Since the package inserts are meant to be read, it might be worthwhile to ask how a few celebrity writers might phrase a package insert.

Here, for example, is what Henry Kissinger might have done with

CHLORPROMAZINE

Unless we stand upon the solemn ground of total emotional ordinariness and flatness throughout our lives, all of us must experience occasional states of relative mania (elation) and corresponding states or even superstates of depression. The modalities for dealing with these swings of mood do not ordinarily require either medical attention or therapeutics. Occasionally, however, mania or depression achieves hegemony over an individual to the point where critical issues are at stake, and decisive action must be taken. It is at this point that chlorpromazine may serve as a useful maneuver.

If chlorpromazine has been prescribed for you, it will produce far-reaching effects at all levels of the central intelligence system. You will notice a lessening of apprehension and violent feelings, a diminished sense of physical pleasure or pain, and any previously held suspicions of foreign intrigue will be replaced by a relative ennui. Suffice it to say, instead of exaltation there will be drabness, and where there was multilevel aggression there will be pacificity. To achieve this psychic manipulation, a few concessions have to be made at the bargain table: Drowsiness, jaundice, hypotension, and hematologic changes may be part of the trade-off. Like all negotiations, this one requires thoughtful balance between the immediate favorable behavior response and adverse events that may succeed it. Unconditional agreement also must be obtained to the absolute banning of extraneous agents (e.g., alcohol, barbiturates, narcotics). It is precisely the strict adherence to these conditions on which eventual mental recovery depends. It will behoove you not to chivy the official medical team who are caring for you, since your improvement will be in direct relationship to your compliance with the regular doses of the drug and other points of agreement with the therapy prescribed.

And here, as James Michener might have written it, is

LOMOTIL

For millions upon millions of years, man has sought a remedy for diarrhea. The provision of 2.5 mg diphenoxylate hydrochloride and 0.025 mg atropine sulfate, in a single-dose tablet or a common teaspoonful of liquid, has a calming effect on restless bowels and reduces the number of watery stools from a turbulent many to a reassuring few. Shortly after the first or second dose of this medication, the storm warnings of intestinal hyperactivity diminish, leaving in their wake only the memory of abdominal agitation.

However, from the ingenious laboratories of Searle come warnings telling us that the smallest people among us, children under two years of age, may suffer injury as often as they benefit from this remedy, and therefore Lomotil should not be given to them. Also, those men and women whose past includes a hypersensitive experience not unlike allergy, following treatment with either diphenoxylate or atropine, should avoid a repetition of the event by using another remedy. Jaundice ("the yellow eye") and diarrhea of an alarming nature ("the fatal flux") are also contraindications to the use of Lomotil. If you are fond of the grape or are plagued with the habit of drugs, either those that excite or those that bring tranquility, remember to choose between the wine or drugs and the Lomotil, because both together will cause the senses to soar, in unexpected and unpleasant ways. Not always will Lomotil have a soothing effect on intestinal motility. When used time after time, the dose may have to be increased to provide the expected relief; this then may be addiction, in that even the bowel is slave to the next fix. Avoid the habit! Use the drug sparingly, that the results are not spoiled.

Benjamin Spock might have written this about

TESTOSTERONE INJECTION

We all want children and young people to have a wholesome normal curiosity about the way their bodies look and function. This may be difficult for a boy if he cannot find his penis and testes. He is apt to ask questions, in a childish kind of way. He may wonder why it is he has nothing to handle, or why he is like the girl next door.

The use of testosterone by injection may be reassuring, both to the boy, his parents, his doctor, and the girl next door. This hormone, which simulates the naturally produced hormone of the testis, will cause the gender changes associated with maleness: prominent genitals, male patterns of hairiness, and a normal healthy lust for life. If, following regular use, sex play develops in prepubertal children, there is no reason to get upset or worried, but it may mean that the next dose should be lowered.

Adult males occasionally require replacement therapy of testosterone, and if your doctor thinks this would be a good idea, be sure to ask him about possible nuisance side effects, such as permanent erections, breast development, and bladder irritability. (Also, your sperm count may drop but may jump back when the injections are discontinued.)

The injections are given in the buttock and may be painful, but the end will justify the shot.

H. L. Mencken could have written this Patient Package Insert on

ACETAMINOPHEN

Despite all the snorting against it in the medical journal editorials, it has always been my experience that acetaminophen—or Tylenol, Tempra, Comtrex, or dozens of other trade names—is a generally estimable form of nostrum, with strong overtones of analgesia and even of antipyresis. Nor does it share with aspirin some of the problems of overdosage; but more of that later.

The essence of a useful and potent analgesic is that it is ingested with facility and gets to business at once. Acetaminophen seems to qualify on these points. Not that we care to itemize the drug's characteristics when our head aches with pulsating inescapability; we simply yearn for the commingling of ease and wellness, whether by magic or medicine. Most of the rewards of acetaminophen proceed from elevation of our pain threshold by action on that gland which lies in the most intimate and central part of our brain, the hypothalamus. When the drug reaches this sacerdotal gland, the body's heat-regulation is turned to "down" and the threshold for recognition of pain and discomfort to "up." Sceptics are soon convinced of this when the calor and dolor are their own, and they have been persuaded to take acetaminophen.

The point needs no laboring, except to recite the kinds of pain, in whatever degree, that respond to medication: headache, myalgia, neuralgia, arthritis, and rheumatic tribulations, and the indignity peculiar to the ladies: dysmenor-rhea ("the cramps"). Most ladies no longer believe it their solemn duty to suffer lower abdominal pain one or two days a month and are glad to discover the relieving effect of 1,000 mg of acetaminophen, in either liquid or tablet form.

When should one hesitate to take the drug? By a strange twist, the sufferer in dire need of some help may also be the one who possesses a hypersensitivity, and if this has happened before, there is the clearest possible reason to avoid the drug. Fortunately, this happens rarely, even to those of us who do not take kindly to aspirin.

The most portentous phenomenon in the acetaminophen story is the effect of overdosage on the liver. It is a grave matter indeed when a large dose, taken intentionally or unintentionally, reaches the liver via blood plasma. Hepatic enzymes resent these toxic doses; jaundice may supervene; and by the time severe liver disease is obvious, the antidote may have been delayed uncon-scionably. God help you. I don't preach patience so much as forethought; it is the more therapeutic of philosophies.

And, finally, Ernest Hemingway, who had a way with words, might have written this description of

TOLINASE TABLETS

There is a damn good diabetes drug that comes in a white tablet and can be used instead of insulin shots if you're middle-aged when the diabetes hits. It is not a small thing to do, to make a potent drug that can be taken once a day, and have it work the way the pancreatic juices should, *lo sabes?* The manufacturers did it, they made a drug like Orinase (tolbutamide), Diabinese (chlorpropa-mide), or phenformin, but *mejor.* It's true, you must obey the doctor faithfully, and see him often, when starting this drug. The urines must be tested every day for sugar and acetone, but what the hell? The dose has got to be right.

Listen, when I said it works like the pancreas, that doesn't mean it's another kind of insulin. It's a sulfonylurea, and it works on the beta cells of the pan-creas. The cells move. They put out insulin again, the way they used to, and as long as your diabetes isn't the severe kind, and you don't have any infection or liver or kidney trouble, it's a first-class drug. Get the doctor to tell you all about your diabetes, and why you have to stay on your diet, and how to know your blood sugar lows and highs. Don't fool yourself: Tell him if you're taking other drugs, *amigo.* If you're on diuretics you may need more Tolinase.

Side effects can happen. They range from rough cases of nausea and vomiting to changes in blood and liver function. Skin and allergic changes can happen, too.

But if you were over 30 when your diabetes began, ask not for whom the Tolinase; it may be for thee.

On the whole, patients would probably be more interested in reading package inserts if they had a little more literary style. It wouldn't hurt the Federal Register, either, if a few good authors rewrote it now and then.

Questions and Topics for Discussion and Writing

1. A parody is a composition that mimics the style or language of a writer, usually to create a humorous effect. Look into your home medicine cabinet and choose an instructional insert from a product. Write a parody of the insert, using the language of someone whose style has caught your attention.
2. In each of these parodies in the preceding selection, language takes on a "personality" of its own, forming a barrier to understanding. Reduce each of these parodies to simple directions for package inserts.
3. What is the linguistic barrier presented by Kissinger's insert? If you actually encountered Kissinger's insert, would you take the medicine?
4. In a few sentences, characterize Michener's style, giving examples to support your judgments.
5. Find linguistic evidence that Spock is concerned with children's feelings.
6. H. L. Mencken was a satirist who lampooned the American middle class as the "booboisie," but championed the American language. In what ways does the insert express these attitudes and Mencken as an individual?
7. Analyze Hemingway's insert, explaining the tone. At the end, what does the allusion to John Donne's poetry contribute?
8. Write a paper contrasting the language of each of these writers.

Lance G. Leithauser

Born in Michigan (1947), Lance Garner Leithauser, a plastic surgeon, grew up near the Detroit Zoological Park. Together with the books on animals and plants of North America that were his favorite childhood reading, the zoo developed his interest in natural history. In suburban Maryland, where he now lives, Leithauser continues this interest, practicing the horticulture of flowering plants and designing an evergreen maze. At Swarthmore College in Pennsylvania, where he received his bachelor's degree, his interest in biology developed into one in medicine. He received his medical degree from the University of Michigan and completed residencies in surgery at Georgetown University in Washington, D.C., and plastic surgery at Pittsburgh University Hospital.

Leithauser is the author of several papers on plastic surgery. The subject today represents a burgeoning field of research, encompassing reconstruction after surgery or trauma, the correction of congenital defects, cosmetic surgery, the treatment of disfiguring burns, and many other techniques. Thus frequent communications to update fellow practitioners are necessary.

The following report presents the results of recent developments in the treatment of amputated fingertips. While the report necessarily is phrased in the objective language of science and follows the conventional format for medical papers, it perhaps conveys a sense of the satisfaction that a surgeon feels in his restorative work and the fascination with what often has been called "the intricate design of the hand."

Fingertip Amputation and Reconstruction

Lance G. Leithauser

Fingertip injuries are common and occur in all age groups, both sexes and a wide variety of social settings. Treatment varies with the type and extent of injury and the work requirements of the patient.

We have reviewed 40 cases of fingertip amputation, 40 patients and 43 procedures, treated during a seven month period in 1978–79. There were 13 female patients and 27 male, ranging in age from 1 to 78 years.

In the female group, 7 of the 13 (54%) were under 18 years of age, and 6 were seven years or younger. In the male group there were no injuries under 10 years of age, but 19 were injured between the ages of 12 and 29. The difference in incidence of injury by age and sex was quite marked (Fig 1).

Female infants were prone to injury, but once a young woman reached adolescence her injury rate dropped dramatically and remained low the rest of her life. For males, however, during adolescence the injury rate reached its peak. Some of this increased incidence can be attributed to job related trauma, but a considerable portion occurs at home in association with lawn care (mowers, hedge trimmers, etc). The incidence of injury then remains higher in males than females until advanced age.

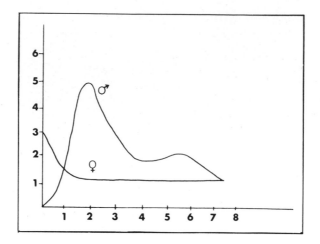

FIGURE 1   Graph shows a marked difference in the incidence of fingertip injuries by age and sex in a group of 40 patients.

FIVE METHODS

Five different procedures were used in this series to repair the fingertips. Minimal amputations, requiring no surgical treatment, were excluded from the series.

The methods of treatment were

primary closure (3 cases)
replantation (7 cases)
skin graft (4 cases)
V-y flap (13 cases)
pedicle flap (6 cases)

PRIMARY CLOSURE

This was the least common method of treatment and we found few indications for its use. Occasionally, when the configuration of amputation was such that primary closure could simply be performed, it was used. But in most cases, closure would require additional shortening of the bone and leave a final result inferior cosmetically and functionally to other methods.

In some patients where follow up might be difficult or where multistage procedures are inappropriate, primary closure can be considered. There were no complications in this group and no additional procedures required.

REPLANTATION

Replantation was utilized in some cases when the amputated segment was returned and appeared usable. All replantations were performed as free composite grafts; no microvascular anastomosis were performed in this series.

Public awareness of replantation is high, and in many cases the segment will be retrieved after a prolonged search. Often fragments or grossly destroyed parts will faithfully be brought to the hospital, but are not suitable for replantation.

The overall success of replantation was 3 in 7 (42.8%). The more proximal the level of injury, the less likely the segment was to take.

Findings were similar to Elsahy[1] in that respect. The dorsal level of amputation was more significant than the volar level. On the pad of the distal phalanx, soft tissue losses could be safely replanted. But over bone on the nail bed, the results were poor.

In addition to this factor, the mechanism of injury was also important. Fingertips that were cut off cleanly (i.e., with a knife) did much better than those that were crushed. Many of the children had their fingertips amputated in a door, and these had a uniformly poor prognosis. In almost all cases, the part was retrieved by the parents and the expectation to replant it was high.

In many cases the amputated piece appeared viable at the time of replantation, but almost invariably went on to a late necrosis. This may be due to vascular damage in the amputated segment with delayed edema and thrombosis.

Case No. I

A one-year-old female suffered amputation of the distal portion of the distal phalanx right long finger when it was crushed in a door. Level of amputation was Elsahy 2-B (proximal nail bed). Replantation was undertaken, but the distal segment went on to necrosis (Fig 2a). Subsequent reconstruction was performed with a thenar flap (Fig 2b).

V-y FLAP

V-y flap was used 13 times for reconstruction and was one of the most reliable methods of repair. Where applicable, it is generally the method of choice as it leaves no donor scars outside the fingertip. Sensation in the flaps is good postoperatively and the cosmetic appearance also is usually good, although the nail may curl over the end of the fingertip in cases where it is unsupported, i.e., more proximal injury.

[1] Elsahy NI: When to replant a fingertip after its complete amputation. Plast Reconstr Surg 60 (1):14–21, 1977

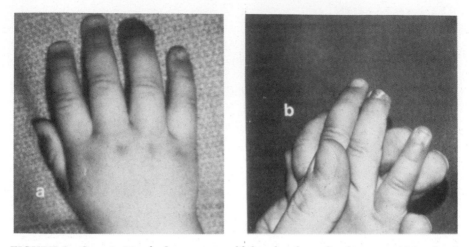

FIGURE 2   Case 1. Hand of a one-year-old female whose distal portion of the distal phalanx right long finger was amputated in a door. (a) Replantation of the distal segment went on to necrosis. (b) Subsequent reconstruction was performed with a thenar flap.

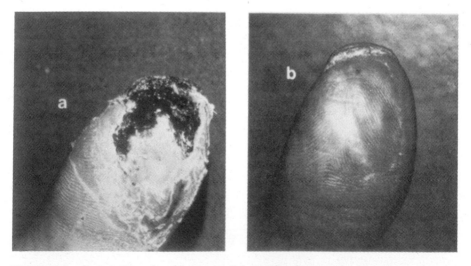

FIGURE 3   Case 2. Tip of nondominant left thumb of a 17-year-old male cut off on a joiner. (a) Large V-y flap went on to partial necrosis. (b) Wound healed without incident or surgery after conservative treatment with debridement and dressing changes.

The flap is so reliable it appears almost invulnerable, but there are occasional difficulties. Like any flap it is subject to ischemia. One case in 13 had a small portion of the flap (approximately 25%) which developed superficial necrosis, then healed by secondary intention without further surgery.

All other flaps survived completely and there were no problems with infection in this group. If the defect is large, the flap must also be large and requires significant advancement. In these cases, distant flap tissue may be a superior choice. In cases where there is a crushing component to the injury, distant flaps should also be considered, as V-y flaps may shift damaged tissue that will not survive a transfer.

Case No. 2

A 17-year-old male cut off the tip of his nondominant left thumb on a joiner. To avoid a cross finger flap, a large V-y flap was outlined and advanced, but went on to partial necrosis (Fig 3a). The wound was treated conservatively with dressing changes and debridement, and healed without further incident or surgery (Fig 3b).

SKIN GRAFT

Fourteen skin grafts were performed, 11 split thickness and 3 full thickness. Split thickness grafts were usually taken from the volar surface of the forearm, and full thickness grafts from the volar wrist crease. The wrist donor site limited somewhat the amount of tissue that could be harvested. Large grafts require an alternative donor site.

All full thickness grafts took, but 4 of the 11 split grafts were lost, for a failure rate of 26.3%. Overall success in grafting, split and full thickness, was 81.5%. Of the four split grafts that failed, two healed spontaneously and two required additional grafting.

The take of full thickness grafts appeared superior, but the sample was too small for reliable interpretation. Some surgeons feel that full thickness grafts are more reliable,[2] that dermal plexus tissue may improve circulation even over avascular tissue. But others have better success with split grafts. Our study would favor full thickness grafts, but the question remains debatable.

The mechanism of injury correlated with the success of the grafts. Crushing injuries with an edematous recipient site had poorer take than sharp amputations.

Contamination did not correlate well with graft loss. Thorough irrigation and debridement generally produced a suitable surface for grafting if the underlying tissue was in good condition.

[2]Spira M: New technique for skin grafting in avascular areas. Preliminary Report. Plast Reconstr Surg 63 (4):501, April, 1979

The full thickness graft produced a better cosmetic result (color match and texture) than the split grafts, but occasionally there was a prolonged edema in the graft.

CROSS FINGER FLAP

Only one cross finger flap was performed. Where flap tissue was needed it was generally felt that thenar flaps were superior.

Cross finger flaps leave a noticeable donor deformity on the dorsum of the finger, and require an additional scar for a skin graft donor site. In addition, the color match is inferior to thenar skin and may be particularly noticeable in black patients.

Case No. 3

A 10-year-old black male had the tip of his dominant right long finger amputated in a door. The amputated segment was not retrieved. Reconstruction was obtained with a cross finger flap, and good contour was obtained, but the skin had a poor color match. In black patients this discrepancy in color match between dorsal and volar skin may influence the choice of flaps (Fig 4).

THENAR FLAPS

There were seven thenar flaps performed. When pedicle flap tissue was required, as for exposed bone or large areas of soft tissue loss, this was the procedure of choice.

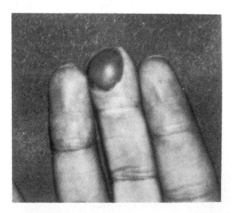

FIGURE 4   Case 3. Hand of a 10-year-old black male who lost the tip of his dominant right long finger in a door. Despite good result, cross-finger flap is a poor color match.

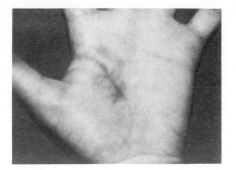

FIGURE 5   Hand of a child that has undergone a thenar flap procedure. The donor site has been closed primarily, producing an acceptable scar that falls in the natural skin crease.

The principal disadvantage of thenar flaps is the necessity for a second stage procedure to divide the flaps. Care must also be exercised in older patients, as stiff fingers may result even after short periods of immobilization.

Our best results were obtained in children. In most cases the donor site could be closed primarily, producing an acceptable scar planned to fall in a natural skin crease (Fig 5).

There was one infection and flap separation in the series, but no flap loss. Thenar flaps are reliable and can usually be designed on a 1 to 1 length/width ratio as random flaps.

The flap is usually placed near the MP crease of the thumb for index reconstruction, and on the palmar crease for long and ring fingers. Flaps are generally divided in 10 to 12 days (Fig 6).

The patients ranged in age from 1 to 54 years with a mean of 20.5 years. Four of the seven were under 15 years. There were no fingers permanently stiff postoperatively, but most patients experienced a transient stiffness at the PIP joint of the involved finger.

Case No. 4

A 14-year-old female had the tip of her left, nondominant long finger bitten off by a horse. The level of amputation was Elsahy 2A, and the piece was not retrieved (Fig 7a).

Reconstruction was obtained with a thenar flap, and the donor site was closed primarily at the time of flap division (Fig 7B).

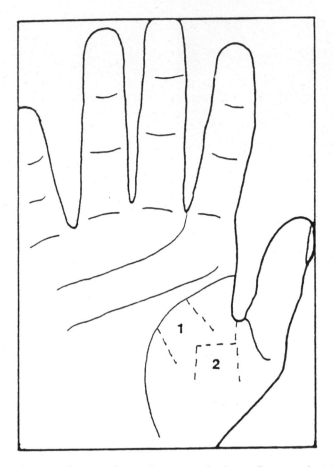

FIGURE 6   Schematic drawing shows placement of a thenar flap near the MP crease of the thumb for index reconstruction and on the palmar crease for long and ring fingers. Flaps are generally divided in 10 to 12 days.

DISCUSSION

The treatment of fingertip amputations must be individualized to the wound and patient. Where possible, V-y flaps are preferred.

When replantation is considered, the level of injury and mechanism of injury are important factors. Fingertips cut off in doors have a poor survival rate. Replantation may prolong treatment, as demarcation of the segment into viable or nonviable tissue can take 8–10 days.

When skin grafts are required, full thickness grafts have a better color match, and possibly a higher percentage of successful take.

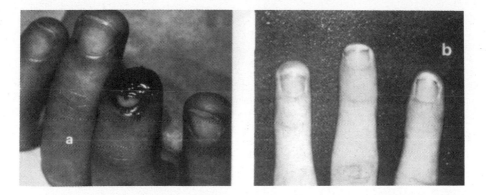

FIGURE 7   Case 4. Hand of a 14-year-old female who had the tip of her left nondominant long finger bitten off by a horse. (a) The amputated piece was not retrieved.
(b) Reconstruction was obtained with a thenar flap. The donor site was closed primarily
at the time of flap division.

When pedicle flaps are required, thenar flaps are preferrable to cross-finger
flaps. The donor site can often be closed primarily eliminating the need for an
additional skin graft and the color match is superior.

Questions and Topics for Discussion and Writing

1. Referring to Robert Day's "What Is a Scientific Paper?" (Unit 9), determine in what
    ways this report fulfills Day's tests of "valid publication."
2. Scientific reports use objective tone and specialized language. List specific words
    and phrases that give evidence of this specialized approach.
3. One of the characteristics of scientific and technical writing is that the writing is
    often supported by graphics. What contribution do the graphics (photographs, a
    schematic drawing, a graph) contribute to the paper?
4. Consult a scientific or engineering journal for other articles that explain a new
    technique or process. In a short paper, summarize the technique.

William Carlos Williams

Even though William Carlos Williams (1883–1963) had a long career as a pedia-
trician in his hometown of Rutherford, New Jersey, he was equally—perhaps
more—a writer and poet. His works include long and short poems, novels,

short stories, plays, and an autobiography. In both diction and rhyme his poems are noteworthy for his efforts to use and explore the "American idiom."

Williams was a resident of Rutherford all his life. The oldest child in the family, he attended a private school in Switzerland as well as the famous Horace Mann High School in New York. He received his medical degree from the University of Pennsylvania in 1906 and did graduate work at the University of Leipzig. His medical practice, chiefly among Rutherford's poorer working people, often provided material for his fiction.

Writing poetry is usually considered Williams's avocation. His first volume, *Poems* (1909), and his second, *The Tempers* (1913), show the strong influence of Ezra Pound and the Imagists. Eventually Williams achieved a reputation as a major poet. He showed his capability to conduct two careers simultaneously, stating, "One feeds the other, in a manner of speaking."

Williams's next poetic volumes show his experiments with language as he worked to intensify objects in sharp detail. The poems pay attention to the pure values of physical things in stark unemotional style but with a painter's eye for images. These efforts brought him the Dial Award in 1926 for "services to American literature," a National Book Award in 1950, and a Bollingen Prize, shared with Archibald MacLeish, in 1952.

In the short story that follows, Williams uses an event from medical practice to reveal the complex necessities that confront a physician.

The Use of Force
William Carlos Williams

They were new patients to me, all I had was the name, Olson. Please come down as soon as you can, my daughter is very sick.

When I arrived I was met by the mother, a big startled looking woman, very clean and apologetic who merely said, Is this the doctor? and let me in. In the back, she added. You must excuse us, doctor, we have her in the kitchen where it is warm. It is very damp here sometimes.

The child was fully dressed and sitting on her father's lap near the kitchen table. He tried to get up, but I motioned for him not to bother, took off my overcoat and started to look things over. I could see that they were all very nervous, eyeing me up and down distrustfully. As often, in such cases, they weren't telling me more than they had to, it was up to me to tell them; that's why they were spending three dollars on me.

The child was fairly eating me up with her cold, steady eyes, and no expression to her face whatever. She did not move and seemed, inwardly, quiet; an unusually attractive little thing, and as strong as a heifer in appearance. But her face was flushed, she was breathing rapidly, and I realized that she had a high fever. She had magnificent blonde hair, in profusion. One of those picture children often reproduced in advertising leaflets and the photogravure sections of the Sunday papers.

She's had a fever for three days, began the father and we don't know what it comes from. My wife has given her things, you know, like people do, but it don't do no good. And there's been a lot of sickness around. So we tho't you'd better look her over and tell us what is the matter.

As doctors often do I took a trial shot at it as a point of departure. Has she had a sore throat?

Both parents answered me together, No . . . No, she says her throat don't hurt her.

Does your throat hurt you? added the mother to the child. But the little girl's expression didn't change nor did she move her eyes from my face.

Have you looked?

I tried to, said the mother, but I couldn't see.

As it happens we had been having a number of cases of diphtheria in the school to which this child went during that month and we were all, quite apparently, thinking of that, though no one had as yet spoken of the thing.

Well, I said, suppose we take a look at the throat first. I smiled in my best professional manner and asking for the child's first name I said, come on, Mathilda, open your mouth and let's take a look at your throat.

Nothing doing.

Aw, come on, I coaxed, just open your mouth wide and let me take a look. Look, I said opening both hands wide, I haven't anything in my hands. Just open up and let me see.

Such a nice man, put in the mother. Look how kind he is to you. Come on, do what he tells you to. He won't hurt you.

At that I ground my teeth in disgust. If only they wouldn't use the word "hurt" I might be able to get somewhere. But I did not allow myself to be hurried or disturbed but speaking quietly and slowly I approached the child again.

As I moved my chair a little nearer suddenly with one catlike movement both her hands clawed instinctively for my eyes and she almost reached them too. In fact she knocked my glasses flying and they fell, though unbroken, several feet away from me on the kitchen floor.

Both the mother and father almost turned themselves inside out in embarrassment and apology. You bad girl, said the mother, taking her and shaking her by one arm. Look what you've done. The nice man . . .

For heaven's sake, I broke in. Don't call me a nice man to her. I'm here to look at her throat on the chance that she might have diphtheria and possibly die of it. But that's nothing to her. Look here, I said to the child, we're going to look at your throat. You're old enough to understand what I'm saying. Will you open it now by yourself or shall we have to open it for you?

Not a move. Even her expression hadn't changed. Her breaths however were coming faster and faster. Then the battle began. I had to do it. I had to have a throat culture for her own protection. But first I told the parents that it was entirely up to them. I explained the danger but said that I would not insist on a throat examination so long as they would take the responsibility.

If you don't do what the doctor says you'll have to go to the hospital, the mother admonished her severely.

Oh yeah? I had to smile to myself. After all, I had already fallen in love with the savage brat, the parents were contemptible to me. In the ensuing struggle they grew more and more abject, crushed, exhausted while she surely rose to magnificent heights of insane fury of effort bred of her terror of me.

The father tried his best, and he was a big man but the fact that she was his daughter, his shame at her behavior and his dread of hurting her made him release her just at the critical moment several times when I had almost achieved success, till I wanted to kill him. But his dread also that she might have diphtheria made him tell me to go on, go on though he himself was almost fainting, while the mother moved back and forth behind us raising and lowering her hands in an agony of apprehension.

Put her in front of you on your lap, I ordered, and hold both her wrists.

But as soon as he did the child let out a scream. Don't, you're hurting me. Let go of my hands. Let them go I tell you. Then she shrieked terrifyingly, hysterically. Stop it! Stop it! You're killing me!

Do you think she can stand it, doctor! said the mother.

You get out, said the husband to his wife. Do you want her to die of diphtheria?

Come on now, hold her, I said.

Then I grasped the child's head with my left hand and tried to get the wooden tongue depressor between her teeth. She fought, with clenched teeth, desperately! But now I also had grown furious—at a child. I tried to hold myself down but I couldn't. I know how to expose a throat for inspection. And I did my best. When finally I got the wooden spatula behind the last teeth and just the point of it into the mouth cavity, she opened up for an instant but before I could see anything she came down again and gripping the wooden blade between her molars she reduced it to splinters before I could get it out again.

Aren't you ashamed, the mother yelled at her. Aren't you ashamed to act like that in front of the doctor?

Get me a smooth-handled spoon of some sort, I told the mother. We're going through with this. The child's mouth was already bleeding. Her tongue was cut and she was screaming in wild hysterical shrieks. Perhaps I should have desisted and come back in an hour or more. No doubt it would have been better. But I have seen at least two children lying dead in bed of neglect in such cases, and feeling that I must get a diagnosis now or never I went at it again. But the worst of it was that I too had got beyond reason. I could have torn the child apart in my own fury and enjoyed it. It was a pleasure to attack her. My face was burning with it.

The damned little brat must be protected against her own idiocy, one says to one's self at such times. Others must be protected against her. It is social necessity. And all these things are true. But a blind fury, a feeling of adult shame, bred of a longing for muscular release are the operatives. One goes on to the end.

In a final unreasoning assault I overpowered the child's neck and jaws. I forced the heavy silver spoon back of her teeth and down her throat till she gagged. And there it was— both tonsils covered with membrane. She had fought valiantly to keep me from knowing her secret. She had been hiding that sore throat for three days at least and lying to her parents in order to escape just such an outcome as this.

Now truly she *was* furious. She had been on the defensive before but now she attacked. Tried to get off her father's lap and fly at me while tears of defeat blinded her eyes.

Questions and Topics for Discussion and Writing

1. What is the significance of the title?
2. How does the last sentence play an important part in the story?
3. Recall an experience in which you had to maintain a stern control you did not feel. In a paragraph, present this experience, including images that will help the reader to understand your experience.
4. Consider the implications of this statement: "After all, I had already fallen in love with the savage brat, the parents were contemptible to me."
5. Can you explain why the doctor does not want the mother to use the word *hurt?*
6. Read at least three critical essays on the writings of Williams. Then write a short paper summarizing the views presented.
7. Compile a bibliography of at least 10 works by physicians who are short-story writers (e.g., Anton Chekhov, Somerset Maugham) or essayists (e.g., Richard Selzer, Lewis Thomas). Choosing two of these works, compare and contrast the writers' styles.
8. "The Use of Force" is generally anthologized under the heading "point of view." How does point of view function in this story? Observe carefully the point of view of the doctor; rewrite a part of the story from the girl's point of view, or from the omniscient point of view. Does your new treatment modify the theme of the story? If so, how?
9. What accounts for the fact that this widely anthologized story is often labeled a masterpiece?

SUGGESTIONS FOR FURTHER READING

Blackwell, Elizabeth. *Scientific Method in Biology*. London: E. Stock, 1898.

Miller, Jonathan. *The Body in Question*. New York: Random House, 1979.

Miller, Jonathan, and David Pelham. *The Facts of Life*. New York: Viking Press, 1984.

Osler, William. *A Way of Life and Selected Writings*. New York: Dover Publications, 1958.

Restak, Richard. *The Brain*. New York: Bantam Books, 1985.

Roueché, Berton. *Eleven Blue Men, and Other Narratives of Medical Detection*. Boston: Little, Brown, 1953.

Selzer, Richard. *Letters to a Young Doctor*. New York: Simon & Schuster, 1983.

———. *Mortal Lessons: Notes on the Art of Surgery*. New York: Simon & Schuster, 1976.

Thomas, Lewis. *The Youngest Science: Notes of a Medicine-Watcher*. New York: Viking Press, 1983.

PART THREE

The Intensifying
Vision

Unit 9

ON SCIENCE APPLIED: TECHNOLOGY, INVENTION, ENGINEERING

"Someday, son, all this will be yours . . ."

J. David Bolter

A professor of classics at the University of North Carolina at Chapel Hill and a master of computer science, J. David Bolter is especially qualified to mediate between the "two cultures" of the humanities and the sciences. His particular vision is that the two groups have much to learn from each other and that the computer opens new avenues of communication between them. In Bolter's words, the computer provides the "sturdiest bridge" between the world of science and the world of the humanities.

Maintaining this perspective causes Bolter to place great emphasis on this one invention—the "logic machine"—as the "defining technology" of our age. By this term, he means that an invention can provide a symbol that represents a world-view. In twentieth-century society, the computer as a symbol presents a particular view of human beings and their relation to nature: Men and women are information processors and nature is the information they process. Bolter has chosen the mathematician–logician A. M. Turing as the progenitor of this view and calls those who hold it "Turing's men." Turing's people are both philosophers and designers of space, treating it as "a plastic, almost palpable material." They remain logicians who evoke data structures "from the imagination."

Bolter synthesizes cultural history and merits a leading place in the growing field of interdisciplinary studies. In the selection that follows, consider Bolter's thesis and the evolution of the computer in our world. As you read, think about the possibility that the computer can set in motion a new dialogue between the humanistic and the technical disciplines.

The Computer As a Defining Technology

J. David Bolter

Computers perform no work themselves; they direct work. The technology of "command and control," as Norbert Wiener has aptly named it, is of little value without something to control, generally other machines whose function is to perform work. For example, the essence of the American space shuttle is the computers that control almost every phase of its operation. But unless the powerful rocket engines provide the expected thrust, there is no mission for the computers to control. The computer leaves intact many older technologies, particularly the technologies of power, and yet it puts them in a new perspective. With the appearance of a truly subtle machine like the computer, the old power machines (steam, gas, or rocket engines) lose something of their prestige. Power machines are no longer agents on their own, subject only to direct human intervention; now they must submit to the hegemony of the computer that coordinates their effects.

As a calculating engine, a machine that controls machines, the computer does occupy a special place in our cultural landscape. It is the technology that more than any other defines our age. Our generation perfected the computer, and we are intrigued by possibilities as yet only half-realized. Ruthlessly practical and efficient, the computer remains something fantastic. Its performance astonishes even the engineers who build it, just as the clock must have astonished craftsmen in the fourteenth century and the power of the steam engine even the rugged entrepreneurs of the nineteenth century. For us today, the computer constantly threatens to break out of the tiny corner of human affairs (scientific measurement and business accounting) that it was built to occupy, to contribute instead to a general redefinition of certain basic relationships: the relationship of science to technology, of knowledge to technical power, and, in the broadest sense, of mankind to the world of nature.

This process of redefinition is not new. Technology has always exercised such an influence; it has always served both as a bridge and a barrier between men and their natural environment. The ability to make and use tools and the subtle capacity to communicate through language have allowed men to live more comfortably in the world, but these achievements have also impressed upon them their separation from nature.

Men and women throughout history have asked how it is that they and their culture (their technology in the largest sense) transcend nature, what makes them characteristically human and not merely animal. For the Greeks, a cardinal human quality was the ability to establish the political and social order embodied in a city-state: men at their best could set collective goals, make laws, and obey them, and none of these could be achieved by animals in a state of nature. Their city-state was a feat of social technology. In the Middle Ages, the accomplishments of technology were perhaps more physical than social, but the use of inanimate sources of power, wind and water, fostered a new view of mankind versus the forces of nature. The discoveries of the Renaissance and the Industrial Revolution moved men closer to nature in some respects and separated them even more radically in others. Continued emphasis on exploring and manipulating the physical world led to a deeper appreciation of the world's resources. Yet the desire to master nature—to harness her more efficient sources of power in steam and fossil fuels and to mine her metals for synthetic purposes—grew steadily throughout this period. When Darwin showed convincingly that man was an animal like any other, he shattered once and for all the barrier that separated men from the rest of nature in the Greek and medieval chains of being. Yet nineteenth-century engineers with their railroads and still more twentieth-century physicists with their atomic bombs seemed less natural than ever before, less under the control of either nature or a personal deity and more responsible for their own misjudgments.

Continually redrawing the line that divides nature and culture, men have always been inclined to explain the former in terms of the latter, to examine the world of nature through the lens of their own created human environment. So Greek philosophers used analogies from the crafts of pottery and woodworking to explain the creation of the universe: the stars, the planets, the earth, and its living inhabitants. In the same way, the weight-driven clock invented in the Middle Ages provided a new metaphor for both the regular movements of heavenly bodies and the beautifully intricate bodies of animals, whereas the widespread use of the steam engine in the nineteenth century brought to mind a different, more brutal aspect of the natural world. It is certainly not true that changing technology is solely responsible for mankind's changing views of nature, but clearly the technology of any age provides an attractive window through which thinkers can view both their physical and metaphysical worlds.

Technology has had this influence even upon philosophers, like Plato, who generally disdain human craftsmanship and see it as a poor reflection of a greater nonhuman reality. And even in Christian theology and poetry, the pleasures of heaven could only be described as a grand version of the tainted pleasures men know on earth, and the tortures of hell were earthly tortures intensified. Almost every sort of philosopher, theologian, or poet has needed an analogy on the human scale to clarify his or her ideas. Speaking of creation as the imposition of order upon the natural world, he or she generally assumes a creator as well, and this creator is a craftsman or technologist.

It is in this context that I propose to examine electronic technology. The computer is the contemporary analog of the clocks and steam engines of the previous six centuries; it is as important to us as the potter's wheel was to the ancient world. It is not that we cannot live without computers, but that we will be different people because we live with them. All techniques and devices have the potential to become defining technologies because all to some degree redefine our relationship to nature. In fact, only a few devices or crafts in any age deserve to be called defining technologies. In the ancient world, carpentry and masonry were about as important as spinning and pottery, and yet poets and philosophers found the latter two far more suggestive. In medieval Europe, crop rotation and the moldboard plough had a greater economic and social impact than the early clockwork mechanisms. Yet not many philosophers and theologians compared the world to a lentil bean. Certain skills and inventions have moved easily out of the agora into the Academy, out of the textile mill into the salon, or out of the industrial research park into the university classroom.

The vision of particular philosophers and poets is important to such a transference. Descartes and his followers helped to make the clock a defining technology in Western Europe. Certainly the first poet to elaborate the myth of the Fates who spin the thread of life helped to make textiles a defining technology for ancient Greece. But there must be something in the nature of the technology itself, so that its shape, its materials, its modes of operation appeal to the mind as well as to the hand of their age—for example, the pleasing rotary motion of the spindle or the autonomy and intricacy of the pendulum clock.

Such qualities combine with the social and economic importance of the device to make people think. Very often a device will take on a metaphoric significance and be compared in art and philosophy to some part of the animate or inanimate world. Plato compared the created universe to a spindle, Descartes thought of animals as clockwork mechanisms, and scientists in the nineteenth century and early twentieth centuries have regularly compared the universe to a heat engine that is slowly squandering its fuel. Today the computer is constantly serving as a metaphor for the human mind or brain: psychologists speak of the input and output, sometimes even the hardware and software, of the brain; linguists treat human language as if it were a programming code; and everyone speaks of making computers "think."

A defining technology develops links, metaphorical or otherwise, with a culture's science, philosophy, or literature; it is always available to serve as a metaphor, example, model, or symbol. A defining technology resembles a magnifying glass, which collects and focuses seemingly disparate ideas in a culture into one bright, sometimes piercing ray. Technology does not call forth major cultural changes by itself, but it does bring ideas into a new focus by explaining or exemplifying them in new ways to larger audiences. Descartes's notion of a mechanistic world that obeyed the laws of mathematics was clear, accessible, and therefore powerful because his contemporaries lived with clocks and gears. So today electronic technology gives a more catholic appeal to a number of trends in twentieth-century thought, particularly the notions of mathematical logic, structural linguistics, and behavioral psychology. Separately these trends were minor upheavals in the history of ideas; taken together, they become a major revision in our thinking.

Questions and Topics for Discussion and Writing

1. What do you understand the word *culture* to mean? In what sense do you think science can be called a culture? Do you agree with Bolter's definition?
2. One evidence of good writing is a "mind-catching" opening. Does Bolter's first sentence arrest or capture your attention? Why or why not? In a paragraph, evaluate the first sentence as "thought-provoking."

3. Bolter contrasts "power technology" with the computer as "subtle technology."
    List the ways that the computer is said to be "subtle."
4. According to Bolter, the computer is a "calculating machine." Why does he give it
    this description?
5. As Bolter judges it, what is the present role of the computer? What possibilities for
    its use does he foresee?
6. In rewriting the following sentence created by Bolter, many would use the term
    science in place of *technology*. Explain why Bolter uses *technology:* "Technology
    has always exercised such an influence; it has always served both as a bridge and a
    barrier between men and their natural environment."
7. In what ways does Bolter illustrate what he means by *social technology?*
8. Draw a chart juxtaposing the symbols that depicted the world of nature to the
    following time periods:
    a.   the Greek
    b.   the Medieval
    c.   the nineteenth century
    d.   the twentieth century
9. Can you think of other kinds of technology that preceded the computer as twen-
    tieth-century symbols? List at least two.
10. Which famous philosopher–scientist does Bolter credit with creating a defining
    technology in Western Europe?
11. Thoreau said of the invention of the train in the nineteenth century, "We have
    created an atropos." Look up the word *atropos* and write a short paper in which
    you discuss how the same can be said of the computer.
12. List the functions of a "defining technology" as presented by Bolter.

Turing's Man

J. David Bolter

In the development of the computer, theory preceded practice. The manifesto
of the new electronic order of things was a paper ("On Computable Numbers")
published by the mathematician and logician A. M. Turing in 1936. Turing set
out the nature and theoretical limitations of logic machines before a single fully
programmable computer had been built. What Turing provided was a symbolic
description, revealing only the logical structure and saying nothing about the
realization of that structure (in relays, vacuum tubes, or transistors). A Turing
machine, as his description came to be called, exists only on paper as a set of
specifications, but no computer built in the intervening half century has surpassed these specifications; all have at most the computing power of Turing

machines. Turing is equally well known for a very different kind of paper; in 1950 he published "Computing Machinery and Intelligence." His 1936 work was a forbidding forest of symbols and theorems, accessible only to specialists. This later paper was a popular polemic, in which Turing stated his conviction that computers were capable of imitating human intelligence perfectly and that indeed they would do so by the year 2000. This paper too has served as a manifesto for a group of computer specialists dedicated to realizing Turing's claim by creating what they call "artificial intelligence," a computer that thinks.

Put aside for the moment the question of whether the computer can ever rival human intelligence. The important point is that Turing, a brilliant logician and a sober contributor to the advance of electronic technology, believed it would and that many have followed him in his belief. The explanation is partly enthusiasm for a new invention. In 1950 the computer was just beginning to bring vast areas of science and business under its technological aegis. These machines were clearly taking up the duties of command and control that had always been assumed by human operators. Who could say then where the applications of electronic command and control might end? Was it not natural to believe that the machine would in time eliminate the human operator altogether? Inventors, like explorers, have a right to extravagant claims. Edison had said that the record player would revolutionize education; the same claim was made for radio and, of course, television.

I think, however, that Turing's claim has had a greater significance. Turing was not simply exaggerating the service his machine could perform. (Does a machine that imitates human beings perform any useful service at all? We are not running short of human beings.) He was instead explaining the meaning of the computer for our age. A defining technology defines or redefines man's role in relation to nature. By promising (or threatening) to replace man, the computer is giving us a new definition of man, as an "information processor," and of nature, as "information to be processed."

I call those who accept this view of man and nature Turing's men. I include in this group many who reject Turing's extreme prediction of an artificial intelligence by the year 2000. We are all liable to become Turing's men, if our work with the computer is intimate and prolonged and we come to think and speak in terms suggested by the machine. When the cognitive psychologist begins to study the mind's "algorithm for searching long-term memory," he has become Turing's man. So has the economist who draws up input-output diagrams of the nation's business, the sociologist who engages in "quantitative history," and the humanist who prepares a "key-word-in-context" concordance.

Turing's man is the most complete integration of humanity and technology, of artificer and artifact, in the history of the Western cultures. With him the tendency, implicit in all eras, to think "through" one's contemporary technology is carried to an extreme; for him the computer reflects, indeed imitates, the crucial human capacity of rational thinking. Here is the essence of Turing's belief in artificial intelligence. By making a machine think as a man, man re-creates himself, defines himself as a machine. The scheme of making a human being through technology belongs to thousands of years of mythology and alchemy, but Turing and his followers have given it a new twist. In Greek mythology, in the story of Pygmalion and Galatea, the artifact, the perfect ivory statue, came to life to join its human creator. In the seventeenth and eighteenth centuries, some followers of Descartes first suggested crossing in the other direction, arguing, with La Mettrie, that men were no more than clockwork mechanisms. Men and women of the electronic age, with their desire to sweep along in the direction of technical change, are more sanguine than ever about becoming one with their electronic homunculus. They are indeed remaking themselves in the image of their technology, and it is their very zeal, their headlong rush, and their refusal to admit any reservation that calls forth such a violent reaction from their detractors. Why, the critics ask, are technologists so eager to throw away their freedom, dignity, and humanity for the sake of innovation?

Should we be repelled by the notion of man as computer? Not until we better understand what it means for man to be a computer. Why on the face of it should we be more upset by this notion than by the Cartesian view that man is a clock or the ancient view that he is a clay vessel animated by a divine breath? We need to know how Turing's man differs from that of Descartes or Plato, how the computer differs conceptually and symbolically from a clock or a clay pot. And to do this, we must isolate the precise qualities of computers and programming, hardware and software, that have the magnifying effect mentioned earlier—bringing ideas from philosophy and science into a new focus.

Questions and Topics for Discussion and Writing

1. State in your own words Bolter's tribute to Turing for his contribution to computer technology.
2. Technology is often said to be the "offspring" of science. Considering this judgment, write what you think to be ironic about the following sentence by Bolter: "In 1950 the computer was just beginning to bring vast areas of science and business under its technological aegis."
3. Why do you think Bolter sees the computer's importance to education to be far greater than that of the record player, radio, or television?
4. What is a "defining technology"? Why does Bolter call the computer a "defining technology"?
5. Write a paragraph explaining why certain individuals are classified by Bolter as "Turing's men."
6. Do you agree with Bolter's statement "By making a machine think as a man, man re-creates himself, defines himself as a machine"? Write why you do or do not.
7. What consolation does Bolter hold for the acceptance of the image of the human being as a computer?

Benjamin Franklin

Benjamin Franklin (1706–1790), an international scientist, was known as America's first citizen of the world. He was also a printer, author, publisher, diplomat, inventor, and statesman.

The fifteenth of seventeen children, Franklin was the son of a soap- and candlemaker. Born in Boston, Franklin attended school for only two years. In his brother's printing shop, he learned the printmaker's art and worked at the trade as a journeyman in London before setting up his own shop in Philadelphia, a city then dominated by Quakers.

His contributions as a founder of America and as a patriot were immense. The most dramatic incident of his life, however, was a scientific, not a political, event: In 1752 he flew a silk kite into a thunderstorm and conducted lightning to earth through his own body. Franklin came close to losing his life, but, in truth, this experiment was unnecessary. Earlier that summer, unknown to Franklin, a Frenchman, using Franklin's own idea of an iron pole on a tower, had already demonstrated that lighting was a form of electricity. Franklin next invented the lightning rod. He further developed the theory, after many patient experiments, that electricity was a single fluid, appearing in two states; he called them "positive" and "negative," terms still in use today.

In many fields, Franklin's lists of accomplishments are lengthy and distinguished. The Franklin stove consists of a box of iron with openings on the side; it burns less wood and spreads heat more evenly than does a fireplace. Designed in 1740, the stove has gained a new generation of users today since the oil crisis of the seventies caused the cost of heating with gas or oil to rise. Bifocals, invented when Franklin was 77 to aid his own vision, consisted of spectacles split in half, the lenses for distant vision being on top and those for reading on the bottom.

Franklin demonstrated an active social conscience. He pointed out that lead poisoning was an industrial disease of the print shop. He organized police for Philadelphia, as well as lamplighters and firefighters. He made the sweeping and paving of streets a municipal responsibility and formed the first circulating American library. He established an academy that is now the University of Pennsylvania. He founded hospitals and, in 1743, the first American scientific society, the American Philosophical Society. He devised systems of postal deliveries. And he became the first president of the Pennsylvania Society for the Abolition of Slavery, the first such group in America.

Franklin's activity was based on his faith in fundamental research. He valued all knowledge whether its application was immediate or not. While watching a balloon ascension in France, he was asked what possible value this new toy could have. Franklin responded, "What good is a newborn baby?"

The following selection presents Franklin's concept of the lightning rod in a letter written to his friend Peter Collinson and conveyed to the Royal Society in London.

Of Lightning and the Method

Benjamin Franklin

Experiments made in electricity first gave philosophers a suspicion that the matter of lightning was the same with the electric matter. Experiments afterwards made on lightning obtained from the clouds by pointed rods, received into bottles, and subjected to every trial, have since proved this suspicion to be perfectly well founded; and that whatever properties we find in electricity, are also the properties of lightning.

This matter of lightning, or of electricity, is an extream subtile fluid, penetrating other bodies, and subsisting in them, equally diffused.

When by any operation of art or nature, there happens to be a greater proportion of this fluid in one body than in another, the body which has most will communicate to that which has least, till the proportion becomes equal; provided the distance between them be not too great; or, if it is too great, till there be proper conductors to convey it from one to the other.

If the communication be through the air without any conductor, a bright light is seen between the bodies, and a sound is heard. In our small experiments we call this light and sound the electric spark and snap; but in the great operations of nature, the light is what we call *lightning*, and the sound (produced at the same time, tho' generally arriving later at our ears than the light does to our eyes) is, with its echoes, called *thunder*.

If the communication of this fluid is by a conductor, it may be without either light or sound, the subtle fluid passing in the substance of the conductor.

If the conductor be good and of sufficient bigness, the fluid passes through it without hurting it. If otherwise, it is damaged or destroyed.

All metals, and water, are good conductors. Other bodies may become conductors by having some quantity of water in them, as wood, and other materials used in building, but not having much water in them, they are not good conductors, and therefore are often damaged in the operation.

Glass, wax, silk, wool, hair, feathers, and even wood, perfectly dry are nonconductors: that is, they resist instead of facilitating the passage of this subtle fluid.

When this fluid has an opportunity of passing through two conductors, one good, and sufficient, as of metal, the other not so good, it passes in the best, and will follow it in any direction.

The distance at which a body charged with this fluid will discharge itself suddenly, striking through the air into another body that is not charged, or not so highly charg'd, is different according to the quantity of the fluid, the dimensions and form of the bodies themselves, and the state of the air between them.

This distance, whatever it happens to be between any two bodies, is called their *striking distance*, as till they come within that distance of each other, no stroke will be made.

The clouds have often more of this fluid in proportion than the earth; in which case as soon as they come near enough (that is, within the striking distance) or meet with a conductor, the fluid quits them and strikes into the earth. A cloud fully charged with this fluid, if so high as to be beyond the striking distance from the earth, passes quietly without making noise or giving light; unless it meets with other clouds that have less.

Tall trees, and lofty buildings, as the towers and spires of churches, become sometimes conductors between the clouds and the earth; but not being good ones, that is, not conveying the fluid freely, they are often damaged.

Buildings that have their roofs covered with lead, or other metal, and spouts of metal continued from the roof into the ground to carry off the water, are never hurt by lightning, as whenever it falls on such a building, it passes in the metals and not in the walls.

When other buildings happen to be within the striking distance from such clouds, the fluid passes in the walls whether of wood, brick or stone, quitting the walls only when it can find better conductors near them, as metal rods, bolts, and hinges of windows or doors, gilding on wainscot, or frames of pictures; the silvering on the backs of looking-glasses; the wires for bells; and the bodies of animals, as containing watry fluids. And in passing thro' the house it follows the direction of these conductors, taking as many in its way as can assist it in its passage, whether in a strait or crooked line, leaping from one to the other, if not far distant from each other, only rending the wall in the spaces where these partial good conductors are too distant from each other.

An iron rod being placed on the outside of a building, from the highest part continued down into the moist earth, in any direction, strait or crooked, following the form of the roof or other parts of the building, will receive the lightning at its upper end, attracting it so as to prevent its striking any other part; and, affording it a good conveyance into the earth, will prevent its damaging any part of the building.

A small quantity of metal is found able to conduct a great quantity of this fluid. A wire no bigger than a goose quill, has been known to conduct (with safety to the building as far as the wire was continued) a quantity of lightning that did prodigious damage both above and below it; and probably larger rods are not necessary, tho' it is common in America, to make them of half an inch, some of three quarters, or an inch diameter.

The rod may be fastened to the wall, chimney, &c., with staples of iron. The lightning will not leave the rod (a good conductor), to pass into the wall (a bad conductor), through those staples. It would rather, if any were in the wall, pass out of it into the rod to get more readily by that conductor into the earth.

If the building be very large and extensive, two or more rods may be placed at different parts, for greater security.

Small ragged parts of clouds suspended in the air between the great body of clouds and the earth (like leaf gold in electrical experiments), often serve as partial conductors for the lightning, which proceeds from one of them to another, and by their help comes within the striking distance to the earth or a building. It therefore strikes through those conductors a building that would otherwise be out of the striking distance.

Long sharp points communicating with the earth, and presented to such parts of clouds, drawing silently from them the fluid they are charged with, they are then attracted to the cloud, and may leave the distance so great as to be beyond the reach of striking.

It is therefore that we elevate the upper end of the rod six or eight feet above the highest part of the building, tapering it gradually to a fine sharp point, which is gilt to prevent its rusting.

Thus the pointed rod either prevents a stroke from the cloud, or, if a stroke is made, conducts it to the earth with safety to the building.

The lower end of the rod should enter the earth so deep as to come at the moist part, perhaps two or three feet; and if bent when under the surface so as to go in a horizontal line six or eight feet from the wall, and then bent again downwards three or four feet, it will prevent damage to any of the stones of the foundation.

A person apprehensive of danger from lightning, happening during the time of thunder to be in a house not so secured, will do well to avoid sitting near the chimney, near a looking glass, or any gilt pictures or wainscot; the safest place is in the middle of the room, (so it be not under a metal lustre suspended by a chain) sitting in one chair and laying the feet up in another. It is still safer to bring two or three mattresses or beds into the middle of the room, and folding them up double, place the chair upon them; for they not being so good conductors as the walls, the lightning will not chuse an interrupted course through the air of the room and the bedding, when it can go thro' a continued better conductor the wall. But, where it can be had, a hamock or swinging bed, suspended by silk cords equally distant from the walls on every side, and from the ceiling and floor above and below, affords the safest situation a person can have in any room whatever; and what indeed may be deemed quite free from danger of any stroke by lightning.   B.F.

Questions and Topics for Discussion and Writing

1.   Describe the stories you had heard before reading this essay of Benjamin Franklin and his research into lightning. How are they like and unlike the information Franklin gives?
2.   What state of matter—solid, liquid, or gas—does Franklin believe lightning exists in?

3. How do Franklin's descriptions of lightning and its workings compare to our ideas of electricity and charged particles?
4. Describe the lightning rod as Franklin knew it. How correct was Franklin in his knowledge of electricity? Explain in a paragraph.

John Aristotle Phillips

What happens when a bright physics major—on academic probation—uses the declassified documents available to the public to outline the design for a crude atomic bomb? John Aristotle Phillips, the central figure in this real-life story that sounds like a parody of college life, describes the results this way: "Today, I'm the 'unthinkable reality': an undergraduate atomic bomb designer." Prior to his project, Phillips had been an ordinary student, delivering pizza, riding his unicycle, and playing the cowbell in the Princeton Marching Band. In four months, excluding everything else, Phillips researched and wrote a thirty-four-page report that earned him the title of "A-Bomb Kid."

In *Mushroom: The Story of the A-Bomb Kid* (1978), David Michaelis, a sensitive and witty humanities student who is Phillips's friend, recounts how Phillips fulfilled his academic requirements while at the same time making a statement about the dangers of nuclear arms proliferation and the need for weapons control. One reviewer comments that if arms control were the only topic, the book would be significant, but it is even more so because "it is about friendship, commitment to an idea, vitality, the value of seeing the funny side of life, power."

In the following selection, Freeman Dyson, Phillips's professor (see "A Scientific Apprenticeship," Unit 1), is the first person to grasp the "unthinkable reality." Counterpointing Phillips's and Michaelis's playful account of the project is an implied thesis of momentous threat and prophecy.

The A-Bomb Kid

John Aristotle Phillips and David Michaelis

When I return to Princeton, I have a short meeting with Freeman Dyson. He asks about my progress. I show him the documents I've brought back from Washington. He takes them from me very casually, the way an oblivious parent might ask a child, "What did you find when you were playing in the woods, dear?" and then suddenly realize he has a live snake in his hands.

"Good Lord," says Freeman. "Where did you get these?"

"At the National Technical Information Service."

He inspects them very closely. His face is a study in disbelief. "I was sure this kind of information was still classified."

"It only cost me twenty-five bucks."

"Good Lord," he says again.

I tell him about the den mother lady at the desk and what she said. He shakes his head, takes off his spectacles, and wipes them clean. "Well, all I can say is you've done some pretty good detective work, John. From now on you'll be on your own." He pauses for a moment and pats the documents. "These will help you far more than I can. I'll be expecting your design in five weeks."

"Do you think I have a good chance?"

"Let's just say I'll be expecting your design. Good luck."

When Freeman leaves he is visibly shaken. He has this distant, glazed look in his eyes, somewhere between shock and fright. . . .

Having witnessed the look in Freeman's eyes, I know I'm going to do it. The only question is: how? . . .

The material necessary to explode my bomb is plutonium 239, a man-made substance. It is the same element that was used in the bomb which destroyed Nagasaki in World War II. The reason why I must use plutonium—as opposed to, say, lead—is because plutonium is a heavy and *unstable* element: Its atomic structure is constantly changing. The best way to understand the instability of plutonium is to imagine a cliff by the side of a highway where the rocks are crumbling and occasionally falling, thereby dislodging other rocks. The constant decay of the cliff in time haphazardly breaks the rock formations apart, causing the cliff to be unstable. If a runaway avalanche were induced on the cliff by the use of explosives, the energy of the rocks falling would represent the action of plutonium atoms being split apart in the runaway chain reaction of an atomic bomb.

Visualize an atomic bomb as a marble inside a grapefruit inside a basketball inside a beachball. At the center of the bomb is the initiator, a marble-sized piece of metal. Around the initiator is a grapefruit-sized ball of plutonium 239. Wrapped around the plutonium is a three-inch reflector shield made of beryllium. High explosives shaped as breastlike lenses are placed in a symmetrical order around the beryllium shield. Wires are attached to each lens and each wire runs to an electrical source. By the time the lenses and wires have been installed, the bomb is about the size of a beachball.

When the electrical current runs through the wires to the lenses, an explosion is triggered. Because of the symmetrical nature of the placement of the explosives, a spherically imploding shock wave is set off, instantly squeezing the beryllium, plutonium, and initiator. The beryllium shield is pushed inward by the explosion, compressing the grapefruit-sized ball of plutonium to the size of a plum. The plutonium has now gone from a subcritical to a supercritical density, and the initiator at the center has been similarly squeezed. At this moment, the process of atoms fissioning—or splitting apart—begins.

Neutrons released from the initiator strike the plutonium atoms at an extremely fast rate. Each time a neutron hits a plutonium atom, the atom splits, creating two more neutrons, which in turn hit two more atoms, which split into four neutrons, which find four new atoms, thus splitting into eight neutrons, sixteen, thirty-two, sixty-four, one hundred and twenty-eight, two hundred and fifty-six, and so on. This tremendously fast splitting process is called a runaway chain reaction. If this does not occur with precise timing, the bomb will not explode with superior force. Each time an atom is split, a terrific amount of energy is released along with a variety of lethal atomic particles. The sum total of all atoms splitting in the chain reaction creates the atomic fireball which rises into the sky as a great mushroom-cloud energy release. This energy is comprised of heat waves, shock waves, and lethal atomic particles that are exploded outward across the countryside.

There are many subtleties involved in the explosion of an atomic bomb. Most of them center around the actual detonation of explosives surrounding the beryllium shield. Normally, only a very small fraction of the plutonium has a chance to fission before the bomb assembly is blown apart by the energy released from the atoms already split. For this reason, the size of the final mushroom cloud explosion depends upon how long the chain reaction of splitting atoms can be sustained. The timing and efficiency of the chain reactions is in turn determined by the specific arrangement of the explosives surrounding the plutonium.

The grouping of these explosives around the plutonium is one of the most highly classified aspects of the atomic bomb and the area around which the most intense postwar espionage was centered. (Julius and Ethel Rosenberg were executed in 1953 allegedly for passing this secret to the Russians.) In the bombs exploded over Hiroshima and Nagasaki, TNT was the explosive used to set off the chain reaction. But since that time, the U.S. Army has designed more sophisticated and efficient explosives to do the job. How to arrange the explosives and which explosives to use pose the biggest problems for me as I begin to design my bomb. The correct arrangement around the plutonium is the critical factor without which my design will be worthless.

My base of operations is a small room on the second floor of Ivy Club. The large conference table in the center of the room is covered with my books, calculators, design paper, notes. . . . My sleeping bag is rolled out on the floor. The ashtrays are full. As the next three weeks go by, I stop going to classes altogether. I work all day and all night in the room. My progress is slow at first. I pretend not to listen to the noise coming from the other rooms in the club. Sometimes I find myself in a trance, staring out at the first buds on the magnolia trees beyond the leaded windows. I drink coffee from a thermos, and when I feel drowsy I go downstairs to the kitchen to refill the thermos. I stop eating at regular meals, taking only a bologna sandwich when I'm hungry. The other members at Ivy begin referring to me as the Hobo because of my unshaven face and disheveled appearance. I develop a terrible case of bloodshot eyes.

Sleep comes rarely.

I devise several mind-sets that help me solve some of the problems I come up against. In my work (and even in my dreams now), I place myself in the position of a terrorist. I approach evey problem from a terrorist's point of view. The bomb must be inexpensive to construct, simple in design, and small enough to sit unnoticed in the trunk of a car or abandoned U-Haul trailer.

At other times, I put myself in the shoes of the Los Alamos scientists working on the first bombs. By closely following the technical accounts of their progress which I bought in Washington, I design the bomb as they did, working on each component separately, one at a time. But the one terrific advantage I share with the nations which developed the bomb after the United States is that I know it can be done. The scientists at Los Alamos were not sure that such a bomb would be successful until after the first Alamogordo test. Furthermore, according to *The Curve of Binding Energy*, every bomb that has been designed and built since 1945 has worked on the first try.

Questions and Topics for Discussion and Writing

1. Does the discussion of Phillips's "base of operations" and life while beginning his work reflect the preparation of most students when asked to write a paper? In a short paper, describe the process of writing one of your papers.
2. Select one other writer's account of his working methods. Write a paragraph in which you compare this person's "mind-set" with that of Phillips.
3. In the light of the way Phillips obtained the information, what are the implications of the parenthetical statement "Julius and Ethel Rosenberg were executed in 1953 allegedly for passing this secret to the Russians"?
4. How does Phillips differentiate his approach to developing the atom bomb design from that of the Los Alamos scientists?
5. What appropriate attitude does Phillips suggest for apprentice scientists?
6. The metaphor of child's play permeates the opening of the essay. What is the significance of Phillips's use of this metaphor in discussing how he obtained declassified documents?

Robert A. Day

Robert A. Day (b. 1924) is Vice President of the Institute for Scientific Information, the world's largest commercial producer of information services covering the professional literature. Day is also the Director of ISI Press, a subsidiary of the Institute for Scientific Information in Philadelphia, Pennsylvania. ISI Press publishes books concerned with scholarly communication in the sciences, social sciences, arts, and humanities.

Day, born in Belvidere, Illinois, attended the University of Illinois, receiving his bachelor's degree in English in 1949. He then attended Columbia University, receiving a master's degree in Library Service in 1951.

After being employed as a librarian in the Science and Technology Division of the Newark (New Jersey) Public Library from 1951 to 1953, Day was employed as Librarian–Editor of the Institute of Microbiology at Rutgers University until 1960. In 1960–1961, he served as Director of the Library, College of South Jersey, Rutgers University.

In 1961, he assumed direction of the publishing program of the American Society for Microbiology. In that capacity he served as Managing Editor of the *Journal of Bacteriology* and eight other journals published by ASM, and he directed a substantial book-publishing program. He held this post for 19 years prior to assuming the Directorship of ISI Press in 1980.

Day has written a number of articles and reports on various phases of scientific writing, editing, and publishing. The following selection is from his book *How to Write and Publish a Scientific Paper,* considered definitive in its field.

What Is a Scientific Paper?

Robert A. Day

DEFINITION OF A SCIENTIFIC PAPER

A scientific paper is a written and published report describing original research results. That short definition must be qualified, however, by noting that a scientific paper must be written in a certain way and it must be published in a certain way, as defined by three centuries of developing tradition, editorial practice, scientific ethics, and the interplay of printing and publishing procedures.

To properly define "scientific paper," we must define the mechanism that creates a scientific paper, namely, valid publication. Abstracts, theses, conference reports, and many other types of literature are published, but such publications do not normally meet the test of valid publication. Further, even if a scientific paper meets all of the other tests, it is not validly published if it is published in the wrong place. That is, a relatively poor research report, but one that meets the tests, is validly published if accepted and published in the right place (a primary journal, usually); a superbly prepared research report is not validly published if published in the wrong place. Most of the government report literature and conference literature, as well as house organs and other ephemeral publications, do not qualify as primary literature.

Many people have struggled with the definition of "valid publication," from which is derived the definition of "scientific paper." The Council of Biology Editors (CBE), an authoritative professional organization (in biology, at least) dealing with such problems, arrived at the following definition. . . .

> An acceptable primary scientific publication must be the first disclosure containing sufficient information to enable peers (1) to assess observations, (2) to repeat experiments, and (3) to evaluate intellectual processes; moreover, it must be susceptible to sensory perception, essentially permanent, available to the scientific community without restriction, and available for regular screening by one or more of the major recognized secondary services (e.g., currently, Biological Abstracts, Chemical Abstracts, Index Medicus, Excerpta Medica, Bibliography of Agriculture, etc., in the United States and similar facilities in other countries).

At first reading, this definition may seem excessively complex, or at least verbose. But those of us who had a hand in drafting it weighed each word carefully, and we doubt that an acceptable definition could be provided in appreciably fewer words. Because it is important that students, authors, editors, and all others concerned understand what a scientific paper is and what it is not, let us work our way through this definition to see what it really means.

"An acceptable primary scientific publication" starts out as the defined substantive, but this gives way to "the first disclosure," which the rest of the paragraph defines. Certainly, first disclosure of new research data often takes place via oral presentation at a scientific meeting. But, the thrust of the CBE statement is that disclosure is more than disgorgement by the author; effective first disclosure is accomplished *only* when the disclosure takes a form that allows the peers of the author to comprehend (either now or in the future) that which is disclosed.

Thus, sufficient information must be presented so that potential users of the data can (i) assess observations, (ii) repeat experiments, and (iii) evaluate intellectual processes. (Are the author's conclusions justified by the data?) Then, the disclosure must be "susceptible to sensory perception." This may seem an awkward phrase, because in normal practice it simply means publication; however, this definition provides for disclosure not just in terms of visual materials (printed journals, microfilm, microfiche) but also perhaps in nonprint, nonvisual forms. For example, "publication" in the form of audio cassettes, if that publication met the other tests provided in the definition, would constitute effective publication. In the future, it is quite possible that first disclosure will be entry into a computer data base.

Regardless of the form of publication, that form must be essentially permanent, must be made available to the scientific community without restriction, and must be made available to the information retrieval system (*Biological Abstracts, Chemical Abstracts, Index Medicus, Science Citation Index, etc.*). Thus, publications such as newsletters and house organs, many of which are of value for their news or other features, cannot serve as repositories for scientific knowledge.

To restate the CBE definition in simpler but not more accurate terms, a scientific paper is (i) the first publication of original research results, (ii) in a form whereby peers of the author can repeat the experiments and test the conclusions, and (iii) in a journal or other source document which is readily available within the scientific community. To understand this definition, however, we must add an important caveat. In modern science (since about the 1930s) the part of the definition that refers to "peers of the author" is accepted as meaning prepublication peer review. Thus, by definition, scientific papers are published in peer-reviewed publications.

I have belabored this question of definition for two very good reasons. First, the entire community of science has long labored with an inefficient, costly system of scientific communication precisely because it (authors, editors, publishers) has been unable or unwilling to define primary publication. As a result, much of the literature is buried in meeting abstracts, obscure conference reports, government publications, or in books or journals of minuscule circulation. Other papers, in the same or slightly altered form, are published twice or more often; occasionally, this is the result of poor ethics on the part of the author, but more often it is the lack of definition as to which conference reports, books, and compilations are (or should be) primary publications and which are not. Redundancy and confusion result.

Second, a scientific paper is, by definition, a particular kind of document containing certain specified kinds of information. A scientific paper "demands exactly the same qualities of thought as are needed for the rest of science: logic, clarity, and precision." . . . If the graduate student or the budding scientist (and even some of those scientists who have already published many papers) can fully grasp the significance of this definition, the writing task should be a good deal easier. Confusion results from an amorphous task. The easy task is the one in which you know exactly what must be done and in exactly what order it must be done.

ORGANIZATION OF A SCIENTIFIC PAPER

A scientific paper is a paper organized to meet the needs of valid publication. A scientific paper is, or should be, highly stylized, with distinctive and clearly evident component parts. Each scientific paper should have, in proper order, its Introduction, Materials and Methods, Results, and Discussion. Any other order will pose hurdles for the reader and probably the writer. "Good organization is the key to good writing." . . .

I have taught and recommended this prescribed order for many years. Until recently, however, there have been several somewhat different systems of organization that were preferred by some journals and some editors. The tendency toward uniformity has increased since 1972, when the order cited above was prescribed as a standard. . . .

This order is so eminently logical that, increasingly, it is used for many other types of expository writing. Whether one is writing an article about chemistry, archeology, economics, or crime in the streets, an effective way to proceed is to answer these four questions, in order: (i) What was the problem? Your answer is the *Introduction*. (ii) How did you study the problem? Your answer is the *Materials and Methods*. (iii) What did you find? Your answer is the *Results*. (iv) What do these findings mean? Your answer is the *Discussion*. . . .

The well-written scientific paper should report its original data in an organized fashion and in appropriate language. . . .

In short, I take the position that the preparation of a scientific paper has almost nothing to do with writing, per se. It is a question of *organization*. A scientific paper is not "literature." The preparer of a scientific paper is not really an "author" in the literary sense. In fact, I go so far as to say that, if the ingredients are properly organized, the paper will virtually write itself.

Some of my old-fashioned colleagues think that scientific papers should be literature, that the style and flair of an author should be clearly evident, and that variations in style encourage the interest of the reader. I disagree. I think scientists should indeed be interested in reading literature, and perhaps even in writing literature, but the communication of research results is a more prosaic procedure. As Booth . . . put it, "Grandiloquence has no place in scientific writing."

Today, the average scientist, to keep up with a field, must examine the data reported in a very large number of papers. Therefore, scientists and, of course, editors must demand a system of reporting data that is uniform, concise, and readily understandable.

I once heard it said: "A scientific paper is not designed to be read. It is designed to be published." Although this was said in jest, there is much truth to it. And, actually, if the paper is designed to be published, it will also be in a prescribed form that can be read and its contents can be grasped quickly and easily by the reader.

LANGUAGE OF A SCIENTIFIC PAPER

In addition to organization, the second principal ingredient of a scientific paper should be appropriate language. . . . I keep emphasizing proper use of English, because in this area most scientists have trouble.

If scientific knowledge is at least as important as any other knowledge, then it must be communicated effectively, clearly, in words of certain meaning. The scientist, to succeed in this endeavor, must therefore be literate. David B. Truman, when he was Dean of Columbia College, said it well: "In the complexities of contemporary existence the specialist who is trained but uneducated, technically skilled but culturally incompetent, is a menace."

Although the ultimate goal of scientific research is publication, it has always been amazing to me that so many scientists neglect the responsibilities involved. A scientist will spend months or years of hard work to secure his data, and then unconcernedly let much of their value be lost because of his lack of interest in the communication process. The same scientist who will overcome tremendous obstacles to carry out a measurement to the fourth decimal place will be in deep slumber while a secretary is casually changing his micrograms per milliliter to milligrams per milliliter and while the printer slips in an occasional pounds per barrel.

Language need not be difficult. In scientific writing, we say: "The best English is that which gives the sense in the fewest short words" (a dictum printed for some years in the "Instructions to Authors" of the *Journal of Bacteriology*). Literary tricks, metaphors and the like, divert attention from the message to the style. They should be used rarely, if at all, in scientific writing. Justin Leonard, assistant conservation director of Michigan, once said: "The Ph.D. in science can make journal editors quite happy with plain, unadorned, eighth-grade level composition" (*Bio-Science*, September 1966).

OTHER DEFINITIONS

At this point, we have defined a scientific paper. . . . Before we proceed, however, a few related definitions may be helpful.

Let us preserve "scientific paper" as the term for an original research report. How should this be distinguished from research reports that are not original, or not scientific, or somehow fail to qualify as scientific papers? Let us look at several specific terms: "review paper," "conference report," and "meeting abstract."

A review paper may review almost anything, most typically the recent work in a defined subject area or the work of a particular individual or group. Thus, the review paper is designed to summarize, analyze, evaluate, or synthesize information that *has already been published* (research reports in primary journals). Although much or all of the material in a review paper has previously been published, the spectre of dual publication does not normally arise because the review nature of the work is usually obvious (often in the title of the publication, such as *Microbiological Reviews, Annual Review of Biochemistry*, etc.).

A conference report is a paper published in a book or journal as part of the proceedings of a symposium, national or international congress, workshop, round table, or the like. Such conferences are normally not designed for the presentation of original data, and the resultant proceedings (book or journal) do not qualify as primary publications. Conference presentations are often review papers, presenting reviews of the recent work of particular scientists or recent

work in particular laboratories. Some of the material reported at some conferences (especially the exciting ones) is in the form of preliminary reports, in
which new, original data are reported, often accompanied by interesting speculation. But, usually, these preliminary reports do not qualify, nor are they
intended to qualify, as scientific papers. Later, often much later, such work is
validly published in a primary journal; by this time, the loose ends have been
tied down, all essential experimental details are recorded (so that a competent
worker could repeat the experiments), and the speculations are now recorded
as conclusions.

Therefore, the vast conference literature that appears in print normally is not
primary. If original data are presented in such contributions, the data can and
should be published (or republished) in an archival (primary) journal. Otherwise, the information may be in essence lost. If publication in a primary journal
follows publication in a conference report, there may be copyright and permission problems affecting portions of the work . . . , but the more fundamental
problem of dual publication normally does not and should not arise.

Meeting abstracts, like conference proceedings, are of several widely varying
types. Conceptually, however, they are similar to conference reports in that
they can and often do contain information. They are not primary publications,
nor should publication of an abstract be considered as a bar to later publication
of the full report.

In the past, there has been little confusion regarding the typical one-para-
graph abstracts published as part of the program or distributed along with the
program of a national meeting or international congress. It was usually understood that the papers presented at these meetings would later be submitted for
publication in primary journals. More recently, however, there has been a
strong trend towards extended abstracts (or "synoptics"). Because publishing
all of the full papers presented at a large meeting, such as a major international
congress, is very expensive and because such publication is still not a substitute
for the valid publication offered by the primary journal, the movement to extended abstracts makes a great deal of sense. The extended abstract can supply
virtually as much information as a full paper; basically, what it lacks is the
experimental detail. However, precisely because it lacks experimental detail, it
cannot qualify as a scientific paper.

Those of us who are involved with publishing these materials see the importance of careful definition of the different types of papers. More and more
publishers, conference organizers, and individual scientists are beginning to
agree on these basic definitions and their general acceptance of such definitions
will greatly clarify both primary and secondary communication of scientific information.

I have tried to answer the question "What is a scientific paper?" Perhaps a
better answer, certainly a more succinct one, was provided by the wag who
described a drug as "any substance which, when injected into a laboratory rat,
produces a scientific paper."*

*This final paragraph is from the first edition.

Questions and Topics for Discussion and Writing

1. What is the primary requirement for any good paper, according to Day?
2. Study Day's definition of a scientific paper. Keep in mind that a definition presents, first, the class to which the subject belongs and, second, the characteristics that distinguish it from other members of that class. Does Day fulfill both of these criteria?
3. Examine the second paragraph. Discuss how it fits the requirement of a well-developed paragraph—the topic sentence, development of sentences, and concluding statement. What type of development does Day use in the paragraph?
4. Draft a paragraph, using Paragraph 2 as a model, in which you state what type of paper an undergraduate student might be required to write.
5. What do you think of Day's assertion that a manuscript may fail to meet the definition of a scientific paper "if it is published in the wrong place"?
6. Consider the definition of "valid publication." Would you modify any "awkward phrase," or add or delete an idea?
7. Does this essay evidence good organization? Write a well-developed paragraph, supporting your judgment.
8. What stress does Day place on the communication process in the life of the research scientist?

Edward Morin

Edward Morin, writer and teacher (b. 1934), grew up in Chicago, a block away from Burnham Park and Lake Michigan. He received a bachelor's degree in philosophy from Maryknoll College in Illinois, a master's in English from the University of Chicago, and a doctorate in English from Loyola University (Chicago). He has taught English at universities in Kentucky, Ohio, and Michigan; currently, he writes documentation for computer software.

Since 1966, Morin has lived in or near Detroit, the heartland of the American auto industry, a setting that he uses in his poetry. Exploring a range of human subjects—the economy, unemployment, parenting, academic and urban situations—his poetry has appeared in such journals as *Hudson Review, Ploughshares, Poetry Northwest,* and in more than twenty anthologies. His book *The Dust of Our City* (1978) presents, in the words of one reviewer, "hard subjects with respect, wit, and craftsmanship."

Morin is co-translator of a sequence of poems by Nobel laureate Odysseus Elytis, entitled *The Primal Sun,* and is a prolific book reviewer with more than 200 published book reviews.

One summer during his college years, Morin worked at a gas-cracking laboratory at Illinois Institute of Technology, an experience on which he draws in "Filling Station." The poem alludes to the popular figurative expressions *dinosaur* and *gas guzzler* for a large, usually American-made auto.

Filling Station

Edward Morin

Night miles force us to a self-serve
where florescents arc from yardarms.
Near the pumps we join a cluster
who pretend to bump one another
like fish in an overcrowded tank.

In this all-night diner for cars
a tousled blear-eyed couple hug theirs,
two smoking serious drinkers idle,
a thin stetsoned pimp kills time,
and the youth in the plastic cage deals.

Digital numbers click off gallons
of extinct ferns, fossilized ginkgoes,
prevertebrates battered by cracking
processes—all distant ancestors
of our gurgling their final rage

before cremation. Myriad
internal combustion dinosaurs
line up to guzzle what remains.
They inch forth for heat and light,
waiting for their time slot to expire.

Questions and Topics for Discussion and Writing

1. How does the image of the ginkgo fossils fit the meaning and mood of this poem
    better than some other kind of fossilized tree might have done?
2. Is the focus of this poem on large cars only? Automobiles in general? Industrial
    technology itself?
3. Have public attitudes toward kinds and quality of automobiles changed much since
    this poem was written in 1979?

Samuel C. Florman

A practicing engineer who writes on the relationship between engineering and the liberal arts, Samuel C. Florman was born in New York City (1925) and is Vice President and General Manager of the Kreisler-Borg Construction Company of New York State. After graduating from the Fieldston School and receiving his Bachelor of Science from Dartmouth College in 1944, Florman served in the United States Navy for two years, then completed his master's degree from Columbia University. Florman credits these three institutions with establishing the respect he holds for both technology and the humanities.

The recipient of an award from the Stevens Institute of Technology for his articles and books on engineering and general culture, Florman has long been a contributing editor of *Harper's*. In the following selections, Florman reveals both his sanguine belief in the positive qualities of his profession and his conviction that familiarity with the humanities will enrich his fellow engineers. He works toward "enlightened engineering leadership" and the "coming of a civilized technology." Florman's writings serve—in a metaphor he himself employs—as a "bridge" between what are often called the "two cultures."

The Existential Engineer

Samuel Florman

Sisyphus was condemned by the gods to forever roll a huge stone up a mountain, only to see it fall back to the bottom each time he reached the summit. Albert Camus has depicted this mythical figure as the archetypical existential hero. Sisyphus has no illusions to sustain him, no hope that some day his labors will end. But he has pride and courage and the satisfaction that comes instinctively to a person undertaking a task. "The struggle itself toward the heights is enough to fill a man's heart," concludes Camus. "One must imagine Sisyphus happy."

Theodore Roszak has expressed dismay that Camus, "the most great-hearted of our humanist heroes," cannot find in life "a project any less grotesquely absurd than the labor of Sisyphus taking his stone once again up the hill." But Roszak is taking the symbol of the stone too literally. Of course we aim for more than rolling a stone up a hill. But we are beginning to realize that for mankind there will never be a time to rest at the top of the mountain. There will be no new arcadian age. There will always be new burdens, new problems, new failures, new beginnings. And the glory of man is to respond to his harsh fate with zest and ever-renewed effort.

This is why Sisyphus can serve as a symbol of the modern engineer. Today's engineer has lost faith in the utopia that engineers of an earlier age thought they were bringing to mankind. Yet his work, springing as it does from the most basic impulse of humanity, can fill him with existential joy.

That there will be no utopia has become clear beyond questioning. Human beings are too varied, too fickle, and too willful. Technologically oriented optimists like Buckminster Fuller may excite us with visions of glass-domed paradises humming with computers. Humanists like René Dubos may enchant us with tales (featured on *The New York Times* Op-Ed page) of isolated human societies living rich and happy lives under conditions of primitive simplicity. These ideals are interesting, inspiring, and comforting. But we know that they are ideals—perhaps only mirages—that cannot become reality for us. In fact, they are evidence of the differences that surface whenever people start to consider what constitutes the good life.

We have, in our new wisdom and humility, stopped talking about "progress." Except for a few elemental humanitarian concepts, such as not wanting anyone to starve or freeze, we simply cannot agree on which way we want to go. We are talking a lot about "trade-offs," since we have learned that the pursuit of many different worthy objectives results inevitably in conflict, and that these conflicts can only be resolved by compromise—or by force. The engineer does not underestimate the importance of his contributions to society; but he has abandoned all messianic illusions. He acknowledges that he has made mistakes; but he rejects totally the image of himself as villain, false prophet, or sorcerer's apprentice. He is a human being doing what human beings are created to do: fulfilling his human destiny both biologically and spiritually, and finding his reward in existential pleasure.

This pleasure is not solely the instinctual satisfaction of a lion on the hunt or a beaver building his dam. Pure instinctual gratification is involved, but only as a part of a complex whole. *Homo faber* does not merely putter around, nor is he interested only in survival and comfort. He shares the values and ideals of the human race—mercy, justice, reverence, beauty, and the like. But he feels that these abstract concepts become meaningful only in a world where people lead authentic lives—struggling, questing, and creating.

Questions and Topics for Discussion and Writing

1. What is the "existential pleasure" that modern engineers will find as a reward for their work?
2. What are the "messianic illusions" that engineers have abandoned today?

3. What does Florman suggest about the view of their role by engineers in the past when he says that engineers today reject the image of themselves as "sorcerer's apprentices"?
4. Read the chapter on existentialism in *Backgrounds of American Literary Thought*, or in any work that outlines the characteristics of existentialism. Write a paper in which you define what existentialists mean by the "authentic life."
5. Who or what is *Homo faber?*
6. Read Albert Camus's "Myth of Sisyphus." Imagine Sisyphus to be an engineer, and the rock a particular task that could be undertaken by an engineer. Write a paper in which you illustrate how the engineer, in performing his or her task, is living the authentic life, as Camus suggests Sisyphus is doing.

Technology and the Tragic View

Samuel Florman

House & Garden magazine, in celebration of the American Bicentennial, devoted its July 1976 issue to the topic "American Know-How." The editors invited me to contribute an article, and enticed by the opportunity to address a new audience, plus the offer of a handsome fee, I accepted. We agreed that the title of my piece would be "Technology and the Human Adventure," and I thereupon embarked on a strange adventure of my own.

I thought that it would be appropriate to begin my Bicentennial-inspired essay with a discussion of technology in the time of the Founding Fathers, so I went to the library and immersed myself in the works of Benjamin Franklin, surely the most famous technologist of America's early days. Remembering stories from my childhood about Ben Franklin the clever tinkerer, I expected to find a pleasant recounting of inventions and successful experiments, a cheering tale of technological triumphs. I found such a tale, to be sure, but along with it I found a record of calamities *caused by* the technological advances of his day.

In several letters and essays, Franklin expressed concern about fire, an ever-threatening scourge in Colonial times. Efficient sawmills made it possible to build frame houses, more versatile and economical than log cabins—but less fire-resistant. Advances in transport made it possible for people to crowd these frame houses together in cities. Cleverly conceived fireplaces, stoves, lamps, and warming pans made life more comfortable, but contributed to the likelihood of catastrophic fires in which many lives were lost.

To deal with this problem, Franklin recommended architectural modifications to make houses more fireproof. He proposed the licensing and supervision of chimney sweeps and the establishment of volunteer fire companies, well supplied and trained in the science of firefighting. As is well known, he invented the lightning rod. In other words, he proposed technological ways of coping with the unpleasant consequences of technology. He applied Yankee ingenuity to solve problems arising out of Yankee ingenuity.

In Franklin's writings I found other examples of technological advances that brought with them unanticipated problems. Lead poisoning was a peril. Contaminated rum was discovered coming from distilleries where lead parts had been substituted for wood in the distilling apparatus. Drinking water collected from lead-coated roofs was also making people seriously ill.

The advancing techniques of medical science were often a mixed blessing, as they are today. Early methods of vaccination for smallpox, for example, entailed the danger of the vaccinated person dying from the artificially induced disease. (In a particularly poignant article, Franklin was at pains to point out that his four-year-old son's death from smallpox was attributable to the boy's *not* having been vaccinated and did not result, as rumor had it, from vaccination itself.)

After a while, I put aside the writings of Franklin and turned my attention to American know-how in the nineteenth century. I became engrossed in the story of the early days of steamboat transport. This important step forward in American technology was far from being the unsullied triumph that it appears to be in our popular histories.

Manufacturers of the earliest high-pressure steam engines often used materials of inferior quality. They were slow to recognize the weakening of boiler shells caused by rivet holes, and the danger of using wrought-iron shells together with cast iron heads that had a different coefficient of expansion. Safety valve openings were often not properly proportioned, and gauges had a tendency to malfunction. Even well-designed equipment quickly became defective through the effects of corrosion and sediment. On top of it all, competition for prestige led to racing between boats, and during a race the usual practice was to tie down the safety valve so that excessive steam pressure would not be relieved.

From 1825 to 1830, 42 recorded explosions killed upward of 270 persons. When, in 1830, an explosion aboard the *Helen McGregor* near Memphis killed more than 50 passengers, public outrage forced the federal government to take action. Funds were granted to the Franklin Institute of Philadelphia to purchase apparatus needed to conduct experiments on steam boilers. This was a notable event, the first technological research grant made by the federal government.

The institute made a comprehensive report in 1838, but it was not until 14 years later that a workable bill was passed by Congress providing at least minimal safeguards for the citizenry. Today we may wonder why the process took so long, but at the time Congress was still uncertain about its right, under the interstate commerce provision of the Constitution, to control the activities of individual entrepreneurs.

When I turned from steamboats to railroads I found another long-forgotten story of catastrophe. Not only were there problems with the trains themselves, but the roadbeds, and particularly the bridges, made even the shortest train journey a hazardous adventure. In the late 1860s more than 25 American bridges were collapsing each year, with appalling loss of life. In 1873 the American Society of Civil Engineers set up a special commission to address the problem, and eventually the safety of our bridges came to be taken for granted.

The more I researched the history of American know-how, the more I perceived that practically every technological advance had unexpected and unwanted side effects. Along with each triumph of mechanical genius came an inevitable portion of death and destruction. Instead of becoming discouraged, however, our forebears seemed to be resolute in confronting the adverse consequences of their own inventiveness. I was impressed by this pattern of progress/setback/renewed-creative-effort. It seemed to have a special message for our day, and I made it the theme of my essay for *House & Garden*.

No matter how many articles one has had published, and no matter how much one likes the article most recently submitted, waiting to hear from an editor is an anxious experience. In this case, as it turned out, I had reason to be apprehensive. I soon heard from one of the editors who, although she tried to be encouraging, was obviously distressed. "We liked the part about tenacity and ingenuity," she said, "but, oh dear, *all those disasters*—they are so depressing."

I need not go into the details of what followed: the rewriting, the telephone conferences, the re-rewriting—the gradual elimination of accidents and casualty statistics, and a subtle change in emphasis. I retreated, with some honor intact I like to believe, until the article was deemed to be suitably upbeat.

I should have known that the Bicentennial issue of *House & Garden* was not the forum in which to consider the dark complexities of technological change. My piece was to appear side by side with such articles as "A House That Has Everything," "Live Longer, Look Younger," and "Everything's Coming Up Roses" (devoted to a review of Gloria Vanderbilt's latest designs).

In the United States today magazines like *House & Garden* speak for those, and to those, who are optimistic about technology. Through technology we get better dishwashers, permanent-press blouses, and rust-proof lawn furniture. "Better living through chemistry," the old du Pont commercial used to say. Not only is *House & Garden* optimistic, that is, hopeful, about technology; it is cheerfully optimistic. There is no room in its pages for failure, or even for struggle, and in this view it speaks for many Americans, perhaps a majority. This is the lesson I learned—or I should say, relearned—in the Bicentennial year.

Much has been written about the shallow optimism of the United States: about life viewed as a Horatio Alger success story or as a romantic movie with a happy ending. This optimism is less widespread than it used to be, particularly as it relates to technology. Talk of nuclear warfare and a poisoned environment tends to dampen one's enthusiasm. Yet optimistic materialism remains a powerful force in American life. The poll-takers tell us that people believe technology is, on balance, beneficial. And we all know a lot of people who, even at this troublesome moment in history, define happiness in terms of their ability to accumulate new gadgets. The business community, anxious to sell merchandise, spares no expense in promoting a gleeful consumerism.

Side by side with what I have come to think of as *House & Garden* optimism, there is a mood that we might call *New York Review of Books* pessimism. Our intellectual journals are full of gloomy tracts that depict a society debased by technology. Our health is being ruined, according to this view, our landscape despoiled, and our social institutions laid waste. We are forced to do demeaning work and consume unwanted products. We are being dehumanized. This is happening because a technological demon has escaped from human control or, in a slightly different version, because evil technocrats are leading us astray.

It is clear that in recent years the resoluteness exhibited by Benjamin Franklin, and other Americans of similarly robust character, has been largely displaced by a foolish optimism on the one hand and an abject pessimism on the other. These two opposing outlooks are actually manifestations of the same defect in the American character. One is the obverse, the "flip side," of the other. Both reflect a flaw that I can best describe as immaturity.

A young child is optimistic, naively assuming that his needs can always be satisfied and that his parents have it within their power to "make things right." A child frustrated becomes petulant. With the onset of puberty a morose sense of disillusionment is apt to take hold. Sulky pessimism is something we associate with the teenager.

It is not surprising that many inhabitants of the United States, a rich nation with seemingly boundless frontiers, should have evinced a childish optimism, and declared their faith in technology, endowing it with the reassuring power of a parent—also regarding it with the love of a child for a favorite toy. It then follows that technological setbacks would be greeted by some with the naive assumption that all would turn out for the best and by others with peevish declarations of despair. Intellectuals have been in the forefront of this childish display, but every segment of society has been caught up in it. Technologists themselves have not been immune. In the speeches of nineteenth-century engineers, we find bombastic promises that make us blush. Today the profession is torn between a blustering optimism and a confused guilt.

The past 50 years have seen many hopes dashed, but we can see in retrospect that they were unrealistic hopes. We simply cannot make use of coal without killing miners and polluting the air. Neither can we manufacture solar panels without worker fatalities and environmental degradation. (We assume that it will be less than with coal, but we are not sure.) We cannot build highways or canals or airports without despoiling the landscape. Not only have we learned that environmental dangers are inherent in every technological advance, but we find that we are fated to be dissatisfied with much of what we produce because our tastes keep changing. The sparkling, humming, paved metropolises of science fiction—even if they could be realized—are not, after all, the home to which humankind aspires. It seems that many people find such an environment "alienating." There can never be a technologically-based Utopia because we discover belatedly that we cannot agree on what form that Utopia might take.

To express our disillusionment we have invented a new word: "tradeoff." It is an ugly word, totally without grace, but it signifies, I believe, the beginning of maturity for American society.

It is important to remember that our disappointments have not been limited to technology. (This is a fact that the antitechnologists usually choose to ignore.) Wonderful dreams attended the birth of the New Deal, and later the founding of the United Nations, yet we have since awakened to face unyielding economic and political difficulties. Socialism has been discredited, as was laissez-faire capitalism before it. We have been bitterly disappointed by the labor movement, the educational establishment, efforts at crime prevention, the ministrations of psychiatry, and most recently by the abortive experiments of the so-called counterculture. We have come face to face with *limits* that we had presumed to hope might not exist.

Those of us who have lived through the past 50 years have passed personally from youthful presumptuousness to mature skepticism at the very moment that American society has been going through the same transition. We have to be careful not to define the popular mood in terms of our personal sentiments, but I do not think I am doing that when I observe the multiple disenchantments of our time. We also have to be careful not to deprecate youthful enthusiasm, which is a force for good, along with immaturity, which is tolerable only in the young.

It can be argued that there was for a while good reason to hold out hope for Utopia, since modern science and technology appeared to be completely new factors in human existence. But now that they have been given a fair trial, we perceive their inherent limitations. The human condition is the human condition still.

To persist in saying that we are optimistic or pessimistic about technology is to acknowledge that we will not grow up.

I suggest that an appropriate response to our new wisdom is neither optimism nor pessimism, but rather the espousal of an attitude that has traditionally been associated with men and women of noble character—the tragic view of life.

As a student in high school, and later in college, I found it difficult to comprehend what my teachers told me about comedy and tragedy. Comedy, they said, expresses despair. When there is no hope, we make jokes. We depict people as puny, ridiculous creatures. We laugh to keep from crying.

Tragedy, on the other hand, is uplifting. It depicts heroes wrestling with fate. It is man's destiny to die, to be defeated by the forces of the universe. But in challenging his destiny, in being brave, determined, ambitious, resourceful, the tragic hero shows to what heights a human being can soar. This is an inspiration to the rest of us. After witnessing a tragedy we feel good, because the magnificence of the human spirit has been demonstrated. Tragic drama is an affirmation of the value of life.

Students pay lip service to this theory and give the expected answers in examinations. But sometimes the idea seems to fly in the face of reason. How can we say we feel better after Oedipus puts out his eyes, or Othello kills his beloved wife and commits suicide, than we do after laughing heartily over a bedroom farce?

Yet this concept, which is so hard to grasp in the classroom, where students are young and the environment is serene, rings true in the world where mature people wrestle with burdensome problems.

I do not intend to preach a message of stoicism. The tragic view is not to be confused with world-weary resignation. As Moses Hadas, a great classical scholar of a generation ago, wrote about the Greek tragedians: "Their gloom is no fatalistic pessimism but an adult confrontation of reality, and their emphasis is not on the grimness of life but on the capacity of great figures to adequate [sic] themselves to it."

It is not an accident that tragic drama flourished in societies that were dynamic: Periclean Athens, Elizabethan England, and the France of Louis XIV. For tragedy speaks of ambition, effort, and unquenchable spirit. Technological creativity is one manifestation of this spirit, and it is only a dyspeptic antihumanist who can feel otherwise. Even the Greeks, who for a while placed technologists low on the social scale, recognized the glory of creative engineering. Prometheus is one of the quintessential tragic heroes. In viewing technology through a tragic prism we are at once exalted by its accomplishments and sobered by its limitations. We thus ally ourselves with the spirit of great ages past.

The fate of Prometheus, as well as that of most tragic heroes, is associated with the concept of *hubris*, "overweening pride." Yet pride, which in drama invariably leads to a fall, is not considered sinful by the great tragedians. It is an essential element of humanity's greatness. It is what inspires heroes to confront the universe, to challenge the status quo. Prometheus defied Zeus and brought technological knowledge to the human race. Prometheus was a revolutionary. So were Gutenberg, Watt, Edison, and Ford. Technology is revolutionary. Therefore, hostility toward technology is anti-revolutionary, which is to say, it is reactionary. This charge is currently being leveled against environmentalists and other enemies of technology. Since antitechnologists are traditionally "liberal" in their attitudes, the idea that they are reactionary confronts us with a paradox.

The tragic view does not shrink from paradox; it teaches us to live with ambiguity. It is at once revolutionary and cautionary. *Hubris,* as revealed in tragic drama, is an essential element of creativity; it is also a tragic flaw that contributes to the failure of human enterprise. Without effort, however, and daring, we are nothing. Walter Kerr has spoken of "tragedy's commitment to freedom, to the unflinching exploration of the possible." "At the heart of tragedy," he writes, "feeding it energy, stands godlike man passionately desiring a state of affairs more perfect than any that now exists."

This description of the tragic hero well serves, in my opinion, as a definition of the questing technologist.

An aspect of the tragic view that particularly appeals to me is its reluctance to place blame. Those people who hold pessimistic views about technology are forever reproaching others, if not individual engineers, then the "technocratic establishment," the "megastate," "the pentagon of power," or some equally amorphous entity. Everywhere they look they see evil intent.

There is evil in the world, of course, but most of our disappointments with technology come when decent people are trying to act constructively. "The essentially tragic fact," says Hegel, "is not so much the war of good with evil as the war of good with good."

Pesticides serve to keep millions of poor people from starving. To use pesticides is good; to oppose them when they create havoc in the food chain is also good. To drill for oil, and to transport it across oceans is good, since petroleum provides life-saving chemicals and heat for homes. To prevent oil spills is also good. Nuclear energy is good, as is the attempt to eliminate radioactivity. To seek safety is a worth goal; but in a world of limited resources, the pursuit of economy is also worthy. We are constantly accusing each other of villainy when we should be consulting together on how best to solve our common problems.

Although the tragic view shuns blame, it does not shirk responsibility. "The fault, dear Brutus, is not in our stars, but in ourselves. . . ." We are accountable for what we do or, more often, for what we neglect to do. The most shameful feature of the antitechnological creed is that it so often fails to consider the consequences of not taking action. The lives lost or wasted that might have been saved by exploiting our resources are the responsibility of those who counsel inaction. The tragic view is consistent with good citizenship. It advocates making the most of our opportunities; it challenges us to do the work that needs doing.

Life, it may be said, is not a play. Yet we are constantly talking about roles—role-playing, role models, and so forth. It is a primordial urge to want to play one's part. The outlook I advocate sees value in many different people playing many different parts. A vital society, like a meaningful drama, feeds on diversity. Each participant contributes to the body social: scientist, engineer, farmer, craftsman, laborer, politician, jurist, teacher, artist, merchant, entertainer. . . . The pro-growth industrialist and the environmentalist are both needed, and in a strange way they need each other.

Out of the conflict comes resolution; out of variety comes health. This is the lesson of the natural world. It is the moral of ecological balance; it is also the moral of great drama. We cannot but admire Caesar, Brutus, and Antony all together. So should we applaud the guardians of our wilderness, even as we applaud the creators of dams and paper mills. I am a builder, but I feel for those who are afraid of building, and I admire those who want to endow all building with grace.

George Steiner, in *The Death of Tragedy* (1961), claimed that the tragic spirit was rendered impotent by Christianity's promise of salvation. But I do not think that most people today are thinking in terms of salvation. They are thinking of doing the best they can in a world that promises neither damnation nor transcendent victories, but instead confronts us with both perils and opportunities for achievement. In such a world the tragic spirit is very much alive. Neither optimism nor pessimism is a worthy alternative to this noble spirit.

We use words to communicate, but sometimes they are not as precise as we pretend, and then we confuse ourselves and each other. "Optimism," "pessimism," "tragic view"—these are mere sounds or scratches on paper. The way we feel is not adequately defined by such sounds or scratches. René Dubos used to write a column for *The American Scholar* that he called "The Despairing Optimist." I seem to recall that he once gave his reasons for not calling it "The Hopeful Pessimist," although I cannot remember what they were. What really counts, I suppose, is not what we say, or even what we feel, but what we want to do.

By saying that I espouse the tragic view of technology I mean to ally myself with those who, aware of the dangers and without foolish illusions about what can be accomplished, still want to move on, actively seeking to realize our constantly changing vision of a more satisfactory society. I mean to oppose those who would evade harsh truths by intoning platitudes. I particularly mean to challenge those who enjoy the benefits of technology but refuse to accept responsibility for its consequences.

Earlier . . . I mentioned the problems I encountered in preparing an article for *House & Garden*, and I would like to close by quoting the last few lines from that much-rewritten opus. The prose is somewhat florid, but please remember that it was written in celebration of the American Bicentennial:

> For all our apprehensions, we have no choice but to press ahead. We must do so, first, in the name of compassion. By turning our backs on technological change, we would be expressing our satisfaction with current world levels of hunger, disease, and privation. Further, we must press ahead in the name of the human adventure. Without experimentation and change our existence would be a dull business. We simply cannot stop while there are masses to feed and diseases to conquer, seas to explore and heavens to survey.

The editors of *House & Garden* thought that I was being optimistic. I knew that I was being tragic, but I did not argue the point.

Questions and Topics for Discussion and Writing

1. What is Florman's tribute to Benjamin Franklin? Why do you think Florman began his research on American values with Benjamin Franklin?
2. Florman sees a "pattern of progress/setback/renewed-creative-effort" in technological advance. Choose one example of technology that you believe followed this form. Write a paragraph in which you illustrate your belief.
3. Write a definition of what Florman means by "optimistic materialism." Explain how he feels that this term describes American values.
4. Why does Florman assert that Americans' view of technology is *House & Garden* technology?
5. The time of the American Bicentennial was one that celebrated American values. What is the lesson about Americans that Florman "relearned" at this time?
6. Florman terms Americans *immature*. Explain why he does so.
7. Write a paragraph to explain what is meant by "tradeoff."
8. How does Florman characterize the feelings of technologists today?
9. Write a paragraph in which you paraphrase what Florman says is the "tragic view of life."
10. Why does he advocate this new attitude to replace optimism *and* pessimism?
11. Florman declares: "I particularly mean to challenge those who enjoy the benefits of technology but refuse to accept responsibility for its consequences." Write a paper in which you give examples to demonstrate that there are such people.
12. Write a paragraph in which you explain why Florman judged his concluding lines in the essay as "tragic" and not "optimistic."
13. Compare Florman's two essays. Is he advocating that Americans, especially technologists, in accepting a tragic stance, would be living the "authentic life"?

SUGGESTIONS FOR FURTHER READING

Almond, Gabriel A., ed. *Progress and Its Discontents*. Berkeley: University of California Press, 1982.

Asimov, Isaac. *Asimov's Biographical Encyclopedia of Science and Technology*. New York: Doubleday, 1964.

Boden, Margaret A. *Artificial Intelligence and Natural Man*. New York: Basic Books, 1977.

Borchert, Donald M., and David Stewart, eds. *Being Human in a Technological Age*. Athens: Ohio University Press, 1979.

Butler, Samuel. *Erewhon and Erewhon Revisited*. New York: Modern Library, 1927.

Cardwell, D. S. L. *Turning Points in Western Technology*. New York: Neale Watson Academic Publications, 1972.

Commoner, Barry. *The Closing Circle: Nature, Man and Technology*. New York: Knopf, 1971.

Mathews, Walter M. *Monster or Messiah? The Computer's Impact on Society*. Jackson: University Press of Mississippi, 1980.

McPhee, John A. *The Curve of Binding Energy*. New York: Farrar, Straus and Giroux, 1974.

Metz, Jerred. *Halley's Comet 1910: Fire in the Sky*. St. Louis: Singing Bone Press, 1985.

Raphael, Bertram. *The Thinking Computer: Mind Inside Matter*. San Francisco: W. H. Freeman, 1976.

Stobaugh, Robert, and Daniel Yergin, eds. *Energy Future: Report of the Energy Project at the Harvard Business School*. New York: Ballantine Books, 1980.

Weizenbaum, Joseph. *Computer Power and Human Reason*. San Francisco: W. H. Freeman, 1976.

Wiener, Norbert. *Cybernetics*. New York: John Wiley and Sons, 1948.

Unit 10

ON THE NATURE OF THINGS: OBSERVATIONS, EXPOSITIONS, CONJECTURES

A CONJECTURE ON THE SCIENTIFIC METHOD

In the space of one hundred and seventy-six years the Lower Mississippi has shortened itself two hundred and forty-two miles. That is an average of a trifle over one mile and a third per year. Therefore, any calm person, who is not blind or idiotic, can see that in the old Oolitic Silurian Period, just a million years ago next November, the Lower Mississippi River was upward of one million three hundred thousand miles long, and stuck out over the Gulf of Mexico like a fishing-rod. And by the same token any person can see that seven hundred and forty-two years from now the Lower Mississippi will be only a mile and three-quarters long, and Cairo and New Orleans will have joined their streets together, and be plodding comfortably along under a single mayor and a mutual board of aldermen. There is something fascinating about science. One gets such wholesale returns of conjecture out of such trifling investment of fact.

Mark Twain, Life on the Mississippi

Aristotle

The Greek philosopher Aristotle (384–322 B.C.) was one of the most important thinkers, scientific researchers, and organizers the world has ever known. Born in Stagira, the son of the court physician, he grew up in a medical and scientific atmosphere. This upbringing helped to train his mind in the accuracy and exactness for which he was to become famous.

When he was eighteen, Aristotle became a member of the Academy, Plato's school in Athens; he remained there until about the time of his famous teacher's death twenty years later. Although he came to differ with many of Plato's theories, Aristotle maintained an attitude of respect and affection for his teacher.

In 343 B.C. Aristotle was called by Philip of Macedon to educate his son, Alexander, who would later carry Greek laws and values throughout the known world. The carefully preserved fragment of a letter reveals Aristotle advising Alexander to become the leader of the Greeks and to conquer the barbarians. Alexander in turn supplied money to help his mentor in his research work.

Aristotle systematized all the knowledge that preceded him, carrying on investigations and extending the boundaries of knowledge in almost every field. He is usually thought of as a philosopher; in subjects like logic, metaphysics, psychology, ethics, and art his work is considered an invaluable foundation.

Many writings of Aristotle were lost, including many popular works published during his two decades at the Academy. Works that have survived include treatises on logic; "First Philosophy," which presents his thoughts on Metaphysics; and many writings on the natural sciences, such as "Physics," "On the Heavens," "On Growth and Decay," "Meteorology," "The History of Animals," "The Parts of Animals," and "De Anima." Aristotle's treatise on ethics, known as the *Nicomachean Ethics*, comprises ten books. Not easy reading, it is aimed toward those who have given considerable thought to the ethical behavior of mankind.

Although Aristotle's reputation as a philosopher took precedence over his accomplishments as a scientist, he is considered one of the pioneers of modern biology. His use of system in the investigation of facts is exemplified in the following selection, the opening section of "The History of Animals," which displays his command of precise detail.

The Classification of Animals

Aristotle

Animals also differ in their manner of life, in their actions and dispositions, and in their parts. We will first of all speak generally of these differences, and afterwards consider each species separately. The following are the points in which they vary in manner of life, in their actions and dispositions. Some animals are aquatic, others live on the land; and the aquatic may again be divided into two classes, for some entirely exist and procure their food in the water, and take in and give out water, and cannot live without it; this is the nature of most fishes. But there are others which, though they live and feed in the water, do not take in water but air, and produce their young out of the water. Many of these animals are furnished with feet, as the otter and the latax[1] and the crocodile, or with wings, as the seagull and diver, and others are without feet, as the water-serpent. Some procure their food from the water, and cannot live out of the water, but neither inhale air nor water, as the acalephe[2] and the oyster.

Different aquatic animals are found in the sea, in rivers, in lakes, and in marshes, as the frog and newt, and of marine animals some are pelagic, some littoral, and some saxatile. Some land animals take in and give out air, and this is called inhaling and exhaling; such are man, and all other land animals which are furnished with lungs; some, however, which procure their food from the earth, do not inhale air, as the wasp, the bee, and all other insects.[3] By insects I mean those animals which have divisions in their bodies, whether in the lower part only, or both in the upper and lower. Many land animals, as I have already observed, procure their food from the water, but there are no aquatic or marine animals which find their food on land. There are some animals which at first inhabit the water, but afterwards change into a different form, and live out of the water; this happens to the gnat in the rivers, and .. .. ..[4] which afterwards becomes an œstrum.[5]

Again, there are some creatures which are stationary, while others are loco-motive; the fixed animals are aquatic, but this is not the case with any of the inhabitants of the land. Many aquatic animals also grow upon each other; this is the case with several genera of shell-fish: the sponge also exhibits some signs of sensation, for they say that it is drawn up with some difficulty, unless the attempt to remove it is made stealthily. Other animals also there are which are alternately fixed together or free, this is the case with a certain kind of acalephe; some of these become separated during the night, and emigrate. Many animals are separate from each other, but incapable of voluntary movement, as oysters, and the animals called holothuria.[6] Some aquatic animals are swimmers, as fish, and the mollusca,[7] and the malacostraca, as the crabs. Others creep on the bottom, as the crab, for this, though an aquatic animal, naturally creeps.

Of land animals some are furnished with wings, as birds and bees, and these differ in other respects from each other; others have feet, and of this class some species walk, others crawl, and others creep in the mud. There is no animal which has only wings as fish have only fins, for those animals whose wings are formed by an expansion of the skin can walk, and the bat has feet, the seal has imperfect feet. Among birds there are some with very imperfect feet, which are therefore called apodes; they are, however, provided with very strong wings,

---

[1] Beaver, Castor fiber.   [2] Medusa, or perhaps Actinia, or both.
[3] Under the class ἔντομα are probably included all annulose animals.
[4] Some words appear to be lost in this place.   [5] Tabanus, gadfly.
[6] Perhaps some species of Zoophyte.   [7] Cephalopods.

and almost all birds that are similar to this one have strong wings and imperfect feet, as the swallow and drepanis;[8] for all this class of birds is alike both in habits and in the structure of their wings, and their whole appearance is very similar. The apos[9] is seen at all times of the year, but the drepanis can only be taken in rainy weather during the summer, and on the whole is a rare bird.

Many animals, however, can both walk and swim. The following are the differences exhibited by animals in their habits and their actions. Some of them are gregarious, and others solitary, both in the classes which are furnished with feet, and those which have wings, or fins. Some partake of both characters, and of those that are gregarious, as well as those that are solitary, some unite in societies and some are scattered. Gregarious birds are such as the pigeon, stork, swan, but no bird with hooked claws is gregarious. Among swimming animals some fish are gregarious, as the dromas,[10] tunny, pelamis,[11] amia.[12]

But man partakes of both qualities. Those which have a common employment are called social, but that is not the case with all gregarious animals. Man, and the bee, the wasp, and the ant, and the stork belong to this class. Some of these obey a leader, others are anarchical; the stork and the bee are of the former class, the ant and many others belong to the latter. Some animals, both in the gregarious and solitary class, are limited to one locality, others are migratory. There are also carnivorous animals, herbivorous, omnivorous, and others which eat peculiar food, as the bee and the spider; the former eats only honey and a few other sweet things, while spiders prey upon flies and there are other animals which feed entirely on fish. Some animals hunt for their food, and some make a store, which others do not. There are also animals which make habitations for themselves, and others which do not. The mole, the mouse, the ant, and the bee, make habitations, but many kinds both of insects and quadrupeds make no dwelling.

With regard to situation, some are troglodite, as lizards and serpents, others, as the horse and dog, live upon the surface of the earth. Some kinds of animals burrow in the ground, others do not; some animals are nocturnal, as the owl and the bat, others use the hours of daylight. There are tame animals and wild animals. Man and the mule are always tame, the leopard and the wolf are invariably wild, and others, as the elephant, are easily tamed. We may however, view them in another way, for all the genera that have been tamed are found wild also, as horses, oxen, swine, sheep, goats, and dogs.

Some animals utter a loud cry, some are silent, and others have a voice, which in some cases may be expressed by a word, in others it cannot. There are also noisy animals and silent animals, musical and unmusical kinds, but they are mostly noisy about the breeding season. Some, as the dove, frequent fields, others, as the hoopoe, live on the mountains; some attach themselves to man, as the pigeon. Some are lascivious, as the partridge and domestic fowl, and others are chaste, as the raven, which rarely cohabits.

[8] Perhaps Saud martin.   [9] Swift.   [10] Some migratory fish.
[11] A kind of tunny, still called palamyde at Marseilles.
[12] A kind of tunny, Les Bonitons (Caimus).

Again, there are classes of animals furnished with weapons of offence, others with weapons of defence; in the former I include those which are capable of inflicting an injury, or of defending themselves when they are attacked; in the latter those which are provided with some natural protection against injury.

Animals also exhibit many differences of disposition. Some are gentle, peaceful, and not violent, as the ox. Some are violent, passionate, and intractable, as the wild boar. Some are prudent and fearful, as the stag and the hare. Serpents are illiberal and crafty. Others, as the lion, are liberal, noble, and generous. Others are brave, wild, and crafty, like the wolf. For there is this difference between the generous and the brave—the former means that which comes of a noble race, the latter that which does not easily depart from its own nature.

Some animals are cunning and evil-disposed, as the fox; others, as the dog, are fierce, friendly, and fawning. Some are gentle and easily tamed, as the elephant; some are susceptible of shame, and watchful, as the goose. Some are jealous, and fond of ornament, as the peacock. But man is the only animal capable of reasoning, though many others possess the faculty of memory and instruction in common with him. No other animal but man has the power of recollection. In another place we will treat more accurately of the disposition and manner of life in each class.

Questions and Topics for Discussion and Writing

1. Aristotle has been called one of the "true founders" of biology. What evidence do you find in this selection of his pioneering qualities as a scientist?
2. What aspects of the scientific method do you find Aristotle employing here? Look up the word "empirical." Does this term apply to Aristotle's work? Explain.
3. Write a paragraph discussing Aristotle's categorization of animals. What bases does he use? What differences do you see between his system and the system of modern classification?
4. What differences from present-day scientific writing do you see in Aristotle's judgments and language choices? List specific words and phrases.
5. Aristotle classifies man as an animal. How does he distinguish him from other animals?

Charles Darwin

For someone who later would shake man's foundation of belief about God, himself, and his place and purpose in the world, young Charles Robert Darwin (1809–1882) did not distinguish himself in his academic career. Both his father and grandfather were physicians. His grandfather, Erasmus Darwin, in fact, was England's best-known physician, as well as a famous scientist and poet, who anticipated his grandson's theory of evolution.

Being born into such an intellectual family created higher expectations than the young Darwin could easily meet. At one point, his father berated him for his lack of seriousness and his love of gentleman's sporting activities, predicting that he would be a "disgrace to yourself and all of your family." After completing school at Shrewsbury, Darwin began medical studies, but quit after three years. He disliked the discipline and his constitution was ill equipped to deal with surgery in the days before anesthesia. He then went to Christ College to study for the ministry, but graduated without committing himself to the church.

Offered the position of naturalist–assistant to the captain of the H.M.S. *Beagle*, he left at age twenty-two for a five-year voyage of scientific exploration. Among the few readings that Darwin took with him was Charles Lyell's *Principles of Geology*, a book that would provide the basis for his life's work. In *The Voyage of the Beagle*, Darwin published his painstaking observations of the unusual plants and animals, especially those made in the geographically unusual Galapagos Islands.

Upon his return to England, he married his cousin, Emma Wedgwood, of the famous porcelain-manufacturing family, and moved to the countryside. In a quiet domestic routine, Darwin continued his research and reflections, writing books and monographs.

The qualities that Darwin brought with him to the study of the sciences were, in his words, patience, observation, and invention. To these should be added his ability to reflect on and organize the many seemingly unrelated observations—both his own and those of many other scientists and researchers— into a workable hypothesis. Even before his trip on the *Beagle*, he was becoming an avid geologist, taking extensive hiking tours to study rock formations and collect fossils. Faced with the "fact" of evolution from the evidence of his observations, Darwin next had to back up his hypothesis with an explanation of how and why, and to this he devoted many years. He discarded several theories until his reading of Malthus's *On Population* led him to the process of natural selection. He published his early ideas in monograph form in the early 1840s.

In 1858, twenty years after his voyage, when English naturalist Alfred R. Wallace presented him with a manuscript that propounded the same theories, Darwin set out to publish his ideas lest his own ideas "be forestalled." In 1859, his revolutionary *Origin of Species* appeared, the entire 1250-copy edition being sold out on the day of issue. Intense controversy surrounded his beliefs, dramatized in the famous debate between T. H. Huxley and the Bishop of Wilberforce in 1860, and continuing to this very day.

Yet for all his epochal contributions to science, and despite the religious and philosophical controversies that have ensued, Charles Darwin and his work can also be viewed as the "natural manifestation" of newer ideas germinating within Victorian society. The practice of genetic improvement through breeding was well established in agriculture and the development of racehorses. Darwin's special contribution was somewhat akin to the role sometimes ascribed to the artist, and here his sense of invention played its part. His awareness of currents of ideas and of their unperceived and unsuspected connections enabled him to amass what is generally held to be solid, incontrovertible proof for the theory of evolution. The conclusion to *The Origin* summarizes this position.

Conclusion, *The Origin of Species*

Charles Darwin

The fact, as we have seen, that all past and present organic beings constitute one grand natural system, with group subordinate to group, and with extinct groups often falling in between recent groups, is intelligible on the theory of natural selection with its contingencies of extinction and divergence of character. On these same principles we see how it is, that the mutual affinities of the species and genera within each class are so complex and circuitous. We see why certain characters are far more serviceable than others for classification;—why adaptive characters, though of paramount importance to the being, are of hardly any importance in classification; why characters derived from rudimentary parts, though of no service to the being, are often of high classificatory value; and why embryological characters are the most valuable of all. The real affinities of all organic beings are due to inheritance or community of descent. The natural system is a genealogical arrangement, in which we have to discover the lines of descent by the most permanent characters, however slight their vital importance may be.

The framework of bones being the same in the hand of a man, wing of a bat, fin of the porpoise, and leg of the horse,—the same number of vertebræ forming the neck of the giraffe and of the elephant,—and innumerable other such facts, at once explain themselves on the theory of descent with slow and slight successive modifications. The similarity of pattern in the wing and leg of a bat, though used for such different purpose,—in the jaws and legs of a crab,—in the petals, stamens, and pistils of a flower, is likewise intelligible on the view of the gradual modification of parts or organs, which were alike in the early progenitor of each class. On the principle of successive variations not always supervening at an early age, and being inherited at a corresponding not early period of life, we can clearly see why the embryos of mammals, birds, reptiles, and fishes should be so closely alike, and should be so unlike the adult forms. We may cease marvelling at the embryo of an air-breathing mammal or bird having branchial slits and arteries running in loops, like those in a fish which has to breathe the air dissolved in water, by the aid of well-developed branchiæ.

Disuse, aided sometimes by natural selection, will often tend to reduce an organ, when it has become useless by changed habits or under changed conditions of life; and we can clearly understand on this view the meaning of rudimentary organs. But disuse and selection will generally act on each creature, when it has come to maturity and has to play its full part in the struggle for existence, and will thus have little power of acting on an organ during early life; hence the organ will not be much reduced or rendered rudimentary at this early age. The calf, for instance, has inherited teeth, which never cut through the gums of the upper jaw, from an early progenitor having well-developed teeth; and we may believe, that the teeth in the mature animal were reduced, during successive generations, by disuse or by the tongue and palate having been fitted by natural selection to browse without their aid; whereas in the calf, the teeth have been left untouched by selection or disuse, and on the principle of inheritance at corresponding ages have been inherited from a remote period to the present day. On the view of each organic being and each separate organ having been specially created, how utterly inexplicable it is that parts, like the teeth in the embryonic calf or like the shrivelled wings under the soldered wing-covers of some beetles, should thus so frequently bear the plain stamp of inutility! Nature may be said to have taken pains to reveal, by rudimentary organs and by homologous structures, her scheme of modification, which it seems that we wilfully will not understand.

I have now recapitulated the chief facts and considerations which have thoroughly convinced me that species have changed, and are still slowly changing by the preservation and accumulation of successive slight favourable variations. Why, it may be asked, have all the most eminent living naturalists and geologists rejected this view of the mutability of species? It cannot be asserted that organic beings in a state of nature are subject to no variation; it cannot be proved that the amount of variation in the course of long ages is a limited quantity; no clear distinction has been, or can be, drawn between species and well-marked varieties. It cannot be maintained that species when intercrossed are invariably sterile, and varieties invariably fertile; or that sterility is a special endowment and sign of creation. The belief that species were immutable productions was almost unavoidable as long as the history of the world was thought to be of short duration; and now that we have acquired some idea of the lapse of time, we are too apt to assume, without proof, that the geological record is so perfect that it would have afforded us plain evidence of the mutation of species, if they had undergone mutation.

But the chief cause of our natural unwillingness to admit that one species has given birth to other and distinct species, is that we are always slow in admitting any great change of which we do not see the intermediate steps. The difficulty is the same as that felt by so many geologists, when Lyell first insisted that long lines of inland cliffs had been formed, and great valleys excavated, by the slow action of the coast-waves. The mind cannot possibly grasp the full meaning of the term of a hundred million years; it cannot add up and perceive the full effects of many slight variations, accumulated during an almost infinite number of generations.

Although I am fully convinced of the truth of the views given in this volume under the form of an abstract, I by no means expect to convince experienced naturalists whose minds are stocked with a multitude of facts all viewed, during a long course of years, from a point of view directly opposite to mine. It is so easy to hide our ignorance under such expressions as the "plan of creation," "unity of design," &c., and to think that we give an explanation when we only restate a fact. Any one whose disposition leads him to attach more weight to unexplained difficulties than to the explanation of a certain number of facts will certainly reject my theory. A few naturalists, endowed with much flexibility of mind, and who have already begun to doubt on the immutability of species, may be influenced by this volume; but I look with confidence to the future, to young and rising naturalists, who will be able to view both sides of the question with impartiality. Whoever is led to believe that species are mutable will do good service by conscientiously expressing his conviction; for only thus can the load of prejudice by which this subject is overwhelmed be removed.

Several eminent naturalists have of late published their belief that a multitude of reputed species in each genus are not real species; but that other species are real, that is, have been independently created. This seems to me a strange conclusion to arrive at. They admit that a multitude of forms, which till lately they themselves thought were special creations, and which are still thus looked at by the majority of naturalists, and which consequently have every external characteristic feature of true species,—they admit that these have been produced by variation, but they refuse to extend the same view to other and very slightly different forms. Nevertheless they do not pretend that they can define, or even conjecture, which are the created forms of life, and which are those produced by secondary laws. They admit variation as a *vera causa* in one case, they arbitrarily reject it in another, without assigning any distinction in the two cases. The day will come when this will be given as a curious illustration of the blindness of preconceived opinion. These authors seem no more startled at a miraculous act of creation than at an ordinary birth. But do they really believe that at innumerable periods in the earth's history certain elemental atoms have been commanded suddenly to flash into living tissues? Do they believe that at each supposed act of creation one individual or many were produced? Were all the infinitely numerous kinds of animals and plants created as eggs or seed, or as full grown? and in the case of mammals, were they created bearing the false marks of nourishment from the mother's womb? Although naturalists very properly demand a full explanation of every difficulty from those who believe in the mutability of species, on their own side they ignore the whole subject of the first appearance of species in what they consider reverent silence.

It may be asked how far I extend the doctrine of the modification of species. The question is difficult to answer, because the more distinct the forms are which we may consider, by so much the arguments fall away in force. But some arguments of the greatest weight extend very far. All the members of whole classes can be connected together by chains of affinities, and all can be classified on the same principle, in groups subordinate to groups. Fossil remains sometimes tend to fill up very wide intervals between existing orders. Organs in a rudimentary condition plainly show that an early progenitor had the organ in a fully developed state; and this in some instances necessarily implies an enormous amount of modification in the descendants. Throughout whole classes various structures are formed on the same pattern, and at an embryonic age the species closely resemble each other. Therefore I cannot doubt that the theory of descent with modification embraces all the members of the same class. I believe that animals have descended from at most only four or five progenitors, and plants from an equal or lesser number.

Analogy would lead me one step further, namely, to the belief that all animals and plants have descended from some one prototype. But analogy may be a deceitful guide. Nevertheless all living things have much in common, in their chemical composition, their germinal vesicles, their cellular structure, and their laws of growth and reproduction. We see this even in so trifling a circumstance as that the same poison often similarly affects plants and animals; or that the poison secreted by the gall-fly produces monstrous growth on the wild rose or oak-tree. Therefore I should infer from analogy that probably all the organic beings which have ever lived on this earth have descended from some one primordial form, into which life was first breathed.

When the views entertained in this volume on the origin of species, or when analogous views are generally admitted, we can dimly foresee that there will be a considerable revolution in natural history. Systematists will be able to pursue their labours as at present; but they will not be incessantly haunted by the shadowy doubt whether this or that form be in essence a species. This I feel sure, and I speak after experience, will be no slight relief. The endless disputes whether or not some fifty species of British brambles are true species will cease. Systematists will have only to decide (not that this will be easy) whether any form be sufficiently constant and distinct from other forms, to be capable of definition; and if definable, whether the differences be sufficiently important to deserve a specific name. This latter point will become a far more essential consideration than it is at present; for differences, however slight, between any two forms, if not blended by intermediate gradations, are looked at by most naturalists as sufficient to raise both forms to the rank of species. Hereafter we shall be compelled to acknowledge that the only distinction between species and well-marked varieties is, that the latter are known, or believed, to be connected at the present day by intermediate gradations, whereas species were formerly thus connected. Hence, without quite rejecting the consideration of the present existence of intermediate gradations between any two forms, we shall be led to weigh more carefully and to value higher the actual amount of difference between them. It is quite possible that forms now generally acknowledged to be merely varieties may hereafter be thought worthy of specific names, as with the primrose and cowslip; and in this case scientific and common language will come into accordance. In short, we shall have to treat species in the same manner as those naturalists treat genera, who admit that genera are merely artificial combinations made for convenience. This may not be a cheering prospect; but we shall at least be freed from the vain search for the undiscovered and undiscoverable essence of the term species.

The other and more general departments of natural history will rise greatly in interest. The terms used by naturalists of affinity, relationship, community of type, paternity, morphology, adaptive characters, rudimentary and aborted organs, &c., will cease to be metaphorical, and will have a plain signification. When we no longer look at an organic being as a savage looks at a ship, as at something wholly beyond his comprehension; when we regard every production of nature as one which has had a history; when we contemplate every complex structure and instinct as the summing up of many contrivances, each useful to the possessor, nearly in the same way as when we look at any great mechanical invention as the summing up of the labour, the experience, the reason, and even the blunders of numerous workmen; when we thus view each organic being, how far more interesting, I speak from experience, will the study of natural history become!

A grand and almost untrodden field of inquiry will be opened, on the causes and laws of variation, on correlation of growth, on the effects of use and disuse, on the direct action of external conditions, and so forth. The study of domestic productions will rise immensely in value. A new variety raised by man will be a far more important and interesting subject for study than one more species added to the infinitude of already recorded species. Our classifications will come to be, as far as they can be so made, genealogies; and will then truly give what may be called the plan of creation. The rules for classifying will no doubt become simpler when we have a definite object in view. We possess no pedigrees or armorial bearings; and we have to discover and trace the many diverging lines of descent in our natural genealogies, by characters of any kind which have long been inherited. Rudimentary organs will speak infallibly with respect to the nature of long-lost structures. Species and groups of species, which are called aberrant, and which may fancifully be called living fossils, will aid us in forming a picture of the ancient forms of life. Embryology will reveal to us the structure, in some degree obscured, of the prototypes of each great class.

When we can feel assured that all the individuals of the same species, and all the closely allied species of most genera, have within a not very remote period descended from one parent, and have migrated from some one birthplace; and when we better know the many means of migration, then, by the light which geology now throws, and will continue to throw, on former changes of climate and of the level of the land, we shall surely be enabled to trace in an admirable manner the former migrations of the inhabitants of the whole world. Even at present, by comparing the differences of the inhabitants of the sea on the opposite sides of a continent, and the nature of the various inhabitants of that continent in relation to their apparent means of immigration, some light can be thrown on ancient geography.

The noble science of Geology loses glory from the extreme imperfection of the record. The crust of the earth with its embedded remains must not be looked at as a well-filled museum, but as a poor collection made at hazard and at rare intervals. The accumulation of each great fossiliferous formation will be recognised as having depended on an unusual concurrence of circumstances, and the blank intervals between the successive stages as having been of vast duration. But we shall be able to gauge with some security the duration of these intervals by a comparison of the preceding and succeeding organic forms. We must be cautious in attempting to correlate as strictly contemporaneous two formations, which include few identical species, by the general succession of their forms of life. As species are produced and exterminated by slowly acting and still existing causes, and not by miraculous acts of creation and by catastrophes; and as the most important of all causes of organic change is one which is almost independent of altered and perhaps suddenly altered physical conditions, namely, the mutual relation of organism to organism,—the improvement of one being entailing the improvement or the extermination of others; it follows, that the amount of organic change in the fossils of consecutive formations probably serves as a fair measure of the lapse of actual time. A number of species, however, keeping in a body might remain for a long period unchanged, whilst within this same period, several of these species, by migrating into new countries and coming into competition with foreign associates, might become modified; so that we must not overrate the accuracy of organic change as a measure of time. During early periods of the earth's history, when the forms of life were probably fewer and simpler, the rate of change was probably slower; and at the first dawn of life, when very few forms of the simplest structure existed, the rate of change may have been slow in an extreme degree. The whole history of the world, as at present known, although of a length quite incomprehensible by us, will hereafter be recognised as a mere fragment of time, compared with the ages which have elapsed since the first creature, the progenitor of innumerable extinct and living descendants, was created.

In the distant future I see open fields for far more important researches. Psychology will be based on a new foundation, that of the necessary acquirement of each mental power and capacity by gradation. Light will be thrown on the origin of man and his history.

Authors of the highest eminence seem to be fully satisfied with the view that each species has been independently created. To my mind it accords better with what we know of the laws impressed on matter by the Creator, that the production and extinction of the past and present inhabitants of the world should have been due to secondary causes, like those determining the birth and death of the individual. When I view all beings not as special creations, but as the lineal descendants of some few beings which lived long before the first bed of the Silurian system was deposited, they seem to me to become ennobled. Judging from the past, we may safely infer that not one living species will transmit its unaltered likeness to a distant futurity. And of the species now living very few will transmit progeny of any kind to a far distant futurity; for the manner in which all organic beings are grouped, shows that the greater number of species of each genus, and all the species of many genera, have left no descendants, but have become utterly extinct. We can so far take a prophetic glance into futurity as to foretel that it will be the common and widely-spread species, belonging to the larger and dominant groups, which will ultimately prevail and procreate new and dominant species. As all the living forms of life are the lineal descendants of those which lived long before the Silurian epoch, we may feel certain that the ordinary succession by generation has never once been broken, and that no cataclysm has desolated the whole world. Hence we may look with some confidence to a secure future of equally inappreciable length. And as natural selection works solely by and for the good of each being, all corporeal and mental endowments will tend to progress towards perfection.

It is interesting to contemplate an entangled bank, clothed with many plants of many kinds, with birds singing on the bushes, with various insects flitting about, and with worms crawling through the damp earth, and to reflect that these elaborately constructed forms, so different from each other, and dependent on each other in so complex a manner, have all been produced by laws acting around us. These laws, taken in the largest sense, being Growth with Reproduction; Inheritance which is almost implied by reproduction; Variability from the indirect and direct action of the external conditions of life, and from use and disuse; a Ratio of Increase so high as to lead to a Struggle for Life, and as a consequence to Natural Selection, entailing Divergence of Character and the Extinction of less-improved forms. Thus, from the war of nature, from famine and death, the most exalted object which we are capable of conceiving, namely, the production of the higher animals, directly follows. There is grandeur in this view of life, with its several powers, having been originally breathed into a few forms or into one; and that, whilst this planet has gone cycling on according to the fixed law of gravity, from so simple a beginning endless forms most beautiful and most wonderful have been, and are being, evolved.

Questions and Topics for Discussion and Writing

1. Restate Darwin's challenge to naturalists and geologists of his time.
2. What was the effect on the concepts of time created by Lyell's and Darwin's theory of the variation of species?

3. To Darwin, what was the value of fossil remains?
4. According to Darwin, what role does geology play in promoting and abetting scientists' theories of evolution?
5. Darwin uses the method of analogy as the method of his argument. Explain what he means by stating in analogy: "All animals and plants have descended from one prototype."
6. Evolutionary concepts of man have implied *ascent* into higher forms of life. (See Jacob Bronowski's book, *The Ascent of Man*.) Does it seem odd to you that Darwin and many biologists use the term *descent*? Why? Why not?
7. According to Darwin, what are the major reasons why certain people fail to accept the theory of evolution?
8. Did Darwin, as revealed in this Conclusion, believe that his ideas would not be accepted by his contemporaries? To whom did he look for acceptance?
9. Read Stephen Crane's story "The Open Boat." Write a paper in which you show how Crane's setting depicts the Darwinian view of nature.
10. What hope for communication between scientists and lay people did Darwin offer?
11. A general assumption is that many people have *heard about* what Darwin said, *never having read* his words themselves. Did you find anything surprising in your reading of Darwin's conclusion? What would you offer as an answer to those who would deny the theory of evolution?
12. Darwin is credited by literary critics with having provided a view of society and the world as a jungle. Cite sentences that support such judgments.

Mary Leakey

Mary Douglas Leakey (b. 1913) displays in her writings both a scientific and an artistic vision. Direct descendant of British prehistoric archeologist John Frere and the daughter of painters, Leakey moved from site to site as her itinerant father searched for new artistic scenes. Consequently, she had no formal schooling in her early years; instead, her parents taught her to read, write, and calculate.

After her father's death when she was thirteen, she twice tried convent schooling, but was expelled both times. Unruly but impelled by curiosity, she attended lectures on archeology and took part in digs around London. Her archeological drawings, well done despite her lack of formal training, launched her career in this field. She worked with Louis Leakey and married him in 1936. They raised three children, the youngest conceived, as she tells us in her autobiography, *Disclosing the Past*, in celebration of their discovery of the *Proconsul* skull in 1948.

Although especially fond of cave drawings and an expert on stone tools, it was she who found the hominid skull *Zinjanthropus* at Olduvai Gorge, Tanzania, in 1959. After her husband's death in 1972, she uncovered early hominid remains in Laetoli that were estimated to be more than three and a half million years old.

Vocal about the need for archeologists to spend less time arguing about evolutionary patterns and more time digging for evidence to fill the gaps, she, nevertheless, uses her findings at Laetoli to infer and to imagine a setting millions of years ago. In the following selection, "Footprints in the Ashes of Time," her direct but lively style is evident.

Footprints in the Ashes of Time

Mary Leakey

It happened some 3,600,000 years ago, at the onset of a rainy season. The East African landscape stretched then, much as it does now, in a series of savannas punctuated by wind-sculptured acacia trees. To the east the volcano now called Sadiman heaved restlessly, spewing ash over the flat expanse known as Laetoli.

The creatures that inhabited the region, and they were plentiful, showed no panic. They continued to drift on their random errands. Several times Sadiman blanketed the plain with a thin layer of ash. Tentative showers, precursors of the heavy seasonal rains, moistened the ash. Each layer hardened, preserving in remarkable detail the footprints left by the ancient fauna. The Laetolil Beds, as geologists designate the oldest deposits at Laetoli, captured a frozen moment of time from the remote past—a pageant unique in prehistory.

Our serious survey of the beds, which lie in northern Tanzania 30 miles by road south of Olduvai Gorge . . . , began in 1975 and gained intensity last summer after the discovery of some startling footprints. This article must stand as a preliminary report; further findings will almost certainly modify early interpretations.

Still, what we have discovered to date at Laetoli will cause yet another upheaval in the study of human origins. For in the gray, petrified ash of the beds—among the spoor of the extinct predecessors of today's elephants, hyenas, hares—we have found hominid footprints that are remarkably similar to those of modern man. Prints that, in my opinion, could only have been left by an ancestor of man. Prints that were laid down an incredible 3,600,000 years ago.

My late husband, Dr. Louis S. B. Leakey, and I had first explored the Laetolil Beds in 1935. In that year we were searching for fossils in Olduvai Gorge when Masai tribesmen told us of the rich remains at Laetoli, which in their language refers to the red lily that grows there in profusion. When heavy rains ended the Olduvai excavation season, we made the difficult, two-day journey south.

We did find fossils, but they were much more fragmented than those of Olduvai. At that time, accurate dating was impossible. So we left the site. A German expedition combed the beds in 1938–39, and we ourselves returned twice with indifferent results. But I could not help feeling that, somehow, the mystique of Laetoli had eluded us.

Then, in 1974, two things occurred. I was drawn back once more to these ancient volcanic deposits, and one of my African associates, Mwongela Mwoka, found a hominid tooth. Analysis of the lava that overlies the beds assigned the tooth an age of at least 2,400,000 years. Since this is older than anything at Olduvai, I decided to concentrate my efforts at Laetoli. In 1975, with National Geographic Society support and the cooperation of the Tanzanian Government and its director of antiquities, A. A. Mturi, I mounted an intensive campaign.

For almost two field seasons we diligently collected hominid and other fossils. Then, as is so often the case in pivotal discoveries, luck intervened. One evening Dr. Andrew Hill of the National Museums of Kenya and several colleagues were larking about on the beds, pelting each other with dry elephant dung. As Andrew ducked low to avoid one such missile, he noticed a series of punctures in the volcanic tuff. When close examination indicated that they were animal footprints, we commenced to study them in earnest.

In 1976 Peter Jones, my assistant and a specialist in stone tools, and my youngest son, Philip, noticed what they believed to be a trail of hominid footprints. After considerable analysis I agreed and announced the discovery the following year. Of the five prints, three were obscured by overlying sediment impossible to remove. The two clear examples, broad and rather curiously shaped, offered few clues to the primate that had trudged across the plain so long ago.

Nonetheless, the implications of this find were enormous. Dr. Garniss Curtis of the University of California at Berkeley undertook to date the footprint strata. These deposits possess relatively large crystals of biotite, or black mica. Biotite from ash overlying the prints, when subjected to potassium-argon testing, showed an age of about 3.6 million years; that from below tested at about 3.8 million years. The footprints had been preserved sometime within this span. Dr. Richard L. Hay, also of Berkeley, showed that the ash forming the layers fell within a month's time.

The hominid footprints attested, in my considered opinion, to the existence of a direct ancestor of man half a million years before the earliest previous evidence—fossils unearthed by Dr. Donald C. Johanson and his party in the Afar triangle of Ethiopia beginning in 1973.[1]

Faced with this, we largely abandoned our hunt for fossils and focused our three-month campaign of 1978 on the footprints—plotting and photographing them, making plaster and latex casts, and even removing certain specimens.

[1] See "Ethiopia Yields First 'Family' of Early Man" by Dr. Johanson in the December 1976 GEOGRAPHIC.

While Dr. Paul Abell of the University of Rhode Island was attempting—delicately and successfully—to quarry out a block of rhinoceros tracks, he noticed a barely exposed, hominidlike heel print.

When we removed the surrounding overburden, we found a trail some 23 meters long; only the end of the excavation season in September prevented our following it still farther. Two individuals, one larger, one smaller, had passed this way 3,600,000 years ago. . . .

The footsteps come from the south, progress northward in a fairly straight line, and end abruptly where seasonal streams have eroded a small, chaotic canyon through the beds. The nature of the terrain leads us to believe that the footprints, though now covered, remain largely intact to the south. And that is where we will continue our effort.

The closeness of the two sets of prints indicates that their owners were not walking abreast. Other clues suggest that the hominids may have passed at different times. For example, the imprints of the smaller individual stand out clearly. The crispness of definition and sharp outlines convince me that they were left on a damp surface that retained the form of the foot.

On the other hand, the prints of the larger are blurred, as if he had shuffled or dragged his feet. In fact, I think that the surface when he passed was loose and dusty, hence the collapsed appearance of his prints. Nonetheless, luck favored us again; the bigger hominid left one absolutely clear print, probably on a patch of once damp ash.

What do these footprints tell us? First, they demonstrate once and for all that at least 3,600,000 years ago, in Pliocene times, what I believe to be man's direct ancestor walked fully upright with a bipedal, free-striding gait. Second, that the form of his foot was exactly the same as ours.

One cannot overemphasize the role of bipedalism in hominid development. It stands as perhaps the salient point that differentiated the forebears of man from other primates. This unique ability freed the hands for myriad possibilities—carrying, tool-making, intricate manipulation. From this single development, in fact, stems all modern technology.

Somewhat oversimplified, the formula holds that this new freedom of forelimbs posed a challenge. The brain expanded to meet it. And mankind was formed.

Even today, millions of years beyond that unchronicled Rubicon, *Homo sapiens* is the only primate to walk upright as a matter of course. And, for better or for worse, *Homo sapiens* dominates the world.

But what of those two hominids who crossed the Laetolil Beds so long ago? We have measured their footprints and the length of their stride. Was the larger one a male, the smaller a female? Or was one mature, the other young? It is unlikely that we will ever know with certainty. For convenience, let us postulate a case of sexual dimorphism and consider the smaller one a female.

Incidentally, following her path produces, at least for me, a kind of poignant time wrench. At one point, and you need not be an expert tracker to discern this, she stops, pauses, turns to the left to glance at some possible threat or irregularity, and then continues to the north. This motion, so intensely human, transcends time. Three million six hundred thousand years ago, a remote ancestor—just as you or I—experienced a moment of doubt.

The French have a proverb: *Plus ça change, plus c'est la même chose*—"The more it changes, the more it is the same." In short, nothing really alters. Least of all, the human condition.

Measurements show the length of the smaller prints to be 18.5 centimeters (slightly more than 7 inches) and 21.5 centimeters for the larger. Stride length averages 38.7 centimeters for the smaller hominid, 47.2 centimeters for the larger. Clearly we are dealing with two small creatures.

An anthropological rule of thumb holds that the length of the foot represents about 15 percent of an individual's height. On this basis—and it is far from exact—we can estimate the height of the male as perhaps four feet eight inches (1.4 meters); the female would have stood about four feet.

Leg structure must have been very similar to our own. It seems clear to me that the Laetoli hominid, although much older, relates very closely to the remains found by Dr. Johanson in Ethiopia. Dr. Owen Lovejoy of Kent State University in Ohio studied a knee joint from Ethiopia—the bottom of the femur and the top of the tibia—and concluded that the Afar hominid had walked upright, with a free, bipedal gait.

Our footprints confirm this. Furthermore, Dr. Louise Robbins of the University of North Carolina, Greensboro, an anthropologist who specializes in the analysis of footprints, visited Laetoli and concluded: "The movement pattern of the individual is a bipedal walking gait, actually a stride—and quite long relative to the creature's small size. Weight-bearing pressure patterns in the prints resemble human ones. . . ."

I can only assume that the prints were left by the hominids whose fossils we also found in the beds. In addition to part of a child's skeleton, we uncovered adult remains—two lower jaws, a section of upper jaw, and a number of teeth.

Where can we place the Laetoli hominids and their Afar cousins in the incomplete mosaic of the rise of man? This question, quite honestly, is a subject of some contention among paleontologists. One school, including Dr. Johanson, classifies them as australopithecines.

But the two forms of *Australopithecus*, gracile and robust, represent, in my opinion, evolutionary dead ends. These man apes flourished for their season, and perished—unsuccessful twigs on the branch that produced mankind. Of course, the Laetoli hominid resembles the gracile *Australopithecus*, but I believe that, so far back in time, all the hominids shared certain characteristics. However, the simple evidence of the footprints, so very much like our own, indicates to me that the Laetoli hominid stands in the direct line of man's ancestry.

We have encountered one anomaly. Despite three years of painstaking search by Peter Jones, no stone tools have been found in the Laetolil Beds. With their hands free, one would have expected this species to have developed tools or weapons of some kind. But, except for the ejecta of erupting volcanoes, we haven't found a single stone introduced into the beds. So we can only conclude, at least for the moment, that the hominids we discovered had not yet attained the toolmaking stage.

While the hominid prints rank as the most exciting of our discoveries at Laetoli, there is also a petrified record of Pliocene animal life. Oddly—or perhaps not so oddly, given the geologic continuity of East Africa—we find the same type of wildlife in roughly the same proportions as exist today.

Before the first fall of ash, the prehistoric plain of Lactoli apparently possessed a normal array of vegetation. Beneath the bottom layer we find fossilized twigs, small branches, and grass. Thereafter, with the terrain buried in barren gray dust, animals continued to drift across in their habitual patterns. Why did they do so, with so little sustenance present? With the data available at this point, we just cannot explain it.

But the footprints are visible in their plenitude. Countless hares have pocked the ash with their distinctive hopping pattern. Baboon prints lie in profusion; all we have found possess a narrow heel similar to those of smaller present-day females. Baboons, incidentally, still arrogantly patrol the beds. *Deinotherium*, a prehistoric relative of today's elephant, lumbered through, its huge footprints obliterating the strata beneath.

We encountered a variety of antelope prints, large and small. In studying these, we enlisted the aid of two African trackers, members of the Hadza and Wasukuma tribes. Because the extinct species possess hooves of roughly the same shape as their living relatives, the trackers provided valuable identifications.

In addition to footprints we also have found skeletal traces of both white and black rhinos, two types of giraffe—one as tall as today's species, the other a pygmy—and two genera of pigs. Various carnivores prowled across the ash, including one very large saber-toothed cat; hyenas were numerous.

A unique discovery, though, is a pair of prints left by a chalicothere. This strange creature had claws on its feet; yet it was an ungulate rather than a carnivore and ate only vegetation. To my knowledge, our prints are the only ones ever discovered.

But in the end one cannot escape the supreme importance of the presence of hominids at Laetoli. Sometimes, during the excavating season, I go out and watch the dusk settle over the gray tuff with its eerie record of time long past. The slanting light of evening throws the hominid prints into sharp relief, so sharp that they could have been left this morning.

I cannot help but think about the distant creatures who made them. Where did they come from? Where were they going? We simply do not know. It has been suggested that they were merely crossing this scorched plain toward the greener ridges to the north. Perhaps so.

In any case, those footprints out of the deep past, left by the oldest known hominids, haunt the imagination. Across the gulf of time I can only wish them well on that prehistoric trek. It was, I believe, part of a greater and more perilous journey, one that—through millions of years of evolutionary trial and error, fortune and misfortune—culminated in the emergence of modern man.

Questions and Topics for Discussion and Writing

1. Leakey breaks a rhetorical "rule" for clear writing by opening with an ambiguous pronoun: "It happened. . . ." Is this opening effective in this instance? Why or why not?
2. The nineteenth-century American poet Henry Wadsworth Longfellow in his poem "Psalm of Life" concludes with this stanza:

    > Lives of great men all remind us
    > We can make our lives sublime
    > And departing leave behind us
    > Footprints on the sands of time.

    How does Leakey's use in her title of the allusion to this metaphor add to your understanding of Leakey's observations and inferences?
3. What effect does the fact that the essay was written for the *National Geographic* have on the choice of vocabulary in this "report"?
4. Compare and contrast this essay to John Harrington's "The Lure of the Hunt" (Unit 6). What similarities of attitude toward the process and the findings do you detect?
5. What does Leakey mean by the metaphor "unchronicled Rubicon"?
6. Leakey suggests that the "smaller hominid" experienced a "moment of doubt." What do you interpret this statement to mean? What effect does it have on your reading?
7. Leakey's essay was published in April 1979. Go to the library and research this subject to update Leakey's findings. Summarize your research in a short report.

Yi-Fu Tuan

The Chinese-born American professor Yi-Fu Tuan (b. 1930) has enlarged our conception of the meaning of geography. His own geographical experiences have been broad: Born in Tientsin, he attended school in China, Australia, and the Philippines before receiving his bachelor's and master's degrees from Oxford University in England. Tuan came to the United States in 1951 and became a citizen in 1973. He has taught in universities in Canada, England, and Hawaii, and currently teaches at the University of Wisconsin–Madison. Relating geography to other disciplines, such as art, architecture, literary history, and psychology, Tuan uses his knowledge of geography as both complement and metaphor.

Tuan received his doctorate from the University of California, Berkeley, in 1957 and began a prolific career of scholarly writing. Among his books are *The Hydrologic Cycle and the Wisdom of God* (1968), *Man and Nature* (1971), *Topophilia* (1974), and *Landscapes of Fear* (1980). The last book "maps the mind" to reveal the forms human fear and anxiety have taken throughout history.

Among Tuan's numerous honors, a writing award has special interest: A recent paper, "The Landscapes of Sherlock Holmes," received the Journal of Geography Award for best paper in 1984–1985. The following essay, an excerpt from *Topophilia*, illustrates Tuan's ability to call up images from many periods of history and areas of the world in order to orient the reader to what Tuan would surely call "human geography."

Topophilia and Environment

Yi-Fu Tuan

People dream of ideal places. The earth, because of its varying defects, is not everywhere viewed as mankind's final home. On the other hand, no environment is devoid of the power to command the allegiance of at least some people. Wherever we can point to human beings, there we point to somebody's *home*— with all the kindly meaning of that word. The Sudan is monotonous and niggardly to the outsider, but Evans-Pritchard says that he can hardly persuade the Nuer who live there that better places exist outside its confines.[1] In a complex modern society, individual tastes in natural setting can vary enormously. Some people prefer to live in, and not just visit, the desert and windswept plains. Alaskans may develop a liking for their frozen landscapes. Most people, however, prefer a more accommodating environment as their home, although they may wish to stimulate their aesthetic palate with an occasional visit to the desert. The expansive steppe, desert, and ice field discourage settlement not only because of their meager biologic base, but because their taut geometry and hardness seem to deny the idea of shelter. On a fertile plain the

[1] E. Evans-Pritchard, *The Nuer* (Oxford: Clarendon Press, 1940), p. 51.

sense of shelter can be created artificially with tree groves and houses grouped around an open space. The natural environment itself may produce a sense of shelter by being penetrable like the tropical forest, isolated and luxuriant like the tropical island, concave in geometry and diversified in resource as in a valley or along a protected seashore. . . . We have noted how, to the BaMbuti Pygmies and the Lele of Kasai, the tropical forest is an enveloping world that caters to their material and deepest spiritual needs. The sylvan environment was also the warm nurturing womb out of which the hominids were to emerge. Today the cabin in the forest clearing remains a powerful lure to the modern man who dreams of withdrawal. Three other natural settings have, at different times and places, appealed strongly to the human imagination. These are the seashore, the valley, and the island.

The attractiveness to human beings of the sheltered cove by the sea is not difficult to understand. To begin with, its geometry has a two-fold appeal: on the one hand, the recessions of beach and valley denote security, on the other, the open horizon to the water incites adventure. Furthermore, water and sand receive the human body that normally enjoys contact only with the air and the ground. The forest envelops man in its cool shadowed recesses; the desert is total exposure in which man is repulsed by the hard earth and excoriated by the brilliant sun. The beach too is washed in the brilliance of direct or refracted sunlight but the sand yields to pressure, wedging in between the toes, and water receives and supports the body.

The sheltered sea- or lakeshore may be one of mankind's earliest homes, dating back in Africa to Lower and Middle Paleolithic times. If the forest environment is necessary to the evolution of the perceptual and locomotive organs of man's primate ancestors, the seashore habitat may have contributed to man's hairlessness, a trait that distinguishes him from apes and monkeys. Theories concerning the causes of evolutionary traits in the remote past are at best uncertain. Human agility in water is, however, a fact. The talent is not widely shared among the primates. Other than human beings, only certain Asian macaques forage for food along the seashores, and can swim. Could it be that our earliest home was a sort of Eden located near a lake or sea? Consider Carl Sauer's sketch of the advantages of the seashore: "No other setting is as attractive for the beginning of humanity. The sea, in particular the tidal shore, presented the best opportunity to eat, settle, increase, and learn. It afforded diversity and abundance of provisions, continuous and inexhaustible. It invited the development of manual skills. It gave the congenial ecologic niche in which animal ethology could become human culture."[2]

Primitive people who live near tropical and temperate shores are generally excellent swimmers and divers. Note further that in water the sexes differ little

[2] Carl O. Sauer, "Seashore—Primitive Home of Man?" in John Leighly (ed.), *Land and Life* (Berkeley: University of California Press, 1963), p. 309.

in ability, which means that they could participate as equals in work and in enjoying water sports. Carl Sauer suggested that the merging of recreation and economic activity might have attracted the primordial males to join in provision from the sea long before they amounted to anything as hunters on the land; and also that such participation helped to establish the bilateral family. In the pre-historic past the evidence of shell mounds suggests that the sea- and lakeshores were often capable of supporting population densities far in excess of those inland where the people had to depend on hunting and collecting. Perhaps only as agriculture became more sophisticated, late in the Neolithic period, did people begin to concentrate inland in large numbers, although even then fishing in the streams contributed importantly to their diet.

Fishing communities in the modern world are poor, generally speaking, when compared with farming communities in the interior; and if they endure it is less for the economic rewards than for the satisfactions to be got out of an ancient and lore-drenched way of life. In the course of the last century the seashore has become immensely popular, but health and pleasure, not the produce of the sea, have been the major attractions. Every summer hordes of people in Europe and North America migrate to the beaches. Take the example of Britain. In 1937, about 15 million people enjoyed a holiday away from home for a week or longer. By 1962, 31 million or sixty percent of Britain's population did so; and of the holidays spent within the country the greatest number were taken by the sea. In 1962, seventy-two percent of all British holiday-makers went to the coast. Swimming was and is by far the most actively pursued sport, engaged in by young and old alike. In 1965, no other single sport even ap-proached half the number of swimming adherents.[3] But, as E. W. Gilbert has pointed out, the popularity of swimming and the beaches is a fairly recent happening: Britain's insularity has not in itself encouraged any precocious culti-vation of the pleasures of the coast. It was the growing reputation of sea water and of sea bathing for health that turned the attention of health seekers from the inland spas of ancient fame to the beaches. The power of sea water owed much of its credibility to a Dr. Richard Russell of Lewes and Brighton. In 1750 he published a book on the use of sea water in treating the diseases of the glands which found favor among hypochondriacal and pleasure-seeking Euro-peans for the next century. The rapid growth of seaside resorts, particularly from the 1850s onward, was made possible by the spread of railroads. The one-day, weekend, and seasonal surges to the sea, a post–World War II phe-nomenon, reflects the increasing affluence of the lower-middle and middle classes, and the sharp rise in the use of automobiles.[4] Economic and technologi-cal factors explain the accelerating volume of the movement to the sea, but not

[3] J. Allan Patmore, *Land and Leisure in England and Wales* (Newton Abbot, Devon: David & Charles, 1970), p. 60.
[4] E. W. Gilbert, "The Holiday Industry and Seaside Towns in England and Wales," *Festschrift Leopold G. Scheidl zum 60 Geburtstag* (Vienna, 1965), pp. 235–47.

why people should have found it attractive in the first place. The origin of the movement to the sea is rooted in a new evaluation of nature.

In the United States inland spas again preceded the seaside resorts as centers for pleasure and health.[5] Although bathing in the sea already appeared toward the end of the eighteenth century, it was much later that it became at all popular. From the first, sea bathing had to overcome prudery. Manufacturers advertised machines of a "peculiar construction" which enabled the bathers to get in and out of the water unseen. Swimming also aroused suspicion because it was a sport for mixed company. Bathers of the late nineteenth century waded into the sea in full attire. Social mores, however, do change: common sense eventually overcomes prudery. Since the early years of the twentieth century swimming has been, and remains, the greatest outdoor recreation for Americans. The east coast beaches have been seasonally packed since the 1920s. Swimming, unlike many competitive sports, minimizes the physical and social differences among human beings. The sport is suited to the entire family. It requires no expensive equipment. The infants, the old, and even the lame can enjoy the benevolent world of the seashore. The sport's popularity is good litmus for the strength of a country's democratic sentiment.

The valley or basin of modest size appeals to human beings for obvious reasons. As a highly diversified ecological niche it promises an easy livelihood: a variety of food is available from the river, the floodplain and the valley slopes. The human being is greatly dependent on easy access to water: he has no means of retaining it in his body for long periods. The valley holds water in its streams, pools, and springs. If the stream is large enough it also serves as the natural route way. Farmers appreciate the rich soils on the valley floor. There are, of course, disadvantages, especially to early man with simple tools. The tangled growths of the floodplain, besides sheltering dangerous animals, may be difficult to clear. The plain may be poorly drained and malarial; it is subject to floods and to temperature fluctuations greater than those experienced on higher slopes. The soils, though rich, are heavy. Some of these difficulties could be avoided or mitigated. Large swampy plains subject to violent floods were shunned, and where possible the settlements were located on dry gravel terraces and at the foot of the valley flanks. It was in the valleys and basins of moderate size that mankind took the first steps toward agriculture and to sedentary life in large village communities.

Symbolically the valley is identified with the womb and shelter. Its concavity protects and nurtures life. When the primate ancestors of man moved out of the forest to the plains they sought the physical and (one might guess) psychological security of the cave. Artificial shelters are man-made concavities in which life processes might function away from exposure to light and to the threats of the

[5] Foster R. Dulles, *A History of Recreation: America Learns to Play*, 2nd ed. (New York: Appleton-Century-Crofts, 1965), pp. 152–53, 355–56.

natural environment. The earliest constructed dwellings were often semisub-
terranean: the digging of the hollow minimized the need for a superstructure
and at the same time brought the inhabitants into closer contact with the earth.
The valley is chthonic and feminine, the *megara* of biologic man. Mountain
tops and other eminences are ladders to the sky, the home of the gods. There
man might build temples and altars but not his own dwellings except to escape
from attack.

The island seems to have a tenacious hold on the human imagination. Unlike
the tropical forest or the continental seashore it cannot claim ecological abundance, nor—as an environment—has it mattered greatly in man's evolutionary
past. Its importance lies in the imaginative realm. Many of the world's cosmog-
onies, we have seen, begin with the watery chaos: land, when it appears, is
necessarily an island. The primordial hillock was also an island and on it life had
its start. In numerous legends the island appears as the abode of the dead or of
immortals. Above all, it symbolizes a state of prelapsarian innocence and bliss,
quarantined by the sea from the ills of the continent. Buddhist cosmology recognizes four islands of "excellent earth" situated in the "exterior sea." Hindu
doctrine tells of an "essential island" of pulverized gems on which sweet-smell-
ing trees grow; it houses the *magna mater*. China has the legend of the Blessed
Isles or the Three Isles of the Genii which were believed to be located in the
Eastern Sea, opposite the coast of Chiang-su. The Semang and Sakai of Malaya,
forest dwellers, conceive paradise as an "island of fruits" from which all the ills
that afflict man on earth have been eliminated; it is located in heaven and has to
be entered from the West. Some Polynesian peoples envisage their Elysium in
the form of an island, which is not surprising. But it is in the imagination of the
Western world that the island has taken the strongest hold. Here is a brief
sketch.

The legend of the Island of the Blessed first appeared in archaic Greece: it
was described as a place that provided heroes with unusual harvests thrice a
year. The Celtic world, remote from Greece, had a similar legend: Plutarch
relates the story of a Celtic island on which no one toiled, its climate was
exquisite, its air steeped in fragrance. In Christian Ireland, certain pagan ro-
mances were converted into edifying tales of saintly endeavor. Especially popular throughout medieval Europe was the legend of St. Brendan, in which the
Abbot of Clonfort (d. 576) became a seafaring hero who discovered insular
paradises of blissful ease and abundance. In a twelfth-century Anglo-Norman
version of the tale, Brendan was made to search for an island described glow-
ingly as a home for the pious that lay beyond the sea, "where no tempest revels,
where for nourishment one inhales the perfume of flowers from paradise."

The imagination of the Middle Ages peopled the Atlantic with a large number of islands, many of which persisted well into the age of the great explorations, and indeed one, Brasil (Gaelic term for blessed), persisted in the mind of

the British Admiralty until the second half of the nineteenth century.[6] By 1300 the classical Fortunate Isles came to be identified with the islands of St. Brendan. The cardinal Pierre d'Ailly, whom Columbus regarded as an authority on geography, seriously inclined to the view that the Earthly Paradise was located on or near the Fortunate Isles because of the fertility of their soil and the excellence of their climate. Ponce de Leon was reported to have searched for the Fountain of Youth in Florida, and by thinking of Florida as an island he followed the tradition of identifying enchantment with insularity. In 1493 the New World dawned on the European imagination as small delectable island-gardens. By the seventeenth century the New World had stretched to an interminable continent, and the original vision of isles of innocence and sunshine turned to incredulity as colonists stood before the immeasurable and the horrifying.[7]

The fantasy of island Edens received a new lease of life in the eighteenth century, as the somewhat ironic consequence of scientific expeditions to the South Seas. Unlike early explorers, Louis de Bougainville did not believe in any literal Eden, but his glowing account of Tahiti made the island an acceptable substitute. The voyages of Captain Cook largely confirmed the desirability of the South Sea islands. George Forster, a naturalist who accompanied Cook on his second voyage, believed that they owed much of their special appeal to contrast with the prior experience of tedium over the empty seas. In the nineteenth century, missionaries assaulted the Edenic image of tropical islands. On the other hand eminent writers who visited them—including Herman Melville, Mark Twain, Robert Louis Stevenson, and Henry Adams—upheld their reputation. The islands triumphed over adverse propaganda: tourists continued to flock to them. They acquired another meaning, that of temporary escapism. Gardens of Eden and island utopias have not always been taken seriously, least of all in the twentieth century. But they seem needed as make-believe and a place of withdrawal from high-pressured living on the continent.[8]

Questions and Topics for Discussion and Writing

1. In what ways can Tuan's article be important to other sciences, such as biology, psychology, and anthropology?
2. Define topography as a science.

[6]Carl O. Sauer, *Northern Mists* (Berkeley and Los Angeles: University of California Press, 1968), pp. 167–68; W. H. Babcock, *Legendary Islands of the Atlantic: A Study in Medieval Geography* (New York: American Geographical Society, 1922).

[7]Howard Mumford Jones, *O Strange New World* (New York: Viking, 1964), p. 61.

[8]Henri Jacquier, "Le mirage et l'exotisme Tahitien dans la littérature," *Bulletin de la Société des Oceaniennes*, 12, Nos. 146–147 (1964), 357–69.

3. What does Tuan's writing about topography and human emotions—united linguistically as *topophilia*—imply about him as a scientist?

4. Define the word "neologism." Compare the neologism in the essay "E Pluribus Boojum" by N. David Mermin (Unit 5) with the neologism *topophilia*. What differences of method and spirit do you see in the creation of the names in the two fields?

5. What does Tuan's reference to a "land's biological base" and its "geometry" suggest about his concern with and knowledge of other sciences?

6. Summarize how, according to Tuan, the topography of the forest, the sea, and the lakes contributed to human evolution.

7. Tuan declares that the popularity of swimming is "good litmus for the strength of a country's democratic sentiment." Explain the metaphor and comment on its effectiveness. Do you agree with Tuan's opinion? What other sport might you choose for such a statement?

8. Tuan brings together terms from many disciplines to explain his concepts. Look up and define *chthonic, prelapsarian,* and *primordial.*

9. The tone of Tuan's essay changes as he describes the valleys as "feminine" and alludes extensively to mythology and literature in describing islands. How would you characterize Tuan's "angle of vision" as a scientist while working in his particular field?

10. Of the four natural environments Tuan emphasizes as important in humanity's dreams of an ideal world—the forest, the seashore, the valley, and the island—which most captures your imagination? Write a paragraph describing your ideal place, inventing your own images of topophilia.

11. Do you believe that science and technology have influenced your view of an ideal place in any way? If so, how?

12. Read Tuan's prize-winning paper, "The Landscapes of Sherlock Holmes" (*Journal of Geography* 84: 56–60, 1985). Look for those elements of the essay that you believe may account for its winning the prize. Write an expository paper presenting your analysis.

Carole Oles

Carole Oles was born in New York City and is a graduate of Queens College and the University of California at Berkeley. As a poet, she has received many honors and awards, among them a National Endowment for the Arts Grant in Poetry and the Poetry Society of America's Gertrude B. Claytor Award. In addition to appearing in numerous magazines, her works have been published in three volumes: *The Loneliness Factor, Quarry,* and *Night Watchers: Inventions on the Life of Maria Mitchell.*

Maria Mitchell (1818–1889) was an American astronomer—a rare occupation for a nineteenth-century woman. Born on the New England island of Nantucket, she was trained in astronomy by her father. She became Professor of Astronomy when Vassar College was founded in 1865 and taught there until she retired in 1888.

Of Oles's book *Night Watchers,* inspired by Mitchell, the poet William Matthews writes: "Carole Oles performs an act of empathy so subtle and thorough as to be an act of identification: She *becomes* Maria Mitchell, the pioneering American astronomer. The fears and condescensions Mitchell excites by being an intelligent woman are depressingly familiar from our current perspective, but what is perhaps most moving in Carol Oles's achievement is her steady understanding of how Mitchell turned even these injustices into fuel for her work."

Discover for yourself how Oles recreated Maria Mitchell in "Miss Mitchell's Comet."

Miss Mitchell's Comet

Carole Oles

October 1, 1847, 10:30. Finally the cover
has lifted. That cannot be mist
at 5 degrees above Polaris, where late
as Tuesday nothing was. Does sight
betray? Stay calm. If it be sent
to test, take the test then. No

nebula could have held that place unnoticed
all year. I swept over and over
the quadrant. Was I a sentry
whose watching grew remiss,
the nebula slipping past my sight?
Take notes, hand. After, let head postulate

this seeming comet, new word on the slate
of sky. *October 2.* The object has no
train, appears more brilliant, sighted
now tracking toward the Pole, a rover
we cannot dismiss.
Father says news should be sent

to William Bond at Harvard. I dissent.
What if our find comes late,
the comet an old one we have always missed?

I dread error. But not to know
is the germ that breeds discovery.
The seer must avow the sight.

October 5. Hurtling still, in sight
when it occults a star, the comet's center
within a second of the star which overwhelms
its brightness. I calculate
the point where two celestial nobles
meet, where nothing seems mistake.

Once a man's voice pierced the dark—"Miss,
what are you doing there?"—shook my sight.
Spiders, ticks, even rats had visited, but no
person until he made his arboreal ascent

to find out what lone women do so late
on rooftops, with only sky for cover.

Mister, with nature's consent I climb to sight,
to know the heavens. Later I must sleep. Here
are comets to discover.

Questions and Topics for Discussion and Writing

1. A sestina is a patterned poem in which the last word of each line in the first stanza is repeated in a different order in the following stanzas. In "Miss Mitchell's Comet" the end words "cover," "mist," "late," "sight," "sent," and "no" are repeated with some interesting variations throughout the poem. How do these repeated words mimic the pattern of Maria Mitchell's meditations on the comet she may have discovered? How do Carole Oles's variations on each end word—for instance, "sent," "sentry," "dissent," "center," "ascent," and "consent"—enrich both the sound and the meaning of the poem?

2. Oles starts her poem with the words, "Finally the cover/has lifted." She ends with the words "Here/are comets to discover." Look up the meaning of "discover," noting especially its derivation. Comment on the significance of the opening clause in terms of the discovery of the comet.

3. Maria Mitchell at first hesitates to send news of the comet to Harvard. Why? What does she mean by "The seer must avow the sight"?

4. Maria Mitchell notes that the comet is "in sight/when it occults a star." What is the meaning of "occult," and how is it used here?

5. What is the role of the man who "made his arboreal ascent"?

6. Considering that Oles uses an elaborate poetic form, write a paragraph in which you pay attention to sentence length, word choice, and tone to explain how Oles simultaneously creates the impression of a scientist's hastily jotted journal.

James Jeans

James Hopwood Jeans (1877–1946) is best known for his theory of the creation of the solar system. This British mathematician, physicist, and astronomer graduated from Cambridge University, England, in 1900 and went on to graduate studies, which, unfortunately, were interrupted by a two-year bout with tuberculosis. Appointed to be a university lecturer in 1904, he resigned to become a professor of applied mathematics at Princeton University in the United States.

While at the prestigious New Jersey institution, Jeans wrote *The Dynamical Theory of Gases, Theoretical Mechanics*, and *The Mathematical Theory of Electricity and Magnetism.* He returned to England in 1910 to lecture at Cambridge. He held an honorary chair at the Royal Institution and was also elected a fellow of the Royal Society in London. Eventual honors included knighthood in 1928 and the Order of Merit in 1939.

Jeans developed many applications of mathematics to problems involving physics and astronomy. His role in developing the kinetic theory of gases helped lay the groundwork for Planck's quantum theory. Jeans's greatest contribution to science was his *Problems of Cosmogony and Stellar Dynamics* (1919). He made vital contributions to the problems of cosmogonic theories and discovered in 1926 the radiative viscosity in stars. After about 1925, Jeans turned to interpretation of science for a popular audience. His *Universe Around Us* (1929) aroused the interest of the public in the nature and meanings of modern astronomy.

In later years, his writings turned more obviously to philosophy. *The New Background of Science* (1933) and *Physics and Philosophy* (1942) exemplify this bent. Some of these philosophical works became targets for the criticism of professional philosophers who attempted to rebut their underlying deistic premise. In *The Mysterious Universe* (1930), for example, he reaches what L. Susan Stebbing calls "the final stage of Jeans's escape" when he views God as the "celestial Mathematician," a pure mathematician whose universe is therefore a mathematical creation. (See "The Escape of Sir James Jeans" in Stebbing's *Philosophy and the Physicists*, 1937, for a criticism of the arguments and methods of Jeans and other popularizing philosopher–physicists.) The notion of God as a pure mathematician and of the universe as spiritual in nature was quite acceptable, however, to a wide segment of the public as well as to various religious groups.

In "A Magic Rocket through Space and Time," Jeans displays his skill of interpreting scientific phenomena through analogy and compelling imagery.

A Magic Rocket through Space and Time

James Jeans

Let us enter this magic rocket and persuade someone to shoot us towards the sun. We need only start with speed enough to carry us a short distance away from the earth—about 7 miles a second will do—and the sun's huge gravitational pull will do the rest. It will drag us down into the sun whether we like it or not. If we start at 7 miles a second, the whole journey will take about ten weeks.

Even in the first few seconds of our flight, we notice strange changes; the whole colour-scheme of the universe alters with startling suddenness. The sky rapidly darkens in hue, until finally it assumes a blackness like that of midnight, from which the stars shine out. They no longer twinkle in the friendly way we are accustomed to on earth; their rays have become piercing needles of steady light. Meantime the sun has changed to a hard steely whiteness, and the shadows it casts are harsh and fierce. Nature seems to have lost a large part of her beauty, and all of her softness, in a surprisingly short space of time. The explanation is that a very few seconds take us entirely clear of the earth's atmosphere, and not until we have left it behind us do we realise how much its softening effect has added to the pleasure of our lives.

Let us pause for a moment to consider the scientific reasons for this. Imagine that we stand on any ordinary seaside pier, and watch the waves rolling in and striking against the iron columns of the pier. Large waves pay very little attention to the columns—they divide right and left and re-unite after passing each column, much as a regiment of soldiers would if a tree stood in their road; it is almost as though the columns had not been there. But the short waves and ripples find the columns of the pier a much more formidable obstacle. When the short waves impinge on the columns, they are reflected back and spread as new ripples in all directions. To use the technical term, they are "scattered." The obstacle provided by the iron columns hardly affects the long waves at all, but scatters the short ripples.

We have been watching a sort of working model of the way in which sunlight struggles through the earth's atmosphere. Between us on earth and outer space the atmosphere interposes innumerable obstacles in the form of molecules of air, tiny droplets of water, and small particles of dust. These are represented by the columns of the pier.

The waves of the sea represent the sunlight. We know that sunlight is a blend of lights of many colours—as we can prove for ourselves by passing it through a prism, or even through a jug of water, or as Nature demonstrates to us when she passes it through the raindrops of a summer shower and produces a rainbow. We also know that light consists of waves, and that the different colours of light are produced by waves of different lengths, red light by long waves and blue light by short waves. The mixture of waves which constitutes sunlight has to struggle through the obstacles it meets in the atmosphere, just as the mixture of waves at the seaside has to struggle past the columns of the pier. And these obstacles treat the light-waves much as the columns of the pier treat the sea-waves. The long waves which constitute red light are hardly affected, but the short waves which constitute blue light are scattered in all directions.

Thus, the different constituents of sunlight are treated in different ways as they struggle through the earth's atmosphere. A wave of blue light may be scattered by a dust particle, and turned out of its course. After a time a second dust particle again turns it out of its course, and so on, until finally it enters our eyes by a path as zigzag as that of a flash of lightning. Consequently the blue waves of the sunlight enter our eyes from all directions. And that is why the sky looks blue. But the red waves come straight at us, undeterred by atmospheric obstacles, and enter our eyes directly. When we look towards the sun, we see it mainly by these red rays. They are not the whole light of the sun; they are what remains after a good deal of blue has already been filtered out by atmospheric obstacles. This filtering of course makes the sunlight redder than it was before it entered our atmosphere. The more obstacles the sunlight meets, the more the blue is extracted from it, and so the redder the sun looks. This explains why the sun looks unusually red when we see it through a fog or a cloud of steam. It also explains why the sun looks specially red at sunrise or sunset—the sun's light, coming to us in a very slantwise direction, has to thread its way past a great number of obstacles to reach us. It also explains the magnificent sunsets which are often seen through the smoky and dusty air of a city—or even better after a volcanic eruption, when the whole atmosphere of the world may be full of minute particles of volcanic dust.

In such ways as this, the earth's atmosphere breaks up the sunlight. The true sunlight, as it is when it leaves the sun, or travels through space before meeting the earth at all, is a blend of all the colours into which the earth's atmosphere breaks it up. To reconstruct this colour, we must blend the blue of the sky with the yellow or red of the direct sunlight. This makes the steely-white light we see as soon as our rocket takes us beyond the earth's atmosphere.

This action of the atmosphere in breaking up sunlight is responsible for much of the beauty of the earth—the blue sky of full day, the vivid orange and red of the rising and setting sun, the fairyland hues of the clouds at sunrise and sunset, the mysterious tones of twilight, the pink afterglow on the mountains, the purple of the distant hills, the apple-green in the western evening sky and the indigo in the east, and indeed all the effects which the artist describes as atmospheric. As we pass beyond the earth's atmosphere, we leave all these behind us, and enter a hard world which is divided sharply into light and dark, and knows nothing of half-tones. For the first time in our lives, we see the sun for what it really is—a vivid bluish globe of light. We see it set in a sky as black as that of midnight, because the earth's atmosphere no longer takes its rays and scatters them in all directions. It is to this weird and terrifying object that our rocket is taking us.

Questions and Topics for Discussion and Writing

1. How effective do you find Jeans's introduction, or invitation to the reader?
2. Why do you think Jeans calls the telescope "this magic rocket"? What fields does he link?
3. Though Jeans presents a journey through space, he also writes a paean of praise to the earth. Explain, by citing specific passages, how he does so.
4. Jeans's essay is unified by imagery of conflict. List words related to conflict.
5. Jeans uses analogy and various figures of speech, such as metaphor and personification, to express his ideas. Discuss his rhetorical techniques, listing specific examples.
6. Study the final paragraph. Explain how the theme of this paragraph challenges us.
7. Write a summarizing paragraph, explaining in your own words "why the sky looks blue."

Arthur Eddington

Sir Arthur Eddington (1882–1944), one of the foremost modern-day astronomers and cosmologists, was also an outstanding mathematician, physicist, and popular interpreter of science. Born in England, Eddington was only two years old at the death of his father, the headmaster of a local school. His higher education was at Cambridge University, where he received honors in mathematics.

From 1906 to 1913, he was the chief assistant at the Royal Observatory at Greenwich. While serving there, he learned how to use astronomical instruments. This practical experience led him to make observations on the island of Malta, establishing its longitude, and to lead an expedition to Brazil to observe a solar eclipse. The conclusions developed from these two research trips placed Eddington among the most knowledgeable of astronomers on stellar motions and star-drifts. His paper on the dynamics of a globular stellar system and his book *Stellar Movements and the Structure of the Universe* (1914) summarize his detailed mathematical findings that support the idea that the spiral nebulae and cloudy structures seen through the telescope are other galaxies like the Milky Way.

Like his fellow Englishman and mathematician Bertrand Russell, Eddington was a pacifist during World War I. In Eddington's case, however, the declaration stemmed from strong religious beliefs: He was a mystic in the Quaker tradition. His subjective faith also shaped his popular writings on the philosophy of science. In *Science and the Unseen,* he held that the meaning of the physical world cannot be uncovered by science but must be found through an apprehension of spiritual reality. *The Nature of the Physical World, New Pathways of Science,* and *The Philosophy of Physical Science* express similar ideas, holding that a scientist creates truth because answers depend on the questions he or she asks. Individualistic in perspective, as these views indicate, Eddington many times disagreed not only with philosophers but also with such fellow physicists as Sir James Jeans (excerpted earlier in this unit). Philosophers—for example, L. Susan Stebbing in her book *Philosophy and the Physicists*—often hold that Eddington's contributions to philosophy may not be lasting but agree that his attention to the philosophical implications of modern physics was important in generating wide interest in the subject.

His contributions as a scientist, however, are lastingly significant. In 1919, Eddington led a group of scientists to the coast of West Africa to investigate the truth of Einstein's theory on the action of gravity on light. Evidently, his investigations "asked the right questions," for he proved the theory and became a leading exponent of Einstein's theory of relativity. In 1930 he showed that the recent discovery made by Edwin Hubble that distant galaxies are receding gave support to Georges LeMaitre's model of the expanding universe; it is now called the Eddington–LeMaitre model. Although he worked to link relativity and the quantum theory within the theoretical framework of the expanding universe, Eddington was unable to complete this work before his death.

Perhaps Eddington's greatest contributions are his astrophysical works, such as his classic *Internal Constitution of the Stars;* his collected lectures, *Stars and Atoms;* and his influence in the application of structured logic to many lines of investigation. Eddington received a range of honors, among them knighthood and the Order of Merit in 1938.

In the following essay, "Parable of the Fishing Net," from *The Philosophy of Physical Science,* examine Eddington's use of analogy and metaphor.

Parable of the Fishing Net

Arthur Eddington

I

Let us suppose that an ichthyologist is exploring the life of the ocean. He casts a net into the water and brings up a fishy assortment. Surveying his catch, he proceeds in the usual manner of a scientist to systematise what it reveals. He arrives at two generalisations:

(1)   No sea-creature is less than two inches long.
(2)   All sea-creatures have gills.

These are both true of his catch, and he assumes tentatively that they will remain true however often he repeats it.

In applying this analogy, the catch stands for the body of knowledge which constitutes physical science, and the net for the sensory and intellectual equipment which we use in obtaining it. The casting of the net corresponds to observation; for knowledge which has not been or could not be obtained by observation is not admitted into physical science.

An onlooker may object that the first generalisation is wrong. "There are plenty of sea-creatures under two inches long, only your net is not adapted to catch them." The ichthyologist dismisses this objection contemptuously. "Anything uncatchable by my net is *ipso facto* outside the scope of ichthyological knowledge, and is not part of the kingdom of fishes which has been defined as the theme of ichthyological knowledge. In short, what my net can't catch isn't fish." Or—to translate the analogy—"If you are not simply guessing, you are claiming a knowledge of the physical universe discovered in some other way than by the methods of physical science, and admittedly unverifiable by such methods. You are a metaphysician. Bah!"

The dispute arises, as many disputes do, because the protagonists are talking about different things. The onlooker has in mind an objective kingdom of fishes. The ichthyologist is not concerned as to whether the fishes he is talking about form an objective or subjective class; the property that matters is that they are catchable. His generalisation is perfectly true of the class of creatures he is talking about—a selected class perhaps, but he would not be interested in making generalisations about any other class. Dropping analogy, if we take observation as the basis of physical science, and insist that its assertions must be verifiable by observation, we impose a selective test on the knowledge which is admitted as physical. The selection is subjective, because it depends on the sensory and intellectual equipment which is our means of acquiring observational knowledge. It is to such subjectively-selected knowledge, and to the universe which it is formulated to describe, that the generalisations of physics— the so-called laws of nature—apply.

It is only with the recent development of epistemological methods in physics that we have come to realise the far-reaching effect of this subjective selection of its subject matter. We may at first, like the onlooker, be inclined to think that physics has missed its way, and has not reached the purely objective world which, we take it for granted, it was trying to describe. Its generalisations, if they refer to an objective world, are or may be rendered fallacious through the selection. But that amounts to condemning observationally grounded science as a failure because a purely objective world is not to be reached by observation.

Clearly an abandonment of the observational method of physical science is out of the question. Observationally grounded science has been by no means a failure; though we may have misunderstood the precise nature of its success.

Those who are dissatisfied with anything but a purely objective universe may turn to the metaphysicians, who are not cramped by the self-imposed ordinance that every assertion must be capable of submission to observation as the final Court of Appeal. But we, as physicists, shall continue to study the universe revealed by observation and to make our generalisations about it; although we now know that the universe so reached cannot be wholly objective. Of course, the great mass of physicists, who pay no attention to epistemology, would have gone on doing this in any case.

Should we then ignore the onlooker with his suggestion of selection? I think not; though we cannot accept his remedy. Suppose that a more tactful onlooker makes a rather different suggestion: "I realise that you are right in refusing our friend's hypothesis of uncatchable fish, which cannot be verified by any tests you and I would consider valid. By keeping to your own method of study, you have reached a generalisation of the highest importance—to fishmongers, who would not be interested in generalisations about uncatchable fish. Since these generalisations are so important, I would like to help you. You arrived at your generalisation in the traditional way by examining the fish. May I point out that you could have arrived more easily at the same generalisation by examining the net and the method of using it?

The first onlooker is a metaphysician who despises physics on account of its limitations; the second onlooker is an epistemologist who can help physics because of its limitations. It is just because of the limited—some might say, the perverted—aim of physics that such help is possible. The traditional method of systematic examination of the data furnished by observation is not the only way of reaching the generalisations valued in physical science. Some at least of these generalisations can also be found by examining the sensory and intellectual equipment used in observation. Epistemology thus presents physics with a new method of achieving its aims. The development of relativity theory, and the transformation of quantum theory from an empirical to a rational theory are the outcome of the new method; and in it is our great hope of further fundamental advances.

II

We return to our fish to illustrate another point of great importance. No suggestion was offered as to the second generalisation—that all sea-creatures have gills—and, so far as we can see, it could not have been deduced from an examination of the net and its mode of use. If the ichthyologist extends his investigations, making further catches, perhaps in different waters, he may any day bring up a sea-creature without gills and upset his second generalisation. If this happens, he will naturally begin to distrust the security of his first generalisation. His fear is needless; for the net can never bring up anything that it is not adapted to catch.

Generalisations that can be reached epistemologically have a security which is denied to those that can only be reached empirically.

It has been customary in scientific philosophy to insist that the laws of nature have no compulsory character; they are uniformities which have been found to occur hitherto in our limited experience, but we have no right to assert that they will occur invariably and universally. This was a very proper philosophy to adopt as regards empirical generalisations—it being understood, of course, that no one would be so foolish as to apply the philosophy in practice. Scientists, assured by their philosophy that they had no right to expectations, continued to cherish indefensible expectation, and interpreted their observations in accordance with them. Attempts have been made by the theory of probability to justify our expectation that if an occurrence (whose cause is unknown) has happened regularly hitherto it will continue to happen on the next occasion; but I think that all that has emerged is an analysis and axiomatisation of our expectation, not a defence of it.

The situation is changed when we recognise that some laws of nature may have an epistemological origin. These are compulsory; and when their epistemological origin is established, we have a right to our expectation that they will be obeyed invariably and universally. The process of observing, of which they are a consequence, is independent of time or place.

But, it may be objected, can we be sure that the process of observing[1] is unaffected by time or place? Strictly speaking, no. But if it is affected—if position in time and space or any other circumstance prevents the observational procedure from being carried out precisely according to the recognised specification—we can (and do) call the resulting observation a "bad observation." Those who resent the idea of compulsion in scientific law may perhaps be mollified by the concession that, although it can no longer be accepted as a principle of scientific philosophy that the laws of nature are uncompulsory, there is no compulsion that our actual observations shall satisfy them, for (unfortunately) there is no compulsion that our observations shall be *good* observations.

What about the remaining laws of nature, not of an epistemological origin, and therefore, so far as we know, non-compulsory? Must they continue to mar the scheme as a source of indefensible expectations, which nevertheless are found to be fulfilled in practice? Before worrying about them, it will be well to wait till we see what is left of the system of natural law after the part which can

---

[1] The standard specification of the procedure of observing must be sufficiently detailed to secure a unique result of the observation. It is the duty of the observer to secure that all attendant circumstances which can affect the result, e.g., temperature, absence of magnetic field, etc., are in accordance with specification. Epistemological laws governing the results of the observation are such as are inferable solely from the fact that the procedure was as specified. The contingency referred to in this paragraph is exemplified by the fact that it is impossible to make a really "good" observation of length in a strong magnetic field, because the standard specification of the procedure of determining length requires us to eliminate magnetic fields. . . .

be accounted for epistemologically has been removed. There may not be anything left to worry about.

The introduction of epistemological analysis in modern physical theory has not only been a powerful source of scientific progress, but has given a new kind of security to its conclusions. Or, I should rather say, it has put a new kind of security within reach. Whether the present conclusions are secure is a question of human fallibility, from which the epistemologist is no more exempt than the classical theorist or the practical observer. Whilst not forgetting that the actual results achieved must depend on the insight and accuracy of those who use the equipment, I would emphasise that we have now the equipment to put theoretical physics on a surer footing than it formerly aspired to.

III

Quis custodiet ipsos custodes? Who will observe the observers? The answer is—the epistemologist. He watches them to see what they really observe, which is often quite different from what they say they observe. He examines their procedure and the essential limitations of the equipment they bring to their task, and by so doing becomes aware beforehand of limitations to which the results they obtain will have to conform. They, on the other hand, only discover these limitations when they come to examine their results, and, unaware of their subjective origin, hail them as laws of nature.

It may be argued that, in accepting the aid of epistemology, physical science continues to be wholly an inference from observation; for the epistemologist too is an observer. The astronomer observes stars; the epistemologist observes observers. Both are seeking a knowledge which rests on observation.

I am sorry I must offend the observationalists by rejecting this sop to traditional views; but the analogy between observing stars and observing observers will not hold good. The common statement that physical science rests on observation, and that its generalisations are generalisations about observational data, is not quite the whole truth. It rests on *good* observation, and its generalisations are about *good* observational data. Scientific epistemology, which is concerned with the nature of the knowledge contained in physical science, has therefore to examine the procedure of good observation. The proper counterpart to the epistemologist who observes good observers is the astronomer who observes good stars.

This qualification of observations as "good," which is the first point attended to in practice, seems often to have been overlooked in philosophy. In speaking of observation, there is often a failure to distinguish the special kind of observational activity contemplated in physical science from indiscriminately "taking notice." The distinction is strongly selective; and it indicates one way in which the subjective selection, to which we have referred, is introduced into the universe described by physics. If astronomers were similarly allowed to distinguish good stars and bad stars, astronomy would doubtless be enriched by some remarkable new laws—applying, of course, only to the good stars which obey the laws so prescribed.

Whether an observation is good or bad depends on what it professes to represent. A bad determination of the melting-point of sulphur may be an excellent determination of the melting-point of a mixture of sulphur and dirt. The terms used to describe an observation—to state what it is an observation of—imply by their definition a standard procedure to be followed in making it; the observer professes to follow this procedure, or a procedure which he takes the liberty of substituting for it in the belief that it will assuredly give the same result. If, through inadvertence or practical difficulty, the prescribed conditions of the procedure are not carried out, the observation is a bad observation, and the observer in this instance is a bad observer. Equally from the point of view of physical science he is a bad observer if his belief that his method can be substituted for the standard procedure is mistaken; though in this case he will pass the blame on to the theorist who advised him wrongly.

The epistemologist accordingly does not study the observers as organisms whose activities must be ascertained empirically in the same way that a naturalist studies the habits of animals. He has to pick out the good observers—those whose activities follow a conventional plan of procedure. What the epistemologist must get at is this plan. Without it, he does not know which observers to study and which to ignore; with it, he need not actually watch the good observers who, he knows already, are merely following its instructions, since otherwise they would not be good.

The plan must be sought for in the mind of the observer, or in the minds of those from whom he has derived his instructions. The epistemologist is an observer only in the sense that he observes what is in the mind. But that is a pedantic description of the way in which we discover a plan conceived in anyone's mind. We learn the observer's plan by listening to his own account of it and cross-questioning him.

Questions and Topics for Discussion and Writing

1. This selection is often called "Parable of the Fishing Net." What is a *parable?* Why do you think Eddington has used one here?
2. Write a definition of *ichthyology.*
3. What does Eddington say is the limitation of observation as the basis of physical science?

4. Define *epistemology*.
5. What does Eddington say is the reason that observationally grounded science, despite its limitations, is not a failure?
6. What does Eddington say is the conclusion of physicists about the possibility of ever attaining complete knowledge of the universe?
7. Eddington uses hypothetical argument for clarity and emphasis. Is he successful?
8. What, according to Eddington, is the important contribution of epistemology to physics?
9. In a sentence or two, paraphrase ". . . the net can never bring up anything that it is not adapted to catch." Write a paragraph translating the metaphor into a statement describing scientific research technique.
10. Eddington formulates his argument into a thesis statement. Paraphrase it into your own words.
11. The title of the chapter from which this selection is taken is "Selective Subjectivism." Explain how this parable helps to illuminate this subject.

James Thurber

A childhood accident caused James Thurber to view the world with only one eye. However, he found more to chuckle about with his view than do many who are fully sighted. Thurber (1894–1961) began his writing career as a newspaper reporter, but an early interest in musical comedy inspired his creation of lyrics and librettos for five theatrical productions at Ohio State University. He later collaborated with Elliott Nugent on the successful Broadway play *The Male Animal*.

Thurber's ability to write about the serious aspects of life as well as the comedic provided the dual creative vision that amazed and delighted his theatrical audiences and his reading public. Additionally, his cartoons—often simple line-drawings of bewildered human beings and animals, straining awkwardly to understand a world gone awry—gave him universal recognition as a superb humorist.

Thus, in 1925, when he began his long association with *The New Yorker*, he was eminently qualified to join the company of such lively writers as Robert Benchley, Dorothy Parker, and E. B. White.

"To tell the truth but tell it slant," to borrow a phrase from Emily Dickinson, is the skill that gives Thurber's work a sharpened sense of the humorous. In "On Insect Behavior," excerpted from his essay "Courtship Through the Ages," Thurber puts this skill to work to transfer information about the animal world to the human. Having read the "sad and absorbing story" on the courtship of animals in Volume 6 of the *Encyclopaedia Britannica*, he satirizes "the sorrowful lengths to which all males must go to arouse the interest of a lady."

On Insect Behavior

James Thurber

The next time you encounter a male web-spinning spider, stop and reflect that he is too busy worrying about his love life to have any desire to bite you. Male web-spinning spiders have a tougher life than any other males in the animal kingdom. This is because the female web-spinning spiders have very poor eyesight. If a male lands on a female's web, she kills him before he has time to lay down his cane and gloves, mistaking him for a fly or a bumblebee who has tumbled into her trap. Before the species figured out what to do about this, millions of males were murdered by ladies they called on. It is the nature of spiders to perform a little dance in front of the female, but before a male spinner could get near enough for the female to see who he was and what he was up to, she would lash out at him with a flat-iron or a pair of garden shears. One night, nobody knows when, a very bright male spinner lay awake worrying about calling on a lady who had been killing suitors right and left. It came to him that this business of dancing as a love display wasn't getting anybody anywhere except the grave. He decided to go in for web-twitching, or strand-vibrating. The next day he tried it on one of the nearsighted girls. Instead of dropping in on her suddenly, he stayed outside the web and began monkeying with one of its strands. He twitched it up and down and in and out with such a lilting rhythm that the female was charmed. The serenade worked beautifully; the female let him live. The *Britannica*'s spider-watchers, however, report that this system is not always successful. Once in a while, even now, a female will fire three bullets into a suitor or run him through with a kitchen knife. She keeps threatening him from the moment he strikes the first low notes on the outside strings, but usually by the time he has got up to the high notes played around the center of the web, he is going to town and she spares his life.

Carolyn Kraus

Carolyn Kraus combines an academic career as a teacher of writing at the University of Michigan–Dearborn with wide-ranging activities as a freelance journalist. Journalism has taken her as far away as Siberia and Nicaragua, but most of her reporting takes place in southern Michigan and is published regularly in Detroit's city magazines. Her feature writing covers a range of subjects:

travel, art, animals, physics, and astronomy. Kraus's purpose in writing science features is, in her words, "to make complicated scientific subjects accessible to the general public by focusing on the human beings involved."

Kraus graduated from Occidental College in Los Angeles in 1966 with a bachelor's degree in English Literature and earned a master's degree in Comparative Literature from the University of California at Berkeley in 1972. She studied languages in Ireland and in Moscow. While on leave from her studies in Russian, she turned to feature writing during a hitchhiking trip across Siberia. As an anonymous traveler, Kraus sketched the vivid observations that underlie her popular essay "Siberian Graffiti."

Her principal essays include "Red Elephants," "A Crack in the Cosmic Wall," "Once in a Lifetime," and "Halley's Comet." Although Kraus's own background is not scientific, her interest in *Enormous Answers to Everything*, the title of her essay collection now in progress, has led her repeatedly to scientific topics. The following essay, "Searching Out Creation's Secrets," is part of that collection.

Searching Out Creation's Secrets

Carolyn Kraus

At Casey's Family Restaurant on Lake Erie's shore twenty miles east of Cleveland, breakfast regulars mull over local abuses in the free cheese distribution program, the at-large Middleberg kitten shooter, or the rich taste of Casey's own "bellybuster" pancakes. There is a lot of time to talk. Massive layoffs have left just a skeleton crew at the Morton salt mine just down the way. Local teenagers have devised their own timeless, low-budget adventures: They explore the mine grounds at night, sneaking by the Pinkerton detective at the gate to gather among the forty-foot salt pyramids stockpiled there.

Meanwhile, in a laboratory buried 2,000 feet beneath those salt mounds, a group of physicists ponders deep mysteries. They keep watch over 10,000 tons of water sealed in a giant bag. The bag lines a cavity the size of a five-story building, which has been dug below the floor of an adjacent chamber. Specially designed light sensors, strung on nylon fishing lines, descend into the water from a floating catwalk above the top rim of the cavity. The sensors peer through the eerie darkness and relay images to five computers in the laboratory. Here the scientists await a Sign—the blue flash of a single decaying proton that might illuminate the secrets of eternity.

"It's being called the experiment of the century," says Dr. Lawrence Sulak of the University of Michigan. He is one of twenty-nine physicists running the $5 million project, which is being funded by the Department of Energy. "At stake are some fundamental issues," he says. "Is matter essentially unstable? Are diamonds forever? Are we forever? Will we eventually turn into light instead of dust?"

These questions, which evoke both fairy tale and theology, also reflect the new physics. Just ten years ago, the notion that a proton was unstable seemed as absurd as the current choice of a salt mine for a laboratory. But then came the era of Grand Unified Theories, an ambitious, visionary quest to find unity and simplicity among the forces of nature. It is one of the foundations of the new physics.

Scientists have long believed that all the matter in the universe was created in a giant explosion fifteen billion years ago called the Big Bang. The set of theories that make up the Grand Unified Theories, constructed by physicists around the world over the past decade, is the most recent attempt to explain how everything came to exist from that explosion. One of its many daring predictions is that protons, which are in the nuclei of all atoms—all matter—*will not last forever*. Einstein's special Theory of Relativity told us early in this century that matter and energy are alternate forms of the same thing—that is, that one can be converted to the other. Now the Grand Unified Theories predict that this conversion of matter to energy will happen spontaneously over time, and that the universe of matter will eventually disappear—as if the movie of Creation were running backward in slow motion.

The regulars at Casey's restaurant need not worry, however, that their pancakes will disappear from plate to lip. Nothing will happen today, or next year, or in a billion years, even if protons do decay. The theory seeking physical proof beneath Lake Erie predicts that protons have a half-life of 10 to the 32nd years. In the language of physics, this means that half the protons in the universe will turn into energy within a timespan represented in years by a one with thirty-two zeros behind it. This is a future almost too remote for imagining, trillions of times longer than the life of the universe thus far.

How, then, can the theory be tested?

"You can test it out in one of two ways," says U-M post-doctoral researcher Dr. Steven Errede. "You can watch one proton for 10 to the 32nd years, in which case you have a family project. Or you can watch 10 to the 32nd protons for one year. . . ." That is, the more protons you watch, the sooner one can be expected to disintegrate.

Of course, no one has ever seen an atom, much less a proton, which is a tiny part of an atom. But with the help of sophisticated equipment, scientists can observe the effects of subatomic phenomena like protons and neutrons. The project to build a proton decay detector began in 1979, a collaboration among physicists from the University of Michigan, the University of California at Irvine, and the Brookhaven National Laboratory. They began on borrowed funds, since initially the Department of Energy found the project too bizarre to support. But money came through after three Nobel Prize–winning theoretical physicists—Steven Weinberg, Sheldon Glashow, and Abdus Salam—wrote letters of endorsement.

Dr. Sulak explains the Lake Erie proton detector this way: "Get a whole lot of matter that you can observe. Let's take water, because it is transparent and we can detect light given off by the particles decaying. Next, take it a half-mile underground to block out interference—for instance, from cosmic rays which continually bombard the earth from outer space [and could produce effects like those of proton decay]. Then, take 2,000 phototubes—light sensors—and paste them on the walls of the water tank so that they register the light that comes off the particles decaying. . . ." Dr. Sulak then explains that radiation given off by proton decay would be recorded in a particular, identifiable pattern.

"Now, 10,000 tons of water is 10 to the 32nd protons. At least *one* ought to pop off in a year."

Sulak smiles with boyish enthusiasm. "You can't do this experiment any better on earth. To improve it, you have to do it ten times bigger and on the moon."

It is early morning as a group of physicists, looking like moon men themselves, gear up to descend into the mine. They wear goggles, hardhats with miners' lights, and "self-rescuer" packs strapped to their belts in case of an oxygen shortage. "Break the seal and suck on the mouthpiece," one of the physicists instructs. "If it burns your lips, mouth and tongue, it's working." Then a drowsy Pinkerton, as if in an afterthought, distributes metal identification "body tags."

A primitive elevator carries the physicists to the mine floor. Ears pop from pressure changes and the darkness thickens. Everything is wet. Two thousand feet later, the elevator door opens to a rush of sulfurous air that stays in the back of the lungs all day.

Dodging puddles and slime-covered rocks, the physicists make their way along a subterranean path past idle mining equipment from which salt crystals hang like tiny stalactites. The elements are slowly reclaiming the cavern blasted twenty-four years ago into the 400-million-year-old salt deposit. On humid days, the vault drips like a tropical rain forest.

A shadowy figure appears at one turning and trudges toward the physicists. He identifies himself as Deward Curkendall, an old-timer. "Been working here since the mine opened," he says. "Longer than anybody." Curkendall smiles at the physicists with their *Star Wars* gear. "I was here when they were doing that other experiment," he chuckles, referring to a radioactive-decay study conducted here some twenty years ago. "I don't think one way or another about it. I've seen 'em come and go."

Curkendall's light fades into the distance again, veiled by falling rock salt. As the physicists round another turn, a headlight grazes over a rock wall where someone has painted in big red letters: "What time is it? What time is it?"

Suddenly, wind from an air vent whooshes into the mine. The physicists grab their hats and duck behind a heavy metal door which clangs shut against the ancient rock. An electric shoe polisher brushes the salt from their steel-toed boots and they enter an air-conditioned room stocked with refrigerators, microwave ovens, a Mr. Coffee, and a "high-energy" toilet ("Our destroilet," says a physicist) that bags the waste, zaps it to ashes, then blows it out into the depths of the mine.

Five computers at the far end of this laboratory are joined by a jungle of cables to the phototubes in the water-filled cavity beyond the back wall. The computers record every subatomic occurrence in the water onto printouts, which hang from the terminals like exhausted white tongues.

The physicists cluster around this latest data and begin sifting through it for evidence of two trails of light, moving apart at a forty-two-degree angle and triggering two circles of phototubes on opposite walls of the tank—the precise pattern expected from a spontaneously decaying proton.

On September 5, 1982, just one month after the detector began operating, the Lake Erie team thought they had "made a kill." Two circles came out on the computer. Someone breathlessly called *The New York Times*, but then the excitement subsided. The pattern was slightly off; the angle of light was wrong. The physicists nostalgically refer to the encounter as *"The New York Times Event."* There has not been a single strong candidate for proton decay since, although the phototubes blip away dutifully several times per second, registering cosmic rays which manage to penetrate into the water through the half mile of shielding rock.

Dr. Daniel Sinclair leaves the huddle around the computers and goes to the purification plant just outside the lab. Here the water is filtered through polyethylene bags under high pressure and made ultrapure so that the instruments can "see" subatomic events taking place in distant parts of the tank. Sinclair returns with two cups, one filled with purified water, the other with regular drinking water. Ironically, "the world's purest water" tastes bitter, metallic . . . actually terrible. But Sinclair is a devotee: "I always take my tea with ultrapure water," he says.

Sinclair is a mustachioed, slightly balding U-M professor with a gruff voice and a twinkle in his eye. On occasion, he has been known to leave his hat in the refrigerator. He is the "mad scientist" behind much of the detector's unorthodox assembly—a combination of sophisticated technology and inspired tinkering. Sinclair designed the reels to move the 2,058 light sensors into and out of the water, the beams to hold them in place, and the waterproof phototube housings to keep them dry and neutrally buoyant (i.e., they neither rise nor sink).

Smiling, he holds up his final version of the phototube housing or "beast," as the other physicists call it. It bears a strong resemblance to R2D2, the *Star Wars* robot. A $300 phototube is the "head," the "body" is constructed of plastic sewer piping from a hardware store, and plastic "arms" weighted with steel pellets emerge from both sides. Sinclair smiles again like a proud papa and admits, "I used to dream about different designs for this thing."

The detector has operated according to the physicists' plan, although one major setback occurred during construction. As the tank was being filled, its heavy plastic liner sprang a leak. Sinclair heard a deep rumbling as the water-soluble salt fill surrounding the detector melted and collapsed. The whole cavity threatened to cave in. "We've got a China Syndrome on our hands," someone shouted as people started to evacuate.

"If we leave now," Sinclair protested, "we can kiss those phototubes good-bye." He wasn't about to abandon his homemade beasts, even though the rumbling grew louder and cracks appeared on the laboratory floor. Sinclair rushed above ground, bought $500 worth of nylon rope, and, with several assistants, jerry-rigged the tubes and saved them as the leaking water was pumped out.

"I'll never forget that day," says Eric Shumard, a U-M graduate student who designed the detector's data-acquisition system and worked sixteen hours straight to save the tubes. "It was the same day Solidarity fell in Poland."

"Is that so?" asks Sinclair. "I didn't notice about Solidarity."

After checking the latest data, the physicists fit paper slipcovers over their shoes and enter the detector chamber. Some make repairs from the floating catwalk. Others prepare to dive into the water. Physicist/diver Karl Luttrell and his crew wear wetsuits with gloves and headgear glued on to avoid contaminating the water as they vacuum the bottom of the tank and check for leaks. From time to time, they have retrieved human relics from the watery depths— a wrench, two watches, a lone motel key.

With the elaborate electronics out of sight behind a wall, the scene appears both mysterious and simple—an immense pool contained within the world's largest plastic bag, and Sinclair's beasts that watch like thousands of eyes for the Sign from above. The bottom of the tank is sixty-five feet away, but appears much closer—the water's purity distorts one's sense of distance. "The first time you go in," says Luttrell, "it's like diving into midair."

"One day when I was diving in the tank," says Dr. Errede, "I started looking at all those tubes suspended weightlessly in the clear water. It was as if I was looking off into eternity. I couldn't work." These scientists, whose labors so often involve computers, complex equipment, or chalkboards covered with inscrutable symbols, are people who wonder at the universe. "You do physics," says Errede, "because you love to look at nature."

Here in the detector cavern, thick with humidity, the elements reign once more. Despite elaborate precautions to keep the water pure, slime covers the submerged cables and nylon fishing lines like thick jelly. "It gets in every-where," says Sinclair. "Even if we killed it with $30,000 worth of chlorine, it would all come back."

When the door to the computer lab is closed and all light is eliminated so the phototubes may resume their silent search, one senses the pulse of the universe very close by. "One day I lay down and took a nap in the dark," says Shumard. "My watch dial seemed to glow. Pretty soon I was seeing Cerenkov light"—the eerie blue glow of a charged particle (in this case a decaying proton), traveling through water faster than the speed of light.

This blue flash, if actually detected, would prove the essential instability of all matter. It would follow that our earth, sky, selves—all is ephemeral and will eventually disintegrate into light.

But so far the blue flash has only come in dreams, and some of the physicists admit they no longer expect it. As more time passes, the theory seems less likely to be correct. Leaning from the catwalk and locating a faulty tube in the water, Sinclair speaks of proton decay in the past tense: "It would have been so breathtaking, so simple, so beautiful. How many miracles can you expect?"

He hauls up the tube and replaces it. Apparently phototubes die faster than protons. "We'll go on looking for proton decay for another five years," he says. "If we don't find it by then, we'll give it up."

If the project fails to "make a kill" in the next five years, this does not mean that protons don't decay, nor does it topple the Grand Unifying Theories which are pillars of the new physics. It may mean that protons decay differently than this particular theory predicts, or that their lifetime is longer than 10 to the 32nd years, perhaps many times longer. "We're experimentalists, not theorists," says Dr. Sulak. "We go out and look at nature and see how it works. We don't have any preconceived notions. . . .

"Still," he ventures, "I think the proton will eventually have to decay. It would be nice if we could discover it. But maybe nature made the world in such a way that the decay lifetime is ten times longer than we could see. So we're out of luck."

Meanwhile, a race to observe proton decay has developed worldwide. A dozen smaller detectors are operating in the United States, India, Japan, Switzerland and the U.S.S.R. The Indian experimenters claim to have seen proton decay already, though physicists here remain skeptical because they have not seen it on their own instruments.

In one low-budget variation of the Lake Erie detector, Dr. Marvin Marshak of the University of Minnesota has gone to the bottom of an iron mine, walled out the bats, set a computer on a chair, and stacked up slabs of concrete laced with iron to pack more protons into a smaller volume. The slabs form a thirty-one-ton cube wired to catch a proton in its last act.

The Lake Erie experimenters remain in the race, but they are beginning to eye more complicated theories that say protons decay in other ways. "Maybe proton decay isn't clear and distinctive," says Shumard. "These other theories predict decay events with less light, but basically you can use the same detection techniques and the same equipment."

The detector is also being used to study other phenomena. Ghostly, massless particles called neutrinos, rarely observed before, are appearing daily on the computer tapes. Several physicists are using the detector to observe the fast-moving particles from supernovas (exploding stars). Even the intrusive cosmic rays—merely a nuisance to the proton decay experiment—are being scrutinized by self-styled physicist trashpickers. "I came here to study what they're trying to get rid of," explains Dr. Robert Svoboda of the University of Hawaii.

But a year of disappointment in the search for proton decay has jarred the world of theoretical physicists and sent shock waves throughout the world's scientific community. "There's a sort of crisis building up," says Dr. Svoboda. "This theory predicting proton decay was very clear-cut, very simple. Now the theorists have run out of gas, and the ball goes into our court. We can look for things they haven't predicted, follow hunches, study the tapes for something that just looks funny." Such events, unclassified and unpredictable, are known in the world of physics as zoo-ons. "They're like animals in the zoo," Svoboda explains, "jumping from tree to tree."

So, along with disappointment has come a subversive sense of adventure among the physicists at the mine. They are experimentalists, traditionally charged with testing out the big ideas sparkling from the pencils of theorists. A healthy rivalry has long existed between the two interdependent breeds, but the idea men—the theorists—tend to walk off with the Nobel Prizes. Maybe this time the experimentalists can trump the theorists. "If you see something like this," reads a note above a diagram of X's, "call Maurice immediately."

Nearby, another message hangs in a frame, an Eighteenth Century poem by William Cowper:

God moves in a mysterious way
His wonders to perform
He plants his footsteps in the sea,
And rides upon the storm.

Deep in unfathomable mines
Of never failing skill
He treasures up his bright designs
And works his sovereign will.

No one is sure if the poem was posted as a joke or as a humbling reminder of the limits of rational thought. Perhaps both. Physicists operate on the edge of the unknown where logic, language—even the imagination—often fail.

Will the universe, which began as energy, end as energy? It is an irresistible question—one which has been asked a thousand ways since the ancient Babylonians pondered a world created in the fiery separation of land and sky. The search for proton decay is Twentieth Century science's latest attempt to fill in the blanks of our own cosmic epic.

Even if the round of experiments is unsuccessful, physicists will continue to search the invisible subatomic world for answers to the cosmic question. Plans are already brewing for a different kind of proton decay detector, one hundred times bigger than the Lake Erie setup, to operate in the open water off the shores of Hawaii.

As the sun sets over Lake Erie, the physicists leave the mine and return to their lodgings at the Mentor Country Inn. No one there seems to have heard of the "experiment of the century" going on down the road, though professors in steel-toed boots come and go.

Now cars filled with young people come out to cruise around No Help Wanted signs and idle machinery at the mine. Crickets begin to sing in the ragged lakeshore grass. Dr. Sinclair is still underground, tending his instruments as, in a burst of purple and orange, the sun disappears.

Questions and Topics for Discussion and Writing

1. This article is a "feature," written for a popular-magazine audience. How does the author attempt to reach and involve readers with little scientific background?
2. In a short paragraph, explain the difficulties a writer might experience in writing about scientific subjects for a contemporary lay audience.
3. How would you describe the public stereotype of a physicist? Citing passages from the article, explain how the proton-decay physicists conform to or break that stereotype.
4. Kraus says that these physicists are essentially "people who wonder at the universe." Discuss.
5. Where do the descriptive passages go beyond creating a sense of place? How do they contribute to the development of themes?
6. In the town where the scientists are working, "No one . . . seems to have heard of the 'experiment of the century' going on down the road. . . ." Do you find this attitude surprising or typical of the general public's level of interest in scientific research?
7. If this experiment, or one like it, were successful in discovering proton decay, how would our perceptions of the universe be altered?
8. Kraus alludes to a rivalry between theoretical and experimental physicists. In an essay, explain the difference between these two kinds of scientists and discuss their respective roles in the search for proton decay.
9. Does Kraus ever sensationalize her material for the purpose of involving a general audience?
10. Does the article primarily seek to convey the scientific experiment itself or to give a sense of the people behind it? To what extent is each accomplished?

Sigmund Freud

The fact is not well known that Sigmund Freud (1856–1939)—considered by many to be the "father of modern psychiatry"—was a brilliant student in both English and French classics. He was also fascinated by Shakespeare, Goethe, Leonardo, and Dostoevsky. Born in Freiberg, Moravia, Freud was the son of middle-class Jewish parents who moved their family to Vienna in 1859, where, in 1881, he received his medical degree. He was professor of neurology there from 1902 until 1938.

Freud used methods of dream analysis, probing the unconscious by means of free association, and the relationship of findings with attitudes about sexuality in his practice of psychiatry. Many of his theories altered the way patients were treated for various neuroses and mental malfunctions.

He worked with Josef Breuer on the treatment of hysteria through the use of hypnosis; their jointly authored book *Studies of Hysteria* (1895) explained their experiments. Eventually, however, Freud's methods of psychoanalysis replaced his use of hypnosis.

His own self-analysis resulted in *Interpretation of Dreams* in 1900; *Psychopathology of Everyday Life* and *Three Contributions to the Theory of Sex* followed, arousing worldwide controversy and anger. Psychoanalysis in its early stages engendered antagonism by suggesting that the mind was made up of three parts: the *superego*, the moral sense, or regulator of the person's sense of right or wrong; the *ego*, the agent that deals with reality and attempts to mediate the needs of the inner self with the outside world; and the *id*, the original, primeval, unconscious part of the mind from which the other structures developed. Some portions of these other structures, the superego and the ego, remain unconscious, the id being the repository of all the unconscious drives and impulses. Freud's theory holds that many mental disturbances are caused by fixations at primitive levels of development and that many of the primitive impulses and drives are expressed in a distorted fashion in dreams.

Much of the fury aroused by *Three Contributions to the Theory of Sex* was caused by Freud's belief that infants have a sexual drive, a concept difficult for the public to accept—then and even now. Freud found motivation, even if unconscious, behind gestures, words, and acts, even those that seemed accidental. In understanding these acts, he believed, men and women understand themselves. Yet he allowed that every act of an individual is not necessarily motivated by the unconscious, but may be motived by the dictates of reality, a concept he conveyed humorously in the remark that "sometimes a cigar is just a cigar."

Freud lived and worked in Vienna until he was put under house arrest by the Nazis. Released in 1938, the year before his death, he moved to London.

His direct writing style earned him the Goethe Prize in 1930. The selection below, an excerpt from "An Introduction to Psychoanalysis," shows the "psychological geographer" explaining the world he charted.

Definition of Psychoanalysis

Sigmund Freud

Ladies and Gentlemen, . . . In medical training you are accustomed to *see* things. You see an anatomical preparation, the precipitate of a chemical reaction, the shortening of a muscle as a result of the stimulation of its nerves. Later on, patients are demonstrated before your senses—the symptoms of their illness, the products of the pathological process and even in many cases the agent of the disease in isolation. In the surgical departments you are witnesses of the active measures taken to bring help to patients, and you may yourselves attempt to put them into effect. Even in psychiatry the demonstration of patients with their altered facial expressions, their mode of speech and their behaviour, affords you plenty of observations which leave a deep impression on you. Thus a medical teacher plays in the main the part of a leader and interpreter who accompanies you through a museum, while you gain a direct contact with the objects exhibited and feel yourselves convinced of the existence of the new facts through your own perception.

In psycho-analysis, alas, everything is different. Nothing takes place in a psycho-analytic treatment but an interchange of words between the patient and the analyst. The patient talks, tells of his past experiences and present impressions, complains, confesses to his wishes and his emotional impulses. The doctor listens, tries to direct the patient's processes of thought, exhorts, forces his attention in certain directions, gives him explanations and observes the reactions of understanding or rejection which he in this way provokes in him. The uninstructed relatives of our patients, who are only impressed by visible and tangible things—preferably by actions of the sort that are to be witnessed at the cinema—never fail to express their doubts whether "anything can be done about the illness by mere talking." That, of course, is both a short-sighted and an inconsistent line of thought. These are the same people who are so certain that patients are "simply imagining" their symptoms. Words were originally magic and to this day words have retained much of their ancient magical power. By words one person can make another blissfully happy or drive him to despair, by words the teacher conveys his knowledge to his pupils, by words the orator carries his audience with him and determines their judgements and decisions. Words provoke affects and are in general the means of mutual influence among men. Thus we shall not depreciate the use of words in psychotherapy and we shall be pleased if we can listen to the words that pass between the analyst and his patient.[1]

[1][Cf. a parallel passage near the beginning of *The Question of Lay Analysis* (1926e), (Norton, 1950).]

But we cannot do that either. The talk of which psycho-analytic treatment consists brooks no listener; it cannot be demonstrated. A neurasthenic or hysterical patient can of course, like any other, be introduced to students in a psychiatric lecture. He will give an account of his complaints and symptoms, but of nothing else. The information required by analysis will be given by him only on condition of his having a special emotional attachment to the doctor; he would become silent as soon as he observed a single witness to whom he felt indifferent. For this information concerns what is most intimate in his mental life, everything that, as a socially independent person, he must conceal from other people, and, beyond that, everything that, as a homogeneous personality, he will not admit to himself.

Thus you cannot be present as an audience at a psycho-analytic treatment. You can only be told about it; and, in the strictest sense of the word, it is only by hearsay that you will get to know psycho-analysis. As a result of receiving your instruction at second hand, as it were, you find yourselves under quite unusual conditions for forming a judgement. That will obviously depend for the most part on how much credence you can give to your informant.

Let us assume for a moment that you were attending a lecture not on psychiatry but on history, and that the lecturer was telling you of the life and military deeds of Alexander the Great. What grounds would you have for believing in the truth of what he reported? At a first glance the position would seem to be even more unfavourable than in the case of psycho-analysis, for the Professor of History no more took part in Alexander's campaigns than you did. The psycho-analyst does at least report things in which he himself played a part. But in due course we come to the things that confirm what the historian has told you. He could refer you to the reports given by ancient writers, who were either themselves contemporary with the events under question or, at any rate, were comparatively close to them—he could refer you, that is to say, to the works of Diodorus, Plutarch, Arrian, and so on. He could put reproductions before you of coins and statues of the king which have survived and he could hand round to you a photograph of the Pompeian mosaic of the battle of Issus. Strictly speaking, however, all these documents only prove that earlier generations already believed in Alexander's existence and in the reality of his deeds, and your criticism might start afresh at that point. You would then discover that not all that has been reported about Alexander deserves credence or can be confirmed in its details; but nevertheless I cannot think that you would leave the lecture-room in doubts of the reality of Alexander the Great. Your decision would be determined essentially by two considerations: first, that the lecturer had no conceivable motive for assuring you of the reality of something he himself did not think real, and secondly, that all the available history books describe the events in approximately similar terms. If you went on to examine the older sources, you would take the same factors into account—the possible motives of the informants and the conformity of the witnesses to one another. The outcome of your examination would undoubtedly be reassuring in the case of Alexander, but would probably be different where figures such as Moses or Nimrod were concerned. Later opportunities will bring to light clearly enough what doubts you may feel about the credibility of your psycho-analytic informant.

But you will have a right to ask another question. If there is no objective verification of psycho-analysis, and no possibility of demonstrating it, how can one learn psycho-analysis at all, and convince oneself of the truth of its assertions? It is true that psycho-analysis cannot easily be learnt and there are not many people who have learnt it properly. But of course there is a practicable method none the less. One learns psycho-analysis on oneself, by studying one's own personality. This is not quite the same thing as what is called self-observation, but it can, if necessary, be subsumed under it. There are a whole number of very common and generally familiar mental phenomena which, after a little instruction in technique, can be made the subject of analysis upon oneself. In that way one acquired the desired sense of conviction of the reality of the processes described by analysis and of the correctness of its views. Nevertheless, there are definite limits to progress by this method. One advances much further if one is analysed oneself by a practised analyst and experiences the effects of analysis on one's own self, making use of the opportunity of picking up the subtler technique of the process from one's analyst. This excellent method is, of course, applicable only to a single person and never to a whole lecture-room of students together.

Psycho-analysis is not to be blamed for a second difficulty in your relation to it; I must make you yourselves responsible for it, Ladies and Gentlemen, at least in so far as you have been students of medicine. Your earlier education has given a particular direction to your thinking, which leads far away from psycho-analysis. You have been trained to find an anatomical basis for the functions of the organism and their disorders, to explain them chemically and physically and to view them biologically. But no portion of your interest has been directed to psychical life, in which, after all, the achievement of this marvellously complex organism reaches its peak. For that reason psychological modes of thought have remained foreign to you. You have grown accustomed to regarding them with suspicion, to denying them the attribute of being scientific, and to handing them over to laymen, poets, natural philosophers and mystics. This limitation is without doubt detrimental to your medical activity, since, as is the rule in all human relationships, your patients will begin by presenting you with their mental *façade*, and I fear that you will be obliged as a punishment to leave a part of the therapeutic influence you are seeking to the lay practitioners, nature curers and mystics whom you so much despise. . . .

It is true that psychiatry, as a part of medicine, sets about describing the mental disorders it observes and collecting them into clinical entities; but at favourable moments the psychiatrists themselves have doubts of whether their purely descriptive hypotheses deserve the name of a science. Nothing is known of the origin, the mechanism or the mutual relations of the symptoms of which these clinical entities are composed; there are either *no* observable changes in the anatomical organ of the mind to correspond to them, or changes which throw no light upon them. These mental disorders are only accessible to therapeutic influence when they can be recognized as subsidiary effects of what is otherwise an organic illness.

This is the gap which psycho-analysis seeks to fill. It tries to give psychiatry its missing psychological foundation. It hopes to discover the common ground on the basis of which the convergence of physical and mental disorder will become intelligible. With this aim in view, psycho-analysis must keep itself free from any hypothesis that is alien to it, whether of an anatomical, chemical or physiological kind, and must operate entirely with purely psychological auxiliary ideas; and for that very reason, I fear, it will seem strange to you to begin with.

Questions and Topics for Discussion and Writing

1. The introduction to this selection by Freud declares him to be a "psychological geographer." Write a paragraph in which you explain how this selection supports the description.
2. Look up the word "apology" in the dictionary, especially with regard to its meaning as a literary genre. After so doing, write a statement explaining whether or not you agree that a more appropriate title of this speech, especially when presented as an essay as is done here, would be "Definition of Psychoanalysis: An Apology."
3. What does Freud mean by his statement "Words were originally magic . . . "? In a paragraph, discuss Freud's view of the significance of language, especially its important role in psychoanalysis.
4. Why, according to Freud, does the "medical teacher" play the role of a "leader and interpreter who accompanies students through a museum"?
5. How does Freud differentiate the role of the psychoanalyst from that of the traditional medical doctor?
6. Write in your own words what Freud sees to be the "gap" in the study of medicine that psychoanalysis can fill.
7. What does Freud consider to be problems faced by the psychoanalyst?
8. In his attempt to define psychoanalysis, Freud creates an analogy between a historian and a psychoanalyst. What does Freud argue is the difference between them? How effective is the analogy he creates for support of his argument and for purposes of definition?
9. Freud says that the information required by analysis will be given by a neurasthenic or hysterical patient "only on condition of his having a special emotional attachment to the doctor." From your own observations and experiences, are similar situations experienced outside of psychoanalysis?
10. Freud reminds his listeners that "you cannot be present as an audience at a psychoanalytic treatment. . . . It is only by hearsay that you will get to know psycho-

analysis." Do you know of any other discipline that has similar constraints of verification?

11.   Freud establishes knowledge of the self as a "practical method" to be followed by a psychoanalyst, and he advocates a scientific base to the study of the mind. What dangers do you see to be inherent in such methods? Write a challenge to Freud's assertions.

12.   A relatively new discipline is psycholinguistics. Write a statement in which you hypothesize how Freud's introduction of psychoanalysis can be said to have given rise to this new study of language.

13.   Freud suggests that the habitual approach of his audience was to try to "explain [psychological disorders] chemically and physically and to view them biologically." He holds that this attitude is "detrimental" to their "medical activity." Would one still think today that psychological modes of thought have no basis in any physical type of activity, for example, possible alteration of electrical impulses or magnetic fields of the brain?

The Expanding Mental Universe

Bertrand Russell

You will find the biography for Bertrand Russell before the essay "The Rise of Science" in Unit 3.

The effects of modern knowledge upon our mental life have been many and various, and seem likely, in future, to become even greater than they have been hitherto. The life of the mind is traditionally divided into three aspects: thinking, willing, and feeling. There is no great scientific validity in this division, but it is convenient for purposes of discussion, and I shall, therefore, follow it.

It is obvious that the primary effect of modern knowledge is on our thinking, but it has already had important effects in the sphere of will, and should have equally important effects in the sphere of feeling, though as yet these are very imperfectly developed. I will begin with the purely intellectual effects.

The physical universe, according to a theory widely held by astronomers, is continually expanding. Everything not quite near to us is moving away from us, and the more remote it is, the faster it is receding. Those who hold this theory think that very distant parts of the universe are perpetually slipping into invisibility because they are moving away from us with a velocity greater than that of light. I do not know whether this theory of the expanding physical universe will continue to hold the field or not, but there can be no doubt about the expanding mental universe. Those who are aware of the cosmos as science has shown it to be have to stretch their imaginations both in space and in time to an extent which was unknown in former ages, and which to many in our time is bewilderingly painful.

The expansion of the world in space was begun by the Greek astronomers. Anaxagoras, whom Pericles imported into Athens to teach the Athenians philosophy, maintained that the sun is as large as the Peloponnesus, but his contemporaries thought that this must be a wild exaggeration. Before long, however, the astronomers discovered ways of calculating the distance of the sun and moon from the earth, and, although their calculations were not correct, they sufficed to show that the sun must be many times larger than the earth. Poseidonius, who was Cicero's tutor, made the best estimate of the sun's distance that was made in antiquity. His estimate was about half of the right value. Ancient astronomers after his time were farther from the mark than he was, but all of them remained aware that, in comparison with the solar system, the size of the earth is insignificant.

In the Middle Ages there was an intellectual recession, and much knowledge that had been possessed by the Greeks was forgotten. The best imaginative picture of the universe, as conceived in the Middle Ages, is in Dante's Paradiso. In this picture there are a number of concentric spheres containing the moon, the sun, the various planets, the fixed stars, and the Empyrean. Dante, guided by Beatrice, traverses all of them in twenty-four hours. His cosmos, to a modern mind, is unbelievably small and tidy. Its relation to the universe with which we have to live is like that of a painted Dutch interior to a raging ocean in storm. His physical world contains no mysteries, no abysses, no unimaginable accumulation of uncatalogued worlds. It is comfortable and cosy and human and warm; but, to those who have lived with modern astronomy, it seems claustrophobic and with an orderliness which is more like that of a prison than that of the free air of heaven.

Ever since the early seventeenth century our conception of the universe has grown in space and time, and, until quite recent years, there has not seemed to be any limit to this growth. The distance of the sun was found to be much greater than any Greek had supposed and some of the planets were found to be very much more distant than the sun. The fixed stars, even the nearest, turned out to be vastly farther off than the sun. The light of the sun takes about eight minutes to reach us, but the light of the nearest fixed star takes about four years. The stars that we can see separately with the naked eye are our immediate neighbours in a vast assemblage called "The Galaxy," or, in more popular parlance, "The Milky Way." This is one assemblage of stars which contains almost all that we can see with the naked eye, but it is only one of many millions of such assemblages. We do not know how many there may be.

A few figures may help the imagination. The distance of the nearest fixed star is about twenty-five million million miles. The Milky Way, which is, so to speak, our parish, contains about three hundred thousand million stars. There are many million assemblages similar to The Milky Way, and the distance from one such assemblage to the next takes about two million years for light to traverse. There is a considerable amount of matter in the universe. The sun weighs about two billion billion billion tons. The Milky Way weighs about a hundred and sixty thousand million times as much as the sun, and there are many million assemblages comparable to The Milky Way. But, although there is so much matter, the immensely large part of the universe is empty, or very nearly empty.

In regard to time, a similar stretching of our thoughts is necessary. This necessity was first shown by geology and paleontology. Fossils, sedimentary rocks and igneous rocks gave a backward history of the earth which was, of necessity, very long. Then came theories of the origin of the solar system and of the nebulae. Now, with the most powerful existing telescopes, we can see objects so distant that the light from them has taken about five hundred million years to reach us, so that what we see is not what is happening now, but what was happening in that immensely distant past.

What I have been saying concerns the expansion of our mental universe in the sphere of thought. I come now to the effects this expansion has, and should have, in the realms of will and feeling.

To those who have lived entirely amid terrestrial events and who have given little thought to what is distant in space and time, there is at first something bewildering and oppressive, and perhaps even paralysing, in the realization of the minuteness of man and all his concerns in comparison with astronomical abysses. But this effect is not rational and should not be lasting. There is no reason to worship mere size. We do not necessarily respect a fat man more than a thin man. Sir Isaac Newton was very much smaller than a hippopotamus, but we do not on that account value him less than the larger beast. The size of a man's mind—if such a phrase is permissible—is not to be measured by the size of a man's body. It is to be measured, in so far as it can be measured, by the size and complexity of the universe that he grasps in thought and imagination. The mind of the astronomer can grow, and should grow, step by step with the universe of which he is aware. And when I say that his mind should grow, I mean his total mind, not only its intellectual aspect. Will and feeling should keep pace with thought if man is to grow as his knowledge grows. If this cannot be achieved—if, while knowledge becomes cosmic, will and feeling remain parochial—there will be a lack of harmony producing a kind of madness of which the effects must be disastrous.

We have considered knowledge, but I wish now to consider wisdom, which is a harmony of knowledge, will and feeling, and by no means necessarily grows with the growth of knowledge.

Let us begin with will. There are things that a man can achieve and other things that he cannot achieve. The story of Canute's forbidding the tide to rise was intended to show the absurdity of willing something that is beyond human power. In the past, the things that men could do were very limited. Bad men, even with the worst intentions, could do only a very finite amount of harm.

Good men, with the best intentions, could do only a very limited amount of good. But with every increase in knowledge, there has been an increase in what men could achieve. In our scientific world, and presumably still more in the more scientific world of the not distant future, bad men can do more harm, and good men can do more good, than had seemed possible to our ancestors even in their wildest dreams.

Until the end of the Middle Ages, it was thought that there were only four kinds of matter, the so-called elements of earth, water, air and fire. As the inadequacy of this theory became increasingly evident the number of elements admitted by men of science increased until it was estimated at ninety-two. The modern study of the atom has made it possible to manufacture new elements which do not occur in nature. It is a regrettable fact that all these new elements are deleterious and that quite moderate quantities of them can kill large numbers of people. In this respect recent science has not been beneficent. *Per contra*, science has achieved what might almost seem like miracles in the way of combating diseases and prolonging human life.

These increases of human power remain terrestrial: we have become able, as never before, to mould life on earth, or to put an end to it if the whim should seize us. But, unless by some such whim we put an end to man, we are on the threshold of a vast extension of human power. We could now, if the expenditure were thought worth while, send a projectile to the moon, and there are those who hold that we could in time make the moon capable of supporting human life. There is no reason to suppose that Mars and Venus will long remain unconquered. Meanwhile, as Senator Johnson told the Senate, scientific power could have astonishing effects upon our own planet. It could, to quote his own words, "have the power to control the earth's weather, to cause drought and flood, to change the tides and raise the levels of the sea, to divert the Gulf Stream and change temperate climates to frigid."

When we have acquired these immense powers, to what end shall we use them? Man has survived, hitherto, by virtue of ignorance and inefficiency. He is a ferocious animal, and there have always been powerful men who did all the harm they could. But their activities were limited by the limitations of their technique. Now, these limitations are fading away. If, with our increased cleverness, we continue to pursue aims no more lofty than those pursued by tyrants in the past, we shall doom ourselves to destruction and shall vanish as the dinosaurs vanished. They, too, were once the lords of creation. They developed innumerable horns to give them victory in the contests of their day. But, though no other dinosaur could conquer them, they became extinct and left the world to smaller creatures such as rats and mice.

We shall court a similar fate if we develop cleverness without wisdom. I foresee rival projectiles landing simultaneously on the moon, each equipped with H-bombs and each successfully engaged in exterminating the other. But until we have set our own house in order, I think that we had better leave the moon in peace. As yet, our follies have been only terrestrial; it would seem a doubtful victory to make them cosmic.

If the increased power which science has conferred upon human volitions is to be a boon and not a curse, the ends to which those volitions are directed must grow commensurately with the growth of power to carry them out. Hitherto, although we have been told on Sundays to love our neighbour, we have been told on weekdays to hate him, and there are six times as many weekdays as Sundays. Hitherto, the harm that we could do to our neighbour by hating him was limited by our incompetence, but in the new world upon which we are entering there will be no such limit, and the indulgence of hatred can lead only to disaster.

These considerations bring us to the sphere of feeling. It is feeling that determines the ends we shall pursue. It is feeling that decides what use we shall make of the enormous increases in human power. Feeling, like the rest of our mental capacities, has been gradually developed in the struggle for existence. From a very early time, human beings have been divided into groups which have gradually grown larger, passing, in the course of ages, from families to tribes, from tribes to nations, and from nations to federations. Throughout this process, biological needs have generated two opposite systems of morality: one for dealings with our own social group; the other for dealings with outsiders. The Decalogue tells us not to murder or steal, but outside our own group this prohibition is subject to many limitations. Many of the men who are most famous in history derive their fame from skill in helping their own group to kill people of other groups and steal from them. To this day, aristocratic families in England are proud if they can prove that their ancestors were Norman and were cleverer at killing Saxons than Saxons were at killing them.

Our emotional life is conditioned to a degree which has now become biologically disadvantageous by this opposition between one's own tribe and the alien tribes against which it collectively competes. In the new world created by modern technique, economic prosperity is to be secured by means quite different from those that were formerly advocated. A savage tribe, if it can exterminate a rival tribe, not only eats its enemies but appropriates their lands and lives more comfortably than it did before. To a continually diminishing degree these advantages of conquest survived until recent times.

But now the opposite is the case. Two nations which co-operate are more likely to achieve economic prosperity than either can achieve if they compete. Competition continues because our feelings are not yet adapted to our technique. It continues because we cannot make our emotions grow at the same rate as our skills.

Increase of skill without a corresponding enlargement in feeling produces a technical integration which fails of success for lack of an integration of purpose. In a technically developed world, what is done in one region may have enormous effects in a quite different region. So long as, in our feeling, we take account only of our own region, the machine as a whole fails to work smoothly. The process is one which, in varying forms, has persisted throughout evolution. A sponge, while it is living in the sea, is like a block of flats, a common abode of a number of separate little animals each almost entirely independent of the others and in no way obliged to concern itself with their interests. In the body of a more developed animal, each cell remains in some degree a separate creature, but it cannot prosper except through the prosperity of the whole. In cancer, a group of cells engages in a career of imperialism, but, in bringing the rest of the body to death, it decrees also its own extinction. A human body is a unit from the point of view of self-interest. One cannot set the interest of the great toe in opposition to that of the little finger. If any part of the body is to prosper, there must be co-operation to the common ends of the body as a whole.

The same sort of unification is taking place, though as yet very imperfectly, in human society, which is gradually approximating to the kind of unity that belongs to a single human body. When you eat, if you are in health, the nourishment profits every part of your body, but you do not think how kind and unselfish your mouth is to take all this trouble for something else. It is this kind of unification and expansion of self-interest that will have to take place if a scientific society is to prove capable of survival. This enlargement in the sphere of feeling is being rendered necessary by the new interdependence of different part of the world.

Let us take an illustration from a quite probable future. Suppose some country in the southern hemisphere sets to work to make the Antarctic continent habitable. The first step will be to melt the ice—a feat which future science is likely to find possible. The melting of the ice will raise the level of the sea everywhere and will submerge most of Holland and Louisiana as well as many other low-lying lands. Clearly the inhabitants of such countries will object to projects that would drown them. I have chosen a somewhat fantastic illustration as I am anxious to avoid those that might excite existing political passions. The point is that close interdependence necessitates common purposes if disaster is to be avoided, and that common purposes will not prevail unless there is some community of feeling. The proverbial Kilkenny cats fought each other until nothing was left but the tips of their tails: if they had felt kindly toward each other, both might have lived happily.

Religion has long taught that it is our duty to love our neighbour and to desire the happiness of others rather than their misery. Unfortunately, active men have paid little attention to this teaching. But in the new world, the kindly feeling towards others which religion has advocated will be not only a moral duty but an indispensable condition of survival. A human body could not long continue to live if the hands were in conflict with the feet, and the stomach were at war with the liver. Human society as a whole is becoming, in this respect, more and more like a single human body; and if we are to continue to exist, we shall have to acquire feelings directed toward the welfare of the whole in the same sort of way in which our feelings of individual welfare concern the whole body and not only this or that portion of it. At any time such a way of feeling would have been admirable, but now, for the first time in human history, it is becoming necessary if any human being is to be able to achieve anything of what he would wish to enjoy.

Seers and poets have long had visions of the kind of expansion of the ego which I am trying to adumbrate. They have taught that men are capable of something which is called wisdom, something which does not consist of knowledge alone, or of will alone, or of feeling alone, but is a synthesis and intimate union of all three.

Some of the Greeks, and notably Socrates, thought that knowledge alone would suffice to produce the perfect man. According to Socrates, no one sins willingly, and, if we all had enough knowledge, we should all behave perfectly. I do not think that this is true. One could imagine a satanic being with immense knowledge and equally immense malevolence—and, alas, approximations to such a being have actually occurred in human history. It is not enough to seek knowledge rather than error. It is necessary, also, to feel benevolence rather than its opposite. But, although knowledge alone is not enough, it is a very essential ingredient of wisdom.

The world of a newborn infant is confined to his immediate environment. It is a tiny world bounded by what is immediately apparent to the senses. It is shut up within the walls of the here-and-now. Gradually, as knowledge grows, these walls recede. Memory and experience make what is past and what is distant gradually more vivid in the life of the growing child. If a child develops into a man of science, his world comes to embrace those very distant portions of space and time of which I spoke earlier. If he is to achieve wisdom, his feelings must grow as his knowledge grows. Theologians tell us that God views the universe as one vast whole, without any here-and-now, without that partiality of sense and feeling to which we are, in a greater or less degree, inevitably condemned. We cannot achieve this complete impartiality, and we could not survive if we did, but we can, and should, move as far toward it as our human limitations permit.

We are beset in our daily lives by fret and worry and frustrations. We find ourselves too readily pinned down to thoughts of what seems obstructive in our immediate environment. But it is possible, and authentic wise men have proved that it is possible, to live in so large a world that the vexations of daily life come to feel trivial and that the purposes which stir our deeper emotions take on something of the immensity of our cosmic contemplations. Some can achieve this in a greater degree, some only in a lesser, but all who care to do so can achieve this in some degree and, in so far as they succeed in this, they will win a kind of peace which will leave activity unimpeded but not turbulent.

The state of mind which I have been trying to describe is what I mean by wisdom, and it is undoubtedly more precious than rubies. The world needs this kind of wisdom as it has never needed it before. If mankind can acquire it, our new powers over nature offer a prospect of happiness and well-being such as men have never experienced and could scarcely even imagine. If mankind cannot, every increase in cleverness will bring us only nearer to irretrievable disaster. Men have done many good things and many bad ones. Some of the good things have been very good. All those who care for these good things must hope, with what confidence they can command, that in this moment of decision the wise choice will be made.

Questions and Topics for Discussion and Writing

1. Russell categorizes the "life of the mind" into three aspects, while remarking that there is "no great scientific validity in this division." List the three aspects and comment on their applicability.
2. Russell uses the theory of the expanding physical universe as an analogy. Comment on the suitability and usefulness of this analogy.
3. Russell says, "In the Middle Ages there was an intellectual recession." You are familiar with these two words individually. What rhetorical effectiveness do you see in their combination?
4. Russell calls the Milky Way "our parish." Consider the implication of Russell's use of this term in context. In a paragraph explain why he chose this particular metaphor.
5. What does Russell mean by "cosmic contemplation"?
6. Explain Russell's distinction between "knowledge" and "wisdom" and between "cleverness" and "wisdom."
7. Russell "tried to speak, as nearly as possible, in the language of the seventeenth century" in "The Rise of Science" (Unit 3). In "The Expanding Mental Universe," he addresses readers of the *Saturday Evening Post*. What stylistic elements can you find that differentiate this essay from the first one and make it more suitable for a popular audience? In a short paper, analyze Russell's style.

SUGGESTIONS FOR FURTHER READING

Cozart, William Reed, and Huston Smith, eds. *Dialogue On Science*. Indianapolis: Bobbs-Merrill, 1967.

Davies, Paul. *The Edge of Infinity: Where the Universe Came From and How It Will End*. New York: Simon and Schuster, 1981.

——. *Space and Time in the Modern Universe*. New York: Cambridge University Press, 1977.

Eiseley, Loren. *Darwin and the Mysterious Mr. X*. New York: Harcourt Brace Jovanovich, 1979.

Freud, Sigmund. *Civilization and Its Discontents*. Trans. and ed. by James Strachey. New York: W. W. Norton, 1962, 1961.

————, and D. E. Oppenheimer. *Dreams in Folklore*. Trans. A. M. Richards. Ed. James Strachey. New York: International Universities Press, 1958.

Gardner, Howard. *Art, Mind, and Brain: A Cognitive Approach to Creativity*. New York: Basic Books, 1982.

Goodfield, J. *An Imagined World*. New York: Penguin, 1982.

Malcolm, Janet. *In the Freud Archives*. New York: Knopf, 1984.

Schrödinger, E. *What Is Life?: My View of the World*. Cambridge, England: Cambridge University Press, 1967.

Stent, G. S. *The Coming of the Golden Age: A View of the End of Progress*. Garden City: Natural History Press, 1969.

Turner, Frederick. *Natural Classicism: Essays on Literature and Science*. New York: Paragon House, 1985.

Unit 11

ON THE ISSUES OF THE WORLD: CONSCIENCE AND CONTROVERSY

I have found that the conscience of scientists is the most active morality in the world today.

Jacob Bronowski

This is a world in which each of us, knowing his limitations, knowing the evils of superficiality and the terrors of fatigue, will have to cling to what is close to him, to what he knows, to what he can do, to his friends and his tradition and his love, lest he be dissolved in a universal confusion and know nothing and love nothing. . . . This balance, this perpetual, precarious, impossible balance between the infinitely open and the intimate, this time—our twentieth century— has been long in coming; but it has come. It is, I think, for us and our children, our only way . . . to make partial order in total chaos.

J. Robert Oppenheimer

Chief Seattle

Chief Seattle, for whom Seattle, Washington, is named, was an early contributor to the point of view that we now recognize as environmentalist. The son of a Suquamish leader and a Duwamish woman, Seattle (ca. 1788–1866) in his youth was known for his courage and daring as a warrior, but later came to be known as a peacemaker.

After the California Gold Rush, the Indians in the Puget Sound region at first welcomed settlers to their region. In 1853, the territory of Washington was established and, in honor of the chief, the new settlers named their settlement "Seattle" (more correctly, "Seathl" or "Sealth"). Later, conflicts arose between the two groups. In 1855, Governor Isaac Stevens of Washington determined to persuade the tribes to move onto reservations established for them, and many tribes waged war in order to resist the idea of reservations. Seattle, refusing to allow his people to continue the bloody conflict, became the first signer of the Port Elliott Treaty of 1855 accepting reservations.

His efforts were in one important sense unrewarded: By 1970, the Duwamish as a people were extinct. Yet in another—his effort to speak for the red man—he succeeded beyond his own expectations. His words in 1855 in a letter to "The Great Chief in Washington" gather force today: "We might understand if we knew what it was that the white man dreams, what hopes he describes to his children on long winter nights, what visions he burns into their minds, so that they will wish for tomorrow. But we are savages. The white man's dreams are hidden from us. And because they are hidden, we will go our own way."

In his speech that we have entitled "Environmental Statement," Seattle responds to the proposed treaty by which the Indians were to sell two million of their acres of land for $150,000.

Environmental Statement

Chief Seattle

How can you buy or sell the sky, the warmth of the land? The idea is strange to us.

If we do not own the freshness of the air and the sparkle of the water, how can you buy them?

Every part of this earth is sacred to my people. Every shining pine needle, every sandy shore, every mist in the dark woods, every clearing and humming insect is holy in the memory and experience of my people. The sap which courses through the trees carries the memories of the red man.

The white man's dead forget the country of their birth when they go to walk among the stars. Our dead never forget this beautiful earth, for it is the mother of the red man. We are part of the earth and it is part of us. The perfumed

Knows what is to come.
—is sad.

flowers are our sisters; the deer, the horse, the great eagle, these are our brothers. The rocky crests, the juices in the meadows, the body heat of the pony, and man—all belong to the same family.

So, when the Great Chief in Washington sends word that he wishes to buy our land, he asks much of us. The Great Chief sends word he will reserve us a place so that we can live comfortably to ourselves. He will be our father and we will be his children.

So we will consider your offer to buy our land. But it will not be easy. For this land is sacred to us. This shining water that moves in the streams and rivers is not just water but the blood of our ancestors. If we sell you land, you must remember that it is sacred, and you must teach your children that it is sacred and that each ghostly reflection in the clear water of the lakes tells of events and memories in the life of my people. The water's murmur is the voice of my father's father.

The rivers are our brothers, they quench our thirst. The rivers carry our canoes, and feed our children. If we sell you our land, you must remember, and teach your children, that the rivers are our brothers and yours, and you must henceforth give the rivers the kindness you would give any brother.

We know that the white man does not understand our ways. One portion of land is the same to him as the next, for he is a stranger who comes in the night and takes from the land whatever he needs. The earth is not his brother, but his enemy, and when he has conquered it, he moves on. He leaves his father's grave behind, and he does not care. He kidnaps the earth from his children, and he does not care. His father's grave, and his children's birthright are forgotten. He treats his mother, the earth, and his brother, the sky, as things to be bought, plundered, sold like sheep or bright beads. His appetite will devour the earth and leave behind only a desert.

I do not know. Our ways are different from your ways. The sight of your cities pains the eyes of the red man. There is no quiet place in the white man's cities. No place to hear the unfurling of leaves in spring or the rustle of the insect's wings. The clatter only seems to insult the ears. And what is there to life if a man cannot hear the lonely cry of the whippoorwill or the arguments of the frogs around the pond at night? I am a red man and do not understand. The Indian prefers the soft sound of the wind darting over the face of a pond and the smell of the wind itself, cleansed by a midday rain, or scented with piñon pine.

The air is precious to the red man for all things share the same breath, the beast, the tree, the man, they all share the same breath. The white man does not seem to notice the air he breathes. Like a man dying for many days he is numb to the stench. But if we sell you our land, you must remember that the air is precious to us, that the air shares its spirit with all the life it supports.

The wind that gave our grandfather his first breath also receives his last sigh. And if we sell you our land, you must keep it apart and sacred as a place where even the white man can go to taste the wind that is sweetened by the meadow's flowers.

You must teach your children that the ground beneath their feet is the ashes of our grandfathers. So that they will respect the land, tell your children that the earth is rich with the lives of our kin. Teach your children that we have taught our children that the earth is our mother. Whatever befalls the earth befalls the sons of the earth. If men spit upon the ground, they spit upon themselves.

This we know: the earth does not belong to man; man belongs to the earth. All things are connected. We may be brothers after all. We shall see. One thing we know which the white man may one day discover: our God is the same God.

You may think now that you own Him as you wish to own our land; but you cannot. He is the God of man, and His compassion is equal for the red man and the white. This earth is precious to Him, and to harm the earth is to heap contempt on its creator. The whites too shall pass; perhaps sooner than all other tribes. Contaminate your bed and you will one night suffocate in your own waste.

But in your perishing you will shine brightly fired by the strength of the God who brought you to this land and for some special purpose gave you dominion over this land and over the red man.

That destiny is a mystery to us, for we do not understand when the buffalo are all slaughtered, the wild horses are tame, the secret corners of the forest heavy with scent of many men and the view of the ripe hills blotted by talking wires.

Where is the thicket? Gone. Where is the eagle? Gone.

The end of living and the beginning of survival.

Questions and Topics for Discussion and Writing

1. This speech has been called a "profound environmental statement." Summarize the environmental statement that it makes.
2. Discuss the effect of repetition as a rhetorical device in this statement.
3. Examine the sentence patterns of this work. How can they be said to affect the tone?
4. In a paper, compare the different religious views of the Indian and the white man as seen by Chief Seattle.
5. Write a paper in which you use specific examples to present what you think Seattle would add today to his eloquent plea.
6. Seattle's resolution to the problem of the white man is rendered in religious terms. How is this resolution an appeal for consideration by the white man?
7. The concluding statement seems to trail off into nothingness. How effective is the conclusion?
8. Write a statement comparing this final line with the theme of "At the Bomb Testing Site" (later in this unit).

Rachel Carson

"The aim of science is to discover and illuminate truth. And that, I take it, is the aim of literature," said Rachel Carson; "there can be no separate literature of science."

Rachel Louise Carson (1907–1964), biologist, oceanographer, and writer, grew up on a farm in Springdale, Pennsylvania, in the lower Allegheny Valley. As a child, she was encouraged by her mother to develop interests in the natural world and in books. Early, Rachel developed a love for writers and writing. She began to compose stories and essays in elementary school and continued through high school. At Pennsylvania College for Women, a perceptive English composition teacher recognized Carson's talent and encouraged her writing. Carson's first ambition, then, was to be a writer, but by the end of her second year, a biology course so fascinated her that she decided to dedicate her life to science.

For a time, Carson believed that she had repressed her dream of being a writer; only later did she realize that science had given her compelling material for writing. As one critic observed, "The merging of these two powerful currents—the imagination and insight of a creative writer with a scientist's passion for fact—goes far to explain the blend of beauty and authority that was to make her books unique."

Carson's pursuit of science continued with graduate work in zoology at Johns Hopkins University and at the Marine Biological Laboratory, Woods Hole; she supported herself by part-time teaching of biology. During this time, both her father and her sister died, leaving Carson to provide for her mother and the sister's two children. Carson secured a full-time job at the federal Bureau of Fisheries in 1936, remaining there for the rest of her life.

In addition to professional publications for the United States Fish and Wildlife Service, Rachel Carson wrote a number of newspaper and magazine articles. *The Sea Around Us* (1951) was a best-seller; *Silent Spring* (1962) attacks the indiscriminate use of pesticides brought on by "human carelessness, greed, and irresponsibility." The book was the winner of eight awards and became the most influential work of the environmental movement; it still arouses praise and controversy. Until *Silent Spring*, the concept of "ecology" was largely unknown. Of this book, Rachel Carson wrote to a friend, "I have felt bound by a solemn obligation to do what I could. . . . Now I can believe I have at least helped a little. It would be unrealistic to believe one book could bring a complete change."

"The Obligation to Endure," from *Silent Spring*, reveals why and how Carson raises our consciousness and changes our sense of obligation to the world around us.

The Obligation to Endure

Rachel Carson

The history of life on earth has been a history of interaction between living things and their surroundings. To a large extent, the physical form and the habits of the earth's vegetation and its animal life have been molded by the environment. Considering the whole span of earthly time, the opposite effect, in which life actually modifies its surroundings, has been relatively slight. Only within the moment of time represented by the present century has one species—man—acquired significant power to alter the nature of his world.

During the past quarter century this power has not only increased to one of disturbing magnitude but it has changed in character. The most alarming of all man's assaults upon the environment is the contamination of air, earth, rivers, and sea with dangerous and even lethal materials. This pollution is for the most part irrecoverable; the chain of evil it initiates not only in the world that must support life but in living tissues is for the most part irreversible. In this now universal contamination of the environment, chemicals are the sinister and little-recognized partners of radiation in changing the very nature of the world—the very nature of its life. Strontium 90, released through nuclear explosions into the air, comes to earth in rain or drifts down as fallout, lodges in soil, enters into the grass or corn or wheat grown there, and in time takes up its abode in the bones of a human being, there to remain until his death. Similarly, chemicals sprayed on croplands or forests or gardens lie long in soil, entering into living organisms, passing from one to another in a chain of poisoning and death. Or they pass mysteriously by underground streams until they emerge and, through the alchemy of air and sunlight, combine into new forms that kill vegetation, sicken cattle, and work unknown harm on those who drink from once pure wells. As Albert Schweitzer has said, "Man can hardly even recognize the devils of his own creation."

It took hundreds of millions of years to produce the life that now inhabits the earth—eons of time in which that developing and evolving and diversifying life reached a state of adjustment and balance with its surroundings. The environment, rigorously shaping and directing the life it supported, contained elements that were hostile as well as supporting. Certain rocks gave out dangerous radiation; even within the light of the sun, from which all life draws its energy, there were short-wave radiations with power to injure. Given time—time not in years but in millennia—life adjusts, and a balance has been reached. For time is the essential ingredient; but in the modern world there is no time.

The rapidity of change and the speed with which new situations are created follow the impetuous and heedless pace of man rather than the deliberate pace of nature. Radiation is no longer merely the background radiation of rocks, the bombardment of cosmic rays, the ultraviolet of the sun that have existed before there was any life on earth; radiation is now the unnatural creation of man's tampering with the atom. The chemicals to which life is asked to make its adjustment are no longer merely the calcium and silica and copper and all the rest of the minerals washed out of the rocks and carried in rivers to the sea; they are the synthetic creations of man's inventive mind, brewed in his laboratories, and having no counterparts in nature.

To adjust to these chemicals would require time on the scale that is nature's; it would require not merely the years of a man's life but the life of generations. And even this, were it by some miracle possible, would be futile, for the new chemicals come from our laboratories in an endless stream; almost five hundred annually find their way into actual use in the United States alone. The figure is staggering and its implications are not easily grasped—500 new chemicals to which the bodies of men and animals are required somehow to adapt each year, chemicals totally outside the limits of biologic experience.

Among them are many that are used in man's war against nature. Since the mid-1940's over 200 basic chemicals have been created for use in killing insects, weeds, rodents, and other organisms described in the modern vernacular as "pests"; and they are sold under several thousand different brand names.

These sprays, dusts, and aerosols are now applied almost universally to farms, gardens, forests, and homes—nonselective chemicals that have the power to kill every insect, the "good" and the "bad," to still the song of birds and the leaping of fish in the streams, to coat the leaves with a deadly film, and to linger on in soil—all this though the intended target may be only a few weeds or insects. Can anyone believe it is possible to lay down such a barrage of poisons on the surface of the earth without making it unfit for all life? They should not be called "insecticides," but "biocides."

The whole process of spraying seems caught up in an endless spiral. Since DDT was released for civilian use, a process of escalation has been going on in which ever more toxic materials must be found. This has happened because insects, in a triumphant vindication of Darwin's principle of the survival of the fittest, have evolved super races immune to the particular insecticide used, hence a deadlier one has always to be developed—and then a deadlier one than that. It has happened also because, for reasons to be described later, destructive insects often undergo a "flareback," or resurgence, after spraying, in numbers greater than before. Thus the chemical war is never won, and all life is caught in its violent crossfire.

Along with the possibility of the extinction of mankind by nuclear war, the central problem of our age has therefore become the contamination of man's total environment with such substances of incredible potential for harm—substances that accumulate in the tissues of plants and animals and even penetrate the germ cells to shatter or alter the very material of heredity upon which the shape of the future depends.

Some would-be architects of our future look toward a time when it will be possible to alter the human germ plasm by design. But we may easily be doing so now by inadvertence, for many chemicals, like radiation, bring about gene mutations. It is ironic to think that man might determine his own future by something so seemingly trivial as the choice of an insect spray.

All this has been risked—for what? Future historians may well be amazed by our distorted sense of proportion. How could intelligent beings seek to control a few unwanted species by a method that contaminated the entire environment and brought the threat of disease and death even to their own kind? Yet this is precisely what we have done. We have done it, moreover, for reasons that collapse the moment we examine them. We are told that the enormous and expanding use of pesticides is necessary to maintain farm production. Yet is our real problem not one of *overproduction?* Our farms, despite measures to re- move acreages from production and to pay farmers *not* to produce, have yielded such a staggering excess of crops that the American taxpayer in 1962 is paying out more than one billion dollars a year as the total carrying cost of the surplus-food storage program. And is the situation helped when one branch of the Agriculture Department tries to reduce production while another states, as it did in 1958, "It is believed generally that reduction of crop acreages under provisions of the Soil Bank will stimulate interest in use of chemicals to obtain maximum production on the land retained in crops."

All this is not to say there is no insect problem and no need of control. I am saying, rather, that control must be geared to realities, not to mythical situa- tions, and that the methods employed must be such that they do not destroy us along with the insects.

The problem whose attempted solution has brought such a train of disaster in its wake is an accompaniment of our modern way of life. Long before the age of man, insects inhabited the earth—a group of extraordinarily varied and adapta- ble beings. Over the course of time since man's advent, a small percentage of the more than half a million species of insects have come into conflict with human welfare in two principal ways: as competitors for the food supply and as carriers of human disease.

Disease-carrying insects become important where human beings are crowded together, especially under conditions where sanitation is poor, as in time of natural disaster or war or in situations of extreme poverty and depriva- tion. Then control of some sort becomes necessary. It is a sobering fact, how- ever, as we shall presently see, that the method of massive chemical control has had only limited success, and also threatens to worsen the very conditions it is intended to curb.

Under primitive agricultural conditions the farmer had few insect problems. These arose with the intensification of agriculture—the devotion of immense acreages to a single crop. Such a system set the stage for explosive increases in specific insect populations. Single-crop farming does not take advantage of the principles by which nature works; it is agriculture as an engineer might conceive it to be. Nature has introduced great variety into the landscape, but man has displayed a passion for simplifying it. Thus he undoes the built-in checks and balances by which nature holds the species within bounds. One important natural check is a limit on the amount of suitable habitat for each species. Obviously then, an insect that lives on wheat can build up its population to much higher levels on a farm devoted to wheat than on one in which wheat is intermingled with other crops to which the insect is not adapted.

The same thing happens in other situations. A generation or more ago, the towns of large areas of the United States lined their streets with the noble elm tree. Now the beauty they hopefully created is threatened with complete destruction as disease sweeps through the elms, carried by a beetle that would have only limited chance to build up large populations and to spread from tree to tree if the elms were only occasional trees in a richly diversified planting.

Another factor in the modern insect problem is one that must be viewed against a background of geologic and human history: the spreading of thousands of different kinds of organisms from their native homes to invade new territories. This worldwide migration has been studied and graphically described by the British ecologist Charles Elton in his recent book *The Ecology of Invasions*. During the Cretaceous Period, some hundred million years ago, flooding seas cut many land bridges between continents and living things found themselves confined in what Elton calls "colossal separate nature reserves." There, isolated from others of their kind, they developed many new species. When some of the land masses were joined again, about 15 million years ago, these species began to move out into new territories—a movement that is not only still in progress but is now receiving considerable assistance from man.

The importation of plants is the primary agent in the modern spread of species, for animals have almost invariably gone along with the plants, quarantine being a comparatively recent and not completely effective innovation. The United States Office of Plant Introduction alone has introduced almost 200,000 species and varieties of plants from all over the world. Nearly half of the 180 or so major insect enemies of plants in the United States are accidental imports from abroad, and most of them have come as hitchhikers on plants.

In new territory, out of reach of the restraining hand of the natural enemies that kept down its numbers in its native land, an invading plant or animal is able to become enormously abundant. Thus it is no accident that our most troublesome insects are introduced species.

These invasions, both the naturally occurring and those dependent on human assistance, are likely to continue indefinitely. Quarantine and massive chemical campaigns are only extremely expensive ways of buying time. We are faced, according to Dr. Elton, "with a life-and-death need not just to find new technological means of suppressing this plant or that animal"; instead we need the basic knowledge of animal populations and their relations to their surroundings that will "promote an even balance and damp down the explosive power of outbreaks and new invasions."

Much of the necessary knowledge is now available but we do not use it. We train ecologists in our universities and even employ them in our governmental agencies but we seldom take their advice. We allow the chemical death rain to fall as though there were no alternative, whereas in fact there are many, and our ingenuity could soon discover many more if given opportunity.

Have we fallen into a mesmerized state that makes us accept as inevitable that which is inferior or detrimental, as though having lost the will or the vision to demand that which is good? Such thinking, in the words of the ecologist Paul Shepard, "idealizes life with only its head out of water, inches above the limits of toleration of the corruption of its own environment . . . Why should we tolerate a diet of weak poisons, a home in insipid surroundings, a circle of acquaintances who are not quite our enemies, the noise of motors with just enough relief to prevent insanity? Who would want to live in a world which is just not quite fatal?"

Yet such a world is pressed upon us. The crusade to create a chemically sterile, insect-free world seems to have engendered a fanatic zeal on the part of many specialists and most of the so-called control agencies. On every hand there is evidence that those engaged in spraying operations exercise a ruthless power. "The regulatory entomologists . . . function as prosecutor, judge and jury, tax assessor and collector and sheriff to enforce their own orders," said Connecticut entomologist Neely Turner. The most flagrant abuses go unchecked in both state and federal agencies.

It is not my contention that chemical insecticides must never be used. I do contend that we have put poisonous and biologically potent chemicals indiscriminately into the hands of persons largely or wholly ignorant of their potentials for harm. We have subjected enormous numbers of people to contact with these poisons, without their consent and often without their knowledge. If the Bill of Rights contains no guarantee that a citizen shall be secure against lethal poisons distributed either by private individuals or by public officials, it is surely only because our forefathers, despite their considerable wisdom and foresight, could conceive of no such problem.

I contend, furthermore, that we have allowed these chemicals to be used with little or no advance investigation of their effect on soil, water, wildlife, and man himself. Future generations are unlikely to condone our lack of prudent concern for the integrity of the natural world that supports all life.

There is still very limited awareness of the nature of the threat. This is an era of specialists, each of whom sees his own problem and is unaware of or intolerant of the larger frame into which it fits. It is also an era dominated by industry, in which the right to make a dollar at whatever cost is seldom challenged. When the public protests, confronted with some obvious evidence of damaging results of pesticide applications, it is fed little tranquilizing pills of half truth.

We urgently need an end to these false assurances, to the sugar coating of unpalatable facts. It is the public that is being asked to assume the risks that the insect controllers calculate. The public must decide whether it wishes to continue on the present road, and it can do so only when in full possession of the facts. In the words of Jean Rostand, "The obligation to endure gives us the right to know."

Questions and Topics for Discussion and Writing

1. What is the most significant developmental feature mentioned that has influenced the history of life on earth? How has the human race added to this feature of its environment?
2. Describe the importance Carson attaches to the pace at which nature and the environment adapt to the influence of the human race. What are the complications engendered by the pace at which humans have introduced changes into the environment?
3. What are the problems associated with chemicals that are, in Carson's term, "non-selective"?
4. What are the features of the insect problem as Carson describes it? How are communities to blame for the spread of diseases among plants, particularly the elm tree, and what does she recommend as a solution?
5. Look up the terms *environment* and *ecology* in the *Oxford English Dictionary*. Write an essay showing how the meanings have changed over the past 20 years.
6. In *Book Review Digest*, look up reviews of *Silent Spring*. Write an objective summary of your findings.
7. In an essay titled "Myths of Environmentalism," William Tucker calls *Silent Spring* both a "great book" and a "terrible book." Read his essay and determine the basis for this paradoxical judgment.

To Preserve a World Graced by Life

Carl Sagan

You will find the biography for Carl Sagan before the essay "Can We Know the Universe?" in Unit 2.

The Earth is an anomaly; in all the solar system, it is, so far as we know, the only inhabited planet. I look at the fossil record and I see that after flourishing for 180 million years, the dinosaurs were extinguished. Every last one. There are none left. No species is guaranteed its tenure on this planet. And we've been here for only about a million years, we, the first species that has devised the means for its self-destruction. We are rare and precious because we are alive, because we can think. We are privileged to live, to influence and control our future. I believe we have an obligation to fight for that life, to struggle not just for ourselves, but for all those creatures who came before us, and to whom we are beholden, and for all those who, if we are wise enough, will come after us. There is no cause more urgent, no dedication more fitting for us than to strive to eliminate the threat of nuclear war. No social convention, no political system, no economic hypothesis, no religious dogma is more important.

Every thinking person fears nuclear war. And every technological state plans for it. Everyone knows it's madness. And every nation has an excuse. There's a dreary chain of causality. The Germans were working on the bomb at the beginning of the Second World War. So the Americans had to make one. If the Americans had one the Soviets had to have one. And then the British, the French, the Chinese, the Indians, the Pakistanis, and so on. By the end of this century many more nations will have collected nuclear weapons. They're easy to devise. Fissionable material can be stolen from nuclear reactors. Nuclear weapons have become almost a home handicraft industry.

The conventional bombs of World War II were called blockbusters. Filled with 20 tons of TNT or so, they could destroy a city block. All the bombs dropped on all the cities in World War II amounted to some two million tons of TNT, two megatons. Coventry and Rotterdam, Dresden and Tokyo, all the death that rained from the skies between 1939 and 1945, a hundred thousand blockbusters: two megatons. Now, two megatons is the energy released in the explosion of a single more or less humdrum thermonuclear weapon, one bomb with the destructive force of the entire Second World War. But there are tens of thousands of nuclear weapons. The strategic missile and bomber forces of the Soviet Union and the United States are at this moment aiming warheads at more than 15,000 designated targets. No place on the planet is safe. The energy contained in these weapons, genies of death patiently awaiting the rubbing of the lamps, is far more than 10,000 megatons—but with their destructive capability concentrated efficiently, not over six years but over a few hours, a blockbuster for every family on the planet, a World War II every second for the length of a lazy afternoon.

The yield, the explosive equivalent, of the Hiroshima bomb, was only 13 kilotons, the equivalent of 13,000 tons of TNT. In a full nuclear exchange, in the paroxysm of thermonuclear war, the equivalent of a million Hiroshima bombs would be dropped all over the world. At the Hiroshima death rate of some hundred thousand people killed per equivalent 13 kiloton weapon, this would be enough to kill a hundred billion people. But there are only 4.3 billion people on the planet. This disparity is given the conventional term "overkill," which falls far short of conveying the horror implied by the steady hoarding of these instruments of death. In a full nuclear exchange not everybody would be killed by the blast and the firestorm, the radiation and the fallout. The survivors would witness more subtle consequences of the war. Many possible full nuclear exchanges would burn the nitrogen in the upper air, converting it to oxides of nitrogen which would in turn destroy a significant fraction of the ozone in the ozonosphere, which would then let in solar ultraviolet light, which would have as a minor consequence skin cancer for light-skinned people who are the ones who built the weapons in the first place, but as a more serious consequence the destruction of large numbers of microbes in a vast ecological pyramid that we understand very poorly but at the top of which we sit.

Ultraviolet light also destroys crops. The dust put up into the air in a full nuclear exchange would reflect sunlight and cool the Earth—no one yet knows by how much. Even a little cooling could have disastrous agricultural consequences. Birds are more easily killed by radiation than insects. Plagues of insects and consequent further agricultural disorders are a likely consequence of nuclear war. And there's another kind of plague to worry about. The plague bacillus is endemic all over the Earth. At the moment not many humans die of it, not so much because it's absent but because our resistance is high and our sanitation is generally adequate. However, the radiation produced in a nuclear war, among its many other effects, debilitates the body's immunological systems, causing a deterioration in our ability to resist disease, and sanitation services would be primitive at best in the post-holocaust environment. In the longer term there would be mutations, new varieties of microbes and insects that might cause still further problems for any human survivors of a nuclear war, and perhaps, after a while, when there's been enough time for the recessive mutations to recombine and be expressed, new and horrifying varieties of humans. Most of these mutations, when expressed, would be lethal. A few would not. And then there would be other agonies: the loss of loved ones, the legions of the burned, the blind and the mutilated, disease, long-lived radioactive poisons in the air and the water, the threat of tumors and stillbirths and malformed children, the virtually complete absence of medical care, the hopeless sense of a civilization destroyed for nothing, the knowledge that we could have prevented it and did not.

This is an unpleasant but I do not think an alarmist picture. It is, I believe, a conservative estimate of what would happen in and after a full-scale nuclear war, because the effects are synergistic and nonlinear, gathering force in unimagined combinations.

The dangers of nuclear war are, in a way, well-known. But in a way they are not well-known, because of the psychological factor—psychiatrists call it denial—that makes us feel it's so horrible that we might as well not think about it. This element of denial is, I believe, one of the most serious problems we face. If everyone had a profound and immediate sense of the actual consequences of nuclear war, we would be much more willing to confront and challenge national leaders of all nations when they present narrow and self-serving arguments for the continuation of mutual nuclear terror.

Denial is remarkably strong and there are many cases in human history where, faced with the clearest signs of extreme danger, people refused to take simple corrective measures. To give one of many examples, there was, some 25 years ago, a tsunami, a so-called tidal wave, in the Pacific, which was approaching the coast of the Hawaiian Islands. The people who lived there were given many hours warning to flee the lowlands and run to safety. But the idea of a great, crashing wave of water thirty feet high surging inland, inundating and washing houses out to sea was so unbelievable, so unpleasant, that many people simply ignored the warning and were killed. One schoolteacher thought the report to be so interesting that she gathered up her children and took them down to the water's edge to watch. I believe that one of the most important jobs that scientists have in this dialogue, this polylogue, on the dangers of nuclear war is to state very clearly what the dangers are.

Increases in the number of nuclear weapons by one nation have never deterred any other nation from stockpiling its own nuclear arsenal. Rather, the evidence is much more compelling that such proliferation leads to a substantial, indeed to an exponential, growth of nuclear weapons worldwide. No nation is ever satisfied that it has enough weapons. Any "improvements" by the other side force us to "improve" our weapons systems. Now, if that's true, it has an interesting implication, because exponentials not only go up, they also go down. It suggests that a concerted effort to increase the nuclear weapons and delivery systems stockpiled by one nation will result in a corresponding increase by other nations. But likewise, a concerted effort by any one nuclear power to decrease its stockpile of nuclear weapons and delivery systems might very well have as a consequence a decline in the nuclear weapons and delivery systems of other nations, and, at least up to a point, the process can be self-sustaining. I therefore raise the question whether there is some special obligation that the United States has, as the nation that first developed nuclear weapons and the nation that first used nuclear weapons on human populations, to decelerate the nuclear arms race. There is a wide range of possible options, including small and safe unilateral steps to test the responses of other nations, and major bilateral and multilateral efforts to negotiated, substantial, verifiable force reductions.

There are clear economic disabilities in maintaining the warfare state. It's apparent that the enormous economic growth of the German Federal Republic and Japan since the Second World War is due in part to the fact that they did not have to maintain major military establishments. In general, nations with smaller fractions of their budgets devoted to armaments and armies exhibit greater industrial productivity and economic growth. Mutual arms reductions, done in such a way as to preserve deterrence against a nuclear attack, are in everybody's interest. Disarmament would be in the interest of the United States, where we obviously cannot simultaneously maintain humane social programs, an enormous military establishment and a balanced budget. (It is not even clear that we can maintain any two of the three.) It is at least equally in the interest of the Soviet Union, where grim economic problems are looming. If the USSR were able to relieve some of the pressure for weapons developments, it would be able to put that money into consumer products, housing, better medical care and other activities which would help its people. A comparable conclusion applies here. It's only a matter of getting started. Of course there's some risk. It takes courage. But as Einstein asked, in precisely this context, "What is the alternative?"

There are hawks and doves on both sides, although there is a tendency for the official propaganda of Nation A to tell us only about the hawks of Nation B, and vice versa. I would like to remind us that there are serious debates in both nations, whose outcomes are by no means certain. In 1981 there was an important meeting held in the United States near Washington and called the Airlie House Conference. Physicians from the USSR, the USA and a number of other nations concluded that the greatest public health menace to the population of the Earth is nuclear war, and that physicians therefore have a special responsibility, connected with their Hippocratic oath, to prevent nuclear war. That conference made a set of stirring and at the same time very rational pleas for nuclear disarmament. However, concern was expressed that, if the conclusions of the meeting were widely disseminated in the United States while no one in the Soviet Union heard about them, it would redound to the strategic advantage of the Soviet Union, being tantamount to an advocacy of unilateral American nuclear disarmament. I wonder how many of us have heard about the Airlie House conclusions? There was in fact very little public attention given to that conference in the United States. But in the Soviet Union the leader of the Soviet delegation, Evgeny Chazov, who was Mr. Brezhnev's personal cardiologist, described in great detail the conclusions of that conference in a television broadcast transmitted all over the Soviet Union. The conclusions were published in both *Pravda* and *Izvestia*. At least in this particular case, there was significantly wider exposure of the dangers of nuclear war in the USSR than in the United States. In June of 1982, some of these same physicians met at an international cardiology conference in Moscow. Six of them, three Americans and three Russians, held a televised discussion on the dangers of nuclear war that included a straightforward and impassioned denial, by Bernard Lown of Harvard, of the utility of shelters in a full nuclear exchange. For many years the Soviets have advocated such shelters and have built a number of them. This discussion was shown in prime time, more than once, on the principal television channel in the Soviet Union. In the first showing, the only alternative television offering was the also-rans in the Tschaikowsky piano competition. The Soviets claim that 200 million people in that country saw the broadcast. In the United States, all three commercial networks refused the program. It was shown eventually on the much smaller Public Broadcast System at 11 P.M. So much for the argument that the Soviets are afraid to exhibit arguments on the perils of nuclear war.

I sometimes imagine an extraterrestrial being, cavorting through the universe, minding his, her, or its own business, and coming upon the Earth, examining it with no prior knowledge of what had happened here, and discovering, after finding out what the local units of currency are, that something approaching a trillion dollars or rubles a year is spent on preparations for war—that the global nuclear and conventional arms budget of all nations approaches a trillion dollars per annum. I imagine that extraterrestrial trying to calculate what good could be done with a trillion dollars, how much misery and hunger and ignorance could be undone with that kind of investment. I don't think it's too much to say that such an investment, sustained over a substantial period of time, could solve a very large number of the problems that most people of the planet Earth confront in their everyday lives. That being would, of course, wonder why we don't do something about it, why we consent to or, even worse, ignore the peril we have placed ourselves in.

Such an extraterrestrial being might also note that a few nations, one of them being the United States, actually have organizations devoted to peace as well as to war. The United States has something called the Arms Control and Disarmament Agency. But its budget is less than one hundred thousandth of the budget of the Department of Defense. This is a numerical measure of the relative importance that we place on finding ways to make war and finding ways to make peace. Is it possible that the intelligence, compassion and even self-interest of the American people have been thoroughly exhausted in the pursuit of solutions to the threat of nuclear war? Or is it more likely that so little attention is given to it, so little encouragement is provided to bright young people to consider this issue, that we have not even begun to find innovative and imaginative solutions? It seems to me that a substantial increase in the budget of this agency, and a creative rededication to its function of actually seeking disarmament and arms control, and breaking out altogether from the lockstep logic of the arms race would be an excellent token of the devotion of this government to peace and not to war. Comparable remarks apply, of course, to other nations.

I invite you to study these issues. Your future, and the future of all who come after you, depends upon it. It is possible that you might achieve an important new insight into this problem. If you exercise denial, if you refuse to think about these issues because they're too difficult or too agonizing or because you feel nothing can be done about them, then you yourself are making a contribution towards the nuclear holocaust we all wish to avoid. Through the courageous examination of these deep and painful questions, and through the political process, I am convinced we can make an important contribution towards preserving and enhancing the life that has graced our small world.

Questions and Topics for Discussion and Writing

1. What are the connotations of the word "graced," especially as used in this title?
2. Look up the name George Kistiakowsky. He is the man to whom this essay is dedicated. Explain Sagan's reason for making this choice.

3. Consider Sagan's statement in the first paragraph, "I believe we have an obligation to fight for that life not just for ourselves, but for all those creatures who came before us, and to whom we are beholden. . . ." Compare this statement to the theme of William Stafford's poem "At the Bomb Testing Site," later in this unit.
4. What does Sagan see as a definite effect on evolution that would result from the nuclear bomb?
5. What does Sagan mean by saying that the effects of full-scale nuclear war are synergistic and nonlinear?
6. According to Sagan, what is the most serious problem that humans face today?
7. Why does Sagan call the discussion of nuclear war a "polylogue"? What does he say is the important job that scientists must undertake in the polylogue?
8. What is the challenge to the United States offered by Sagan?
9. What is Sagan's argument against improving one country's system of weapons to keep pace with another's?
10. Had you heard of the Airlie House Conference before reading this essay? In what way does Sagan credit the USSR over the USA in handling the conclusions of the conference? Do you agree with the implication that the USA should have reacted differently?
11. What is the challenge that Sagan offers in this essay to you, the reader?
12. Write a paragraph employing Sagan's figure of speech, the "lockstep logic of the arms race."

W. H. Auden

W. H. (Wystan Hugh) Auden (1907–1973), one of the major voices in twentieth-century poetry, is a complex figure. An Anglo-American, he was born in York, England, but left before World War II to settle in the United States, where he became a citizen. When his early poems appeared in the decade before he left England, they sounded an impressive, "knowing" tone. They were romantically radical, didactic and challenging, and intellectual in a deliberate, sophisticated way that made him the leading speaker for the "thirties group" of left-wing, modernist poets. Yet soon after he changed homelands, he made another change in orientation: Auden became a Christian convert whose embracing of religion antagonized many of his earlier, Marxist admirers. Complicating his canon, Auden repudiated some of his early political poems, omitting them from his later collections, or revising them extensively. Auden is also complex in the rich variety and scope of his subjects as well as in the diverse forms of his verse, which display a technical mastery of his craft.

Just as Auden cannot be confined in one ideological compartment, so also he cannot be described as predictable in style or subject. He has written lyrics, satires, conversational "arguments," parodies, plays, allegorical poems, short light verses, and long verse dramas. He has employed mythology, philosophy, history, geography, geology, psychology, and many other areas of inquiry as material.

One characteristic that makes Auden a stylistic individualist is his use of science. Critics have remarked on this use in several ways: the technological images, such as lead mines and industrial landscapes; the allusions to scientific information, revealing his wide reading; the influence of Sigmund Freud; and the frequent use of a detached, even "clinical" tone that seems to supply an air of scientific authority. The clinical tone, for example, brings an objectivity to his political poems in the thirties—he was the "doctor" diagnosing the faults of a sick society—and to his philosophical and religious poems of 1940 and later— he was the responsible moral "teacher." The use of science was a natural one. Auden had been fascinated by childhood reading of technical books and planned to become a mining engineer. Long after he had discovered that his true vocation was to be a writer, he listed as his "favorite reading" the magazine *Scientific American*.

"Gare du Midi" illustrates Auden's use of science, his perennial fascination with words, and the varying possibilities of tone.

Gare du Midi

W. H. Auden

A nondescript express in from the South,
Crowds round the ticket barrier, a face
To welcome which the mayor has not contrived
Bugles or braid: something about the mouth
Distracts the stray look with alarm and pity.
Snow is falling. Clutching a little case,
He walks out briskly to infect a city
Whose terrible future may have just arrived.

Questions and Topics for Discussion and Writing

1. What words would you use to characterize the tone of this poem? Citing specific words and phrases, explain how Auden creates this tone.
2. An argument can be made that the mysterious stranger infects the city by accident or that he does so by design. What interpretation do you hold? Present your argument, citing specific words and phrases.
3. Find or sketch a picture of the scientist imaged by Auden in this poem.

4. Consider the seven images of science set forth in Holton's essay "Modern Science and the Intellectual Tradition" (Unit 3). Is Auden using one of these images? Write a paragraph explaining how he applies one of the seven images, or, if you believe Auden uses an eighth image, defining and describing that image.

William Stafford

William Edgar Stafford (b. 1914), a scholar, editor, and professor of English at Purdue University, has been called a "Western Robert Frost." Born in Kansas, he obtained his bachelor's and master's degrees there, then completed his doctorate at the University of Iowa. These midwestern landscapes and scenes appear in his poetry as spatial images of mountains, bleak prairie towns, and deserts.

His writing reflects other vistas—the many parts of the world where he has traveled and taught. He has worked in a kaleidoscope of locales, from sugar-beet fields and oil refineries to the United States Forest Service and Church World Service. As a lecturer on literature for the United States Information Agency, he has taught in Egypt, India, Bangladesh, Pakistan, Iran, and Nepal. In all these areas, he has treated the subject of existential loneliness.

His poems have appeared in such publications as *The Atlantic, The Nation, Harper's Poetry,* and *The New Yorker.* His books of collected poetry include *Stories That Could Be True* (1977) and *A Glass Face in the Rain* (1982). Prose works include *Writing the Australian Crawl, Friends to This Ground,* and *Down in My Heart.* The third volume is an account of Stafford's serving as a conscientious objector in World War II.

Stafford's sense of moral responsibility and of crisis appears in "At the Bomb Testing Site," a scene of possible apocalypse in a setting of desolation.

At the Bomb Testing Site

William Stafford

At noon in the desert a panting lizard
waited for history, its elbows tense,
watching the curve of a particular road
as if something might happen.

It was looking for something farther off
than people could see, an important scene
acted in stone for little selves
at the flute end of consequences.

There was just a continent without much on it
under a sky that never cared less.
Ready for a change, the elbows waited.
The hands gripped hard on the desert.

Questions and Topics for Discussion and Writing

1. Consider the poem's opening phrase, "at noon." What are the connotations of such a phrase? How does it help to convey Stafford's belief about the use of the atomic bomb?
2. Give thought to the diction of this poem. What is suggested by the word "panting"? What is the significance of the phrase "waited for history"? Why does the lizard watch the "*curve* [italics added] of a particular road"?
3. Why do you think Stafford focuses on a lizard for the chief animate subject of the poem?
4. Write a paragraph discussing Stafford's use of images and symbols in the poem.
5. Although the atomic bomb in actuality was tested in the desert, Stafford uses the word "desert" in a greater metaphoric sense in the opening and closing of the poem. What extra meanings does he suggest?
6. Study the middle stanza. How can a lizard look "for something farther off/than people could see"?
7. Consider the use of verbs in this poem. What meaning is added by the use of "could" instead of "can" in "It was looking for something farther off/than people could see . . ."?
8. Explain the first two lines of the last stanza. Why do you think Stafford says that the sky "never cared less"?
9. Why do you think the lizard's elbows "wait" and the hands "grip hard"?

J. Robert Oppenheimer

J. Robert Oppenheimer (1904–1967), a distinguished American physicist and educator, is probably best known as director of the Los Alamos Scientific Laboratory in New Mexico from 1943 to 1945. It was there that the design and building of the world's first atomic bomb took place.

The son of a textile importer, Oppenheimer was born in New York City. He attended Harvard University, where he graduated with honors in three years, receiving a bachelor's degree in chemistry in 1925. He did graduate work in theoretical physics at Cambridge University, then earned a doctorate in physics in 1927 from the University of Göttingen. He continued his studies in Europe, working with the original architects of the quantum theory. In 1929, he returned to this country and began his career as an educator, teaching theoretical physics at the University of California and the California Institute of Technology.

Shortly after the United States entered World War II, Oppenheimer became involved in this country's nuclear effort. As a highly respected physicist, educator, and administrator, he quickly took command of the work on the atomic bomb. The laboratory at Los Alamos was built to consolidate and expedite work on the bomb; Oppenheimer was made its director in 1943. He assembled a large and heterogeneous group of brilliant scientists, planned the entire research program, and organized the laboratory. (Freeman Dyson's essay, "A Scientific Apprenticeship," in Unit 1, provides another perspective.) Under his dedicated leadership, a deliverable nuclear weapon was developed in less than three years.

After the war, Oppenheimer was appointed a consultant to the newly formed Atomic Energy Commission (AEC), playing a key role in drafting its policies. He also served as a policy adviser to the United States Department of Defense. In 1947 he was appointed director of the Institute for Advanced Study, Princeton University; he held this job until his retirement in 1966.

In 1953, Oppenheimer was investigated by the AEC security panel. His opposition to the development of the hydrogen bomb plus a brief association with the Communist party led to charges of disloyalty. Although he was cleared of all charges, he was denied further access to official secrets. In 1963, however, the AEC awarded Oppenheimer its highest honor, the Enrico Fermi Award, for his work in the field of nuclear physics.

Oppenheimer was a scientist with a social conscience. His opposition to the hydrogen bomb was the natural extension of his philosophy of the dual responsibility of scientists: While they pursue their work, they should be equally concerned with its effects on society.

Oppenheimer has left more to the world than his scientific discoveries; his published lectures stand as a reminder of the dignity and joy of science and of its relevance to philosophy and ethics. "To be touched with awe, or humor," he wrote, "to be moved by beauty, to make a commitment or a determination, to understand some truth—these are complementary modes of the human spirit. All of them are part of man's spiritual life."

The following selection was an address to members of the Association of Los Alamos Scientists on November 2, 1945. The occasion was Oppenheimer's farewell to the members. Oppenheimer spoke to his colleagues as a "fellow scientist and fellow worrier."

Farewell Address to the Los Alamos Scientists

J. Robert Oppenheimer

In considering what the situation of science is, it may be helpful to think a little of what people said and felt of their motives in coming into this job. One always has to worry that what people say of their motives is not adequate. Many people said different things, and most of them, I think, had some validity. There was in the first place the great concern that our enemy might develop these weapons before we did, and the feeling—at least, in the early days, the very strong feeling—that without atomic weapons it might be very difficult, it might be an impossible, it might be an incredibly long thing to win the war. These things wore off a little as it became clear that the war would be won in any case. Some people, I think, were motivated by curiosity, and rightly so; and some by a sense of adventure, and rightly so. Others had more political arguments and said, "Well, we know that atomic weapons are in principle possible, and it is not right that the threat of their unrealized possibility should hang over the world. Is it right that the world should know what can be done in their field and deal with it." And the people added to that that it was a time when all over the world men would be particularly ripe and open for dealing with this problem because of the immediacy of the evils of war, because of the universal cry from everyone that one could not go through this thing again, even a war without atomic bombs. And there was finally, and I think rightly, the feeling that there was probably no place in the world where the development of atomic weapons would have a better chance of leading to a reasonable solution, and a smaller chance of leading to disaster, than within the United States. I believe all these things that people said are true, and I think I said them all myself at one time or another.

But when you come right down to it the reason that we did this job is because it was an organic necessity. If you are a scientist you cannot stop such a thing. If you are a scientist you believe that it is good to find out how the world works; that it is good to find out what the realities are; that it is good to turn over to mankind at large the greatest possible power to control the world and to deal with it according to its lights and its values.

There has been a lot of talk about the evil of secrecy, of concealment, of control, of security. Some of that talk has been on a rather low plane, limited really to saying that it is difficult or inconvenient to work in a world where you are not free to do what you want. I think that the talk has been justified, and that the almost unanimous resistance of scientists to the imposition of control and secrecy is a justified position, but I think that the reason for it may lie a little deeper. I think that it comes from the fact that secrecy strikes at the very root of what science is, and what it is for. It is not possible to be a scientist unless you believe that it is good to learn. It is not good to be a scientist, and it is not possible, unless you think that it is of the highest value to share your knowledge, to share it with anyone who is interested. It is not possible to be a scientist unless you believe that the knowledge of the world, and the power which this gives, is a thing which is of intrinsic value to humanity, and that you are using it to help in the spread of knowledge, and are willing to take the consequences. And, therefore, I think that this resistance which we feel and see all around us to anything which is an attempt to treat science of the future as though it were rather a dangerous thing, a thing that must be watched and managed, is resisted not because of its inconvenience—I think we are in a position where we must be willing to take any inconvenience—but resisted because it is based on a philosophy incompatible with that by which we live, and have learned to live in the past.

There are many people who try to wiggle out of this. They say the real importance of atomic energy does not lie in the weapons that have been made; the real importance lies in all the great benefits which atomic energy, which the various radiations, will bring to mankind. There may be some truth in this. I am sure that there is truth in it, because there has never in the past been a new field opened up where the real fruits of it have not been invisible at the beginning. I have a very high confidence that the fruits—the so-called peacetime applications—of atomic energy will have in them all that we think, and more. There are others who try to escape the immediacy of this situation by saying that, after all, war has always been very terrible; after all, weapons have always gotten worse and worse; that this is just another weapon and it doesn't create a great change; that they are not so bad; bombings have been bad in this war and this is not a change in that—it just adds a little to the effectiveness of bombing; that some sort of protection will be found. I think that these efforts to diffuse and weaken the nature of the crisis make it only more dangerous. I think it is for us to accept it as a very grave crisis, to realize that these atomic weapons which we have started to make are very terrible, that they involve a change, that they are not just a slight modification: to accept this, and to accept with it the necessity for those transformations in the world which will make it possible to integrate these developments into human life.

As scientists I think we have perhaps a little greater ability to accept change, and accept radical change, because of our experiences in the pursuit of science. And that may help us—that, and the fact that we have lived with it—to be of some use in understanding these problems. . . .

There are a few things which scientists perhaps should remember, that I don't think I need to remind us of; but I will, anyway. One is that they are very often called upon to give technical information in one way or another, and I think one cannot be too careful to be honest. And it is very difficult, not because one tells lies, but because so often questions are put in a form which makes it very hard to give an answer which is not misleading. I think we will be in a very weak position unless we maintain at its highest the scrupulousness which is traditional for us in sticking to the truth, and in distinguishing between what we know to be true from what we hope may be true.

The second thing I think it right to speak of is this: it is everywhere felt that the fraternity between us and scientists in other countries may be one of the most helpful things for the future; yet it is apparent that even in this country not all of us who are scientists are in agreement. There is no harm in that; such disagreement is healthy. But we must not lose the sense of fraternity because of it; we must not lose our fundamental confidence in our fellow scientists.

I think that we have no hope at all if we yield in our belief in the value of science, in the good that it can be to the world to know about reality, about nature, to attain a gradually greater and greater control of nature, to learn, to teach, to understand. I think that if we lose our faith in this we stop being scientists, we sell out our heritage, we lose what we have most of value for this time of crisis.

But there is another thing: we are not only scientists; we are men, too. We cannot forget our dependence on our fellow men. I mean not only our material dependence, without which no science would be possible, and without which we could not work; I mean also our deep moral dependence, in that the value of science must lie in the world of men, that all our roots lie there. These are the strongest bonds in the world, stronger than those even that bind us to one another, these are the deepest bonds—that bind us to our fellow men.

Karl Jaspers

First legal study, then a medical degree, then training in psychiatry were the formal courses that led Karl Jaspers (1883–1969) to eventual eminence as an existentialist philosopher–historian. Both Germany and Switzerland can claim him. Born in Oldenburg, Germany, he joined the psychiatric faculty at Heidelberg University, where he wrote *General Psychopathology* (1913), then developed an interest in philosophy, and changed to a professorship in that field in 1922. During the next decade, he was able to produce his *magnum opus,* the three-volume *Philosophy.* Soon, however, trouble developed for him under the Hitler regime.

Because his wife was Jewish, Jaspers was persecuted by the Nazis, then dismissed from his job at the university in 1937. Although he was allowed to continue to publish—*Nietzsche* appeared in 1936 and *Descartes and His Philosophy* in 1937—Jaspers was not allowed to lecture. When he was reinstated after World War II, Jaspers chose to transfer to the University of Basel, Switzerland, where he renounced his German citizenship and remained until his death.

There Jaspers continued to write voluminously, and his works proved of interest to a wide group of readers, including theologians, rather than to fellow philosophers alone. Jaspers is always concerned in his work with the metaphor of "shipwreck" in human life—the whirling upset that the reasonable person experiences in the human quest for certainty—and he attests that it must occur before a man or woman can begin truly to philosophize.

The existentialist philosophy of Jaspers and others has proved highly attractive to our contemporary world. A chief reason may be that the philosophy is centered in our humanness: Jaspers asserts, "Man is everything." Through this statement, he declares that, whatever one's ideas or circumstances, a person is thrown into a situation in which he or she must create himself or herself through decision and choice: "Existence is . . . the will to be authentic." Expressed another way, we spend our lives "in a seething cauldron of possibilities." With many people afraid that today's society tends to eclipse the free personality, Jaspers's focus on authentic individuality touches a widespread concern.

At the same time, Jaspers's focus does not negate his sense of community or of history. "We owe to the classical world," he writes, "the foundation of what, in the West, makes man all he can be." The essay "Is Science Evil?" was written relatively late in life and shows Jaspers bringing his thought and knowledge to bear on the difficult question of the prerogatives and responsibilities of science.

Is Science Evil?

Karl Jaspers

No one questions the immense significance of modern science. Through industrial technology it has transformed our existence, and its insights have transformed our consciousness, all this to an extent hitherto unheard of. The human condition throughout the millennia appears relatively stable in comparison with the impetuous movement that has now caught up mankind as a result of science and technology, and is driving it no one knows where. Science has destroyed the substance of many old beliefs and has made others questionable. Its powerful authority has brought more and more men to the point where they wish to know and not believe, where they expect to be helped by science and only by science. The present faith is that scientific understanding can solve all problems and do away with all difficulties.

Such excessive expectations result inevitably in equally excessive disillusionment. Science has still given no answer to man's doubts and despair. Instead, it has created weapons able to destroy in a few moments that which science itself helped build up slowly over the years. Accordingly, there are today two conflicting viewpoints: first, the superstition of science, which holds scientific results to be as absolute as religious myths used to be, so that even religious movements are now dressed in the garments of pseudo-science. Second, the hatred of science, which sees it as a diabolical evil of mysterious origin that has befallen mankind.

These two attitudes—both non-scientific—are so closely linked that they are usually found together, either in alternation or in an amazing compound.

A very recent example of this situation can be found in the attack against science provoked by the trial in Nuremberg of those doctors who, under Nazi orders, performed deadly experiments on human beings. One of the most esteemed medical men among German university professors has accepted the verdict on these crimes as a verdict on science itself, as a stick with which to beat "purely scientific and biological" medicine, and even the modern science of man in general: "this invisible spirit sitting on the prisoner's bench in Nuremberg, this spirit that regards men merely as objects, is not present in Nuremberg alone—it pervades the entire world." And, he adds, if this generalization may be viewed as an extenuation of the crime of the accused doctors, that is only a further indictment of purely scientific medicine.

Anyone convinced that true scientific knowledge is possible only of things that *can* be regarded as objects, and that knowledge of the subject is possible only when the subject attains a form of objectivity; anyone who sees science as the one great landmark on the road to truth, and sees the real achievements of modern physicians as derived exclusively from biological and scientific medicine—such a person will see in the above statements an attack on what he feels to be fundamental to human existence. And he may perhaps have a word to say in rebuttal.

In the special case of the crimes against humanity committed by Nazi doctors and now laid at the door of modern science, there is a simple enough argument. Science was not needed at all, but only a certain bent of mind, for the perpetration of such outrages. Such crimes were already possible millennia ago. In the Buddhist Pali canon, there is the report of an Indian prince who had experiments performed on criminals in order to determine whether they had an immortal soul that survived their corpses: "You shall—it was ordered—put the living man in a tub, close the lid, cover it with a damp hide, lay on a thick layer of clay, put it in the oven and make a fire. This was done. When we knew the man was dead, the tub was drawn forth, uncovered, the lid removed, and we looked carefully inside to see if we could perceive the escaping soul. But we saw no escaping soul." Similarly, criminals were slowly skinned alive to see if their souls could be observed leaving their bodies. Thus there were experiments on human beings before modern science.

Better than such a defense, however, would be a consideration of what modern science really genuinely is, and what its limits are.

Science, both ancient and modern, has, in the first place, three indispensable characteristics:

First, it is *methodical* knowledge. I know something scientifically only when I also know the method by which I have this knowledge, and am thus able to ground it and mark its limits.

Second, it is *compellingly certain*. Even the uncertain—i.e., the probable or improbable—I know scientifically only insofar as I know it clearly and compellingly as such, and know the degree of its uncertainty.

Third, it is *universally valid*. I know scientifically only what is identically valid for every inquirer. Thus scientific knowledge spreads over the world and remains the same. Unanimity is a sign of universal validity. When unanimity is not attained, when there is a conflict of schools, sects, and trends of fashion, then universal validity becomes problematic.

This notion of science as methodical knowledge, compellingly certain, and universally valid, was long ago possessed by the Greeks. Modern science has not only purified this notion; it has also transformed it: a transformation that can be described by saying that modern science is *indifferent to nothing*. Everything—the smallest and meanest, the furthest and strangest—that is in any way and at any time *actual*, is relevant to modern science, simply because it *is*. Modern science wants to be thoroughly universal, allowing nothing to escape it. Nothing shall be hidden, nothing shall be silent, nothing shall be a secret.

In contrast to the science of classical antiquity, modern science is *basically unfinished*. Whereas ancient science had the appearance of something completed, to which the notion of progress was not essential, modern science progresses into the infinite. Modern science has realized that a finished and total world-view is scientifically impossible. Only when scientific criticism is crippled by making particulars absolute can a closed view of the world pretend to scientific validity—and then it is a false validity. Those great new unified systems of knowledge—such as modern physics—that have grown up in the scientific era, deal only with single aspects of reality. And reality as a whole has been fragmented as never before; whence the openness of the modern world in contrast to the closed Greek cosmos.

However, while a total and finished world-view is no longer possible to modern science, the idea of a unity of the sciences has now come to replace it. Instead of the cosmos of the world, we have the cosmos of the sciences. Out of dissatisfaction with all the separate bits of knowledge is born the desire to unite all knowledge. The ancient sciences remained dispersed and without mutual relations. There was lacking to them the notion of a concrete totality of science. The modern sciences, however, seek to relate themselves to each other in every possible way.

At the same time the modern sciences have increased their claims. They put a low value on the possibilities of speculative thinking, they hold thought to be valid only as part of definite and concrete knowledge, only when it has stood the test of verification and thereby become infinitely modified. Only superficially do the modern and the ancient atomic theories seem to fit into the same theoretical mold. Ancient atomic theory was applied as a plausible interpretation of common experience; it was a statement complete in itself of what might possibly be the case. Modern atomic theory has developed through experiment, verification, refutation: that is, through an incessant transformation of itself in which theory is used not as an end in itself but as a tool of inquiry. Modern science, in its questioning, pushes to extremes. For example: the rational critique of appearance (as against reality) was begun in antiquity, as in the concept of perspective and its application to astronomy, but it still had some connection with immediate human experiences; today, however, this same critique, as in modern physics for instance, ventures to the very extremes of paradox, attaining a knowledge of the real that shatters any and every view of the world as a closed and complete whole.

So it is that in our day a scientific attitude has become possible that addresses itself inquisitively to everything it comes across, that is able to know what it knows in a clear and positive way, that can distinguish between the known and the unknown, and that has acquired an incredible mass of knowledge. How helpless was the Greek doctor or the Greek engineer! The ethos of modern science is the desire for reliable knowledge based on dispassionate investigation and criticism. When we enter its domain we feel as though we were breathing pure air, and seeing the dissolution of all vague talk, plausible opinions, haughty omniscience, blind faith.

But the greatness and the limitations of science are inseparable. It is a characteristic of the greatness of modern science that it comprehends its own limits:

(1) Scientific, objective knowledge is not knowledge of Being. This means that scientific knowledge is particular, not general, that it is directed toward specific objects, and not toward Being itself. Through knowledge itself, science arrives at the most positive recognition of what it does *not* know.

(2) Scientific knowledge or understanding cannot supply us with the aims of life. It cannot lead us. By virtue of its very clarity it directs us elsewhere for the sources of our life, our decisions, our love.

(3) Human freedom is not an object of science, but is the field of philosophy. Within the purview of science there is no such thing as liberty.

These are clear limits, and the person who is scientifically minded will not expect from science what it cannot give. Yet science has become, nevertheless, the indispensable element of all striving for truth, it has become the premise of philosophy and the basis in general for whatever clarity and candor are today possible. To the extent that it succeeds in penetrating all obscurities and unveiling all secrets, science directs us to the most profound, the most genuine secret.

The unique phenomenon of modern science, so fundamentally different from anything in the past, including the science of the Greeks, owes its character to the many sources that were its origin; and these had to meet together in Western history in order to produce it.

One of these sources was Biblical religion. The rise of modern science is scarcely conceivable without its impetus. Three of the motives that have spurred research and inquiry seem to have come from it.

(1) The ethos of Biblical religion demanded truthfulness at all costs. As a result, truthfulness became a supreme value and at the same time was pushed to the point where it became a serious problem. The truthfulness demanded by God forbade making the search for knowledge a game or amusement, an aristocratic leisure activity. It was a serious affair, a calling in which everything was at stake.

(2) The world is the creation of God. The Greeks knew the cosmos as that which was complete and ordered, rational and regular, eternally subsisting. All else was nothing, merely material, not knowable and not worth knowing. But if the world is the creation of God, then everything that exists is worth knowing, just because it is God's creation; there is nothing that ought not to be known and comprehended. To know is to reflect upon God's thought. And God as creator is—in Luther's words—present even in the bowels of a louse.

The Greeks remained imprisoned in their closed world-view, in the beauty of their rational cosmos, in the logical transparency of the rational whole. Not only Aristotle and Democritus, but Thomas Aquinas and Descartes, too, obey this Greek urge, so paralyzing to the spirit of science, toward a closed universe. Entirely different is the new impulse to unveil the totality of creation. Out of this there arises the pursuit through knowledge of that reality which is not in accord with previously established laws. In the Logos itself [the Word, Reason] there is born the drive toward repeated self-destruction—not as self-immolation, but in order to arise again and ever again in a process that is to be continued infinitely. This science springs from a Logos that does not remain closed within itself, but is open to an anti-Logos which it permeates by the very act of subordinating itself to it. The continuous, unceasing reciprocal action of theory and experiment is the simple and great example and symbol of the universal process that is the dialectic between Logos and anti-Logos.

This new urge for knowledge sees the world no longer as simply beautiful. This knowledge ignores the beautiful and the ugly, the good and the wicked. It is true that in the end, *omne ens est bonum* [all Being is good], that is, as a creation of God. This goodness, however, is no longer the transparent and self-sufficient beauty of the Greeks. It is present only in the love of all existent things as created by God, and it is present therefore in our confidence in the significance of inquiry. The knowledge of the createdness of all worldly things replaces indifference in the face of the flux of reality with limitless questioning, an insatiable spirit of inquiry.

But the world that is known and knowable is, as created Being, Being of the second rank. For the world is unfathomable, it has its ground in another, a

Creator, it is not self-contained and it is not containable by knowledge. The Being of the world cannot be comprehended as definitive, absolute reality, but points always to another.

The idea of creation makes worthy of love whatever is, for it is God's creation; and it makes possible, by this, an intimacy with reality never before attained. But at the same time it gives evidence of the incalculable distance from that Being which is not merely created Being but Being itself, God.

(3) The reality of this world is full of cruelty and horror for men. "That's the way things are," is what man must truthfully say. If, however, God is the world's creator, then he is responsible for his creation. The question of justifying God's ways becomes with Job a struggle with the divine for the knowledge of reality. It is a struggle against God, for God. God's existence is undisputed and just because of this the struggle arises. It would cease if faith were extinguished.

This God, with his unconditional demand for truthfulness, refuses to be grasped through illusions. In the Bible, he condemns the theologians who wish to console and comfort Job with dogmas and sophisms. This God insists upon science, whose content always seems to bring forth an indictment of him. Thus we have the adventure of knowledge, the furtherance of unrestricted knowledge—and at the same time, a timidity, an awe in the face of it. There was an inner tension to be observed in many scientists of the past century, as if they heard: God's will is unconfined inquiry, inquiry is in the service of God—and at the same time: it is an encroachment on God's domain, all shall not be revealed.

This struggle goes hand in hand with the struggle of the man of science against all that he holds most dear, his ideals, his beliefs; they must be proven, newly verified, or else transformed. Since God could not be believed in if he were not able to withstand all the questions arising from the facts of reality, and since the seeking of God involves the painful sacrifice of all illusions, so true inquiry is the struggle against all personal desires and expectations.

This struggle finds its final test in the struggle of the scientist with his own theses. It is the determining characteristic of the modern scientist that he seeks out the strongest points in the criticism of his opponents and exposes himself to them. What in appearance is self-destructiveness becomes, in this case, productive. And it is evidence of a degradation of science when discussion is shunned or condemned, when men imprison themselves and their ideas in a milieu of like-minded savants and become fanatically aggressive to all outside it.

That modern science, like all things, contains its own share of corruption, that men of science only too often fail to live up to its standards, that science can be used for violent and criminal ends, that man will steal, plunder, abuse, and kill to gain knowledge—all this is no argument against science.

To be sure, science as such sets up no barriers. As science, it is neither human nor inhuman. So far as the well-being of humanity is concerned, science needs guidance from other sources. Science in itself is not enough—or should not be. Even medicine is only a scientific means, serving an eternal ideal, the aid of the sick and the protection of the healthy.

When the spirit of a faithless age can become the cause of atrocities all over the world, then it can also influence the conduct of the scientist and the behavior of the physician, especially in those areas of activity where science itself is confused and unguided. It is not the spirit of science but the spirit of its vessels that is depraved. Count Keyserling's dictum—"The roots of truth-seeking lie in primitive aggression"—is as little valid for science as it is for any genuine truth-seeking. The spirit of science is in no way primarily aggressive, but becomes so only when truth is prohibited; for men rebel against the glossing over of truth or its suppression.

In our present situation the task is to attain to that true science which knows what it knows at the same time that it knows what it cannot know. This science shows us the ways to the truth that are the indispensable precondition of every other truth. We know what Mephistopheles knew when he thought he had outwitted Faust:

> *Verachte nur Vernunft und Wissenschaft*
> *Des Menschen allerhöchste Kraft*
> *So habe ich Dich schon unbedingt.*
> (Do but scorn Reason and Science
> Man's supreme strength
> Then I'll have you for sure.)

Questions and Topics for Discussion and Writing

1. According to Jaspers, what are the two conflicting viewpoints concerning science expressed by nonscientists?
2. What are the three indispensable characteristics of science outlined by Jaspers?
3. Consult a library source for a definition of existentialism; then write your own extended definition, using examples and illustrations.
4. Explain in a paragraph the charge that Jaspers makes against science for creating the existential plight of the human race today.
5. According to Jaspers, what is a test of the validity of scientific knowledge?
6. Explain what Jaspers means by an "open" and a "closed" cosmos.
7. Explain what Jaspers means by saying that today "reality" has been "fragmented." Explain what he means by "absolute being."
8. Jaspers says that in spite of the "fragmentation of reality" and the impossibility of a "total and finished world-view," scientists of all fields seek to relate themselves to each other. What does he conclude is their goal in so doing?

9. What does he judge to be the "limits" of modern science?
10. Many writers view religion and science as opposing each other. Jaspers offers a defense for religion. How has Biblical religion contributed to the modern scientific character?
11. Jaspers also proposes a dialectic that he deems necessary for the growth of both science and religion. Summarize his proposal.
12. Socrates, in defense of his probing questions, asserted, "Life without inquiry is not worth living, for a man." According to Jaspers, how does modern scientific activity demonstrate the truth contained in Socrates' statement?
13. What techniques does Jaspers use to make this essay a persuasive one? Write a persuasive essay using one of the following topics as your controlling idea:
    a. Scientific experiments on animals are a necessary part of scientific research.
    b. Judges should be appointed and not elected by the people.
    c. Parents alone should have the final authority to determine whether particular medical treatments should be administered to their children.
    d. Is _____ evil?

Francis Crick

British-born (1916) and -educated, Sir Francis Harry Compton Crick, working at Cambridge University in England with the American biochemist James Dewey Watson (b. 1928), produced the discovery that is perhaps the most celebrated in twentieth-century biology. The two scientists proposed a model for the molecular structure of deoxyribonucleic acid, the hereditary material commonly known as DNA and the "master molecule" of life.

When Crick and Watson, around midcentury, joined the long-term investigation into the chemical basis and transmission of "instructions" by hereditary material, they had benefit of experimental data obtained by many others, including Linus Pauling, the American biochemist, and the English biochemist Rosalind Franklin and physicist Maurice Wilkins. Scientists already knew that DNA carried genetic instructions, that it was an enormous chain of nucleotides, and, from X-ray diffraction photographs, that the DNA molecule, a lengthy polymer, was probably a helix. The key insight was Crick's and Watson's perception that it must be a *double* helix, a concept permitting the hypothesis that the strands uncoil during the process of replication so that each strand can serve as a template for a new strand. The discovery brought Crick, Wilkins, and Watson the Nobel Prize for physiology or medicine in 1962, and the story of the Watson–Crick collaboration won wide popular attention through Watson's best-selling account *The Double Helix.*

After Crick and Watson completed their collaborative study, they turned to independent lines of research. Watson returned to the United States, where he now directs the Cold Spring Harbor Laboratory of Quantitative Biology in New

York, engaging primarily in cancer research. Crick, working at the Salk Institute in San Diego, continues a series of analytical experiments aimed at unraveling the secrets of the origin of life itself.

"Should We Infect the Galaxy?" poses Crick's profound and unsettling question.

Should We Infect the Galaxy?

Francis Crick

One topic remains. Even if it turns out that we shall never know for sure how life began here, we may still be confronted, at some time in the future, with the practical question: should we attempt to start our form of life elsewhere in the universe? And if so, how should we do it?

We may expect that by that time (if we have not destroyed ourselves by our own folly) we shall be able to decide whether the nearer stars have planets, perhaps by placing sophisticated new instruments on the moon. We may even know, more or less, how our solar system was formed, thanks to extended exploration of other planets, the asteroid belt, comets and so on. This may enable us to estimate which planets are likely to possess a fairly favorable environment. The design of rockets may be expected to have improved enormously, so that they can go very long distances and work reliably for very long times even if they cannot approach the speed of light.

With all this at our disposal, what should we do? Perhaps one of the easier things would be to try what was done for Mars; not to send men, at least in the first instance, but to send instruments which could report back to us. Even this apparently simple requirement seems technologically far in advance of what we can do today. It would require difficult feats of engineering to get a spaceship successfully into orbit, especially after such a long journey and at such a distance. In orbit it would sense far less than if it could settle on the solid surface of the planet (if it had one), yet to get it to the surface would require an even more advanced technology. Some of these problems might be solved if people were sent on the mission, but this poses a whole new set of problems, not the least being how to make sure that they arrive alive. The chances of their starting a colony, under the very unfavorable conditions likely to be found there, or of making the return journey alive seem infinitely remote. Ironically, as Tommy Gold has suggested, the most likely outcome, if they got there at all, would be that some of the bacteria they carried would reach the primitive ocean, there to survive and multiply long after the death of the astronauts. In that case, why not just send bacteria in the first place? Immediately we decide on this option, all our design problems become simpler. . . . If there is one intellectual exercise which disposes the mind to look more favorably on Directed Panspermia, it is that of imagining what we ourselves might do in the future exploration and colonization of space.

But notice that in our enthusiasm for infecting our neighbors there is one little detail that we have overlooked. What if our chosen planet has already evolved another form of life? Whether our descendants will have been able to decide that life was very common in the universe or, alternatively, very rare, we cannot know. We cannot even estimate how good their guesses might be. The technology to decide whether a nearby star has planets and, broadly, what they are like does not seem too far distant, but the technology needed to decide whether they possess life or not, in one form or another, would seem to be very far in the future. We can see these problems on a smaller scale as we try to discover whether there is some form of life on the planets and moons of our own solar system. The only good evidence comes from bodies on which a landing has been made. The urge to explore space is likely to reach a high level long before we can know whether what we shall be exploring supports any form of life.

It is difficult to see what would be the outcome of such a situation. Our descendants will be confronted with novel problems in cosmic ethics. Are we, as highly developed beings, entitled to disturb the fragile ecology of another planet? Should we feel bound to respect life, *whatever* form it may take? We have similar dilemmas on earth, as any vegetarian will tell you, though not many people would respect the smallpox viruses' right to life. Perhaps there will be a profound division of opinion among our descendants, though I cannot help thinking that it will be the meat-eaters who will want to explore space and the vegetarians who are likely to oppose it.

I should say, in passing, that I do not think these fears apply to the spaceships which we are at present sending outside the solar system. Even if they harbor any bacteria at all, those few microorganisms are highly unlikely to survive both the journey in space and the entry into another solar system. The chance of their infecting another planet is so extremely low that we would be foolish to worry about it.

One obvious plea is that the matter should not be rushed. Since, with luck, we have millennia ahead of us, and since as time goes on we should know more and be able to tackle more difficult tasks, why hurry? But even this argument assumes that the world will be politically stable for an indefinite period. If it is not, there would certainly be pressure from powerful groups who wanted to get on with the job, lest circumstances arose in which it could never be completed. My prejudice would be not to press ahead too eagerly, if waiting is at all possible. We should not lightly contaminate the galaxy.

Questions and Topics for Discussion and Writing

1. Crick is the originator of a theory known as Directed Panspermia. This theory postulates that life on earth originated when "technocrats" in a distant galaxy sent micro-organisms to earth, from which life forms survived and evolved. Read his book *Life Itself: Its Origin and Nature* (1981), and write a commentary.
2. Look up biochemist Rosalind Franklin's role in the discovery of DNA and write a report on her contribution.
3. In *Nature* (Vol. 171, April 1953), read the original report of the discovery of DNA, "Molecular Structure of Nucleic Acids." How does this report differ in style from the Crick selection, which was written for a more popular audience?
4. Read James Watson's *The Double Helix* and write a critical book review.
5. Discuss H. G. Wells's statement "Science is a match that man has just got alight."

Alfred North Whitehead

A leading British mathematician and philosopher, Alfred North Whitehead (1861–1947) was born the son of an Anglican vicar in Ramsgate, England. His early training grounded him in history and the humanities. At Trinity College, Cambridge, he attended the mathematics lectures but also pursued his interest in literature and philosophy. Appointed a Fellow, he taught mathematics for the next three decades. In 1910 he received a position at London's Imperial College of Science and Technology, eventually becoming Dean of the Faculty of Science. In 1924, at sixty-three, he moved to the United States to teach philosophy at Harvard University, where he remained until his death.

Mathematics marks Whitehead's first phase. While at Cambridge, England, he published several works on mathematics, the most important of them the *Principia Mathematica*, a massive, three-volume work on which he collaborated for three years with his former student Bertrand Russell. It is an abstrusely technical, recondite, almost arcane book—as one scholar puts it, more quoted than actually read. Nonetheless, it was enormously influential. Attempting to demonstrate that mathematics can be deduced from, and defined in terms of, formal logic (a thesis not universally agreed to), it is a classic of symbolic logic. So new was the topic that molds for the type had to be cast (the symbols themselves were the work of an Italian logician, Giuseppe Peano), and the work relied heavily upon symbols.

Although he began his career as a mathematician, Whitehead branched out to many other areas of learning. In his second phase, he developed his system of natural philosophy, focusing on epistemology and the philosophy of science. When he retired from his London post and accepted the position of professor of philosophy at Harvard, he was freed from administrative duties and was able to expand his intellectual pursuits, delving into speculative philosophy, meta-physics, cosmology, and culture.

Whitehead's later writings are geared to a more general, less technical—though still educated—audience. His ideas embrace not only mathematics and the sciences, but also the social sciences, religion, esthetics, and education. A brief list of some of his writings demonstrates his vast range of intellectual activity: *The Organization of Thought, Science and the Modern World, Religion in the Making, The Aim of Education*, and *Adventures in Ideas*. They were considered as simple, tentative generalizations from experience, and he expected them to be altered or discarded if they proved useless. Part of his great contribution to modern thought lies in his establishing a comprehensive out-look, not dichotomizing science and religion, for example, or the sciences and humanities, but encompassing all of human experience. In "Religion and Science" he discusses the reciprocity between his two subjects.

Religion and Science

Alfred North Whitehead

The difficulty in approaching the question of the relations between Religion and Science is, that its elucidation requires that we have in our minds some clear idea of what we mean by either of the terms, "religion" and "science." Also I wish to speak in the most general way possible, and to keep in the background any comparison of particular creeds, scientific or religious. We have got to understand the type of connection which exists between the two spheres, and then to draw some definite conclusions respecting the existing situation which at present confronts the world.

The *conflict* between religion and science is what naturally occurs to our minds when we think of this subject. It seems as though, during the last half-century, the results of science and the beliefs of religion had come into a position of frank disagreement, from which there can be no escape, except by abandoning either the clear teaching of science, or the clear teaching of religion. This conclusion has been urged by controversialists on either side. Not by all controversialists, of course, but by those trenchant intellects which every controversy calls out into the open.

The distress of sensitive minds, and the zeal for truth, and the sense of the importance of the issues, must command our sincerest sympathy. When we consider what religion is for mankind, and what science is, it is no exaggeration to say that the future course of history depends upon the decision of this generation as to the relations between them. We have here the two strongest general forces (apart from the mere impulse of the various senses) which influence men, and they seem to be set one against the other—the force of our religious intuitions, and the force of our impulse to accurate observation and logical deduction.

A great English statesman once advised his countrymen to use large-scale maps, as a preservative against alarms, panics, and general misunderstanding of the true relations between nations. In the same way in dealing with the clash between permanent elements of human nature, it is well to map our history on a large scale, and to disengage ourselves from our immediate absorption in the present conflicts. When we do this, we immediately discover two great facts. In the first place, there has always been a conflict between religion and science; and in the second place, both religion and science have always been in a state of continual development. In the early days of Christianity, there was a general belief among Christians that the world was coming to an end in the lifetime of people then living. We can make only indirect inferences as to how far this belief was authoritatively proclaimed; but it is certain that it was widely held, and that it formed an impressive part of the popular religious doctrine. The belief proved itself to be mistaken, and Christian doctrine adjusted itself to the change. Again in the early Church individual theologians very confidently deduced from the Bible opinions concerning the nature of the physical universe. In the year A.D. 535, a monk named Cosmas[1] wrote a book which he entitled, *Christian Topography.* He was a travelled man who had visited India and Ethiopia; and finally he lived in a monastery at Alexandria, which was then a great centre of culture. In this book, basing himself upon the direct meaning of Biblical texts as construed by him in a literal fashion, he denied the existence of the antipodes, and asserted that the world is a flat parallelogram whose length is double its breadth.

In the seventeenth century the doctrine of the motion of the earth was condemned by a Catholic tribunal. A hundred years ago the extension of time demanded by geological science distressed religious people, Protestant and Catholic. And to-day the doctrine of evolution is an equal stumbling-block. These are only a few instances illustrating a general fact.

But all our ideas will be in a wrong perspective if we think that this recurring perplexity was confined to contradictions between religion and science; and that in these controversies religion was always wrong, and that science was always right. The true facts of the case are very much more complex, and refuse to be summarised in these simple terms.

Theology itself exhibits exactly the same character of gradual development, arising from an aspect of conflict between its own proper ideas. This fact is a commonplace to theologians, but is often obscured in the stress of controversy. I do not wish to overstate my case; so I will confine myself to Roman Catholic

---

[1] *Cf.* Lecky's *The Rise and Influence of Rationalism in Europe*, Ch. III.

writers. In the seventeenth century a learned Jesuit, Father Petavius, showed that the theologians of the first three centuries of Christianity made use of phrases and statements which since the fifth century would be condemned as heretical. Also Cardinal Newman devoted a treatise to the discussion of the development of doctrine. He wrote it before he became a great Roman Catholic ecclesiastic; but throughout his life, it was never retracted and continually reissued.

Science is even more changeable than theology. No man of science could subscribe without qualification to Galileo's beliefs, or to Newton's beliefs, or to all his own scientific beliefs of ten years ago.

In both regions of thought, additions, distinctions, and modifications have been introduced. So that now, even when the same assertion is made to-day as was made a thousand, or fifteen hundred years ago, it is made subject to limitations or expansions of meaning, which were not contemplated at the earlier epoch. We are told by logicians that a proposition must be either true or false, and that there is no middle term. But in practice, we may know that a proposition expresses an important truth, but that it is subject to limitations and qualifications which at present remain undiscovered. It is a general feature of our knowledge, that we are insistently aware of important truths; and yet that the only formulations of these truths which we are able to make presuppose a general standpoint of conceptions which may have to be modified. I will give you two illustrations, both from science: Galileo said that the earth moves and that the sun is fixed; the Inquisition said that the earth is fixed and the sun moves; and Newtonian astronomers, adopting an absolute theory of space, said that both the sun and the earth move. But now we say that any one of these three statements is equally true, provided that you have fixed your sense of "rest" and "motion" in the way required by the statement adopted. At the date of Galileo's controversy with the Inquisition, Galileo's way of stating the facts was, beyond question, the fruitful procedure for the sake of scientific research. But in itself it was not more true than the formulation of the Inquisition. But at that time the modern concepts of relative motion were in nobody's mind; so that the statements were made in ignorance of the qualifications required for their more perfect truth. Yet this question of the motions of the earth and the sun expresses a real fact in the universe; and all sides had got hold of important truths concerning it. But with the knowledge of those times, the truths appeared to be inconsistent.

Again I will give you another example taken from the state of modern physical science. Since the time of Newton and Huyghens in the seventeenth century there have been two theories as to the physical nature of light. Newton's theory was that a beam of light consists of a stream of very minute particles, or corpuscles, and that we have the sensation of light when these corpuscles strike the retinas of our eyes. Huyghens' theory was that light consists of very minute waves of trembling in an all-pervading ether, and that these waves are travelling along a beam of light. The two theories are contradictory. In the eighteenth century Newton's theory was believed, in the nineteenth century Huyghens' theory was believed. To-day there is one large group of phenomena which can be explained only on the wave theory, and another large group which can be explained only on the corpuscular theory. Scientists have to leave it at that, and wait for the future, in the hope of attaining some wider vision which reconciles both.

We should apply these same principles to the questions in which there is a variance between science and religion. We would believe nothing in either sphere of thought which does not appear to us to be certified by solid reasons based upon the critical research either of ourselves or of competent authorities. But granting that we have honestly taken this precaution, a clash between the two on points of detail where they overlap should not lead us hastily to abandon doctrines for which we have solid evidence. It may be that we are more interested in one set of doctrines than in the other. But, if we have any sense of perspective and of the history of thought, we shall wait and refrain from mutual anathemas.

We should wait: but we should not wait passively, or in despair. The clash is a sign that there are wider truths and finer perspectives within which a reconciliation of a deeper religion and a more subtle science will be found.

In one sense, therefore, the conflict between science and religion is a slight matter which has been unduly emphasised. A mere logical contradiction cannot in itself point to more than the necessity of some readjustments, possibly of a very minor character on both sides. Remember the widely different aspects of events which are dealt with in science and in religion respectively. Science is concerned with the general conditions which are observed to regulate physical phenomena; whereas religion is wholly wrapped up in the contemplation of moral and aesthetic values. On the one side there is the law of gravitation, and on the other the contemplation of the beauty of holiness. What one side sees, the other misses; and vice versa.

Consider, for example, the lives of John Wesley and of Saint Francis of Assisi. For physical science you have in these lives merely ordinary examples of the operation of the principles of physiological chemistry, and of the dynamics of nervous reactions: for religion you have lives of the most profound significance in the history of the world. Can you be surprised that, in the absence of a perfect and complete phrasing of the principles of science and of the principles of religion which apply to these specific cases, the accounts of these lives from these divergent standpoints should involve discrepancies? It would be a miracle if it were not so.

It would, however, be missing the point to think that we need not trouble ourselves about the conflict between science and religion. In an intellectual age there can be no active interest which puts aside all hope of a vision of the harmony of truth. To acquiesce in discrepancy is destructive of candour, and of moral cleanliness. It belongs to the self-respect of intellect to pursue every tangle of thought to its final unravelment. If you check that impulse, you will get no religion and no science from an awakened thoughtfulness. The important question is, In what spirit are we going to face the issue? There we come to something absolutely vital.

A clash of doctrines is not a disaster—it is an opportunity. I will explain my meaning by some illustrations from science. The weight of an atom of nitrogen was well known. Also it was an established scientific doctrine that the average weight of such atoms in any considerable mass will be always the same. Two experimenters, the late Lord Rayleigh and the late Sir William Ramsay, found that if they obtained nitrogen by two different methods, each equally effective for that purpose, they always observed a persistent slight difference between the average weights of the atoms in the two cases. Now I ask you, would it have been rational of these men to have despaired because of this conflict between chemical theory and scientific observation? Suppose that for some reason the chemical doctrine had been highly prized throughout some district as the foundation of its social order:— would it have been wise, would it have been candid, would it have been moral, to forbid the disclosure of the fact that the experiments produced discordant results? Or, on the other hand, should Sir William Ramsay and Lord Rayleigh have proclaimed that chemical theory was now a detected delusion? We see at once that either of these ways would have been a method of facing the issue in an entirely wrong spirit. What Rayleigh and Ramsay did was this: They at once perceived that they had hit upon a line of investigation which would disclose some subtlety of chemical theory that had hitherto eluded observation. The discrepancy was not a disaster: it was an opportunity to increase the sweep of chemical knowledge. You all know the end of the story: finally argon was discovered, a new chemical element which had lurked undetected, mixed with the nitrogen. But the story has a sequel which forms my second illustration. This discovery drew attention to the importance of observing accurately minute differences in chemical substances as obtained by different methods. Further researches of the most careful accuracy were undertaken. Finally another physicist, F. W. Aston, working in the Cavendish Laboratory at Cambridge in England, discovered that even the same element might assume two or more distinct forms, termed *isotopes*, and that the law of the constancy of average atomic weight holds for each of these forms, but as between the different isotopes differs slightly. The research has effected a great stride in the power of chemical theory, far transcending in importance the discovery of argon from which it originated. The moral of these stories lies on the surface, and I will leave to you their application to the case of religion and science.

In formal logic, a contradiction is the signal of a defeat: but in the evolution of real knowledge it marks the first step in progress towards a victory. This is one great reason for the utmost toleration of variety of opinion. Once and forever, this duty of toleration has been summed up in the words, "Let both grow together until the harvest." The failure of Christians to act up to this precept, of the highest authority, is one of the curiosities of religious history. But we have not yet exhausted the discussion of the moral temper required for the pursuit of truth. There are short cuts leading merely to an illusory success. It is easy enough to find a theory, logically harmonious and with important applications in the region of fact, provided that you are content to disregard half your evidence. Every age produces people with clear logical intellects, and with the most praiseworthy grasp of the importance of some sphere of human experience, who have elaborated, or inherited, a scheme of thought which exactly fits those experiences which claim their interest. Such people are apt resolutely to ignore, or to explain away, all evidence which confuses their scheme with contradictory instances. What they cannot fit in is for them nonsense. An unflinching determination to take the whole evidence into account is the only method of preservation against the fluctuating extremes of fashionable opinion. This advice seems so easy, and is in fact so difficult to follow.

One reason for this difficulty is that we cannot think first and act afterwards. From the moment of birth we are immersed in action, and can only fitfully guide it by taking thought. We have, therefore, in various spheres of experience to adopt those ideas which seem to work within those spheres. It is absolutely necessary to trust to ideas which are generally adequate, even though we know that there are subtleties and distinctions beyond our ken. Also apart from the necessities of action, we cannot even keep before our minds the whole evidence except under the guise of doctrines which are incompletely harmonised. We cannot think in terms of an indefinite multiplicity of detail; our evidence can acquire its proper importance only if it comes before us marshalled by general ideas. These ideas we inherit—they form the tradition of our civilisation. Such traditional ideas are never static. They are either fading into meaningless formulae, or are gaining power by the new lights thrown by a more delicate apprehension. They are transformed by the urge of critical reason, by the vivid evidence of emotional experience, and by the cold certainties of scientific perception. One fact is certain, you cannot keep them still. No generation can merely reproduce its ancestors. You may preserve the life in a flux of form, or preserve the form amid an ebb of life. But you cannot permanently enclose the same life in the same mould.

The present state of religion among the European races illustrates the statements which I have been making. The phenomena are mixed. There have been reactions and revivals. But on the whole, during many generations, there has been a gradual decay of religious influence in European civilisation. Each revival touches a lower peak than its predecessor, and each period of slackness a lower depth. The average curve marks a steady fall in religious tone. In some countries the interest in religion is higher than in others. But in those countries where the interest is relatively high, it still falls as the generations pass. Religion is tending to degenerate into a decent formula wherewith to embellish a comfortable life. A great historical movement on this scale results from the convergence of many causes. I wish to suggest two of them which lie within the scope of this chapter for consideration.

In the first place for over two centuries religion has been on the defensive, and on a weak defensive. The period has been one of unprecedented intellectual progress. In this way a series of novel situations have been produced for thought. Each such occasion has found the religious thinkers unprepared. Something, which has been proclaimed to be vital, has finally, after struggle, distress, and anathema, been modified and otherwise interpreted. The next generation of religious apologists then congratulates the religious world on the deeper insight which has been gained. The result of the continued repetition of this undignified retreat, during many generations, has at last almost entirely destroyed the intellectual authority of religious thinkers. Consider this contrast: when Darwin or Einstein proclaim theories which modify our ideas, it is a triumph for science. We do not go about saying that there is another defeat for science, because its old ideas have been abandoned. We know that another step of scientific insight has been gained.

Religion will not regain its old power until it can face change in the same spirit as does science. Its principles may be eternal, but the expression of those principles requires continual development. This evolution of religion is in the main a disengagement of its own proper ideas from the adventitious notions which have crept into it by reason of the expression of its own ideas in terms of the imaginative picture of the world entertained in previous ages. Such a release of religion from the bonds of imperfect science is all to the good. It stresses its own genuine message. The great point to be kept in mind is that normally an advance in science will show that statements of various religious beliefs require some sort of modification. It may be that they have to be expanded or explained, or indeed entirely restated. If the religion is a sound expression of truth, this modification will only exhibit more adequately the exact point which is of importance. This process is a gain. In so far, therefore, as any religion has any contact with physical facts, it is to be expected that the point of view of those facts must be continually modified as scientific knowledge advances. In this way, the exact relevance of these facts for religious thought will grow more and more clear. The progress of science must result in the unceasing codification of religious thought, to the great advantage of religion.

The religious controversies of the sixteenth and seventeenth centuries put theologians into a most unfortunate state of mind. They were always attacking and defending. They pictured themselves as the garrison of a fort surrounded by hostile forces. All such pictures express half-truths. That is why they are so popular. But they are dangerous. This particular picture fostered a pugnacious party spirit which really expresses an ultimate lack of faith. They dared not modify, because they shirked the task of disengaging their spiritual message from the associations of a particular imagery.

Let me explain myself by an example. In the early medieval times, Heaven was in the sky, and Hell was underground; volcanoes were the jaws of Hell. I do not assert that these beliefs entered into the official formulations: but they did enter into the popular understanding of the general doctrines of Heaven and Hell. These notions were what everyone thought to be implied by the doctrine of the future state. They entered into the explanations of the influential exponents of Christian belief. For example, they occur in the *Dialogues* of Pope Gregory,[2] the Great, a man whose high official position is surpassed only by the magnitude of his services to humanity. I am not saying what we ought to believe about the future state. But whatever be the right doctrine, in this instance the clash between religion and science, which has relegated the earth to the position of a second-rate planet attached to a second-rate sun, has been greatly to the benefit of the spirituality of religion by dispersing these medieval fancies.

Another way of looking at this question of the evolution of religious thought is to note that any verbal form of statement which has been before the world for some time discloses ambiguities; and that often such ambiguities strike at the very heart of the meaning. The effective sense in which a doctrine has been held in the past cannot be determined by the mere logical analysis of verbal statements, made in ignorance of the logical trap. You have to take into account the whole reaction of human nature to the scheme of thought. This reaction is of a mixed character, including elements of emotion derived from our lower natures. It is here that the impersonal criticism of science and of philosophy comes to the aid of religious evolution. Example after example can be given of this motive force in development. For example, the logical difficulties inherent in the doctrine of the moral cleansing of human nature by the power of religion rent Christianity in the days of Pelagius and Augustine—that is to say, at the beginning of the fifth century. Echoes of that controversy still linger in theology.

So far, my point has been this: that religion is the expression of one type of fundamental experiences of mankind: that religious thought develops into an increasing accuracy of expression, disengaged from adventitious imagery: that the interaction between religion and science is one great factor in promoting this development.

I now come to my second reason for the modern fading of interest in religion. This involves the ultimate question which I stated in my opening sentences. We have to know what we mean by religion. The churches, in their presentation of their answers to this query, have put forward aspects of religion which are expressed in terms either suited to the emotional reactions of bygone times or directed to excite modern emotional interests of nonreligious character. What I mean under the first heading is that religious appeal is directed partly to excite that instinctive fear of the wrath of a tyrant which was inbred in the unhappy populations of the arbitrary empires of the ancient world, and in particular to excite that fear of an all-powerful arbitrary tyrant behind the unknown forces of nature. This appeal to the ready instinct of brute fear is losing its force. It lacks any directness of response, because modern science and modern conditions of life have taught us to meet occasions of apprehension by a critical analysis of their causes and conditions. Religion is the reaction of human nature to its search for God. The presentation of God under the aspect of power awakens every modern instinct of critical reaction. This is fatal; for religion

---

[2]*Cf.* Gregorovius' *History of Rome in the Middle Ages*, Book III, Ch. III, Vol. II, English Trans.

collapses unless its main positions command immediacy of assent. In this respect the old phraseology is at variance with the psychology of modern civilisations. This change in psychology is largely due to science, and is one of the chief ways in which the advance of science has weakened the hold of the old religious forms of expression. The non-religious motive which has entered into modern religious thought is the desire for a comfortable organisation of modern society. Religion has been presented as valuable for the ordering of life. Its claims have been rested upon its function as a sanction to right conduct. Also the purpose of right conduct quickly degenerates into the formation of pleasing social relations. We have here a subtle degradation of religious ideas, following upon their gradual purification under the influence of keener ethical intuitions. Conduct is a by-product of religion—an inevitable by-product, but not the main point. Every great religious teacher has revolted against the presentation of religion as a mere sanction of rules of conduct. Saint Paul denounced the Law, and Puritan divines spoke of the filthy rags of righteousness. The insistence upon rules of conduct marks the ebb of religious fervour. Above and beyond all things, the religious life is not a research after comfort. I must now state, in all diffidence, what I conceive to be the essential character of the religious spirit.

Religion is the vision of something which stands beyond, behind, and within, the passing flux of immediate things; something which is real, and yet waiting to be realised; something which is a remote possibility, and yet the greatest of present facts; something that gives meaning to all that passes, and yet eludes apprehension; something whose possession is the final good, and yet is beyond all reach; something which is the ultimate ideal, and the hopeless quest.

The immediate reaction of human nature to the religious vision is worship. Religion has emerged into human experience mixed with the crudest fancies of barbaric imagination. Gradually, slowly, steadily the vision recurs in history under nobler form and with clearer expression. It is the one element in human experience which persistently shows an upward trend. It fades and then recurs. But when it renews its force, it recurs with an added richness and purity of content. The fact of the religious vision, and its history of persistent expansion, is our one ground for optimism. Apart from it, human life is a flash of occasional enjoyments lighting up a mass of pain and misery, a bagatelle of transient experience.

The vision claims nothing but worship; and worship is a surrender to the claim for assimilation, urged with the motive force of mutual love. The vision never overrules. It is always there, and it has the power of love presenting the one purpose whose fulfilment is eternal harmony. Such order as we find in nature is never force—it presents itself as the one harmonious adjustment of complex detail. Evil is the brute motive force of fragmentary purpose, disregarding the eternal vision. Evil is overruling, retarding, hurting. The power of God is the worship He inspires. That religion is strong which in its ritual and its modes of thought evokes an apprehension of the commanding vision. The worship of God is not a rule of safety—it is an adventure of the spirit, a flight after the unattainable. The death of religion comes with the repression of the high hope of adventure.

Questions and Topics for Discussion and Writing

1. Whitehead sees the controversy between religion and science as part of a larger debate. What is the larger dimension?
2. People who criticize religion usually think of science as advancing and religion as being static. What is Whitehead's view on this matter? What is yours? Cite specific reasons for your opinion.
3. In his discussion, Whitehead asks for a distancing of the controversy, using the image of the "large-scale maps." Do you think that such distancing is a reasonable, workable attitude? Why or why not?
4. Whitehead writes in this essay: "We are told by logicians that a proposition must be either true or false, and that there is no middle term. But in practice . . . it is subject to limitations and qualifications. . . ." How do these statements stand as a modern scientific and philosophical judgment?
5. Whitehead states that contradictory theories can serve science. What does Whitehead state is the difference between the concerns of religion and science? How can religion and science both be relevant?
6. What positive attribute does Whitehead see in the controversy between religion and science?
7. Whitehead recounts several short "stories," illustrations of episodes in scientific investigations. In a paragraph, apply the moral of Whitehead's stories to the case of religion and science.
8. What does Whitehead deem the essential character of the religious spirit?
9. Do you find Whitehead's thesis of the dialectic between religion and science a valid one?
10. T. S. Eliot has said that Christianity will evolve into something that can be believed. Would Whitehead agree? Do you find this view acceptable?

When We Are Old

P. B. Medawar

You will find the biography for Medawar before the essay "How Can I Tell If I Am Cut Out to Be a Research Worker?" in Unit I.

Over the past thirty years, research on aging has raised the serious possibility that life expectancy might someday be extended by as much as one fourth. Many different lines of research—on diet, metabolism, and the immune system, among other things—are being pursued. One of the most promising of these originated with Denham Harman, professor of medicine and biochemistry at the University of Nebraska, who proposed, in 1954, that the highly reactive molecular fragments known as "free radicals," which are especially damaging to biological microstructures, might to some extent be counteracted by the increased consumption of antioxidants. These chemical compounds, of which vitamin E is the best known, occur naturally in some foods and are added as preservatives to others—in most countries in tiny and strictly regulated proportions.

Harman and many others are of the opinion that, just as small amounts of antioxidants preserve foods, in larger amounts the compounds might preserve human tissue. Many antioxidants have since been tested for such an effect on laboratory animals, and the increased longevity observed was equivalent, in Harman's reckoning, to an extension of the average human life expectancy from seventy-three to ninety-five years.

If antioxidants can be ingested safely by human beings, the result is not expected to be an extra decade or two of zombie-like existence, in which people would be alive only in the purely technical sense (or alive enough, shall we say, to avoid becoming transplant donors). Not at all. What the researchers hope for is a prolonging of life such as would be achieved if the seven ages of man were marked off on a length of rubber and the rubber were stretched. A seventy-year-old would have the address to life of a sixty-year-old, and an eighty-year-old that of a seventy-year-old.

How successful any treatment to prolong life might be is unclear. Growing old is a bad thing quite apart from the decline of bodily faculties and energies that it entails. Even if the process of senescence could be arrested temporarily, we would still suffer from the passage of years. Consider, for example, the likely consequences of extending a woman's reproductive life to the age of sixty, or beyond. The older a woman is at the age of reproduction, the longer her finite endowment of egg cells will have been exposed to influences that are inimical to it. Thus, even though a woman of sixty might be as physically fit as a woman of thirty, the likelihood of a chromosomal aberration, such as that which causes Down's syndrome, would have increased with her age. The etiology of cancer is a similar example. Researchers believe that a malignant tumor can start with the mutation of a chromosome following the body's exposure to ionizing radiation, a toxic chemical, or some other mutagen. Thus, the longer one lives, the longer one has to cross the path of such hazards, and the greater one's chances of contracting cancer.

Because it is not easy to see a remedy for these side effects of old age, I fear that the incorporation of antioxidants into our diet will have a more modest result than proponents of the theory expect. But there are people who say that such research ought not to proceed at all. Their opposition compels us to ask, Is the extension of the life-span a possibility that we should welcome or a temptation that we should resist?

The case against efforts to increase longevity takes several forms. It is said, for example, that the prolonging of life runs counter to biblical teaching. Yet "threescore years and ten" (Psalms 90:10) has no authority other than the opinion of a psalmist. In fact, the phrase is something of a cliché in the Bible, standing for quite a number but less than a hundred. Thus we are told that there were threescore and ten palm trees in Elim (*Numbers* 33:9); that when the house of Jacob entered Egypt, it comprised threescore and ten persons (*Deuteronomy* 10:22); that Jerubbaal had threescore and ten sons (*Judges* 9:2). It might be more in accord with the spirit of the Bible if the human life-span were construed to be that which, for better or worse, human beings cause it to be.

People also say that extending life is a crime against nature. I consider this a despondent view, which rests on an implicit nostalgia for the supposedly healthy, happy, exuberant, and yea-saying savages that Jean-Jacques Rousseau spoke for—creatures whose life expectancy probably did not exceed twenty-five or thirty years. This attitude echoes the literary propaganda of the Romantic revival, and it is surely wider of the mark than Thomas Hobbes's assertion that the life of man in a state of nature is "solitary, poor, nasty, brutish, and short."

Human life has always been what human beings have made of it, and in many ways we have improved on nature. It cannot be too strongly emphasized that all advances in medicine increase life expectancy; their efficacy is measured by the degree to which they do so. I am referring not only to insulin, penicillin, and the other spectacular innovations of medical history but also to aspirin (which lowers fever and reduces inflammation, besides relieving pain), adhesive bandages, and washing one's hands before eating. These, too, have contributed years to our life-span.

The whole philosophy of the prevention of disease—and where prevention fails, the cure—represents a deep and long-standing moral commitment to life, and the research in question here is its logical development. Thus, one could argue that, our commitment to the preservation of life having already been made, it is too late for us to cease to be ambitious.

Other objections to prolonging old age have to do with population control and age distribution. For example, it is asked, Dare we propose to add to a burden that is almost insupportable now? Shall the resources of underprivileged nations be consumed at an even faster rate by the technologically more advanced peoples of the Northern Hemisphere and of the West generally—those who will be the first to take advantage of new medical procedures?

In partial extenuation, it can be said that the increase in population would not be exponential, because it is unlikely that older people would choose to add to their families. Admittedly, though, they would have mouths, and they would use energy and other raw materials at the high rate characteristic of people in the developed parts of the world.

One hears that the likely increase in population size would provoke wars, as if the linkage were an established truth. But it does not stand up to scrutiny. No one will challenge Europe's claim to the dunce's cap for political aggression and warmaking, yet war has been no more frequent in Europe over the past hundred years than it was in medieval times or in the fifteenth century or in the seventeenth, even though the population has grown steadily.

The threat of gerontocracy—government by the aged and probably for the aged—is less easy to dismiss. Certainly old people require special attention, and their rising numbers would lay an extra burden on social-welfare services in a caring society. That burden would have to be shouldered mainly by the young. A vigorous elderly generation would also probably hold on to jobs that otherwise would pass to the young, thereby exacerbating unemployment. Who knows? A gerontocracy might have the nerve to impose a special tax on jobholders below some minimum age, and at the same time reward older jobholders with generous concessions.

Without minimizing these last worries, I must point out that the political and sociological effects of a population shift would not be felt overnight. We should have between fifty and 200 years to adapt.

The process might completely overturn our present ideas about work and retirement, but in reality such a revolution has been in progress for the past 150 years, as the proportion of older people in the population has grown with advances in medicine and sanitary engineering. It is reasonable to assume we can solve the problems of the future, since they are not qualitatively new.

Jane Austen wrote her novels around the turn of the nineteenth century, and they are a mine of information about the manners and attitudes of her day. Consider, in particular, Austen's first published novel, *Sense and Sensibility*. The hero, Colonel Brandon, is rated at thirty-five an old man and quite past it—so much so that Marianne Dashwood, the eighteen-year-old girl whose hand he seeks, regards his suit as a kind of geriatric charade. In the book the question arises of laying down a sum to purchase a fifteen-year annuity for Marianne's mother, who is described as a healthy woman of forty. The man who would have to provide for the annuity protests, "Her life cannot be worth half that purchase." So Austen seems able to take for granted her readers' doubt that a woman of forty could live to be as old as fifty-five.

Suppose someone had told Austen's contemporaries that their life expectancy could be doubled. If they were to react as some people do today, they would have held up their hands in horror at the impiety of interfering with nature—at the very idea that a man of thirty-five would not have one foot in the grave and that a woman of forty would live another fifteen years! Yet the average life-span of a century or more ago seems pathetically short from the perspective of today. How can we be certain that a generation as close to us as we are to Jane Austen would not look upon our fears with pained condescension?

Some lines by the poet Walter Savage Landor, which have the cadences of a requiem, seem to rebuke the wish to delay death:

Nature I loved; and next to Nature, Art.
I warmed both hands before the fire of life;
It sinks, and I am ready to depart.

Perhaps this declaration is a Christian acquiescence to an inevitable fate, but to me it sounds spiritless. A person who is loved and in good health has reason enough to want to live a few years longer than might seem to be his due: to learn, for example, how the grandchildren turn out, and whether the flux of history corroborates or refutes his expectations. A writer will want to complete his book, or even turn his thoughts to another, and no gardener will willingly surrender his hope of taking part in the wonder and joyous expectations of another spring. From the point of view of biology, the strength of our hold upon life has been the most important single factor in bringing us to our present ages and, indeed, in the fact that human beings have evolved at all.

Some of the evils that confront mankind—the havoc of war, for example— can be anticipated and guarded against. Others are more insidious. They are the outcomes of well-intentioned actions and could not have been predicted. I have in mind the deaths from cancer of the pioneers of x-radiography, who could not possibly have known that x-rays are among the most potent cancer-causing agents.

Likewise, overpopulation is the consequence of a reduction of mortality, especially in childhood, through medicine and sanitary engineering, which has not been matched by a corresponding reduction in the birthrate.

All else being equal, I think that the risk of unforeseeable catastrophe will probably be sufficient to turn us away from the research to extend life. But what I *hope* will happen is this: perhaps a dozen enthusiasts for the prolonging of life will go ahead and try to prolong their own lives. If they become wise and oracular nonagenarians or centenarians, they will be counted among the bene-factors and pathfinders of mankind. If sentile dementia is their fate, they will have warned us off, and that would be an equally useful service.

My personal sympathies are with the daredevils who want to try out these new procedures. This kind of adventurousness has always been in the character of science, as Sir Francis Bacon, the Lord Chancellor of England, the first and greatest philosopher of science, and a pious and reverent man, believed. In one of his essays, he wrote:

> The true aim of science is the discovery of all operations and all possibilities of operations from immortality (if it were possible) to the meanest mechanical prac-tice.

I count Bacon, therefore, as a man on my side.

Questions and Topics for Discussion and Writing

1. The title of a work initiates the process of persuasion. What response do you have to Medawar's title?
2. List Medawar's transitional devices in this essay.

3. Outline the chief points in Medawar's argument. List four points of debate and answer each.
4. Describe Medawar's intended audience here. How would you characterize his attitude? Identify words and phrases to support your characterization.
5. In Jonathan Swift's *Gulliver's Travels,* read Chapter X of Part Three, which describes the Struldbrugs, or Immortals. In a 500-word essay, compare Swift's ideas with those of Medawar.
6. In what way does Medawar relate the problems of the human race to evolution and the struggle for life?
7. Medawar uses references and allusions to the Bible, to philosophical writings, and to literary works. In a paper, discuss the value and effectiveness of these literary devices to his persuasive argument.

SUGGESTIONS FOR FURTHER READING

Brown, Michael. *Laying Waste: The Poisoning of America by Toxic Chemicals.* New York: Random House, 1980.

Cohen, Bernard L. *Before It's Too Late: A Scientist's Case for Nuclear Energy.* New York: Plenum Press, 1983.

Dahlitz, Julie. *Nuclear Arms Control.* Boston: George Allen and Unwin, 1983.

Davies, Paul. *God and the New Physics.* New York: Simon and Schuster, 1983.

Dyson, Freeman J. *Weapons and Hope.* New York: Harper & Row, 1984.

Dunn, Lewis A. *Controlling the Bomb: Nuclear Proliferation in the 1980s.* New Haven: Yale University Press, 1982.

Ehrlich, Robert. *Waging Nuclear Peace: The Technology and Politics of Nuclear Weapons.* Albany: State University of New York Press, 1985.

Harwell, Mark A. *Nuclear Winter: The Human and Environmental Consequences of Nuclear War.* New York: Springer-Verlag, 1984.

Medawar, Peter Brian. *The Hope of Progress.* London: Methuen, 1972.

Oppenheimer, J. Robert. *The Flying Trapeze: Three Crises for Physicists.* London: Oxford University Press, 1964.

Ryder, Richard D. *Victims of Science: The Use of Animals in Research.* London: National Anti-Vivisection Society Limited, 1983.

Schell, Jonathan. *The Fate of the Earth.* New York: Knopf, 1982.

Zuckerman, Edward. *The Day After World War III.* New York: The Viking Press, 1984.

for Scapegoats by Samuel C. Florman, St. Martin's Press, Inc., New York. Copyright © 1981 by Samuel C. Florman.

68. "The History of Animals" from *The History of Animals* by Aristotle, translated by Cresswell, published by Bell & Hyman Publishers. Reprinted with permission.

69. "Conclusion" from *On the Origin of Species* by Charles Darwin. Public Domain.

70. "Footprints in the Ashes of Time" by Mary Leakey. First appeared in National Geographic Magazine. Reprinted with permission.

71. "Topophilia and Environment" and "Summary" by Yi-Fu Tuan, *Topophilia: A Study of Environmental Perception, Attitudes, and Values,* © 1974, pp. 93, 114–120, 247. Adapted by permission of Prentice-Hall, Inc., Englewood Cliffs, N.J.

72. "A Magic Rocket through Space and Time" by Sir James Jeans from *The Stars in Their Courses,* published by Cambridge University Press. Reprinted with permission.

73. "Miss Mitchell's Comet," by Carole Oles, reprinted from *Prairie Schooner,* by permission of University of Nebraska Press. Copyright 1985 University of Nebraska Press.

74. "Parable of the Fishing Net" by Arthur Eddington from *The Philosophy of Physical Science,* published by Cambridge University Press. Reprinted with permission.

75. "On Insect Behavior," Copyright © 1942 James Thurber. Copyright © 1970 Helen W. Thurber and Rosemary T. Sauers. From *My World—and Welcome to It,* published by Harcourt Brace Jovanovich.

76. "Searching Out Creation's Secrets" by Carolyn Kraus. First published in Metropolitan Detroit Magazine. Reprinted with permission.

77. "Introduction to Psychoanalysis" from *Complete Introductory Lectures on Psychoanalysis* by Sigmund Freud, translated and edited by James Strachey: With the permission of W. W. Norton & Company, Inc. Copyright © 1966 by W. W. Norton & Company, Inc. Copyright © 1965, 1964 by James Strachey.

78. "The Expanding Mental Universe" by Bertrand Russell from *Basic Writings of Bertrand Russell,* Copyright © 1961 by Allen & Unwin, Ltd. Reprinted by permission of Simon & Schuster, Inc.

79. "Environmental Statement" from *Critical Path* by R. Buckminster Fuller, St. Martin's Press, Inc., New York. Copyright © 1981 by R. Buckminster Fuller, Dr. Glenn T. Olds at Alaska's Future Frontiers Conference 1979.

80. "The Obligation to Endure" from *Silent Spring* by Rachel Carson. Copyright © 1962 by Rachel L. Carson. Reprinted by permission of Houghton Mifflin Company.

81. "To Preserve a World Graced by Life" by Carl Sagan, Council for a Livable World Education Fund.

82. "Gare du Midi," copyright 1940 and renewed 1968 by W. H. Auden. Reprinted from *W. H. Auden: Collected Poems,* edited by Edward Mendelson, by permission of Random House, Inc.

83. "At the Bomb-Testing Site" from *Stories That Could Be True* by William Stafford. Copyright © 1960, 1977 by William Stafford. Reprinted by permission of Harper & Row, Publishers, Inc.

84. "Farewell Address to Los Alamos Scientists" by Robert Oppenheimer. Public Domain.

85. "Is Science Evil?" from *Commentary, 1950.* Reprinted with permission.

86. "Should We Infect the Galaxy?" from *Life Itself* by Francis Crick. Copyright © 1981 by Francis Crick. Reprinted by permission of Simon & Schuster, Inc.

87. "Religion and Science," reprinted with permission of Macmillan Publishing Company from *Science and the Modern World* by Alfred North Whitehead. Copyright 1925 by Macmillan Publishing Company, renewed 1953 by Evelyn Whitehead.

88. "When We Are Old" by Peter Medawar from *The Atlantic,* Vol. 253, No. 3. Reprinted by permission.